Handbook of Obesity: Etiology and Pathophysiology, Second Edition, *edited by George Bray and Claude Bouchard*

Handbook of Obesity: Clinical Applications, Second Edition, *edited by George Bray and Claude Bouchard*

Genomics and Proteomics in Nutrition

edited by

Carolyn D. Berdanier

The University of Georgia
Athens, Georgia, U.S.A.

Naima Moustaid-Moussa

The University of Tennessee
Knoxville, Tennessee, U.S.A.

CRC Press
Taylor & Francis Group
Boca Raton London New York

CRC Press is an imprint of the
Taylor & Francis Group, an **informa** business

CRC Press
Taylor & Francis Group
6000 Broken Sound Parkway NW, Suite 300
Boca Raton, FL 33487-2742

First issued in paperback 2019

© 2004 by Taylor & Francis Group, LLC
CRC Press is an imprint of Taylor & Francis Group, an Informa business

No claim to original U.S. Government works

ISBN-13: 978-0-8247-5430-3 (hbk)
ISBN-13: 978-0-367-39403-5 (pbk)

Library of Congress Cataloging-in-Publication Data
A catalog record for this book is available from the Library of Congress.

Visit the Taylor & Francis Web site at
http://www.taylorandfrancis.com

and the CRC Press Web site at
http://www.crcpress.com

Preface

Over the last century nutrition science has evolved from an initial inquiry into the life essences to a sophisticated inquiry into how cells and cell systems work. Initially, nutrition scientists wanted to identify those nutrients essential to life. To that end they conducted many feeding studies using purified diets. They studied many species. Over the years they became aware of the tremendous variability between species and within species. One animal might require twice as much of a specific vitamin as another animal of the same age and gender. Nutrition scientists attributed this variability to genetics. Once the essential nutrients were identified, nutrition scientists wanted to know how they functioned as single nutrients. Metabolic pathways were uncovered and the detailed role of each nutrient in metabolism was gradually discovered. One of the more puzzling aspects of nutrition was the reconciliation of the metabolic function of the nutrient with the signs and symptoms of its deficiency disease. The feature of skin lesions, for example, in pellegra did not relate (at first) to the role of niacin in the coenzyme, NAD. It was only after the discovery of DNA and its role in cell replication that it was realized that niacin (as NAD or NADP) was essential to the synthesis of the purine and pyrimidine bases that are the backbone of DNA. Because skin cells have a very short half-life, niacin-deficiency-induced skin lesions really reflected the unmet need for niacin to support the DNA/RNA-dictated cell replication.

Using the techniques of genetics, nutrition scientists soon learned how to define in great detail the mechanism involved in nutrient-dictated gene expression. Further, nutrition scientists were able to identify metabolic processes and disease processes that were nutritionally responsive from the genetic point of view. The drive to understand basic biological mechanisms of nutrient gene interactions has led to two distinct, yet related, approaches in the study of molecular biology: genomics and proteomics. A third approach, metabolomics, links the first two with changes in intracellular metabolites.

Three earlier books on nutrition and gene expression (*Nutrition and Gene Expression*, edited by C. D. Berdanier and J. L. Hargrove (CRC Press, 1993); *Nutrients and Gene Expression: Clinical Aspects*, edited by C. D. Berdanier (CRC Press. 1996); and *Nutrient-Gene Interactions in Health and Disease*, edited by N. Moustaid-Moussa and C. D. Berdanier (CRC Press, 2001) addressed questions of how specific nutrients affected the expression of the genetic material DNA. The present book takes the reader one step further. It explores the question of how genetic expression affects the production of gene products. The many authors of this book have each addressed this issue from very specific points of view. Whether the gene product is a known protein or one that has yet to be identified, the authors have used the techniques of biotechnology to provide new insights into nutrition and metabolism. We the editors thank these authors for their conscientious, careful review of the literature in their particular writing assignment. We hope that you, the reader, will appreciate the progress that has been made relating nutrition, gene expression, and tissue/organ and cell function.

Carolyn D. Berdanier
Naima Moustaid-Moussa

Contents

**PART III: PROTEOMICS: LARGE-SCALE ANALYSIS OF
PROTEIN EXPRESSION AND REGULATION**

Contributors

Gerard Ailhaud   CNRS 6543, Centre de Biochimie, Nice, France

David B. Allison   Section on Statistical Genetics, Department of Biostatistics, University of Alabama at Birmingham, Birmingham, Alabama

Maya Belghazi   Institut National de la Recherche Agronomique (INRA), Laboratory of Mass Spectrometry for Proteomics, Nouzilly, France

Brett Campbell   Department of Nutrition and Agricultural Experiment Station, University of Tennessee, Knoxville, Tennessee, U.S.A.

James A. Carroll   Department of Molecular Genetics and Biochemistry, University of Pittsburgh School of Medicine, Pittsburgh, Pennsylvania, U.S.A.

Maria DeLuca   Department of Environmental Health Sciences, University of Alabama at Birmingham, Birmingham, Alabama, U.S.A.

Melissa Derfus   Department of Nutrition and Agricultural Experiment Station, University of Tennessee, Knoxville, Tennessee, U.S.A.

James C. Fleet   Purdue University, West Lafayette, Indiana, U.S.A.

Ivan C. Gerling   University of Tennessee Health Science Center, Memphis, Tennessee, U.S.A.

Richard Giannone University of Tennessee-Oak Ridge National Laboratory Graduate School of Genome Science and Technology, Oak Ridge, Tennessee, U.S.A.

Dawn M. Graunke Pennington Biomedical Research Center, Louisiana State University, Baton Rouge, Louisiana, U.S.A.

Arthur Grider Department of Foods and Nutrition, University of Georgia, Athens, Georgia, U.S.A.

Michael J. Griffin Department of Nutritional Sciences and Toxicology, University of California, Berkeley, California, U.S.A.

Sung Nim Han Jean Mayer USDA Human Nutrition Research Center on Aging at Tufts University, Boston, Massachusetts, U.S.A.

Young-Ran Heo Department of Nutrition and Agricultural Experiment Station, University of Tennessee, Knoxville, Tennessee, U.S.A.

Steven B. Heymsfield New York Obesity Research Center, Columbia University College of Physicians & Surgeons, New York, New York, U.S.A.

Nieves Ibarrola McKusick-Nathans Institute of Genetic Medicine and Department of Biological Chemistry, Johns Hopkins University, Baltimore, Maryland, U.S.A.

Kimberly F. Johnson Duke University Medical Center, Duke University, Durham, North Carolina, U.S.A.

Rashika Joshi Department of Nutrition and Agricultural Experiment Station, University of Tennessee, Knoxville, Tennessee, U.S.A.

Hyoung Yon Kim Department of Nutrition and Agricultural Experiment Station, University of Tennessee, Knoxville, Tennessee, U.S.A.

Jung Han Kim Nutrition Department, University of Tennessee, Knoxville, Tennessee, U.S.A.

Kee-Hong Kim Department of Nutritional Sciences and Toxicology, University of California, Berkeley, California, U.S.A.

Kyoungmi Kim Section on Statistical Genetics, Department of Biostatistics, University of Alabama at Birmingham, Birmingham, Alabama

Suyeon Kim Department of Nutrition and Agricultural Experiment Station, University of Tennessee, Knoxville, Tennessee, U.S.A.

Robert A. Koza Pennington Biomedical Research Center, Louisiana State University, Baton Rouge, Louisiana, U.S.A.

Seth W. Kullman Nicholas School of the Environment, Duke University, Durham, North Carolina, U.S.A.

Simon M. Lin Duke University Medical Center, Duke University, Durham, North Carolina, U.S.A.

Amanda H. McDaniel Monell Chemical Senses Center, Philadelphia, Pennsylvania, U.S.A.

Florence Massiera CNRS 6543, Centre de Biochimie, Nice, France

Clayton E. Mathews Diabetes Institute, Department of Pediatrics, University of Pittsburgh School of Medicine, Pittsburgh, Pennsylvania, U.S.A.

Simin Nikbin Meydani Jean Mayer USDA Human Nutrition Research Center on Aging at Tufts University, and Department of Pathology, Sackler Graduate School of Biomedical Sciences, Tufts University, Boston, Massachusetts, U.S.A.

Orsolya Mezei Department of Biological Sciences, University of Notre Dame, Notre Dame, Indiana, U.S.A.

E. Barry Moser Department of Experimental Statistics, Louisiana State University, Baton Rouge, Louisiana, U.S.A.

Naima Moustaid-Moussa Department of Nutrition and Agricultural Experiment Station, University of Tennessee, Knoxville, Tennessee, U.S.A.

Manabu T. Nakamura Department of Food Science and Human Nutrition, University of Illinois at Urbana-Champaign, Urbana, Illinois, U.S.A.

Jürgen K. Naggert The Jackson Laboratory, Bar Harbor, Maine, U.S.A.

Patsy M. Nishina   The Jackson Laboratory, Bar Harbor, Maine, U.S.A.

Michael J. Pabst   University of Tennessee Health Science Center, Memphis, Tennessee, U.S.A.

Grier P. Page   Section on Statistical Genetics, Department of Biostatistics, University of Alabama at Birmingham, Birmingham, Alabama, U.S.A.

Akhilesh Pandey   McKusick-Nathans Institute of Genetic Medicine and Department of Biological Chemistry, Johns Hopkins University, Baltimore, Maryland, U.S.A.

Andrey Ptitsyn   Pennington Biomedical Research Center, Louisiana State University, Baton Rouge, Louisiana, U.S.A.

Annie Quignard-Boulangé INSERM U465-IFR58, Centre Biomédical des Cordeliers, Paris, France

Chaerkady Raghothama Institute of Bioinformatics, International Technology Park Ltd., Bangalore, India

Danielle R. Reed Monell Chemical Senses Center, Philadelphia, Pennsylvania, U.S.A.

Hélène Rogniaux   Institut National de la Recherche Agronomique (INRA), Unité de Recherche sur les Protéines Végétales et Leurs Interactions, Laboratory of Mass Spectrometry, Nantes, France

Marie-Pierre St-Onge New York Obesity Research Center, Columbia University College of Physicians & Surgeons, New York, New York, U.S.A.

Arnold M. Saxton Department of Animal Science, University of Tennessee, Knoxville, Tennessee, U.S.A.

Neil F. Shay Department of Biological Sciences, University of Notre Dame, Notre Dame, Indiana, U.S.A.

Liang Shi Torrey Mesa Research Institute, Syngenta, San Diego, California, U.S.A.

Steven R. Smith Pennington Biomedical Research Center, Louisiana State University, Baton Rouge, Louisiana, U.S.A.

Jay Snoody University of Tennessee-Oak Ridge National Laboratory Graduate School of Genome Science and Technology, Oak Ridge, Tennessee, U.S.A.

Hei Sook Sul Department of Nutritional Sciences and Toxicology, University of California, Berkeley, California, U.S.A.

Michele Teboul CNRS 6543, Centre de Biochimie, Nice, France

Sumithra Urs Department of Nutrition and Agricultural Experiment Station, University of Tennessee, Knoxville, Tennessee, U.S.A.

Josep A. Villena Department of Nutritional Sciences and Toxicology, University of California, Berkeley, California, U.S.A.

Brynn H. Voy Oak Ridge National Laboratory, Oak Ridge, Tennessee, U.S.A.

Hui Xie Pennington Biomedical Research Center, Louisiana State University, Baton Rouge, Louisiana, U.S.A

Yun Wang The Jackson Laboratory, Bar Harbor, Maine, U.S.A.

Patrick Wortman Department of Nutrition and Agricultural Experiment Station, University of Tennessee, Knoxville, Tennessee, U.S.A.

Bing Zhang University of Tennessee-Oak Ridge National Laboratory Graduate School of Genome Science and Technology, Oak Ridge, Tennessee, U.S.A.

Kui Zhang Section on Statistical Genetics, Department of Biostatistics, University of Alabama at Birmingham, Birmingham, Alabama, U.S.A.

Tong Zhu Torrey Mesa Research Institute, Syngenta, San Diego, California, U.S.A.

Genomics and Proteomics in Nutrition

1

Genetic Modifiers in Rodent Models of Obesity

Yun Wang, Patsy M. Nishina, and Jürgen K. Naggert
The Jackson Laboratory, Bar Harbor, Maine, U.S.A.

ABSTRACT

The identification of single gene mutations leading to obesity in the mouse has greatly enhanced our understanding of the regulation of bodyweight and energy metabolism. Indeed, the discovery of the influence of the leptin/melanocortin pathway on body weight was solely the result of identifying the genetic defects in the mutants, diabetes *(db/db)*, obese *(ob/ob)*, and agouti yellow *(A^y)*. In addition, from such studies it is clear that obesity mutations must reside in a permissive genetic backgound in order to manifest an obese or obese/diabetic phenotype. Such background genes can also modify age of onset, rate of disease progression, or severity of the obesity phenotypes. Background genes that interact with mutant genes are responsible for alterations of specific phenotypes and are called genetic modifier loci. Identification of these modifiers may provide a powerful tool for defining biological pathways that lead from the primary genetic defect to the disease phenotype. Those modifiers that suppress weight gain or progression to non-insulin-dependent diabetes may lead to new therapeutic targets. These targets may be more amenable to manipulation by small molecule drugs than the primary mutant gene product.

1. INTRODUCTION

Individuals affected with the same genetic disorder often differ in their clinical presentation. This effect is evident in the intrafamilial variability observed in weight gain and glycemic status in syndromic diseases such as Bardet-Biedl and Alström syndromes, in which all affected family members carry the same mutation (1,2). Intrafamilial variability in disease phenotypes may be due to environmental influences, genetic modifier loci, or a combination of these factors. In addition, interfamilial variability may be due to allelic differences at specific loci.

Whereas much work has been done on the environmental influences, e.g., diet or exercise on development of obesity and type II diabetes, the role of genetic modifiers is gaining prominence. The phenotypic effects of modifier genes on the manifestation of a primary disease mutation can arise from the modifier's action in the same or in a parallel biological pathway as a disease gene. The effect can be enhancing, causing a more severe mutant phenotype, or suppressive, reducing the mutant phenotype even to the extent of completely restoring the wild-type condition. Modifier genes can also alter the pleiotropy of a given disease, resulting in different combinations of traits. In addition, for any given genetic disorder, alleles of multiple modifier genes may act in combination to create a final, cumulative effect on the observed phenotype. The latter situation may be especially true for complex disease traits such as obesity and type II diabetes, for which it has been extremely difficult to identify underlying genes in the human population.

Studying and identifying genetic modifier loci can yield new insights into the biological pathways in which Mendelian disease genes act and through which they cause disease phenotypes. For example, knowing the molecular basis of a genetic modifier may improve diagnosis and treatment of disease, perhaps by defining a particular subgroup within the disease population. In addition, the identification of modifier genes may lead to new treatments either by providing additional information about the genetic contributions to the phenotype for which treatment may already be available or by pointing to additional steps in a biological pathway that may be more amenable to treatment.

In obesity, environmental influences have typically been emphasized over genetic causes for the phenotype. Examples of modifier genes in human studies are not abundant; their existence can, however, be inferred by the finding of association of obesity subphenotypes with particular alleles of genes that have been implicated in obesity. In animal models of obesity caused by a mutation in a single gene, it can be shown that modifier genes influence phenotypic expression. The fact that most obese individuals do not develop non-insulin-dependent diabetes mellitus (NIDDM), whereas most patients afflicted with NIDDM are obese, can be interpreted to mean that obesity (and obesity genes) are necessary but not sufficient for the development of NIDDM and that NIDDM susceptibility genes may act as modifiers of obesity genes. Better known are the effects of modifier genes in causing non-insulin-dependent type II diabetes in obesity models. In mouse models, such as C57BLKS-Lep^{ob}/Lep^{ob}, the obesity mutation is necessary but not sufficient for the development of diabetes. In this review we will focus on the role of genetic modifiers in rodent models of obesity and diabetes and provide some examples of reported modifier gene action.

2. GENE/GENE INTERACTIONS

The earliest documented gene/gene interaction in an obesity pathway was found between the mouse mutations obese (ob) (3) and diabetes (db) (4). C57BL/6-ob/ob and C57BL/6-db/db mice are hyperphagic, hypometabolic, and massively obese. Because of the phenotypic resemblance of the two mouse strains, Coleman and Hummel undertook a series of parabiosis experiments. Connecting the blood supplies of ob/ob mice with those of wild-type mice caused the ob/ob mice to loose weight. Parabiosis of db/db mice to wild-type mice led to starvation and weight loss in the wild-type mice, while the db/db mice maintained their body weight. And finally, parabiosis of ob/ob and db/db mice led to weight loss in the obese ob/ob mice. Apparently, ob/ob mice were lacking a blood-borne factor that prevented obesity that wild-type mice possessed, and db/db mice could not respond to that factor and overproduced it. From these results, Coleman concluded that the ob locus might encode a satiety hormone and the db locus, its receptor. This prediction was proven correct by the identification of the obese gene as Lep, the adipocyte-secreted hormone, leptin (5), and diabetes as the leptin receptor (Lepr) gene, acting primarily in the hypothalamus (6,7).

The example of the interaction between Lep^{ob} and $Lepr^{db}$ is unusual in that it was not demonstrated by genetic means. More typically, gene/gene interactions are discovered by observing the suppression of a phenotype in double-mutant mice or by the appearance of a phenotype in compound heterozygous animals.

An example for the former is the discovery of an interaction between the coat color mutations yellow (A^y) at the agouti locus on chromosome (Chr) 2 and the nonallelic mahogany ($Atrn^{mg}$) and mahoganoid ($Mgrn1^{md}$) loci on Chrs 2 and 16, respectively. In addition to a yellow coat color, yellow mice develop obesity due to the ectopic expression in the hypothalamus of agouti signal protein (ASP), a melanocortin receptor antagonist normally expressed only in the skin (8,9). In order to determine where mahogany and mahoganoid lie with respect to agouti signaling in a genetic pathway, Miller and colleagues created double-mutant animals (10). They found that homozygosity for either the mahogany or the mahoganoid mutation suppressed the effects of A^y on coat color as well as on obesity, suggesting that mg and md act downstream of agouti to interfere with agouti signaling. Both mutant genes have been identified by positional cloning. The mahogany gene codes for the membrane protein, attractin (11,12), which may act as a low-affinity receptor for the agouti protein (13). Mahoganoid codes for a novel RING-containing protein with E3 ubiquitin ligase activity, which may function in protein turnover (14,15).

In some cases, enough is known about a molecular pathway to test directly for interactions between genes by creating double-mutant mice. Neuropeptide Y (NPY) is an orexigenic peptide that stimulates feeding when injected into the third ventricle of the hypothalamus (16). Administration of leptin suppresses hypothalamic expression and release of NPY; NPY is elevated in leptin-deficient mice (17,18). Mice deficient in NPY, however, show no abnormality in feeding behavior (19) and only a slight increase in body weight (20), suggesting the existence of additional pathways controlling feeding in mice. By generating mice that were deficient for both leptin and NPY, Erickson and coworkers showed that the lack of NPY in these animals attenuated the obesity normally observed in Lep^{ob}/Lep^{ob} mice by reducing their food intake and increasing their energy expenditure (21). This indicated that NPY is a major effector in leptin signaling.

The last example shows, in particular, that a candidate gene approach to gene/gene interactions can provide important confirmation of hypotheses regarding biological pathways. With our rapidly increasing knowledge about gene function, this approach will gain more importance in the future.

3. HUNTING FOR MODIFIER GENES

As seen earlier, the identification of modifier genes can be an important component of understanding biological pathways that lead from a primary mutation to a disease phenotype. Whereas a candidate gene approach is limited by our knowledge of gene function, a reverse genetic approach—going from a phenotype to the causative underlying gene—does not require prior knowledge about the interacting genes, and could be a powerful method for identifying novel pathways. A major stumbling block for the identification of genetic modifiers of obesity and type II diabetes in humans will be the difficulty of mapping these loci in the face of the huge genetic heterogeneity in the human population. And whereas chromosomal localization of modifier loci in large human families segregating for monogenic diseases that cause obesity and type II diabetes may be feasible, a real problem will be moving from a general map position to a narrow enough region that fine physical mapping and gene identification can commence.

Here again, we may be able to use the available monogenic mouse obesity models to gain insight into the pathways that influence obesity and hyperglycemia. The most straightforward approach to mapping modifier genes in the mouse is to carry out crosses between inbred strains that carry the disease-causing mutation and in which a difference in phenotype is observed. A schematic for this approach is shown in Fig. 1, in which the obesity phenotype of mice deficient in leptin (homozygous Lep^{ob}/Lep^{ob} mice) is modified by the C57BL/6J (B6) and BALB/cJ genetic backgrounds (22). B6-Lep^{ob}/Lep^{ob} mice develop an early onset, severe obesity, whereas BALB/cJ-Lep^{ob}/Lep^{ob} mice are reported to be obese but lighter than B6-Lep^{ob}/Lep^{ob}. This suggests the presence of genes in the BALB/cJ background that moderate weight gain. When F1 offspring from a mating between B6-Lep^{ob}/Lep^{ob} and BALB/cJ-+/+ are intercrossed, the modifier genes should segregate in the F2 population. Because a modifier gene itself does not in most cases produce a phenotype, only F2 animals that are homozygous for the primary disease-causing mutation (in the case of a recessively inherited disease) will show phenotypic variation and thus be informative for the analysis. In those F2 animals, standard quantitative trait locus (QTL) analysis methods can be used to map the modifier loci (23). In order not to confuse background QTLs with modifier genes, the F2 animals that do not carry the disease mutation should be examined for variation in the trait of interest. If there is variation, then the background QTLs should also be mapped to distinguish the loci that affect the trait independent of the disease mutation from the true modifier loci. It should be pointed out that the primary mutation does not necessarily have to lead to phenotypic differences in the two parental strains used in the modifier cross. Occasionally, modifier genes are unmasked only by the interaction of the two genetic backgrounds in the segregating F2 population (24). This is the case, for example, for the fat mutation: Although the body weights of C57BLKS-Cpe^{fat}/Cpe^{fat} mice do not differ much from those of HRS-Cpe^{fat}/Cpe^{fat}, in the F2-Cpe^{fat}/Cpe^{fat} population from a (C57BLKS × HRS) F1-$Cpe^{fat}/+$ intercross, body weights vary from normal to severely obese (25).

Although less difficult in the mouse, identification of genetic modifiers of obesity-related traits for which the chromosomal locations have been mapped may still be challenging, especially if more than one gene is contributing to the modification of the phenotype. If a major modifying locus is found (explaining $>40\%$ of the phenotypic variance), then conventional

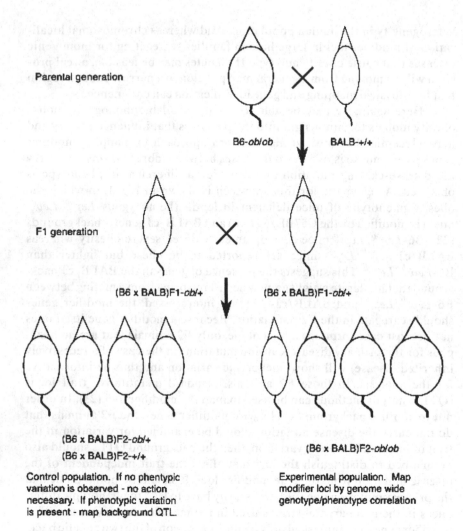

Parental generation

B6-*ob/ob*   BALB-+/+

F1 generation

(B6 x BALB)F1-*ob/+*   (B6 x BALB)F1-*ob/+*

(B6 x BALB)F2-*ob/+*
(B6 x BALB)F2-+/+

(B6 x BALB)F2-*ob/ob*

Control population. If no phentypic
variation is observed - no action
necessary. If phenotypic variation
is present - map background QTL.

Experimental population. Map
modifier loci by genome wide
genotype/phenotype correlation

FIGURE 1   Breeding scheme to carry out a modifier screen.

fine-structure mapping in a large F2 intercross combined with progeny testing can be used to narrow the genetic interval sufficiently to proceed with positional cloning (26). In cases where multiple loci contribute to the phenotypic variance, it may be necessary to construct congenic lines to isolate individual modifier loci (23,27). If the phenotypic effect of the modifier locus in the congenic line is greater than that of the nongenetic variation, then the line can be used in crosses for fine-resolution mapping, as in the case of the major modifier. Once a high-resolution map has been obtained, conventional positional cloning techniques may be applied (26). Currently available methods such as gene expression microarray analysis may be combined with the use of congenic lines to directly identify a misregulated modifier allele, or to point to the misregulation of a pathway in which the modifier gene plays a role (28).

4. OBESITY MODELS SHOWING PHENOTYPIC MODIFICATION

The first recognition of modifier genes in obesity research dates back to the study of the Lep^{ob} and $Lepr^{db}$ mutations on different genetic backgrounds. Coleman and his coworkers noted that B6-ob/ob and B6-db/db mice became obese but remained diabetes-free. However, when placed on the related C57BLKS (BKS) inbred strain background, both mutations led to severe diabetes (29,30). This indicates that the BKS genetic background is diabeto-genic, i.e., BKS carries alleles of diabetes susceptibility genes that are neces-sary but not sufficient for the development of overt diabetes. These diabetes susceptibility alleles have to interact with obesity mutations such as Lep^{ob} and $Lepr^{db}$ to cause hyperglycemia. Although the major diabetes modifiers in BKS have yet to be mapped (31), modifiers of leptin action have been reported in other mouse and rat strains. In addition, obesity and diabetes modifiers have been reported for different obesity mutations.

4.1. Lep and Lepr Mutations

Although the existence of genetic background modifiers affecting glycemic status in the context of leptin receptor mutations was first recognized in mice, the first published mapping studies were carried out in the rat model. The Zucker fatty rat carries a Gln269Pro mutation in the leptin receptor that leads to obesity, hyperinsulinemia, and glucose intolerance. The animals, however, are, normoglycemic (32). In contrast, the same mutation when transferred onto the WKY strain background causes obesity, hyperinsuline-mia, and hyperglycemia (33). Chung and colleagues used this strain differ-ence to map NIDDM susceptibility loci in an F2 intercross between animals of the WKY and 13M strains homozygous for the $Lepr^{fa}$ mutation (34). Significant genotype/phenotype associations were found on rat Chr 1 for pancreatic morphology, on Chr 12 for body weight, and on Chr 16 for plasma glucose levels. It is interesting to note that a number of obesi-ty/diabetes related traits have been mapped to the same region of Chr 1 in other rat models (35) and in the homologous region on mouse Chr 19 (36,37). Whether these loci represent variations in the same gene remains an open question until the genes are cloned. The identification of these loci, however, promises new insights into the reasons for pancreatic failure in type II diabetes.

Similar to the findings for *Lepr* mutations, phenotypes of leptin mutations can also be modified by strain background. Apart from the original observation of hyperglycemia in BKS-Lep^{ob}/Lep^{ob} mice versus normoglycemia in B6-Lep^{ob}/Lep^{ob} [30], modifications of body weight, insulin levels, and glucose levels have been reported in the BALB/cJ (22) and the BTBR strain backgrounds (36). In Lep^{ob}/Lep^{ob} F2 offspring from an intercross of (B6 × BTBR) F1-Lep^{ob}/+ mice, Stoehr et al. were able to map three loci controlling insulin and glucose levels on Chrs 2, 16, and 19 (36). Interestingly, it is the B6 allele on Chr 19 that contributes to increased plasma glucose levels, yet B6-Lep^{ob}/Lep^{ob} mice are protected from diabetes. Susceptibility contributions from an overall resistant background are not uncommon (24), and in this case the resistance of B6 to overt diabetes can be attributed to an interaction between the loci on Chrs 19 and 16. BTBR alleles on Chr 16 are necessary to unmask the deleterious effects of the B6 allele on Chr 19 (36).

4.2. *Tub* Mutation

Mice homozygous for the tubby mutation (*tub*) are a model for sensory loss/obesity syndromes such as Alström syndrome (25,38). Tubby mice develop late onset obesity with insulin resistance, early onset retinal degeneration, and neural hearing loss [39–41]. The tubby phenotype is due to a loss-of-function mutation in the novel *Tub* gene, a member of the small gene family encoding tubby-related proteins (TULPs) [42–44]. The biochemical function of the TULPs is not fully understood. Roles as transcription factors (45), as intermediates in insulin signaling (46), and in intracellular transport [47–49] have been proposed. Identification of genetic modifiers of the different phenotypes observed in tubby mice could provide additional clues to the pathways involved, and so may lead to further insights into TUB function.

The first modifier of a tubby phenotype to be identified was *moth1*, the modifier of tubby hearing 1. In an F1 intercross following a cross between B6-*tub*/*tub* and AKR/J, it was observed that F2 offspring homozygous for the *tub* mutation varied widely in their hearing ability from normal hearing to profound deafness. Hearing was quantified electrophysiologically by measuring auditory brainstem response in the F2-*tub*/*tub* population, and a major QTL, *moth1*, was mapped to Chr 2 (50). In the absence of the *tub* mutation, i.e., in the wild-type B6 strain, this locus has no effect on hearing. Standard positional cloning techniques were used to identify *moth1* as an allele of the gene-encoding microtubule-associated protein 1A (*Map1a*) (26). The B6 *Map1a* allele, associated with hearing loss, carries 12 amino acid alterations and an Ala-Pro repeat length polymorphism compared to the protective AKR allele. It was shown that these polymorphisms lead to a weaker binding of the B6 MAP1A protein than of the AKR variant to the postsynaptic density protein PSD95. That *Mapla* is indeed *moth1* was confirmed by a transgenic rescue experiment showing that B6-*tub/tub* mice carrying a protective 129P2/OlaHsd allele of *Mapla* have nearly normal hearing. MAP1A has been shown to be important in trafficking of vesicles and organelles, and PSD95 is a major component of the synaptic cytoarchitecture. The identification of the *moth1* modifier has, therefore, genetically established that synaptic architecture and intracellular transport are relevant to TUB function.

Although *moth1* has provided more functional information, the findings are compatible with both the transcription factor and the transport hypotheses of TUB function. There are still additional modifiers to be identified that may yield further insight into TUB function. Apart from hearing ability and vision loss, adiposity and plasma levels of glucose, insulin and lipids also show variation in F2 progeny of crosses between B6-*tub/tub* and AKR (25,51), (A. Ikeda, personal communication, 2002). A genome-wide scan using 57 microsatellite markers distributed at about 30-cM intervals was performed on 43 female and 37 male F2-*tub/tub* mice. Several statistically significant and suggestive QTLs have been found for body and fat pad weights as well as for plasma insulin levels on Chr.6 ($p < 4.5 \times 10^{-6}$, $p < 2.6 \times 10^{-6}$, $p < 2.6 \times 10^{-6}$ respectively), for plasma cholesterol on Chr.8 ($p < 2.6 \times 10^{-6}$), and for plasma glucose levels on Chr.4 ($p < 3 \times 10^{-4}$).

4.3. *Cpe^fat* Mutation

The mouse fat mutation is a complex model for obesity and type II diabetes (38). The underlying defect is a mutation in the carboxypeptidase E (*Cpe*) gene (52), which codes for an enzyme responsible for the final proteolytic processing step of prohormone intermediates, such as those for insulin and proopiomelanocortin (53). Because a large number of neuro/endocrine peptides are affected by the *Cpe^fat* mutation, the etiology of obesity and diabetes in the mutant mice is not clear. The identification of modifier genes in this case should point to pathways that are critical for the expression of a particular phenotype.

Cpe^fat is a typical disease gene, i.e., it is necessary but not sufficient for the development of obesity, type II diabetes, and related metabolic disorders. On the HRS/J (HRS) inbred strain background, *Cpe^fat*/*Cpe^fat* mice exhibit early onset hyperinsulinemia followed by postpubertal obesity without hyperglycemia. In contrast, on the C57BLKS/J (BKS) genetic background, *Cpe^fat*/*Cpe^fat* mice become hyperglycemic as well as obese and hyperinsulinemic. In order to map the susceptibility loci responsible for modifying obesity and diabetes associated traits, Cpe^{fat}/Cpe^{fat} male progeny from a large F2 intercross between BKS.HRS-Cpe^{fat}/Cpe^{fat} and HRS-+/+ mice were characterized both genetically and phenotypically. All traits measured—body weight, adiposity, fat pad weights, plasma glucose, insulin, triglycerides, and HDL and non-HDL levels—showed a large variance in the F2 population, indicating the action of modifier loci (24). A genome-wide scan was carried out on 282 male Cpe^{fat}/Cpe^{fat} F2 progeny, and four major modifier QTLs for Cpe^{fat} were detected. Three loci for hyperglycemia (*find2, find1, findc*) were mapped on Chrs 5, 19, and 2, and one locus for adiposity (*final*) on Chr 11 (54). Interestingly, at *find1* it is the HRS allele that contributes to hyperglycemia, indicating that there must be another, as yet unidentified, HRS locus that counteracts diabetes development to maintain normoglycemia in HRS-Cpe^{fat}/Cpe^{fat} mice.

5. MODIFICATION IN DIABETES MODELS

Although obesity appears to be a prerequisite for the common forms of human NIDDM, mutations in genes that are more directly associated with diabetes can also be used to further define the pathways that lead to NIDDM.

Mutations in the insulin receptor can be used to model insulin resistance. Kido et al. showed that the effect of reduced insulin receptor activity in animals heterozygous for an *Insr*-targeted allele are dependent on the genetic background (55). Male 129S6/SvEvTac-$Insr^{tm1Dac}$/+ are hyperinsulinemic and have slightly elevated plasma glucose levels compared to B6-$Insr^{tm1Dac}$/+ mice. Two loci controlling plasma insulin levels were found on Chrs 2 and 10, and both susceptibility alleles are contributions from the resistant B6 strain.

The genetic complexity of type II diabetes comes to light when a compound heterozygous mouse model is created by adding a defect in insulin receptor substrate 1, *Irs1*, to the defect in the insulin receptor in the model described earlier. In this case, it is the doubly mutant B6 mouse that is hyperinsulinemic and diabetic, whereas the same mutations on a 129 strain background cause only a mild elevation in insulin and no hyperglycemia (56). It is possible that the *Irs1* defect unmasks the effects of B6 susceptibility alleles, but in addition to those described in the previous example, other loci are at play. In F2 double-heterozygous mice from a (B6 × 129) F1-$Insr^{tm1Dac}$/+ intercross, Almind et al. mapped one significant and one suggestive locus associated with hyperinsulinemia to Chrs 14 and 12, respectively, and one locus for hyperleptinemia to Chr 7 (57). The last locus acts synergistically with that on Chr 14 to increase hyperinsulinemia and with the Chr 12 locus to increase hyperglycemia.

6. SUMMARY

From the examples given, it is clear that genetic modifiers play a role in the phenotypic variation observed in obesity and type II diabetes. Modifiers have been shown to affect the age of onset, severity, rate of disease progression, and presence or absence of a particular disease phenotype.

Whereas some modifier effects have been localized to a chromosomal region, many have yet to be mapped, and cloning of modifier genes is still a difficult task. In humans, this is in part due to genetic heterogeneity, in terms of both the large number of obesity genes that may exist and the high levels of variation among the genetic backgrounds upon which obesity mutations reside. In addition, modifier effects may occur as a result of several genes modulating a disease phenotype, and unless there is a significant contribution from one locus, they may be difficult to isolate. It can be argued that the approach of studying obesity modifier loci in the mouse and then determining whether those genes play a similar role in humans may be the most efficient means of identifying genetic modifiers. The availability of the complete human and mouse genome sequences is aiding greatly in this quest. Advances in determining the expression patterns of all genes in the genome help to prioritize candidates in the vicinity of mapped modifier genes. Large-scale gene expression analysis using microarrays may identify genes that are coregulated by the modifiers and possibly define novel pathways. Consequently, the rate at which modifiers are identified will increase in the near future.

Although few modifier genes have been identified to date, these have yielded additional information about the pathways in which the primary mutation acts, and have provided new experimental avenues toward understanding the pathological effects of the primary disease genes. Finally, the elucidation of modifier genes associated with attenuation of obesity and prevention of progression to type II diabetes may lead to exploration of new therapeutics aimed at increasing the activity of modifiers in affected patients.

REFERENCES

1. Riise R, Tornqvist K, Wright AF, Mykytyn K, Sheffield VC. The phenotype in Norwegian patients with Bardet-Biedl syndrome with mutations in the BBS4 gene. Arch Ophthalmol 2002; 120:1364–1367.
2. Marshall JD, Ludman MD, Shea SE, Salisbury SR, Willi SM, LaRoche RG, Nishina PM. Genealogy, natural history, and phenotype of Alstrom syndrome in a large Acadian kindred and three additional families. Am J Med Genet 1997; 73:150–161.

3. Ingalls AM, Dickie MM, Snell GD. Obese, a new mutation in the house mouse. Heredity 1950; 41:317.

4. Hummel KP, Dickie MM, Coleman DL. Diabetes, a new mutation in the mouse. Science 1966; 153:1127–1128.

5. Zhang Y, Proenca R, Maffei M, Barone M, Leopold L, Friedman JM. Positional cloning of the mouse obese gene and its human homologue. Nature 1994; 372:425–432.

6. Tartaglia LA, Dembski M, Weng X, Deng N, Culpepper J, Devos R, Richards GJ, Campfield LA, Clark FT, Deeds J, Muir C, Sanker S, Moriarty A, Moore KJ, Smutko JS, Mays GG, Woolf EA, Monroe CA, Tepper RI. Identification and expression cloning of a leptin receptor, OB-R. Cell 1995; 83:1263–1271.

7. Lee GH, Proenca R, Montez JM, Carroll KM, Darvishzadeh JG, Lee JI, Friedman JM. Abnormal splicing of the leptin receptor in diabetic mice. Nature 1996; 379:632–635.

8. Bultman SJ, Michaud EJ, Woychik RP. Molecular characterization of the mouse agouti locus. Cell 1992; 71:1195–1204.

9. Miller MW, Duhl DM, Vrieling H, Cordes SP, Ollmann MM, Winkes BM, Barsh GS. Cloning of the mouse agouti gene predicts a secreted protein ubiquitously expressed in mice carrying the lethal yellow mutation. Genes Dev 1993; 7: 454–467.

10. Miller KA, Gunn TM, Carrasquillo MM, Lamoreux ML, Galbraith DB, Barsh GS. Genetic studies of the mouse mutations mahogany and mahoganoid. Genetics 1997; 146:1407–1415.

11. Gunn TM, Miller KA, He L, Hyman RW, Davis RW, Azarani A, Schlossman SF, Duke-Cohan JS, Barsh GS. The mouse mahogany locus encodes a transmembrane form of human attractin. Nature 1999; 398:152–156.

12. Nagle DL, McGrail SH, Vitale J, Woolf EA, Dussault BJ Jr, DiRocco L, Holmgren L, Montagno J, Bork P, Huszar D, Fairchild-Huntress V, Ge P, Keilty J, Ebeling C, Baldini L, Gilchrist J, Burn P, Carlson GA, Moore KJ. The mahogany protein is a receptor involved in suppression of obesity. Nature 1999; 398:148–152.

13. He L, Gunn TM, Bouley DM, Lu XY, Watson SJ, Schlossman SF, Duke-Cohan JS, Barsh GS. A biochemical function for attractin in agouti-induced pigmentation and obesity. Nat Genet 2001; 27:40–47.

14. Phan LK, Lin F, LeDuc CA, Chung WK, Leibel RL. The mouse mahoganoid coat color mutation disrupts a novel C3HC4 RING domain protein. J Clin Invest 2002; 110:1449–1459.

15. He L, Lu XY, Jolly AF, Eldridge AG, Watson SJ, Jackson PK, Barsh GS, Gunn TM. Spongiform degeneration in mahoganoid mutant mice. Science 2003; 299:710–712.

16. Zarjevski N, Cusin I, Vettor R, Rohner-Jeanrenaud F, Jeanrenaud B. Chronic intracerebroventricular neuropeptide-Y administration to normal rats mimics hormonal and metabolic changes of obesity. Endocrinology 1993; 133:1753–1758.

17. Stephens TW, Basinski M, Bristow PK, Bue-Valleskey JM, Burgett SG, Craft L, Hale J, Hoffmann J, Hsiung HM, Kriauciunas A, MacKellar W, Rosteck PR Jr, Schoner B, Smith D, Tinsley FC, Zhang X-Y, Heiman M. The role of neuro-

peptide Y in the antiobesity action of the obese gene product. Nature 1995; 377:530–532.

18. Schwartz MW, Seeley RJ, Campfield LA, Burn P, Baskin DG. Identification of targets of leptin action in rat hypothalamus. J Clin Invest 1996; 98:1101–1106.

19. Erickson JC, Clegg KE, Palmiter RD. Sensitivity to leptin and susceptibility to seizures of mice lacking neuropeptide Y. Nature 1996; 381:415–421.

20. Segal-Lieberman G, Trombly DJ, Juthani V, Wang X, Maratos-Flier E. NPY ablation in C57BL/6 mice leads to mild obesity and to an impaired refeeding response to fasting. Am J Physiol Endocrinol Metab 2003; 284(6):E1131–9.

21. Erickson JC, Hollopeter G, Palmiter RD. Attenuation of the obesity syndrome of ob/ob mice by the loss of neuropeptide Y. Science 1996; 274:1704–1707.

22. Qiu J, Ogus S, Mounzih K, Ewart-Toland A, Chehab FF. Leptin-deficient mice backcrossed to the BALB/cJ genetic background have reduced adiposity, enhanced fertility, normal body temperature, and severe diabetes. Endocrinology 2001; 142:3421–3425.

23. Moore KJ, Nagle DL. Complex trait analysis in the mouse: The strengths, the limitations and the promise yet to come. Annu Rev Genet 2000; 34:653–686.

24. Leiter EH, Reifsnyder PC, Flurkey K, Partke HJ, Junger E, Herberg L. NIDDM genes in mice: deleterious synergism by both parental genomes contributes to diabetogenic thresholds. Diabetes 1998; 47:1287–1295.

25. Naggert JK, North MA, Nishina PM. Central gene defects causing obesity – fat and tub. In: Bray GA, Ryan DH, eds. Nutrition, Genetics, and Obesity. Baton Rouge, LA: Louisiana State University Press 1999:320–337.

26. Ikeda A, Zheng QY, Zuberi AR, Johnson KR, Naggert JK, Nishina PM. Microtubule-associated protein 1A is a modifier of tubby hearing (moth1). Nat Genet 2002; 30:401–405.

27. Serreze DV, Prochazka M, Reifsnyder PC, Bridgett MM, Leiter EH. Use of recombinant congenic and congenic strains of NOD mice to identify a new insulin-dependent diabetes resistance gene. J Exp Med 1994; 180:1553–1558.

28. Aitman TJ, Glazier AM, Wallace CA, Cooper LD, Norsworthy PJ, Wahid FN, Al-Majali KM, Trembling PM, Mann CJ, Shoulders CC, Graf D, St Lezin E, Kurtz TW, Kren V, Pravenec M, Ibrahimi A, Abumrad NA, Stanton LW, Scott J. Identification of Cd36 (Fat) as an insulin-resistance gene causing defective fatty acid and glucose metabolism in hypertensive rats. Nat Genet 1999; 21: 76–83.

29. Coleman DL, Hummel KP. The influence of genetic background on the expression of the obese (Ob) gene in the mouse. Diabetologia 1973; 9:287–293.

30. Coleman DL, Hummel KP. Symposium IV: diabetic syndrome in animals. Influence of genetic background on the expression of mutations at the diabetes locus in the mouse. II. Studies on background modifiers. Isr J Med Sci 1975; 11:708–713.

31. Mu JL, Naggert JK, Svenson KL, Collin GB, Kim JH, McFarland C, Nishina PM, Levine DM, Williams KJ, Paigen B. Quantitative trait loci analysis for the differences in susceptibility to atherosclerosis and diabetes between inbred mouse strains C57BL/6J and C57BLKS/J. J Lipid Res 1999; 40:1328–1335.

32. Zucker LM, Antoniades HN. Insulin and obesity in the Zucker genetically obese rat "fatty". Endocrinology 1972; 90:1320–1330.

33. Ikeda H, Shino A, Matsuo T, Iwatsuka H, Suzuoki Z. A new genetically obese-hyperglycemic rat (Wistar fatty). Diabetes 1981; 30:1045–1050.

34. Chung WK, Zheng M, Chua M, Kershaw E, Power-Kehoe L, Tsuji M, Wu-Peng XS, Williams J, Chua SC Jr, Leibel RL. Genetic modifiers of Leprfa associated with variability in insulin production and susceptibility to NIDDM. Genomics 1997; 41:332–344.

35. Kim JH, Nishina PM, Naggert JK. Genetic models for non insulin dependent diabetes mellitus in rodents. J Basic Clin Physiol Pharmacol 1998; 9:325–345.

36. Stoehr JP, Nadler ST, Schueler KL, Rabaglia ME, Yandell BS, Metz SA, Attie AD. Genetic obesity unmasks nonlinear interactions between murine type 2 diabetes susceptibility loci. Diabetes 2000; 49:1946–1954.

37. Kim JH, Sen S, Avery CS, Simpson E, Chandler P, Nishina PM, Churchill GA, Naggert JK. Genetic analysis of a new mouse model for non-insulin-dependent diabetes. Genomics 2001; 74:273–286.

38. Ohlemiller KK, Mosinger Ogilvie J, Lett JM, Hughes RM, LaRegina MC, Olson LM. The murine tub (rd5) mutation is not associated with a primary axonemal defect. Cell Tissue Res 1998; 291:489–495.

39. Coleman DL, Eicher EM. Fat (fat) and tubby (tub): two autosomal recessive mutations causing obesity syndromes in the mouse. J Hered 1990; 81:424–427.

40. Heckenlively JR, Chang B, Erway LC, Peng C, Hawes NL, Hageman GS, Roderick TH. Mouse model for Usher syndrome: linkage mapping suggests homology to Usher type I reported at human chromosome 11p15. Proc Natl Acad Sci U S A 1995; 92:11100–11104.

41. Ohlemiller KK, Hughes RM, Mosinger-Ogilvie J, Speck JD, Grosof DH, Silverman MS. Cochlear and retinal degeneration in the tubby mouse. Neuroreport 1995; 6:845–849.

42. Noben-Trauth K, Naggert JK, North MA, Nishina PM. A candidate gene for the mouse mutation tubby. Nature 1996; 380:534–538.

43. Kleyn PW, Fan W, Kovats SG, Lee JJ, Pulido JC, Wu Y, Berkemeier LR, Misumi DJ, Holmgren L, Charlat O, Woolf EA, Tayber O, Brody T, Shu P, Hawkins F, Kennedy B, Baldini L, Ebeling C, Alperin GD, Deeds J, Lakey ND, Culpepper J, Chen H, Glucksmann-Kuis MA, Carlson GA, Duyk GM, Moore KJ. Identification and characterization of the mouse obesity gene tubby: a member of a novel gene family. Cell 1996; 85:281–290.

44. Ikeda A, Nishina PM, Naggert JK. The tubby-like proteins, a family with roles in neuronal development and function. J Cell Sci 2002; 115(Pt 1):9–14.

45. Santagata S, Boggon TJ, Baird CL, Gomez CA, Zhao J, Shan WS, Myszka DG, Shapiro L. G-protein signaling through tubby proteins. Science 2001; 292:2041–2050.

46. Kapeller R, Moriarty A, Strauss A, Stubdal H, Theriault K, Siebert E, Chickering T, Morgenstern JP, Tartaglia LA, Lillie J. Tyrosine phosphorylation of tub and its association with Src homology 2 domain-containing proteins

implicate tub in intracellular signaling by insulin. J Biol Chem 1999; 274:24980–24986.

47. Hagstrom SA, Duyao M, North MA, Li T. Retinal degeneration in tulp1-/- mice: vesicular accumulation in the interphotoreceptor matrix. Invest Ophthalmol Vis Sci 1999; 40:2795–2802.

48. He W, Ikeda S, Bronson RT, Yan G, Nishina PM, North MA, Naggert JK. GFP-tagged expression and immunohistochemical studies to determine the subcellular localization of the tubby gene family members. Brain Res Mol Brain Res 2000; 81:109–117.

49. Hagstrom SA, Adamian M, Scimeca M, Pawlyk BS, Yue G, Li T. A role for the Tubby-like protein 1 in rhodopsin transport. Invest Ophthalmol Vis Sci 2001; 42:1955–1962.

50. Ikeda A, Zheng QY, Rosenstiel P, Maddatu T, Zuberi AR, Roopenian DC, North MA, Naggert JK, Johnson KR, Nishina PM. Genetic modification of hearing in tubby mice: evidence for the existence of a major gene (moth1) which protects tubby mice from hearing loss. Hum Mol Genet 1999; 8:1761–1767.

51. Ikeda A, Naggert JK, Nishina PM. Genetic modification of retinal degeneration in tubby mice. Exp Eye Res 2002; 74:455–461.

52. Naggert JK, Fricker LD, Varlamov O, Nishina PM, Rouille Y, Steiner DF, Carroll RJ, Paigen BJ, Leiter EH. Hyperproinsulinaemia in obese fat/fat mice associated with a carboxypeptidase E mutation which reduces enzyme activity. Nat Genet 1995; 10:135–142.

53. Kim JH, Naggert JK. Prohormone processing and disorders of energy homeostasis. In: Moustaïd-Moussa N, Berdanier CD, eds. Nutrient-Gene Interactions in Health and Disease. Boca Raton, FL: CRC Press, 2001:177–204.

54. Collin GB, Maddatu TP, Sen S, Naggert JK. Genetic modifiers interact with the fat mutation to affect body weight, adiposity, and hyperglycemia. Submitted.

55. Kido Y, Philippe N, Schaffer AA, Accili D. Genetic modifiers of the insulin resistance phenotype in mice. Diabetes 2000; 49:589–596.

56. Kulkarni RN, Almind K, Goren HJ, Winnay JN, Ueki K, Okada T, Kahn CR. Impact of genetic background on development of hyperinsulinemia and diabetes in insulin receptor/insulin receptor substrate-1 double heterozygous mice. Diabetes 2003; 52:1528–1534.

57. Almind K, Kulkarni RN, Lannon SM, Kahn CR. Identification of interactive loci linked to insulin and leptin in mice with genetic insulin resistance. Diabetes 2003; 52:1535–1543.

2

Quantitative Trait Loci for Obesity and Type II Diabetes in Rodents

Jung Han Kim

Nutrition Department, University of Tennessee, Knoxville, Tennessee, U.S.A.

ABSTRACT

Obesity and type II diabetes are two of the most common complex diseases in humans that are influenced by both genetic factors and the environment. These complex diseases are polygenic, implying that multiple genes referred to as susceptibility genes are involved in the onset of these diseases additively or interactively. One approach to dissect the genetic complexity is genome-wide quantitative linkage analysis (or genome-wide scan) using rodent models followed by fine mapping and positional cloning. A genome-wide scan, unlike a candidate gene approach, does not require any previous knowledge about gene functions and can identify chromosomal regions accounting for a quantitatively assessed trait, called quantitative trait loci (QTLs). Various polygenic rodent models mimicking human obesity and type II diabetes have been developed, and these models provide valuable resources to search for genetic factors underlying these metabolic disorders.

1. INTRODUCTION

Obesity and type II diabetes, which often coexist, are very common diseases in humans. An estimated 65 or 31% of U.S. adults are overweight or clinically obese, defined as a body mass index greater than 25 or 30 kg/m², respectively (1). In addition, about 90–95% of diabetes patients suffer type II diabetes (2,3), and currently this accounts for approximately 16 million patients in the United States. (http://www.diabetes.org/main/type2/info/default.jsp.)

The genetic contribution to human obesity and type II diabetes has been revealed through twin, adoption, and family studies, demonstrating that an individual with obese and/or diabetic relatives has a higher risk for being affected by these diseases [4–8].

There are some rare forms of obesity and type II diabetes that are caused by a single gene mutation, and these include mutations of A^y (agouti), *Cpe* (carboxypeptidase E), *Lep* (leptin), *Lepr* (leptin receptor), and *Tub* (tubby) and the maturity-onset diabetes of the young (MODY) [9–11]. To date, six MODYs, MODY1, 2, 3, 4, 5, and 6 have been reported in humans and the responsible genes are hepatocyte nuclear factor (HNF)-4α, glucokinase, HNF-1α, insulin promoter factor-1/pancreas duodenum homeobox-1/islet duodenum homeobox-1, HNF-1β, and NeuroD/BETA2, respectively (11). Mutation in the insulin 2 gene (*Ins2*) has been also reported for one murine MODY in the Akita mouse model (12).

Most common forms of obesity and type II diabetes in humans, however, follow polygenic inheritances: i.e., multiple genes are involved in the development of these diseases (13,14). These multiple genes are referred to as susceptibility genes, reflecting the concept that these genes confer an increased susceptibility to a disease rather than the certainty of developing the disease (14,15).

Animal models including rats and mice have long been an adjunct to human studies, minimizing many difficulties encountered in carrying out genetic studies of obesity and type II diabetes in human populations (16). For example, the capability of genetic and environmental controls, the availability of inbred strains and the ability to generate large experimental cohorts, and the short generation cycle can simplify and facilitate genetic studies. Furthermore, as rodents and humans share basic biological and physiological characteristics, and gene order over large distances has been conserved through evolution, candidate genes or pathways found in rodents can readily be tested in humans (17,18). Indeed, obesity genes known in humans, such as *Lep* and *Lepr*, were discovered in mice first (19,20). This review will discuss common strategies for dissecting genetic factors underlying obesity and type II diabetes using polygenic rodent models and the related genetic studies.

2. STRATEGY TO DEFINE QTLs FOR OBESITY AND TYPE II DIABETES IN RODENTS

2.1. Complex Traits and QTLs

When a one-to-one association between genotype and phenotype does not exist for certain traits, these traits are called complex traits as opposed to simple mendelian traits (21,22). This inconsistency between genotype and phenotype, despite an evidence of strong heredity for the traits, results from the fact that complex traits, unlike single mendelian traits, are determined by multiple factors including genes and environments. Complexity is created by the presence of multiple genes contributing in different degrees to the trait, possible interactions among those genes, and interactions between genes and environments in the determination of the traits (15,23). The list of complex traits can include natural traits such as skin color, wavy hair, height, and behavioral characteristics as well as disease traits [called complex diseases) such as hypertension, obesity, diabetes, alcoholism, and cancer (23–25).

Because of the involvement of multiple genes in controlling the traits, complex traits are also referred as polygenic traits. These traits are usually quantitative or assessed quantitatively, and thus the trait controlling genes (or loci) are called QTLs (21). A review of terminology frequently used in genetics is available (26).

2.2. Dissecting Genetic Factors for Complex Disease Using Polygenic Animal Models

2.2.1. Sources of Polygenic Models

Selected Strains. An often-used approach to create animal models harboring genetic variation is long-term breeding with phenotypic selection (27,28). When individuals in a population differentially express a trait that is heritable (often found in heterogeneous outbred populations), breeding individuals ranking at phenotypic extremes (i.e., heaviest or lightest) can produce offspring that also rank at the extremes (28). Repeating this selective breeding (usually followed by inbreeding) over several generations can fix genetic variants that contribute to the selected trait, creating new lines that possess the extreme phenotypes (9,28).

Standard Inbred Strains. Inbred strains are defined to be homozygous for each gene throughout the genome, and numerous inbred strains, especially for mice, are currently available (25,29,30) (http://www.informatics. jax.org/external/festing/search_form.cgi). Because standard inbred strains are generated via repeated brother-sister mating at random over at least 20 generations, these phenotypic variations among the strains result from naturally occurring gene combinations (28). Indeed, many phenotypic variations including adiposity and glucose metabolism have been reported among existing inbred strains (27) (http://aretha.jax.org/pub-cgi/phenome/mpdcgi?rtn=docs/home). With the currently growing database for standard inbred strains regarding genetic maps (http://www-genome.wi.mit.edu/cgi-bin/rat/gmap_search; http://www-genome.wi.mit.edu/cgi-bin/mouse/index) and phenotypic characterizations (31), these can serve as very valuable resources for biomedical science including complex disease field of obesity and diabetes.

2.3. Mapping of a QTL

Mapping, i.e., identifying the chromosomal location, is the first step to dissect the genetic factors contributing to a trait. Genome-wide QTL linkage analysis (or genome-wide scan) has been a powerful way for comprehensively mapping genetic factors underlying complex diseases (23). Because this approach does not require any molecular knowledge of the traits of interest, it has the potential of discovering new genes or pathways not previously known.

The approach consists of studying whether there is an association (cosegregation) between the genotype at a marker locus and the trait values, and if there is, then this indicates that the marker locus is close by (or linked) to the putative disease QTL (14,26).

2.3.1. Genetic Crosses to Create Segregating Populations

Commonly, QTL mapping is initiated by crossing two different inbred strains that show contrasting phenotype and genotypic variation (9,27). The resultant F1 mice are then intercrossed (sib mating) or backcrossed to one of the parental strains. This will generate F2 or backcross (BC) progeny in which both phenotypes and genotypes segregate unlike in the grandparental and parental generations. The origin of this segregation is the recombination of the DNA segments between the homologous chromosome pairs occurring during meiosis in the production of germ cells from the F1 parent. The recombination events are random and, consequently, individual F2 or BC mice possess a unique combination of progenitor alleles, which gives rise to segregation of genotypes and concurrently phenotypes. Phenotypes can, however, be influenced by environmental factors, such as high and low fat diets (25). An example of a genetic cross is depicted in Fig. 1.

The choice of methods for genetic crosses is reviewed in detail in elsewhere (21,32). In addition to F2 or BC mice, recombinant inbred (RI) strains, which have a fixed genotype, have been used for QTL mapping, and the usage of RI strains is thoroughly reviewed elsewhere (33,34).

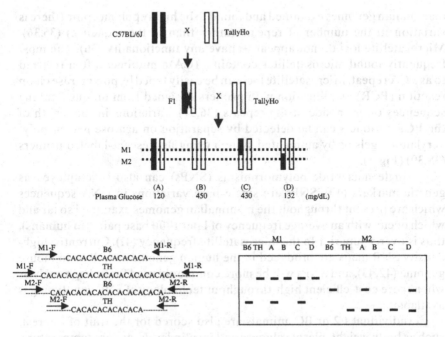

FIGURE 1 Genome-wide scan to identify QTLs controlling plasma glucose levels in the diabetic TallyHo mice. Backcross (BC) mice (shown as A–D) were collected from a cross between F1 (C57BL/6J × TallyHo) and TallyHo mice, genotyped with markers (shown as M1 and M2) throughout the genome, and phenotyped for plasma glucose levels. DNA from the BC mice were PCR amplified with markers (bottom left), run on the agarose gel, and visualized with ethidiumbromide (bottom right). Between M1 and M2, for example, M2 is likely linked to a QTL controlling plasma glucose levels. M1 = marker1; M2 = marker2; F = forward primer for PCR; R = reverse primer for PCR; B6 = C57BL/6J; TH = TallyHo.

2.3.2. Genotyping and Phenotyping of Segregating Populations for Linkage Analysis

A segregating population as described in the preceding paragraphs is then genotyped throughout the entire genome using a series of genetic markers that are polymorphic, i.e., they differ between the two parental strains. Using inbred strains provides a great advantage in that only two alleles of genetic factors originating from each of the crossbred strains segregate in the population, and this makes all polymorphic markers informative for the genotype at all loci (32).

The most commonly used genetic markers are microsatellite repeats, also known as simple sequence repeats, that are present throughout all mammalian genomes examined and found to be highly polymorphic (there is variation in the number of repeats rather than in the sequence) (35,36). Microsatellite loci do not appear to have any functionality (36). The most frequently found microsatellites contain a (CA)n multimer, often referred to as a CA repeat. Microsatellite loci can be easily typed by polymerase chain reaction (PCR) amplification with primers designed from unique flanking sequences on each side of the repeats (36,37). Variations in the length of the PCR products can be detected by separation on agarose gels or poly-acrylamide gels or by automated system using fluorescent-labeled primers (38,39) (Fig. 1).

Single nucleotide polymorphisms (SNPs) can also be employed as genetic markers (40). SNPs are single-base variations in DNA sequences which are present throughout the mammalian genomes examined so far and which occur with an average frequency of 1 per 1000 base pairs (in humans), thus is overly superior to the microsatellite frequency (41). Currently, high-density SNP maps are produced in the human genome as well as in mouse genome (42,43), and SNPs will be more commonly used for linkage analysis when more cost-efficient high-throughput technology of SNP genotyping is available.

Individual F2 or BC animals are also scored for the trait of interest, such as body weight, plasma glucose and insulin levels, or core temperature. Subsequently, using statistical methods, the individual phenotypic scores are examined for correlation with the genotypes of the polymorphic markers (44) (Fig. 2). Regardless of the statistical methods applied, the basic tenet is that markers that are significantly associated with the trait lie close to (are linked to) QTLs that are responsible for variation in the trait of interest. This is based on the assumption that the recombination during meiosis less likely occurs between closely linked loci on the same chromosome, resulting in cotransmission of these closely linked loci (15). With known map positions of genetic markers, the genomic locations of QTLs on the chromosomes can be estimated statistically.

Further details about statistical analysis for QTL mapping including methods, number of markers and animals required, and software available are very well reviewed elsewhere [44–48]. One statistical model for data analysis derived from a cross between outbred strains has been discussed by Nagamine and Haley (49).

2.4. Fine Mapping of a QTL

Once the map position has been determined for a certain disease/trait gene, narrowing the genomic interval (fine mapping) is essential to identifying the actual allelic variants (molecular basis) by positional cloning.

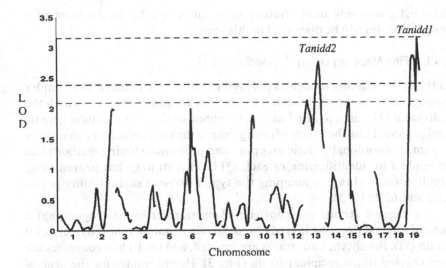

FIGURE 2 Plot of one-dimensional genome-wide scan on 19 autosomes for plasma glucose levels in BC progeny from (B6 × TallyHo) F1 and TallyHo mice using the pseudomarker method. The LOD score is plotted as a function of genome location. The dashed lines represent critical values from permutation tests at the 99, 95, and 90% significance levels. Two QTLs on chromosomes 13 and 19 were significantly linked to the plasma glucose levels and designated *Tanidd1* and *Tanidd2* (TallyHo-associated non-insulin-dependent diabetes), respectively. (Method from Ref. 162; Figure adapted from Ref. 95).

Conventionally, for monogenic traits, this can be done by increasing the number of meioses to collect more crossovers within the previously mapped genomic region. Practically, all animals from an expanded mapping cross are genotyped for markers flanking the locus, and only animals that carry a crossover between the flanking markers are retained and phenotyped. Testing more markers within this genomic interval will then more precisely localize the disease gene. The size of a genetic region determines the amount of time and resources needed for a positional cloning effort, and a smaller critical region greatly simplifies the process.

In the initial mapping, a QTL is usually located within a large interval on the chromosome. Unlike for monogenic traits, fine mapping of a QTL for complex traits is not straightforward because the phenotypes of individual animal are influenced by additional unlinked loci or environmental noise, resulting in difficulties in assigning carrier status in animals carrying recombinations within the QTL interval (22).

Several methods exist that can help to construct a fine map to conduct a practical search for the molecular basis of a specific QTL (21). Among those, the most commonly used strategy of isolating a QTL in the form of a congenic strain will be discussed in this review.

2.4.1. Fine Mapping Using Congenic Strains

In the case of animal models, a polygenic trait can be reduced to a simpler monogenic form through the development of congenic strains in which the individual QTL is separated as a new inbred strain on a common genetic background. Thus, the effects of each gene can then be studied. In this simpler system, conventional genetic mapping and positional cloning methods can be applied for identification of each QTL. This strategy has proven to be highly successful in fine mapping the type I diabetes susceptibility genes, *Idd3* and *Idd10* (50,51).

Congenic strains are constructed by transferring (introgressing) a gene of interest from one inbred (donor) strain to another inbred (recipient) strain (52). Briefly, the two strains are crossed, and the F1 heterozygotes are backcrossed to the recipient strain (Fig. 3). Heterozygotes for the gene of interest (i.e., a QTL) are selected using flanking markers and backcrossed again to the recipient strain. This procedure is repeated for at least ten cycles of backcrossing at which point two heterozygotes are intercrossed to yield offspring that are homozygous for the donor allele. At each backcross, 50% of unlinked loci will become fixed for recipient alleles at random. Consequently, the genetic background becomes progressively enriched with the recipient genome and the donor gene will be "hauled" along with the markers in this procedure. The name of the congenic strain is designated using the format: Recipient.Donor-Introgressed region[differential allele] (53). For example, when a congenic rat is constructed by transferring the *Nidd-11/of* QTL from the OLETF rat to the F344 rat background strain, the congenic rat is designated F344.OLETF-*Nidd1-11/of*.

2.4.2. Positional Cloning of a QTL

Once the position of a QTL has been refined to a small candidate region, a contig of clones from large insert genomic libraries across the minimal region of the QTL needs to be assembled. This can be followed by identification of transcripts within the physical contig, testing for mutations within transcripts that may account for the disease, and verification that a sequence change is not simply a polymorphism and that the mutation identified is indeed the cause of the disease. Technical details including contig-building methods are very well reviewed elsewhere (32,54,55). However, contig assembly may not be necessary anymore with the current availability of high-density sequence databases, and this whole process can be expedited.

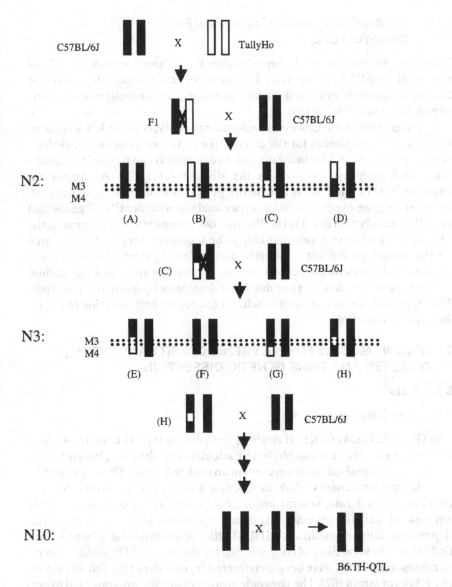

FIGURE 3 Construction of congenic mouse strain carrying diabetes QTL derived from TallyHo (TH) mice in C57BL/6J (B6) background, B6.TH-QTL, by marker-assisted backcrossing. Diabetic TH mouse is crossed to normal B6 mouse, and the resultant F1 mice are backcrossed to B6. Heterozygotes for the QTL are selected using flanking markers (shown as M3 and M4) and backcrossed again to B6. This procedure is repeated for ten cycles of backcrossing at which point two heterozygotes are inter-crossed to yield offspring that are homozygous for the QTL.

2.4.3.   Possible Complementary Strategies to Positional
Cloning of a QTL

Evaluating known candidate genes within the minimal region, called the positional candidate gene test, by sequence comparison between the disease-susceptible strain and normal inbred strains is usually used as a first step in positional cloning (56).

Large-scale gene expression microarray analysis using RNA isolated from likely target tissues for the disease (i.e., liver, adipose tissue, skeletal muscle, pancreas, and brain for obesity and diabetes) can be used to identify genes and pathways involved in the disease (28,57,58). An important question in QTL analysis is how many genes control the trait of interest. Large-scale gene expression microarray analysis will identify all genes that are differentially expressed in the disease state compared to the normal state. The majority of these genes are likely to be secondary targets, i.e., required for the phenotype but not responsible for its development. However, if one or more of the genetic alterations causing the disease also causes disregulation of gene expression, this too should be detectable by microarray analysis. This approach has proven quite useful in the recent identification of a QTL for hypertension (59).

3.   RODENT MODELS FOR POLYGENIC OBESITY AND TYPE II DIABETES AND THEIR GENETIC DISSECTION

3.1.   Rats

3.1.1.   The Goto-Kakizaki Rat

The Goto-Kakizaki (GK) rat strain is a model for type II diabetes without obesity. These rats were established by selective breeding for glucose intolerance from natural variations present in an outbred stock of Wistar rats (60).

In genetic crosses of diabetic GK and normoglycemic Brown-Norway (BN) or Fischer rats, several major and minor QTLs, depending on the amount of genetic variance explained by those loci, for non-insulin-dependent diabetes mellitus (NIDDM)-like phenotypes have been identified (61,62). Several lines of congenic rats for the major QTL, *Niddm1*, on rat chromosome (Chr) 1 have been constructed by transfer of the GK allele onto the F344 rat strain (62). The congenic strains displayed postprandial hyperglycemia, reduced insulin action in adipocytes, and insulin secretory defects (60). Using these congenic strains, Fakhrai-Rad et al. (63) were able to fine map *Niddm1* and identify a candidate gene, insulin-degrading enzyme (*Ide*). A functional defect in the IDE protein has been confirmed by an in vitro assay demonstrating that cells transfected with the GK allele of *Ide*, coding for two amino acid substitutions, exhibited a reduction in insulin-degrading activity by 31%. Further evidence that this reduced enzyme activity is indeed underlying the QTL comes from gene targeting experiments in mice which showed that mice homozygous for an *Ide* null allele are hyperinsulinemic and glucose intolerant (64).

3.1.2. The Otsuka Long-Evans Tokushima Fatty Rat

The Otsuka Long-Evans Tokushima Fatty (OLETF) rat strain was established by selective inbreeding for glucose intolerance from several rats spontaneously exhibiting polyuria, polydipsia, and mild obesity that were observed in an outbred colony of Long-Evans rats (65). A nondiabetic strain of Long-Evans Tokushima Otsuka (LETO) rats was also established from different mating from the same colony of Long-Evans rats.

The OLETF rat strain is characterized by mild obesity, gender- (mostly in males) and age-dependent diabetes, and diabetic nephropathy [66–69]. Diet restriction was effective to prevent or improve diabetes and nephropathy in OLETF rats (68).

Initially, a diabetogenic QTL was mapped on Chr X by analyzing F2 progeny of OLETF and either LETO or F344 rats (70). Subsequently, several genome-wide scan studies were conducted using F2 progeny from intercrosses between OLETF and BN rats or between OLETF and F344 rats as well as BC progeny from a cross between (OLETFxBN) F1 and OLETF rats [71–76]. Through these studies, several QTLs, including *Dmo1* and *Nidd1-11/of*, linked to NIDDM related phenotypes, obesity, and dyslipidemia were detected.

Watanabe et al. (77) and Okuno et al. (78,79) generated congenic rats carrying alleles of *Dmo1* (rat Chr 1) derived from either the BN or the F344 strain in the OLETF background, replacing the susceptibility allele in the OLETF strain with the normal allele. Even though the backcrossing was done only for four generations, a substantial attenuation in the development of obesity, dyslipidemia, and glucose intolerance was observed in these congenic rats confirming the diabetogenic and obesigenic effect of *Dom1* (77,79).

Kose et al. (80) generated 11 lines of congenic rats, F344.OLETF-*Nidd1-11/of*, in the F344 background carrying the OLETF alleles of 11 QTLs (*Nidd1-11/of*) that were identified in a study using F2 progeny of OLETF and F344 rats (72). After five generations of backcrossing to the F344 strain, six lines of the congenic rats containing *Nidd1, 2, 3, 4, 7,* and *10/of* QTLs, respectively, showed glucose intolerance while five other lines did not (80).

3.1.3. The Spontaneously Diabetic Torii Rat

The Spontaneously Diabetic Torii (SDT) rat strain is the most recently (in 1997) established rat model for type II diabetes without obesity. In 1988, five male rats were found to be polyuric and glucosuric in an outbred colony of the Crj:CD(SD) stock (Charles River Japan, Inc., Kanagawa, Japan) of Sprague-Dawley rats, and these affected rats were kept in the Research Laboratories of Torii Pharmaceutical Co., Ltd., Chiba, Japan (81). For 20 generations these rats were brother-sister mated, and the diabetic strain named SDT was established. Like the OLETF rats, the SDT rats exhibit age dependence and gender dimorphism in the development of diabetes, exhibiting 100% incidence in males vs. 33.3% in females at 40 weeks of age (81). The survival rates were, however, almost comparable between males and females. Male SDT rats are characterized by hyperglycemia, hypoinsulinemia, and hypertiglyceridemia, with absence of obesity (81).

Genetic analysis for diabetes QTLs has been performed using BC male progeny obtained from a cross between (BN × SDT) F1 and SDT rats, and three major QTLs, *Gisdt1-3*, linked to glucose intolerance were identified on rat Chrs 1, 2, and X, respectively (82).

3.2. Mice

3.2.1. The Kasukabe Habitat/K Group Mouse

The Kasukabe habitat/K group (KK) mouse strain was one of the inbred strains established from Japanese native mice in 1957 (83). Several years after establishing the strain, the KK mice were found to have diabetes characteristics of glucose intolerance, insulin resistance, hypertrophy and degranulation of pancreatic islets, and nephropathy with moderate obesity (83). The KK mouse strain was also susceptible to diet- or chemical- and genetically (by introducing the A^y allele) induced obesity, which was highly correlated with the severity and incidence of diabetes in this strain (83,84). The KK-A^y/a strain (also called the yellow KK mouse) is a congenic mouse in which the A^y allele at the agouti locus had been introgressed from the B6-A^y strain (85).

A genome-wide search using the 14% phenotypic extremes for adiposity of an F2 progeny obtained from a cross between (KK/HlLt × C57BL6/J) F1 mice revealed two significant QTLs affecting adiposity, *Obq* (obesity QTL) 5 and 6 on Chrs 9 and X, respectively, and two suggestive QTLs on Chrs 18 and 7 (86).

A major QTL linked to fasting glucose levels, *Fgq1*, was mapped on Chr 6 in F2 progeny (a/a genotype) from a cross between F1s of KK-A^y/a and B6-a/a mice (87). In F2 progeny with A^y/a genotype from the same cross, no significant linkage was shown, but some suggestive loci that were not detected in F2-a/a mice accounted for glucose intolerance. These loci, so-called genetic modifiers, possibly interact with A^y. Genetic modifiers are discussed in the chapter by Wang et al. in this volume.

In male BC progeny from a cross between (BALB/c × KK/Ta) F1 and KK/Ta mice, a genome-wide scan analysis identified one major QTL contributing to glucose intolerance, *Igt-1* (impaired glucose tolerance-1), on Chr 6 and two major QTLs accounting for fasting blood glucose levels, *Fbg-1* and *Fbg-2* (fasting blood glucose level-1 and -2), on Chrs 12 and 15, respectively (88). Two additional significant QTLs linked to serum triglyceride levels were detected on Chrs 4 and 8, respectively.

3.2.2. The Nagoya-Shibata-Yasuda Mouse

The Nagoya-Shibata-Yasuda (NSY) mouse strain is a type II diabetes model with moderate obesity, established by selective inbreeding for glucose intolerance in the descendants of streptozotocin-induced diabetic mice from a outbred stock of Jcl:ICR mice (89). Briefly, diabetes was induced in the adult ICR mice by treatment with streptozotocin, and the diabetic mice were then mated. Glucose tolerance was tested in the resultant F1 mice, and the F1 mice with impaired glucose tolerance were selected and mated with siblings. This selection was repeated over several generations. The incidence of diabetes in the NSY mice was age-dependent and significantly higher in males than females (98 vs. 31%) (90). Neither severe obesity nor extreme hyperinsulinemia was observed in these mice, but glucose-stimulated insulin secretion was markedly diminished (90,91).

Three QTLs linked to glucose intolerance, *Nidd1nsy*, *Nidd2nsy*, and *Nidd3nsy*, were localized on Chrs 11, 14, and 6, respectively, in F2 progeny from a cross between F1s of NSY and C3H/He mice (92).

3.2.3. The Tsumura, Suzuki, Obese Diabetes Mouse

The Tsumura, Suzuki, Obese Diabetes (TSOD) mouse strain is an inbred strain originating from six obese male mice with urinary glucose found in a ddY colony (93). The obese male mice were mated with normal females and the resultant offspring were inbred with selection for increased body weight and urinary glucose to establish the obese diabetic strain of TSOD mice (93). From the same pool of initial breeders, a nonobese, nonglycosuric line was also established and named TSNO (Tsumura, Suzuki, Non Obesity) mice (93). Male TSOD mice are characterized by obesity, urinary glucose with hyperphasia and polydipsia, hyperinsulinemia, hyperglycemia, and dyslipidemia as well as hypertrophy of the pancreatic islets (93).

By QTL mapping using F2 progeny from a (TSOD × BALB/cA) F1 intercross, a QTL affecting blood glucose levels during an intraperitoneal glucose tolerance test (IPGTT), *Nidd4*, was localized on Chr 11, and two other QTLs accounting for body weight, *Nidd5* and *Nidd6*, were mapped on Chrs 1 and 2, respectively. *Nidd5* was also linked to blood insulin levels during the IPGTT (94).

3.2.4. The *TallyHo* Mouse

The TallyHo mouse strain was established from several male mice found to be polyuric, glucosuric, hyperglycemic, and hyperinsulinemic in an outbred colony of Theiler Original mice (95). These progenitors were brother-sister mated over several generations with selection for plasma glucose levels >300 mg/dL (nonfasting) in males. The TallyHo mice are characterized by moderate obesity, hyperinsulinemia, hyperlipidemia, and male-biased hyperglycemia.

A genome-wide scan using BC progeny from a cross between F1(B6 × TallyHo) and TallyHo mice revealed two significant QTLs linked to hyperglycemia, *Taniddl* and *Tanidd2* (TallyHo-associated non-insulin-dependent diabetes), on Chrs 19 and 13, respectively (95). The *Taniddl* QTL was also detected in BC progeny from a separate cross between F1(CAST/Ei × TallyHo) and TallyHo mice. The third QTL linked to hyper-glycemia, *Tanidd3*, was also detected in this CAST cross on Chr 16. Two obesity-related QTLs, *Tabw* (TallyHo-associated body weight) and *Tafat* (TallyHo-associated fat), were identified on Chrs 7 and 4, respectively.

3.2.5. The New Zealand Obese Mouse

The New Zealand Obese (NZO) mouse strain is an inbred strain developed from a mixed genetic background colony with selection for obesity (96). The NZO mice are characterized by hyperphasia, insulin resistance, hyperinsulinemia, hyperglycemia (mostly in males), hypertension, hyper-cholesterolemia, and leptin resistance [16,97–99].

Genetic dissections for the polygenic obesity syndrome in NZO mice were conducted using a BC population obtained from a cross between NZO and (NZO × SJL)F1 mice. Two obesity/hyperinsulinemia linked QTLs, *Nobl* and *Nob2* (New Zealand obese1 and 2), on Chrs 5 and 19, respectively (100), and a diabetes QTL, *Nidd/SJL*, on Chr 4 (101) were detected in the BC progeny. Interestingly, there seems to be an interaction between *Nidd/SJL* and *Nobl* accounting for 90% of the variance in plasma glucose levels observed in the BC population, and this interaction was enhanced by a high-fat diet (102). Also, a susceptibility QTL for hyperchole-sterolemia, *Choll/NZO*, was mapped to Chr 5 (103).

Multiple obesity QTLs affecting adiposity, *Obq7-15*, have been localized in an F2 population from an intercross of F1s of NZO and SM (small) mice (104).

During genetic analysis using male F2 intercross progeny from NZO/H1Lt and NON/Lt mice, three QTLs affecting nonfasting glucose and plasma insulin levels, *Niddl*, *Nidd2* and *Nidd3*, were mapped on Chrs 4, 11, and 18, respectively (105). Interestingly, F1 males in this cross have shown a more severe type II diabetes syndrome than the parental strain of NZO mice (105). In a subsequent study, a backcross strategy with the same parental strains was adapted, and several other QTLs affecting obesity, plasma glucose levels, and the diabetes subphenotypes were detected (106). These included QTLs on Chrs 1 (obesity and diabetes), 12 (obesity), and 5 (obesity) that exhibited complex epistatic interactions among themselves or with other genomic regions increasing the severity of the diabesity syndrome.

In an attempt to dissect these genetic complexities, Reifsnyder et al. (107) generated ten lines of interval-directed recombinant congenic (RC) strains in the NON/Lt background (http://www.jax.org/staff/leiter/labsite/type2_genomics.html) based on QTL data from their previous two studies (105,106). All ten RC lines were significantly heavier than the NON parental strain, but not as heavy as the diabetes-prone NZO parental strain. Incidences of diabetes in these RC lines differed among the lines, ranging from 0 to 100%.

The protocol for construction of an RC strain is different in some ways from that for traditional congenic strains described earlier, including the number of backcrosses. Briefly, RC strains are derived by inbreeding after conducting several independent backcrosses (2–3 times) from a donor to a host (recipient) strain, which results in 6.25–12.5% random contribution of the donor genome in each RC strain (108). Phenotypic characterization is followed for the relevant traits in the set of RC strains, and RC strains exhibiting phenotypic differences compared to the host strain are implicated as carrying QTLs originating from the donor strain. Rather than for fine mapping, RC strains can be more useful for the phenotype study of gene/gene interactions (109).

3.2.6. The High and Low Heat Loss Mice

The High (MH) and Low (ML) lines of mice are noninbred lines selectively bred for high heat loss and low heat loss, respectively, from four outbred strains of mice including NIH and ICR (Harlan Sprague-Dawley) and CF-1 and CFW (Charles River) (110,111). Along with this selection, an MC line of unselected controls was also established. Heat loss (kcal/day) measured by individual animal direct calorimetry differed by 37% of the mean between MH and ML mice (112). The MH mice consumed more energy than the ML and the MH mice were leaner than the ML (112). When compared with three common inbred mouse strains, B6, DBA/J, and SWR/J, MH mice exhibited higher heat loss than B6 mice by 26.3% (112). This difference was exaggerated with high-fat diet feeding (112).

Genetic analysis using F2 progeny from an intercross of F1s of MH and B6 mice detected four QTLs controlling heat loss on Chrs 1, 2, 3, and 7, respectively, and the QTL on Chr 1 was confirmed in an F2 population from a separate cross between MH and ML mice (113). A QTL linked to fat pad weight, *Fatq1*, was found on Chr 1, and QTLs linked to body weight were also identified on Chrs 1 (*Wt3q1-2*:3 wk, *Wt6q1-2*: 6wk, *Wt10q1*: 10wk), 3 (*Wt10q2*), 11 (*Wt6q3, Wt10q3*), and 17 (*Wt3q3*) (113).

3.2.7. The M16 Mouse

The M16 mouse strain is an inbred strain established by selection from ICR mice for rapid postweaning (3–6 week) weight gain (114). Compared to unselected controls, the M16 mice exhibited greater body weight gain and hyperphagia (115) as well as greater fat pads as a percentage of body weight (116). The adipose cellularity of M16 mice was characterized by both hypertrophy and hyperplasia (115,117).

By genetic analysis using BC progeny from a cross between M16 and CAST/Ei mice, six QTLs controlling fatness, *Pfat1-6*, were mapped (27).

3.2.8. The DU6 Mouse

The DU6 line of mice was selected for high body weight at six weeks of age from outbred Fzt:DU mice through a systematic crossbreeding scheme with four inbred (CBA/Bln, AB/Bln, C57BL/Bln, and XVII/Bln) strains and four outbred (NMRI orig., Han:NMRI, CFW, and CF1) stocks of mice (118,119). An unselected line of the DUKs was also established from the same base population, and both the DU6 and the DUK mice have been maintained as outbred lines. The DU6 mice have twice the body weight (58 vs. 28 g at 42 days) and three times the fat content of the DUKs mice (120). The DU6 mice were also characterized by hyperleptinemia, hyperinsulinemia, and elevated insulin-like growth factor I (IGF-1) (121).

Nine QTLs for body weight, with the biggest effect of *Bw4* on Chr 11, and eight QTLs for fat accumulation were detected using F2 progeny from an intercross of F1s of the DU6 and the DUKs mice (122).

Recently, a DU6 line has been inbred for four generations generating a partially inbred line of DU6i mice (123). The DU6i mice were crossed to DBA/2 mice to produce F2 progeny and genome-wide QTL mapping was conducted (123). Using this genetic cross, body weight QTLs on Chrs 1, 5, and 7, muscle weight QTLs on Chrs 7, 11, and 12, and abdominal fat weight QTLs on Chrs 4, 5, 7, 11, 12, and 13 were detected, with the largest effects being by *Bw14* on body weight, *Mwq1* on muscle weight, and *Afw9* on abdominal fat weight. These three QTLs were located on Chr 7.

3.2.9. The Large Mouse

The large (LG/J) mouse strain is an inbred strain developed with long-term selection for large body size (124,125). The LG/J mice (albino) are characterized by rapid growth early in life, longer tails, and higher body, liver, and fat pad weights when compared to the SM/J inbred strain (126).

By genetic analysis using F2 progeny of the LG/J and the SM/J mice, Cheverud et al. (127) detected eight QTLs (*Adip1-8*) for adiposity, seven QTLs (*Skl1-7*) for tail length, and four QTLs (*Wt1-4*) for adult body weight.

3.2.10. The Small Mouse

The small (SM/J) mouse strain (white-bellied agouti or black) is an inbred strain established by crossing seven stocks including DBA with selection for small body size at 60 days (128). Although the names sound related, the SM/J and the existing LG/J mice were derived from a separate experiment.

Due to their small body size, the SM/J mouse has been utilized for several genetic studies of obesity. In an intercross population from the SM/J mice and a normal inbred strain of A/J mice, two significant QTLs, *Bwq3* and *Bwq4*, linked to body weight, were detected on Chrs 8 and 18, respectively, as well as five suggestive QTLs (129). The SM/J alleles of both *Bwq3* and *Bwq4* were associated with increase in body weight although the parental SM/J mouse was smaller than the A/J mouse. Exhaustive discussion about parental origins of susceptibility alleles for individual QTL has been avoided in this review. But, oftentimes, one finds that susceptibility alleles of certain QTLs originate from the "low-value" strain (i.e., a small-size animal) rather than from the "high-value" strain (i.e., a large-size animal). This suggests that new combinations of genetic factors in segregating populations, such as F2 or BC populations, can expose susceptibility alleles that were quiescent in parental strains.

3.2.11. The Edinburgh Fat and Lean Mice

The Edinburgh Fat (EF or F) and Edinburgh Lean (EL or L) lines originated from a three-way cross base (two inbred [CBA and JU] and one outbred line [CFLP]) with divergent selection for high and low fat content (130). Initially, the selection index for the first 20 generations was the ratio of gonadal fat pad weight to body weight of 10-week-old males, and subsequently, the index was changed to dry matter content, known to be highly correlated with fat content, of 14-week-old males. Selection was carried out for over 60 generations (131) and the lines are maintained outbred. Average fat contents of EF and EL males were about 21 and 4%, respectively, and EF males were about 11 g heavier at 14 weeks of age than EL males (130).

By genetic analysis using F2 progeny from an intercross of F1s of the EF and the EL mice, Horvat et al. (132) detected four QTLs accounting for fat %, *Fob1-4* (F-line obesity QTL 1–4) on Chrs 2, 12, 15, and X, respectively.

TABLE 1 Polygenic Rodent Models for Obesity and Type II Diabetes and the Related QTLs

Model	Selection	Obesity	Diabetes	Genetic cross		QTLs	Congenics	Gene	Ref.
Rat									
GK	IGT	—	X	F2	F344	Niddm1-3, Niddgk1-6	F344.GK-Niddm1	Ide	(60–63)
OLETF	IGT	X	X	F2	LETO, BN, F344	Odb1-2, Dmo1-3, Obs1-6	OLETF.BN-Dmo1, OLETF.F344-Dmo1	—	(64–80)
				BC	BN	Niddm1-11/of, Dmo1-12	F344.OLETF-Niddm1-11/of		
SDT	—	—	X	BC	BN	Gisdt1-3	—	—	(81,82)
Mouse									
KK	—	Mild	X	F2	B6	Obq5-6, Fgq1	—	—	(83–88)
				BC	BALB	Igt-1, Fbg-1-2			
NSY	IGT	Mild	X	F2	C3H	Nidd1-3nsy	—	—	(89–92)
TSOD	BW & UG	X	X	F2	BALB	Nidd4-6	—	—	(93,94)
TallyHo	HG	X	X	BC	B6, CAST	Tanidd1-3, Tabw, Tafat	—	—	(95)
NZO	Obesity	X	X	BC	SJL	Nob1-2, Nidd/SJL, Chol1/NZO	RC of NON.NZO	—	(96–109)
				F2	SM, NON	Obq7-15, Nidd1-3			
				BC	NON	Chrs 1, 12, 5			

Cross	Pop.	Pheno.	Obesity	Strain	Locus	Ref.
MH		HHL	X	B6	Fatq1	(110–113)
M16	F2	WG	X	ML	Wt3q1-3	(114–117,27)
				ML	Wt6q1-3	
				ML	Wt10q1-3	
DU6	BC	HBW	X	CAST	Pfat1-6	(118–123)
DU6i	F2		X	DUK	Bw4	
	F2		X	DBA	Bw14	
				DBA	Mwq1	
				DBA	Afw9	
LG	F2	HBW	X	SM	Adip1-8	(124–127)
F		HFC	X		Wt1-4	
EPH	F2	HBW	X	EL	Fob1-4	(130,132)
	F2		X	EPL	Chr X (EPL, EPH-Chr X QTL)	(130,133–135)
B6 × SPRET	BC		X		Mob-1-4	(136,137)
NZB × SM	F2		X	SM	Mob-5	(138–141)
EL × 129	F2		X		Obq1-2	(143)
AKR × C57L	F2		X		Obq3-4	(144,145)
B6 × C3H	F2/F3		— (IGT: X)		Chrs 2 & 13 / IGT	(146)
BTBR × B6	F2		— (X)		T2dm1-3	(147–149)
Diet induced						
AKR × SWR	F2		X		Do1-3	(156–158)
B6 × CAST	F2		X		Do3 (B6.CAST-Do3)	(159–161)
					Mob-5-8	

IGT = impaired glucose tolerance; BW = body weight; UG = urinary glucose; HG = hyperglycemia; HHL = high heat loss; WG = rapid post-weaning weight gain; HBW = high body weight; HFC = high fat content; Gisdt1-3 = glucose intolerance SDT1-3; Obq5-6 = obesity QTL5-6; Fgq1 = fasting glucose QTL1; Igt-1 = impaired glucose tolerance-1; Fbg-1-2 = fasting blood glucose-1-2; Taniddf-3 = TallyHo-associated non-insulin-dependent diabetes1-3; Tabw = TallyHo-associated body weight; Tafat = TallyHo-associated fat; Nob1-2 = New Zealand obese1-2; Chol1/NZO = Cholesterol1/NZO; RC = recombinant congenic mice; Mwq1 = muscle weight QTL1; Afw9 = abdominal fat weight 9; Adip1-8 = adipocity1-8; Wt1-4 = weight1-4; Fob1-4 = F-line obesity QTL 1-4; Mob-1-4 = multigenic obesity-1-4; T2dm1-3 = type II diabetes mellitus 1-3; Do1-3 = dietary obese1-3.

3.2.12. The Edinburgh High and Low Mice

The Edinburgh High (EPH) and Edinburgh Low (EPL) lines are divergently selected lines for high and low body weights, respectively, from the same base population as the lines EF and EL described in the preceding paragraph (130). The selection index for the first 20 generations was lean mass in males, and in subsequent generations the index was changed to body weight in both sexes at 10 weeks of age.

Hastings and Veerkamp (133) reported a large additive sex-linked effect (25%) in the body weight divergence between the EPH and the EPL lines. In a follow-up study, a QTL with a significant LOD score was localized to the proximal region of Chr X using F2 progeny from a cross between F1s of the outbred EPH and the inbred EPL mice (134). This QTL was confirmed in a congenic line generated by transfer of the Chr X QTL from the EPH mice into the inbred EPL background (134) and was further fine mapped to the 2cM region (135).

3.2.13. Cross C57BL/6J vs. SPRET Mice

During the genetic studies of atherosclerosis, Fisler et al. (136) noticed varying degrees of obesity among BC animals from a cross between F1(B6 × SPRET) and B6 mice, called BSB, although neither parental inbred strain was obese. The parental strains, however, differed somewhat in carcass lipid %: 7.5 (B6) vs. 2.2 (SPRET) at 3–4 months. BSB progeny exhibited carcass lipid ranging from 0.2 to 49%, body weight from 13 to 49 g, and omental fat pad weight from 0.01 to 2 g in males at 3 months, and similar trends were observed in females.

In the BSB progeny, four QTLs linked to % body fat or fat pad weights, *Mob-1-4* (multigenic obesity-1-4), were detected on Chrs 7, 6, 12, and 15, respectively (137).

3.2.14. Cross New Zealand Black/Bielschowsky NJ vs. SM/J Mice

The New Zealand Black (NZB)/Bielschowsky (Bl) strain was separated and inbred for black coat color from an early generation of the NZO/Bl strain (96). Because of the remarkable difference in plasma lipoprotein concentration between the NZB and the SM strains, these strains are commonly used for studying genetic factors affecting plasma lipid profile and atherosclerosis [138–140].

Multigenic obesity was also observed in F2 progeny from an intercross of F1s of the NZB/BlNJ and the SM/J mice, and a QTL linked to body fat, fat mass, and body weight, *Mob-5*, was detected on Chr 2 in these F2 animals (141).

3.2.15.   Cross EL/Suz vs. 129/Sv Mice

The EL mouse strain (different from the lean line of EL described above) is an inbred strain originating from the Japanese ddY noninbred stock of "Swiss" origin and is known as a model for epilepsy (142). In a preliminary study for epilepsy, apparent segregation for obesity was noticed in an F2 population from an intercross of F1s of EL/Suz and 129/Sv mice, and Taylor and Phillips (143) followed up with genetic analysis for the obesity. By analysis of the leanest 15% and the fattest 15% based on adiposity of the F2 progeny, two major QTLs, *Obq1* and *2*, were detected on Chrs 1 and 7, respectively.

3.2.16.   Cross AKR/J vs. C57L/J Mice

During construction of AKXL RI strains from a cross between the AKR/J and the C57L/J inbred strains, obesity was noticed in mice from several of the partially inbred lines (144).

By genetic analysis using F2 progeny of the AKR/J and the C57L/J mice, two QTLs affecting adiposity, *Obq3* and *4*, were detected on Chrs 2 and 17, respectively (145).

3.2.17.   Cross C57BL/6J vs. C3H/He Mice

When compared to the C3H/He (C3H) inbred mice, B6 mice exhibit impaired glucose tolerance and lower insulin secretion during intra-peritoneal glucose tolerance tests (IPGTT) (146). Genome-wide QTL analysis using 30 F2 and F3 progeny of the B6 and the C3H mice with the highest and lowest 30-minute blood glucose levels during IPGTT, respectively, revealed two QTLs linked to poor glucose tolerance on Chrs 2 and 13, respectively (146).

3.2.18.   Cross BTBR vs. C57BL/6J Mice

The BTBR inbred mouse strain has long been used for chemical mutagenesis rather than metabolic studies (147). However, F1 male offspring from BTBR and B6 mice exhibit elevated fasting insulin levels and profound insulin resistance although neither parental strain showed these metabolic alterations (147).

In an attempt to locate diabetes susceptibility loci, the *ob* obesity allele of the Leptin gene (148) was introduced into the F2 progeny from a (BTBR × B6) F1 intercross by mating BTBR-+/+ mice with B6-*ob*/+ mice, followed by intercrossing the resultant F1-*ob*/+ mice (149). Genome-wide QTL mapping analysis using these F2-*ob*/*ob* progeny with the segregating genetic backgrounds of BTBR and B6 permitted detection of two QTLs modifying diabetes severity, *t2dm1* and *t2dm2* (type II diabetes mellitus

1 and 2), on Chrs 16 and 19, respectively, and of one locus determining fasting plasma insulin levels, *t2dm3*, on Chr 2 (149).

4. THE QTL FOR DIET-INDUCED OBESITY AND TYPE II DIABETES

The incidence of obesity has increased worldwide in parallel with industrialization (150). This epidemic cannot be solely attributed to genetics, because mutations in our genes would not have spread that rapidly (151). However, a significant change has occurred in our diets and life styles. Constantly available, fatty/energy-enriched foods have accompanied industrialization and may now interact with existing genetic susceptibility factors in predisposed individuals, making them more vulnerable to obesity now than in previous times (152–154).

The prevalence of type II diabetes is also growing worldwide, and 29 million people are predicted to be diagnosed with diabetes in the United States alone by year 2050 (3). This increase in prevalence of diabetes may reflect the current obesity epidemics (3).

It has been emphasized that susceptibility genes to common disease, indeed, may play a role as responding factors to environments rather than as etiological factors (155). To be relevant to this multifactorial nature, polygenic animal models for dietary obesity have been developed.

4.1. Mice

4.1.1. Cross AKR/J vs. SWR/J Mice

Among nine inbred mouse strains evaluated, the AKR/J and the SWR/J were the two strains exhibiting the most and the least weight gain, respectively, in response to high-fat diet feeding (156).

Male F2 progeny of the AKR/J and the SWR/J mice were fed with high-fat diets (a high-fat, condensed milk diet, 32% of kilocalories as fat), and linkage analysis with polymorphic markers mapping near known obesity genes, *ob* (*Lep*), *db* (*Lepr*), *tub*, and *fat* (*Cpe*), was conducted for adiposity (157). Significant linkage was found only distal to the *db* locus on Chr 4, and the putative gene was designated as *Do1* (dietary obese 1). In a separate experiment, a genome-wide scan using male F2 animals derived from the same parental strains revealed two major QTLs for adiposity, *Do2* and *Do3*, on Chrs 9 and 15, respectively (157). By genetic analysis using the F2 animals from the 10% tails of the frequency distribution for adiposity, a QTL was also detected on Chr X for body weight and adiposity (158).

4.1.2. Cross CAST/Ei vs. C57BL/6J Mice

When compared to the CAST/Ei (CAST) mice, B6 mice markedly accumulated body fat in response to high-fat diets (159,160).

In an F2 population of male mice from CAST and B6 mice, dietary obesity segregated, and one major QTL (putatively *Do3*) on Chr 15 and two suggestive QTLs on Chrs 2 and 7 for adiposity were detected (159). Congenic mice carrying the Chr 7 QTL derived from CAST mice on the B6 background exhibited 50% lower adiposity than the host (B6) strain when fed with the high-fat diets (161).

Mehrabian et al. (160) also analyzed F2 progeny from an intercross of F1s of CAST and B6 mice fed high fat diets, and three QTLs linked to body fat, *Mob-5, 6,* and *7,* were detected on one chromosome, Chr 2. The fourth QTL for body fat, *Mob-8,* was detected on Chr 9.

5. CONCLUSION

Genome-wide scans using polygenic rodent models has been a powerful strategy to map susceptibility QTLs associated with obesity and type II diabetes. Although, to date, few susceptibility genes have been cloned from these QTLs, the current availability of comprehensive genome sequences and high-throughput functional genomic tools will speed up the search for candidate genes and will lead to the identification of genes underlying QTLs. Fundamental understanding of the gene variants and how they contribute to the susceptibility for obesity and type II diabetes could not be achieved by a single advanced technology or statistical method. A multidisplinary approach merging genetics, genomics, bioinformatics, cell biology, biochemistry, physiology, and nutrition will ultimately lead us to achieve this goal.

ACKNOWLEDGMENTS

I thank Dr. Jürgen K. Naggert at The Jackson Laboratory for his critical review of this manuscript. Manuscript preparation was supported by the College of Education, Health, and Human Science; and Center of Excellence for Genomics and Bioinformatics at the University of Tennessee and the American Heart Association-Scientist Development Award.

REFERENCES

1. Hill JO, Wyatt HR, Reed GW, Peters JC. Obesity and the environment: where do we go from here? Science 2003; 299:853–855.
2. Groop LC, Toumi T. Non-insulin-dependent diabetes mellitus – A collision between thrifty genes and an affluent society. Ann Med 1997; 29:37–53.

3. Boyle JP, Honeycutt AA, Narayan KMV, Hoerger TJ, Geiss LS, Chen H, Thompson TJ. Projection of diabetes burden through 2050. Diabetes Care 2001; 24:1936–1940.

4. Stunkard AJ, Sorensen TI, Hanis C, Teasdale TW, Chakraborty R, Schull WJ, Schulsinger F. An adoption study of human obesity. N Engl J Med 1986; 314:193–198.

5. Stunkard AJ, Foch TT, Hrubec Z. A twin study of human obesity. JAMA 1986; 256:51–54.

6. Koeppen-Schomerus G, Wardle J, Plomin R. A genetic analysis of weight and overweight in 4-year-old twin pairs. Int J Obes Relat Metab Disord 2001; 25:838–844.

7. Elbein SC. The genetics of human noninsulin-dependent (type 2) diabetes mellitus. J Nutr 1997; 127:1891S–1896S.

8. Elbein SC. Perspective: the search for genes for type 2 diabetes in the post-genome era. Endocrinology 2002; 143:2012–2018.

9. Brockmann GA, Bevova MR. Using mouse models to dissect the genetics of obesity. Trends Genet 2002; 18:367–376.

10. Kim JH, Taylor PN, Young D, Karst SY, Nishina PM, Naggert JK. New leptin receptor mutations in mice: Lepr(db-rtnd), Lepr(db-dmpg) and Lepr(db-rlpy). J Nutr 2003; 133:1265–1271.

11. Wang H, Hagenfeldt-Johansson K, Otten LA, Gauthier BR, Herrera PL, Wollheim CB. Experimental models of transcription factor-associated maturity-onset diabetes of the young. Diabetes 2002; 51(Suppl 3):S333–S342.

12. Wang J, Takeuchi T, Tanaka S, Kubo S-K, Kayo T, Lu D, Takata K, Koizumi A, Izumi T. A mutation in the insulin 2 gene induces diabetes with severe pancreatic ß-cell dysfunction in the *Mody* mouse. J Clin Invest 1999; 103: 27–37.

13. Clement K, Boutin P, Froguel P. Genetics of obesity. Am J Pharmacogenomics 2002; 2:177–187.

14. Doria A. Methods for the study of the genetic determinants of diabetes and its complications. Przegl Lek 2000; 57(Suppl 3):7–12.

15. Vink JM, Boomsma DI. Gene finding strategies. Biol Psychol 2002; 61:53–71.

16. Herberg L, Coleman DL. Laboratory animals exhibiting obesity and diabetes syndromes. Metabolism 1977; 26:59–98.

17. Cox RD, Brown SD. Rodent models of genetic disease. Curr Opin Genet Dev 2003; 13:278–83.

18. Nadeau JH. Maps of linkage and synteny homologies between mouse and man. Trends Genet 1989; 5:82–86.

19. Friedman JM. The function of leptin in nutrition, weight, and physiology. Nutr. Rev 2002; 60:S1–S14.

20. Clement K, Vaisse C, Lahlou N, Cabrol S, Pelloux V, Cassuto D, Gourmelen M, Dina C, Chambaz J, Lacorte JM, Basdevant A, Bougneres P, Lebouc Y, Froguel P, Guy-Grand B. A mutation in the human leptin receptor gene causes obesity and pituitary dysfunction. Nature 1998; 392:398–401.

21. Darvasi A. Experimental strategies for the genetic dissection of complex traits in animal models. Nat Genet 1998; 18:19–24.

22. Frankel WN. Taking stock of complex trait genetics in mice. Trends Genet 1995; 11:471–477.
23. Thomson G, Esposito MS. The genetics of complex diseases. Trends Cell Biol 1999; 9:M17–M20.
24. Silver LM. Quantitative traits and polygenic analysis. In: Mouse Genetics. New York: Oxford University Press Inc., 1995:253–263.
25. Grisel JE. Quantitative trait locus analysis. Alcohol Res Health 2000; 24: 169–174.
26. Elston RC. Introduction and overview. Stat Methods Med Res 2000; 9: 527–541.
27. Pomp D. Genetic dissection of obesity in polygenic animal models. Behav Genet 1997; 27:285–306.
28. Phillips TJ, Belknap JK, Hitzemann RJ, Buck KJ, Cunningham CL, Crabbe JC. Harnessing the mouse to unravel the genetics of human disease. Genes Brain Behav 2002; 1:14–26.
29. Silver LM. Laboratory Mice. New York: Oxford University Press Inc., 1995:32–61.
30. Lyon MF, Rastan S, Brown SDM. Genetic Variants and Strains of the Laboratory Mouse. New York: Oxford University Press, 1996.
31. Svenson KL, Bogue MA, Peters LL. Genetic models in applied physiology: invited review: Identifying new mouse models of cardiovascular disease: a review of high-throughput screens of mutagenized and inbred strains. J Appl Physiol 2003; 94:1650–1659.
32. Cicila GT. Strategy for uncovering complex determinants of hypertension using animal models. Curr Hypertens Rep 2000; 2:217–226.
33. Johnson TE, DeFries JC, Markel PD. Mapping quantitative trait loci for behavioral traits in the mouse. Behav Genet 1993; 22:635–653.
34. Williams RW, Gu J, Qi S, Lu L. The genetic structure of recombinant inbred mice: high-resolution consensus maps for complex trait analysis. Genome Biol 2001; 2:1–18.
35. Weber JL, Mary PE. Abundant class of human DNA polymorphism which can be typed using the polymerase chain reaction. Am J Hum Genet 1989; 44:388–396.
36. Silver LM. Laboratory Mice. New York: Oxford University Press Inc., 1995:159–194.
37. O'Brien SJ, Menotti-Raymond M, Murphy WJ, Nash WG, Wienberg J, Stanyon R, Copeland NG, Jenkins NA, Womack JE, Marshall Graves JA. The promise of comparative genomics in mammals. Science 1999; 286: 458–481.
38. Love JM, Knight AM, McAleer MA, Todd JA. Towards construction of a high resolution map of the mouse genome using PCR-analyzed microsatellites. Nucl Acids Res 1990; 18:4123–4130.
39. Iakoubova OA, Olsson CL, Dains KM, Choi J, Kalcheva I, Bentley LG, Cunanan M, Hillman D, Louie J, Machrus M, West DB. Microsatellite marker panels for use in high-throughput genotyping of mouse crosses. Physiol Genomics 2000; 3:145–148.

40. Collins FS, Brooks LD, Chakravarti A. A DNA polymorphism discovery resource for research on human genetic variation. Genome Res 1998; 8: 1229–1231.
41. Brookes AJ. The essence of SNPs. Gene 1999; 234:177–186.
42. Lindblad-Toh K, Winchester E, Daly MJ, Wang DG, Hirschhorn JN, Laviolette JP, Ardlie K, Reich DE, Robinson E, Sklar P, Shah N, Thomas D, Fan JB, Gingeras T, Warrington J, Patil N, Hudson TJ, Lander ES. Large-scale discovery and genotyping of single-nucleotide polymorphisms in the mouse. Nat Genet 2000; 24:381–386.
43. Sachidanandam R, Weissman D, Schmidt SC, Kakol JM, Stein LD, Marth G, Sherry S, Mullikin JC, Mortimore BJ, Willey DL, Hunt SE, Cole CG, Coggill PC, Rice CM, Ning Z, Rogers J, Bentley DR, Kwok PY, Mardis ER, Yeh RT, Schultz B, Cook L, Davenport R, Dante M, Fulton L, Hillier L, Waterston RH, McPherson JD, Gilman B, Schaffner S, Van Etten WJ, Reich D, Higgins J, Daly MJ, Blumenstiel B, Baldwin J, Stange-Thomann N, Zody MC, Linton L, Lander ES, Altshuler D. International SNP map working group. A map of human genome sequence variation contining 1.42 million single nucleotide polymorphisms. Nature 2001; 409:928–933.
44. Broman KW. Review of statistical methods for QTL mapping in experimental crosses. Lab Anim 2001; 30:44–52.
45. Manly KF, Olson JM. Overview of QTL mapping software and introduction to map manager QT. Mamm Genome 1999; 10:327–334.
46. Manly KF, Cudmore RH, Meer JM. Map manager QTX, cross-platform software for genetic mapping. Mamm Genome 2001; 12:930–932.
47. Darvasi A, Soller M. A simple method to calculate resolving power and confidence interval of QTL map location. Behav Genet 1997; 27:125–132.
48. Rapp JP. Genetic analysis of inherited hypertension in the rat. Physiol Rev 2000; 80:135–72.
49. Nagamine Y, Haley CS. Using the mixed model for interval mapping of quantitative trait loci in outbred line crosses. Genet Res Camb 2001; 77:199–207.
50. Denny P, Lord CJ, Hill NJ, Goy JV, Levy ER, Podolin PL, Peterson LB, Wicker LS, Todd JA, Lyons PA. Mapping of the IDDM locus *Idd3* to a 0.35-cM interval containing the interleukin-2 gene. Diabetes 1997; 46:695–700.
51. Podolin PL, Denny P, Lord CJ, Hill NJ, Todd JA, Peterson LB, Wicker LS, Lyons PA. Congenic mapping of the insulin-dependent diabetes (Idd) gene, Idd10, localizes two genes mediating the Idd10 effect and eliminates the candidate Fcgr1. J Immunol 1997; 159:1835–1843.
52. Rogner UC, P Avner. Congenic mice: cutting tools for complex immune disorders. Immunology 2003; 3:243–252.
53. Bennett B. Congenic strains developed for alcohol- and drug-related phenotypes. Pharmacol Biochem Behav 2000; 67:671–681.
54. Rikke BA, Johnson TE. Towards the cloning of genes underlying murine QTLs. Mamm Genome 1998; 9:963–968.
55. Schalkwyk LC, Francis F, Lehrach H. Techniques in mammalian genome mapping. Curr Opin Biotechnol 1995; 6:37–43.

56. Warden CH, Fisler JS. Integrated methods to solve the biological basis of common diseases. Methods 1997; 13:347–357.

57. Bogardus C, Baier L, Permana P, Prochazka M, Wolford J, Hanson R. Identification of susceptibility genes for complex metabolic diseases. Ann N Y Acad Sci 2002; 967:1–6.

58. Doerge RW. Mapping and analysis of quantitative trait loci in experimental populations. Genetics 2001; 3:43–52.

59. Aitman TJ, Glazier AM, Wallace CA, Cooper LD, Norsworthy PJ, Wahid FN, Al-Majali KM, Trembling PM, Mann CJ, Shoulders CC, Graf D, St Lezin E, Kurtz TW, Kren V, Pravenec M, Ibrahimi A, Abumrad NA, Stanton LW, Scott J. Identification of Cd36 (Fat) as an insulin-resistance gene causing defective fatty acid and glucose metabolism in hypertensive rats. Nat Genet 1999; 21:76–83.

60. Goto Y, Suzuki K, Ono T, Sasaki M, Toyota T. Development of diabetes in the non-obese niddm rat (GK rat). Adv Exp Med Biol 1988; 246:29–31.

61. Gauguier D, Froguel P, Parent V, Bernard C, Bihoreau MT, Portha B, James MR, Penicaud L, Lathrop M, Ktorza A. Chromosomal mapping of genetic loci associated with non-insulin dependent diabetes in the GK rat. Nat Genet 1996; 12:38–43.

62. Galli J, Li LS, Glaser A, Ostenson CG, Jiao G, Fakhrai-Rad H, Jacob HJ, Lander ES, Luthamn H. Genetic analysis of non-insulin dependent diabetes mellitus in the GK rat. Nat Genet 1996; 12:31–37.

63. Fakhrai-Rad H, Nikoshkov A, Kamel A, Fernstrom M, Zierath JR, Norgren S, Luthman H, Galli J. Insulin-degrading enzyme identified as a candidate diabetes susceptibility gene in GK rats. Hum Mol Genet 2000; 9:2149–2158.

64. Farris W, Mansourian S, Chang Y, Lindsley L, Eckman EA, Frosch MP, Eckman CB, Tanzi RE, Selkoe DJ, Guenette S. Insulin-degrading enzyme regulates the levels of insulin, amyloid beta-protein, and the beta-amyloid precursor protein intracellular domain in vivo. Proc Natl Acad Sci USA 2003; 100:4162–4167.

65. Kawano K, Hirashima T, Mori S, Kurosumi MYS. A new strain with non-insulin-dependent-diabetes-mellitus, "OLETF". Rat News Lett 1991; 25: 24–26.

66. Kawano K, Hirashima T, Mori S, Natori T. OLETF (Otsuka Long-Evans Tokushima Fatty) rat: a new NIDDM rat strain. Diabetes Res Clin Pract 1994; 24(Suppl):S317–S320.

67. Yagi K, Kim S, Wanibuchi H, Yamashita T, Yamamura Y, Iwao H. Characteristics of diabetes, blood pressure, and cardiac and renal complications in Ostsuka Long-Evans Tokushima Fatty Rats. Hypertension 1997; 29:728–735.

68. Mori S, Kawano K, Hirashima T, Natori T. Relationships between diet control and the development of spontaneous type II diabetes and diabetic nephropathy in OLETF rats. Diabetes Res Clin Pract 1996; 33:145–152.

69. Kawano K, Mori S, Hirashima T, Man Z-W, Natori T. Examination of the pathogenesis of diabetic nephropathy in OLETF rats. J Vet Med Sci 1999; 61:1219–1228.

70. Hirashima T, Kawano K, Mori S, Matsumoto K, Natori T. A diabetogenic gene (ODB-1) assigned to the X-chromosome in OLETF rats. Diabetes Res Clin Pract 1995; 27:91–96.

71. Kanemoto N, Gishigaki G, Miyakita A, Oga K, Okuno S, Tsuji A, Takagi T, Takahashi E-I, Nakamura Y, Watanabe TK. Genetic dissection of "OLETF", a rat model for non-insulin-dependent diabetes mellitus. Mamm Genome 1998; 9:419–425.

72. Moralejo DH, Wei S, Wei K, Weksler-Zangen S, Koike G, Jacob HJ, Hirashima T, Kawano K, Sugiura K, Sasaki Y, Ogino T, Yamada T, Matsumoto K. Identification of quantitative trait loci for non-insulin-dependent diabetes mellitus that interact with body weight in the Otsuka Long-Evans Tokushima Fatty Rat. Proc Assoc Am Physicians 1998; 110:545–558.

73. Watanabe TK, Okuno S, Oga K, Mizoguchi-Miyakita A, Tsuji A, Yamasaki Y, Hishigaki H, Kanemoto N, Takagi T, Takahashi E-I, Irie Y, Nakamura Y, Tanigami A. Genetic dissection of "OLETF", a rat model for non-insulin-dependent diabetes mellitus: quantitative trait locus analysis of (OLETF × BN) × OLETF. Genomics 1999; 58:233–239.

74. Okuno S, Watanabe TK, Ono T, Yamasaki Y, Goto Y, Miyao H, Asai T, Kanemoto N, Oga K, Mizoguchi-Miyakita A, Takagi T, Takahashi E, Nakamura Y, Tanigami A. Genetic determinants of plasma triglyceride levels in (OLETF × BN) × OLETF backcross rats. Genomics 1999; 62:350–355.

75. Sugiura K, Miyake T, Taniguchi Y, Yamada T, Moralejo DH, Wei S, Wei K, Sasaki Y, Matsumoto K. Identification of novel non-insulin-dependent diabetes mellitus susceptibility loci in the Otsuka Long-Evans Tokushima fatty rat by MQM-mapping method. Mamm Genome 1999; 10:1126–31.

76. Ogino T, Wei S, Wei K, Moralejo DH, Kose H, Mizuno A, Shima K, Sasakin Y, Yamada T, Matsumoto K. Genetic evidence for obesity loci involved in the regulation of body fat distribution in obese type 2 diabetes rat, OLETF. Genomics 2000; 70:19–25.

77. Watanabe TK, Okuno S, Ono T, Yamasaki Y, Oga K, Mizoguchi-Miyakita A, Miyao H, Suzuki M, Momota H, Goto Y, Shinomiya H, Hishigaki H, Hayashi I, Asai T, Wakitani S, Takagi T, Nakamura Y, Tanigami A. Single-allele correction of the Dmo1 locus in congenic animals substantially attenuates obesity, dyslipidaemia and diabetes phenotypes of the OLETF rat. Clin Exp Pharmacol Physiol 2001; 28:28–42.

78. Okuno S, Watanabe TK, Ono T, Oga K, Mizoguchi-Miyakita A, Yamasaki Y, Goto Y, Shinomiya H, Momota H, Miyao H, Hayashi I, Asai T, Suzuki M, Harada Y, Hishigaki H, Wakitani S, Takagi T, Nakamura Y, Tanigami A. Effects of Dmo1 on obesity, dyslipidaemia and hyperglycaemia in the Otsuka Long Evans Tokushima Fatty strain. Genet Res 2001; 77:183–190.

79. Okuno S, Kondo M, Yamasaki Y, Miyao H, Ono T, Iwanaga T, Omori K, Okano M, Suzuki M, Momota H, Hishigaki H, Hayashi I, Goto Y, Shinomiya H, Harada Y, Hirashima T, Kanemoto N, Asai T, Wakitani S, Takagi T, Nakamura Y, Tanigami A, Watanabe TK. Substitution of Dmo1 with normal alleles results in decreased manifestation of diabetes in OLETF rats. Diabetes Obes Metab 2002; 4:309–318.

80. Kose H, Moralejo DH, Ogino T, Mizuno A, Yamada T, Matsumoto K. Examination of OlETF-derived non-insulin-dependent diabetes mellitus QTL by construction of a series of congenic rats. Mamm Genome 2002; 13:558–562.

81. Shinohara M, Masuyama T, Shoda T, Takahashi T, Katsuda Y, Komeda K, Kuroki M, Kakehashi A, Kanazawa Y. A new spontaneously diabetic non-obese Torii rat strain with severe ocular complications. Int J Exp Diabetes Res 2000; 1:89–100.

82. Masuyama T, Fuse M, Yokoi N, Shinohara M, Tsujii H, Kanazawa M, Kanazawa Y, Komeda K, Taniguchi K. Genetic analysis for diabetes in a new rat model of nonobese type 2 diabetes, Spontaneously Diabetic Torii rat. Biochem Biophys Res Commun 2003; 304:196–206.

83. Ikeda H, Mouse KK. Diabetes Res Clin Pract 1994; 24(Suppl):S313–S316.

84. Iwatsuka H, Shino A, Suzuoki Z. General survey of diabetic features of yellow KK mice. Endocrinol Jpn 1970; 17:23–25.

85. Nishimura M. Breeding of mice strains for diabetes mellitus. Exp Anim 1969; 18:147–157.

86. Taylor BA, Tarantino LM, Phillips SJ. Gender-influenced obesity QTLs identified in a cross involving the KK type II diabetes-prone mouse strain. Mamm Genome 1999; 10:963–968.

87. Suto J, Matsuura S, Imamura K, Yamanaka H, Sekikawa K. Genetic analysis of non-insulin-dependent diabetes mellitus in KK and KK-Ay mice. Eur J Endocrinol 1998; 139:654–661.

88. Shike T, Hirose S, Kobayashi M, Funabiki K, Shirai T, Tomino Y. Susceptibility and negative epistatic loci contributing to type 2 diabetes and related phenotypes in a KK/Ta mouse model. Diabetes 2001; 50: 1943–1948.

89. Shibata M, Yasuda B. New experimental congenital diabetic mice (N.S.Y. mice). Tohoku J Exp Med 1980; 130:139–142.

90. Ueda H, Ikegami H, Yamato E, Fu J, Fukuda M, Shen G, Kawaguchi Y, Takekawa K, Fujioka Y, Fujisawa T, Nakagawa Y, Hamada Y, Shibata M, Ogihara T. The NSY mouse: a new animal model of spontaneous NIDDM with moderate obesity. Diabetologia 1995; 38:503–508.

91. Hamada Y, Ikegami H, Ueda H, Kawaguchi Y, Yamato E, Nojima K, Yamada K, Babaya N, Shibata M, Ogihara T. Insulin secretion to glucose as well as non-glucose stimuli is impaired in spontaneously diabetic Nagoya-Shibata-Yasuda mice. Metabolism 2001; 50:1282–1285.

92. Ueda H, Ikegami H, Kawaguchi Y, Fujisawa T, Yamato E, Shibata M, Ogihara T. Genetic analysis of late-onset type 2 diabetes in a mouse model of human complex trait. Diabetes 1999; 48:1168–1174.

93. Suzuki W, Iizuka S, Tabuchi M, Funo S, Yanagisawa T, Kimura M, Sato T, Endo T, Kawamura H. A new mouse model of spontaneous diabetes derived from ddY strain. Exp Anim 1999; 48:181–189.

94. Hirayama I, Yi Z, Izumi S, Arai I, Suzuki W, Nagamachi Y, Kuwano H, Takeuchi T, Izumi T. Genetic analysis of obese diabetes in the TSOD mouse. Diabetes 1999; 48:1183–1191.

95. Kim JH, Sen S, Avery CS, Simpson E, Chandler P, Nishina PM, Churchill GA, Naggert JK. Genetic analysis of a new mouse model for non-insulin-dependent diabetes. Genomics 2001; 74(3):273–286.
96. Bielschowsky M, Goodall CM. Origin of inbred NZ mouse strains. Cancer Res 1970; 30:834–836.
97. Ortlepp JR, Kluge R, Giesen K, Plum L, Radke P, Hanrath P, Joost HG. A metabolic syndrome of hypertension, hyperinsulinemia, and hypercholesterolemia in the New Zealand obese (NZO) mouse. Eur J Clin Invest 2000; 30:195–202.
98. Halaas JL, Boozer C, Blair-West J, Fidahusein N, Denton DA, Friedman JM. Physiological response to long-term peripheral and central leptin infusion in lean and obese mice. Proc Natl Acad Sci USA 1997; 94:8878–8883.
99. Igel M, Becker W, Herberg L, Joost HG. Hyperleptinemia, leptin resistance, and polymorphic leptin receptor in the New Zealand obese (NZO) mouse. Endocrinology 1997; 138:4234–4239.
100. Kluge R, Giesen K, Bahrenberg G, Plum L, Ortlepp JR, Joost HG. Quantitative trait loci for obesity and insulin resistance (Nob1, Nob2) and their interaction with the leptin receptor allele (LeprA720T/T1044I) in New Zealand obese mice. Diabetologia 2000; 43:1565–1572.
101. Plum L, Kluge R, Giesen K, Altmüller J, Ortlepp JR, Joost HG. Type-2-diabetes-like hyperglycemia in a backcross model of New Zealand obese (NZO) and SJL mice: characterization of a susceptibility locus on chromosome 4 and its relation with obesity. Diabetes 2000; 49:1590–1596.
102. Plum L, Giesen K, Kluge R, Junger E, Linnartz K, Becker W, Joost HG. Characterization of the diabetes susceptibility locus Nidd/SJL in the New Zealand obese (NZO) mouse: islet cell destruction, interaction with the obesity QTL Nob1, and effect of dietary fat. Diabetologia 2002; 45:823–830.
103. Giesen K, Plum L, Kluge R, Ortlepp J, Joost HG. Diet-dependent obesity and hypercholesterolemia in the New Zealand obese mouse: identification of a quantitative trait locus for elevated serum cholesterol on the distal mouse chromosome 5. Biochem Biophys Res Commun 2003; 304:812–817.
104. Taylor BA, Wnek C, Schroeder D, Phillips SJ. Multiple obesity QTLs identified in an intercross between the NZO (New Zealand obese) and the SM (small) mouse strains. Mamm Genome 2001; 12:95–103.
105. Leiter EH, Reifsnyder PC, Flurkey K, Partke HJ, Junger E, Herberg L. NIDDM genes in mice: deleterious synergism by both parental genomes contributes to diabetic thresholds. Diabetes 1998; 47:1287–1295.
106. Reifsnyder PC, Churchill G, Leiter EH. Maternal environment and genotype interact to establish diabesity in mice. Genome Res 2000; 10:1568–1578.
107. Reifsnyder PC, Leiter EH. Deconstructing and reconstructing obesity-induced diabetes (diabesity) in mice. Diabetes 2002; 51:825–832.
108. Demant P, Hart AA. Recombinant congenic strains – a new tool for analyzing genetic traits determined by more than one gene. Immunogenetics 1986; 24:416–422.
109. Shalom A, Darvasi A. Experimental designs for QTL fine mapping in rodents. Methods Mol Biol 2002; 195:199–223.

110. Nielsen MK, Freking BA, Jones LD, Nelson SM, Vorderstrasse TL, Hussey BA. Divergent selection for heat loss in mice: II Correlated responses in feed intake, body mass, body composition, and number born through fifteen generations. J Anim Sci 1997; 75:1469–1476.

111. Nielsen MK, Jones LD, Freking BA, DeShazer JA. Divergent selection for heat loss in mice: I. Selection applied and direct response through fifteen generations. J Anim Sci 1997; 75:1461–1468.

112. Moody DE, Pomp D, Nielsen MK. Variability in metabolic rate, feed intake and fatness among selection and inbred lines of mice. Genet Res 1997; 70:225–235.

113. Moody DE, Pomp D, Nielsen MK, Van Vleck LD. Identification of quantitative trait loci influencing traits related to energy balance in selection and inbred lines of mice. Genetics 1999; 152:699–711.

114. Hanrahan JP, Eisen EJ, Lagates JE. Effects of population size and selection intensity of short-term response to selection for postweaning gain in mice. Genetics 1973; 73:513–530.

115. Eisen EJ, Leatherwood JM. Adipose cellularity and body composition in polygenic obese mice as influenced by preweaning nutrition. J Nutr 1978; 108:1652–1662.

116. Eisen EJ. Effects of selection for rapid postweaning gain on maturing patterns of fat depots in mice. J Anim Sci 1987; 64:133–147.

117. Robeson BL, Eisen EJ, Leatherwood JM. Adipose cellularity, serum glucose, insulin and cholesterol in polygenic obese mice fed high-fat or high-carbohydrate diets. Growth 1981; 45:198–215.

118. Schuler L. Mouse strain Fzt:DU and its use as model in animal breeding research. Arch Tierz 1985; 28:357–363.

119. Bunger L, Laidlaw A, Bulfield G, Eisen EJ, Medrano JF, Bradford GE, Pirchner F, Renne U, Schlote W, Hill WG. Inbred lines of mice derived from long-term growth selected lines: unique resources for mapping growth genes. Mamm Genome 2001; 12:678–686.

120. Bunger L, Gerrendorfer G, Renne U. Results of long term selection for growth traits in laboratory mice. Proceedings of the 4th World Congress on Genetics Applied to Livestock Production, Univ Edinburgh, Dinburgh, Scotland 1990; 13:321–324.

121. Timtchenko D, Kratzsch J, Sauerwein H, Wegner J, Souffrant WB, Schwerin M, Brockmann GA. Fat storage capacity in growth-selected and control mouse lines is associated with line-specific gene expression and plasma hormone levels. Int J Obes Relat Metab Disord 1999; 23:586–594.

122. Brockmann GA, Haley CS, Renne U, Knott SA, Schwerin M. Quantitative trait loci affecting body weight and fatness from a mouse line selected for extreme high growth. Genetics 1998; 150:369–381.

123. Brockmann GA, Kratzsch J, Haley CS, Renne U, Schwerin M, Karle S. Single QTL effects, epistasis, and pleiotropy account for two-thirds of the phenotypic F2 variance of growth and obesity in DU6i × DBA/2 mice. Genome Res 2000; 10:1941–195.

124. Goodale H. A study of the inheritance of body weight in the albino mouse by selection. J Hered 1938; 29:101–112.

125. Goodale H. Progress report on possibilities in progeny test breeding. Science 1941; 94:442–443.
126. Cheverud JM, Pletscher LS, Vaughn TT, Marshall B. Differential response to dietary fat in large (LG/J) and small (SM/J) inbred mouse strains. Physiol Genomics 1999; 1:33–39.
127. Cheverud JM, Vaughn TT, Pletscher LS, Peripato AC, Adams ES, Erikson CF, King-Ellison KJ. Genetic architecture of adiposity in the cross of LG/J and SM/J inbred mice. Mamm Genome 2001; 12:3–12.
128. MacArthur J. Genetics of body size and related characters. I. Selection of small and large races of the laboratory mouse. Am Nat 1944; 78:142–157.
129. Anunciado RV, Nishimura M, Mori M, Ishikawa A, Tanaka S, Horio F, Ohno T, Namikawa T. Quantitative trait loci for body weight in the intercross between SM/J and A/J mice. Exp Anim 2001; 50:319–324.
130. Bunger L, Hill WG. Inbred lines of mice derived from long-term divergent selection on fat content and body weight. Mamm Genome 1999; 10:645–648.
131. Bunger L, Forsting J, McDonald KL, Horvat S, Duncan J, Hochscheid S, Baile CA, Hill WG, Speakman JR. Long-term divergent selection on fatness in mice indicates a regulation system independent of leptin production and reception. FASEB J 2002; 17:85–87.
132. Horvat S, Bunger L, Falconer VM, Mackay P, Law A, Bulfield G, Keightley PD. Mapping of obesity QTLs in a cross between mouse lines divergently selected on fat content. Mamm Genome 2000; 11:2–7.
133. Hastings IM, Veerkamp RF. The genetic basis of response in mouse lines divergently selected for body weight or fat content. I. The relative contributions of autosomal and sex-linked genes. Genet Res 1993; 62:169–175.
134. Rance KA, Hill WG, Keightley PD. Mapping quantitative trait loci for body weight on the X chromosome in mice I. Analysis of a reciprocal F2 population. Genet Res 1997; 70:117–124.
135. Liu X, Oliver F, Brown SD, Denny P, Keightley PD. High-resolution quantitative trait locus mapping for body weight in mice by recombinant progeny testing. Genet Res 2001; 77:191–197.
136. Fisler JS, Warden CH, Pace MJ, Lusis AJ. BSB: a new mouse model of multigenic obesity. Obesity Res 1993; 1(4):271–280.
137. Warden CH, Fisler JS, Pace MJ, Svenson KL, Lusis AJ. Coincidence of genetic loci for plasma cholesterol levels and obesity in a multifactorial mouse model. J Clin Invest 1993; 92:773–779.
138. Purcell-Huynh DA, Weinreb A, Castellani LW, Mehrabian M, Doolittle MH, Lusis AJ. Genetic factors in lipoprotein metabolism. Analysis of a genetic cross between inbred mouse strains NZB/BINJ and SM/J using a complete linkage map approach. J Clin Invest 1995; 96:1845–1858.
139. Pitman WA, Hunt MH, McFarland C, Paigen B. Genetic analysis of the difference in diet-induced atherosclerosis between the inbred mouse strains SM/J and NZB/BINJ. Arterioscler Thromb Vasc Biol 1998; 18:615–620.

140. Pitman WA, Korstanje R, Churchill GA, Nicodeme E, Albers JJ, Cheung MC, Staton MA, Sampson SS, Harris S, Paigen B. Quantitative trait locus mapping of genes that regulate HDL cholesterol in SM/J and NZB/B1NJ inbred mice. Physiol Genomics 2002; 9:93–102.

141. Lembertas AV, Perusse L, Chagnon YC, Fisler JS, Warden CH, Purcell-Huynh DA, Dionne FT, Gagnon J, Nadeau A, Lusis AJ, Bouchard C. Identification of an obesity quantitative trait locus on mouse chromosome 2 and evidence of linkage to body fat and insulin on the human homologous region 20q. J Clin Invest 1997; 100:1240–1247.

142. Suzuki J, Nakamoto Y. El mouse: a model of sensory precipitating epilepsy. Excerpta Med 1977; 427:81–82.

143. Taylor BA, Phillips SJ. Detection of obesity QTLs on mouse chromosomes 1 and 7 by selective DNA pooling. Genomics 1996; 4:389–398.

144. Taylor BA, Meier H. Mapping the adrenal lipid depletion gene of the AKR/J mouse strain. Genet Res 1975; 26:307–312.

145. Taylor BA, Phillips SJ. Obesity QTLs on mouse chromosomes 2 and 17. Genomics 1997; 43:249–257.

146. Kayo T, Fujita H, Nozaki J, E X, Koizumi A. Identification of two chromosomal loci determining glucose intolerance in a C57BL/6 mouse strain. Comp Med 2000; 50:296–302.

147. Ranheim T, Dumke C, Schueler KL, Cartee GD, Attie AD. Interaction between BTBR and C57BL/6J genomes produces an insulin resistance syndrome in (BTBR × C57BL/6J) F1 mice. Arterioscler Thromb Vasc Biol 1997; 17:3286–3293.

148. Zhang Y, Proenca R, Maffei M, Barone M, Leopold L, Friedman JM. Positional cloning of the mouse obese gene and its human homologue. Nature 1994; 372:425–432.

149. Stoehr JP, Nadler ST, Schueler KL, Rabaglia ME, Yandell BS, Metz SA, Attie AD. Genetic obesity unmasks nonlinear interactions between murine type 2 diabetes susceptibility loci. Diabetes 2000; 49:1946–1954.

150. Popkin BM, Doak CM. The obesity epidemic is a worldwide phenomenon. Nutr Rev 1998; 56:106–114.

151. Hill JO, Peters JC. Environmental contributions to the obesity epidemic. Science 1998; 280:1371–1374.

152. Astrup A, Ryan L, Grunwald GK, Storgaard M, Saris W, Melanson E, Hill JO. The role of dietary fat in body fatness: evidence from a preliminary meta-analysis of ad libitum low-fat dietary intervention studies. Br J Nutr 2000; 83(Suppl 1):S25–S32.

153. Hill JO, Melanson EL, Wyatt HT. Dietary fat intake and regulation of energy balance: implications for obesity. J Nutr 2000; 130:284S–288S.

154. Bray GA, Popkin BM. Dietary fat intake does affect obesity. Am J Clin Nutr 1998; 68:1157–1173.

155. Tiret L. Gene-environment interaction: a central concept in multifactorial diseases. Proc Nutr Soc 2002; 61:457–463.

156. West DB, Boozer CN, Moody DL, Atkinson RL. Dietary obesity in nine inbred mouse strains. Am J Physiol 1992; 262(6 Pt 2):R1025–R1032.
157. West DB, Goudey-Lefevre J, York B, Truett GE. Dietary obesity linked to genetic loci on chromosomes 9 and 15 in a polygenic mouse model. J Clin Invest 1994; 94:1410–1416.
158. York B, Lei K, West DB. Inherited non-autosomal effects on body fat in F2 mice derived from an AKR/J × SWR/J cross. Mamm Genome 1997; 8:726–730.
159. York B, Lei K, West DB. Sensitivity to dietary obesity linked to a locus on chromosome 15 in a CAST/Ei × C57BL/6J F2 intercross. Mamm Genome 1996; 7:677–681.
160. Mehrabian M, Wen PZ, Fisler J, Davis RC, Lusis AJ. Genetic loci controlling body fat, lipoprotein metabolism, and insulin levels in a multifactorial mouse model. J Clin Invest 1998; 101:2485–2496.
161. York B, Truett A A, Monteiro MP, Barry SJ, Warden CH, Naggert JK, Maddatu TP, West DB. Gene-environment interaction: a significant diet-dependent obesity locus demonstrated in a congenic segment on mouse chromosome 7. Mamm Genome 1999; 10:457–462.
162. Sen S, Churchill GA. A statistical framework for quantitative trait mapping. Genetics 2001; 159:371–387.

3

The Human Sweet Tooth and Its Relationship to Obesity

Amanda H. McDaniel and Danielle R. Reed
Monell Chemical Senses Center,
Philadelphia, Pennsylvania, U.S.A.

ABSTRACT

People who have a persistent desire to eat sweets are said to have a "sweet tooth". The term sweet tooth has several meanings in the experimental literature, and the types of behavior purported to represent measures of the human sweet tooth are described and evaluated. The determinants of individual differences among humans in their behavior toward sweetness are largely unknown, but extrapolating from studies in mice and from human families, we predict that these differences may be partially genetic in origin. There is an assumption that preferring sweet foods compared with other types of foods leads to excessive sweet consumption and to obesity, but few studies test this hypothesis directly. Recent advances in our understanding of the molecular biology of sweet sensory systems and metabolism may explain individual differences in the human behavior toward sweetness and provide new avenues for the treatment of nutritional disorders.

1. INTRODUCTION

The term "sweet tooth" has been used widely in both popular culture and in the scientific literature. But what is meant by the term sweet tooth and how do we measure it? When we say that a person has a sweet tooth, we may be thinking of a person who usually prefers to eat a sweet food or beverage rather than one that is savory or salty. Or we might assume that the sweeter a food or beverage is, the more someone with a sweet tooth will prefer it. Because to like sweet foods is seen as a prerequisite to eating too much, the study of the human sweet tooth has usually been undertaken with the goal of understanding how the perception of or preference for sweet foods contributes to overeating and obesity. But the underlying assumptions of this hypothesis—that increased perception of and preference and desire for sugar leads to increased intake of sweet food and drinks—is rarely directly tested.

This review is divided into two sections. In the first section, we assess the ways in which human behavior toward sweets is measured, and the factors that influence it. In the second section, we examine the relationship between the preference for sweet foods, their intake, and the effect on obesity.

2. SWEET TOOTH: MEASUREMENT AND INFLUENCES

2.1. Sensation, Behavior, or Desire?

Sweet is one of the five primary taste qualities, and there are several measures of human perception of sweetness. The lowest concentration at which someone can detect sugar or recognize its sweet quality can be measured. The terms for these measures are detection and recognition thresholds; the detection threshold usually occurs at lower concentrations than recognition because subjects can tell that there is something in a solution before they can identify its quality (1). A second measure of sweetness is how intense above-threshold concentrations of sweetness are perceived to be. For instance, some people may find the sweetness of a commercially available carbonated beverage to be "very strong" but another person might find it to be "weak". This concept is referred to as perceived intensity. The next measure is "liking"—defined as the degree to which the person perceives it as acceptable and desirable when presented with a single stimulus. This measure is sometimes also referred to as "acceptability". Sometimes people have a choice among stimuli and choose the one that is the most acceptable or desirable. These types of measures are referred to as "preference". When the degree of the desire to eat a sweet food or drink is measured, this is referred to as craving. A final and important measure of human behavior toward sweetness is the amount of sugar someone eats when offered a choice of foods or drinks, either in the laboratory or in their daily lives.

There is no agreement within the experimental literature upon a definition of sweet tooth. Sometimes it is assessed using measures of liking or preference (2–9), sometimes by measures of food intake or food selection, and sometimes by measures of the motivation to eat sweet foods (10,11). Most laboratory measures designed to assess sweet tooth use preference measures rather than measures of food intake or desire and motivation. Food intake and food selection outside of the laboratory are hard to measure accurately because of the disinclination of subjects to correctly report the food they eat. Therefore, proxy indices of sugar intake—such as the number of dental caries or the amount of oral bacteria per subject—are sometimes substituted as measures to circumvent report bias (12). Also, asking specific questions about sugar usage, for instance on cereal or in coffee, may elicit accurate responses regarding sugar intake and preference (4,13).

Perhaps one reason that preference is most often measured in human studies is because these methods detect reliable individual differences among subjects (14). Preference measures are also desirable because people can be classified into categories. For instance, some investigators have identified two different response patterns to sucrose solutions, a type I response whereby subjects increase in the liking for sucrose up to a middle range of concentration, followed by a breakpoint after which preference decreases with increasing concentration. This pattern is referred to as an inverted-U shape. The type II response is characterized by increased liking as the concentration increases, but levels off (15). Other investigators have reported similar patterns among subjects (5).

Although laboratory measures of sweet preference are commonly used, they may not predict the preference for other sweeteners (16) or the preference for sweet foods or beverages. Investigators have tried to bridge the gap between preference measures for laboratory stimuli and preference measures for real-world foods and drinks by using mixtures of sugar and milk (17,18) or by adding sugar to simple beverages or foods (9). Finally, some investigators have compared sweet preference measures inside the laboratory to self-reported behaviors outside of the laboratory (8,13).

Human behavior toward sweet may be affected by the degree to which the subjects can perceive the stimuli. There are individual differences in the detection or recognition thresholds for sweetness (19,20), and although rare, there are people who do not perceive a sweet taste from sucrose (21). Therefore, when measuring preference for sucrose at low concentrations, it is important to consider that some people will not be able to perceive the stimulus as well as other people. Thus far in human studies, sweet detection threshold does not predict either how intense higher concentrations are perceived or how much they are liked (22–25). Although in mice there is a relationship between peripheral sensitivity and intake of sweeteners (26), this relationship in humans is unclear, and more focused study is needed.

2.2. Stable and Variable Aspects of Sweet Taste Perception

Individual differences in the response to sweet are present at birth, with some infants responding more positively than others to the taste of sucrose (27), and these individual differences persist as children become young adults (14). However, the same people measured on two occasions, weeks or months apart, have similar but not identical sweet preference, suggesting that sweet preference changes over the short term (3,7,28).

There are effects of race and sex on sweet preference. Americans of African descent prefer higher concentrations and Pima Indians prefer lower concentrations of sugar compared with those of European ancestry (7,13,29–33). However, race differences in sweet preference may be specific to types of foods. For instance, Taiwanese students rate sucrose solutions as more pleasant but sweetened cookies as less pleasant compared with students of European descent (34). Studies of sex differences suggest that male and female infants do not differ in sweet preference (29) but that older boys and men prefer higher concentrations of sweets compared with women (7,11,31,35,36). Although men prefer high concentrations of sweet in their food and drink, studies of food craving in men show they experience less desire to eat sweet foods compared with women (37,38). Sex differences in food craving may be population-specific, however, since women in Egypt did not show elevations in sweet food craving compared with men (38). Week-to-week variations in sex hormone concentrations in women predict changes in sucrose threshold (39) but with equivocal effects on sucrose preference (40,41), and it is not clear to what extent sex hormones account for sex differences in human behavior toward sweet.

In addition to race and sex, age is also a reliable predictor of sweet preference. Children prefer more highly sweetened solutions compared with adults (31,35,42) but see (36). Children may also have lower detection thresholds (23) and lower perceived intensity at high-sucrose concentrations (43) compared with adults, but not all studies agree (35,36,44). Younger people also eat more sugar than do older people (45). Dietary experience alters sweet preference in children; for instance, children fed sweet water like it more than children not fed sweet water (29). Children are less afraid of sugar than other nutrients and even neophobic children will accept sweets (46). Sweet craving changes over the life span, and older women report less craving for sweet food compared with younger women (47).

An immediate but short-lived reduction in the preference for sweet-tasting solutions can be produced by ingesting a sweet solution (48). The reduction of sugar preference immediately after the ingestion of sweet solutions may extrapolate to situations outside the laboratory, such as after a meal. This effect, when measured in the laboratory, is more pronounced in people who are chronic dieters (49,50), is not observed in obese subjects (51), and is influenced by the menstrual cycle (52,53).

2.3. Genes and Genetics

Because family and twin studies have shown modest heritability for sweet intake, sweet perception or preference may be partially due to genetic variation (54). Most studies of sweet preference use sugar or carbohydrate intake as a measure of preference and as measures of food intake collected through diaries. Family and twin studies using other measures of sweet perception and preference are needed to assess more specifically the degree to which these phenotypes are heritable. In considering how and where genetic differences may influence the human behavior toward sweetness, we now discuss recent advances in our understanding of sweet taste biology.

The initial events in the perception of sweet taste occur in taste receptor cells in the tongue, which are found clustered in taste buds in taste papillae. The perception of sweetness intensity is related to the number of papillae (55). The number of taste papillae and taste buds varies widely in humans, and these differences among people may be due to alleles in genes that develop and maintain sensory cells. For at least one genetic disorder (familial dysautonomia), mutations in a single gene (IKBKAP) (56,57) are associated with few or no taste papillae and taste buds (58). It is possible that less harmful alleles of this gene may influence the density of taste buds in otherwise healthy people.

Inside the taste papillae, taste receptor cells produce proteins that participate in sweet taste transduction, and some of these proteins are inserted into the cell membrane to form taste receptors. Two proteins twist together to create a sweet receptor (Fig. 1) (59,60). The names of these proteins are T1R2 and T1R3, for taste receptor family 1, proteins 2 and 3, and the names of the associated genes for these proteins are *Tas1r2* and *Tas1r3*. If T1R3 pairs with the first member of this family, T1R1, the receptor is sensitive to umami, the taste quality of monosodium glutamate and an important flavor principle of Asian cooking.

These sweet and umami receptor genes were discovered through mapping experiments in mice. Inbred mouse strains differ in their intake of saccharin, and the results of breeding experiments suggested that an allele of a single gene was partially responsible for these differences (61). Through

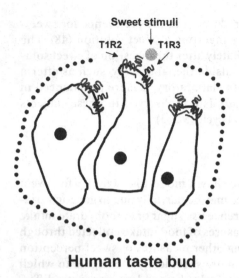

Human taste bud

FIGURE 1 Representation of a human taste bud and taste receptors cells. T1R2 and T1R3 co-localize (and probably dimerize) to create a receptor for sweet stimuli. The receptors are embedded in the apical membrane of the taste receptor cell and stimulate G proteins to initiate a transduction signal inside the cell. Genetic variation in the *Tas1r3* gene (which codes for T1R3 protein) accounts for differences in sweet intake of mice.

positional cloning approaches, this gene was identified and found to be the gene *Tas1r3* (60,62–66). An important advance in our understanding of the behavior of animals toward sweetness was the observation that small changes in the DNA sequence of the mouse *Tas1r3* gene lead to large differences in the consumption of sweetener (67). This reduction of sweetener preference by mice with certain *Tas1r3* alleles is probably due to their reduced ability to perceive the intensity of the sweeteners. Recordings of their peripheral taste nerves suggest that mice with the low-preference *Tas1r3* alleles exhibit lower nerve firing in response to saccharin (26). Furthermore, when the *Tas1r3* gene is eliminated by genetic engineering in mice, the peripheral nerve firing is reduced in response to sweeteners (68).

The pairing of T1R2 and T1R3 does not constitute the only receptor for all sweeteners, however. When the *Tas1r3* gene is knocked out in mice, their ability to detect glucose and maltose is unaffected compared with mice with a normal *Tas1r3* gene (68). Furthermore, the ability to detect other sugars and high-intensity sweeteners is reduced in *Tas1r3* knockout mice, but not absent. Therefore, other receptors or mechanisms exist that signal sweetness in mice, for instance, the remaining partner (T1R2) could act as a taste receptor by itself (69).

If DNA sequence variants have a large effect on the intake of saccharin and other sweeteners in mice, then this may also be true in humans. There is a human counterpart to each of the mouse sweet receptor genes (*TAS1R1*, *TAS1R2*, and *TAS1R3**) (70). Because the peripheral neural responses of humans to sugars predict their verbal reports about the taste of sugars (71), peripheral differences in taste sensitivity may be an important component of the human behavior toward sweetness. There is more variation than appreciated in human perception of sweeteners, and one investigator has even suggested that there is a "different receptor site for each subject" (72) or, in other words, each person may perceive sugars slightly differently. Although the differences in the ability to perceive sweet stimuli has been thought to be of little consequence in human sweet intake and preference, the relationship in mice may stimulate further study of this topic.

Sweet preference may be influenced by genetic variants in the sensory system in humans as it is in mice. However, the appreciation of sweet and the pleasure that it brings to some people may be due to differences in the degree to which they have learned about its rewarding properties. The genes and genetics involved in the perception of the pleasure associated with sugar are not known, but several observations provide clues about which mechanisms may be involved. Sweet preference is increased in opioid addicts compared with healthy subjects (73), and the opioid antagonist naloxone reduces the pleasantness of sucrose (74). Studies suggest that the rewarding aspects of alcohol and sweeteners may also share brain pathways, because alcoholic subjects and their family members may prefer sweeter solutions compared with nonalcoholic subjects (6,75). Therefore, the investigation of genes that participate in the shared brain pathways responsible for the pleasurable effects of drugs and sweeteners is warranted.

3. OBESITY AND SWEET TOOTH

People assume that because increases in sugar consumption in the human diet are associated with a proportional rise in obesity, eating sugar and foods that are sweet is the cause. More specifically, people often hypothesize that if someone has a sweet tooth, it will cause the person to eat sweet food in excess of his or her caloric needs and consequently gain weight. In other words, the sweet tooth is the cause and obesity is the effect. However, an alternative hypothesis is that obesity, per se, may change sweet preference

---

*The protein name for each of the three receptors has the same name in mice and humans (T1R1, T1R2, and T1R3). However, the gene names in the mice (*Tas1r1*, *Tas1r2*, and *Tas1r3*) are lowercase and italic whereas the human gene symbols are in uppercase and italic: *TAS1R1*, *TAS1R2*, *TAS1R3*.

and that metabolism and taste may participate in a feedback loop. Pathways that could influence sweet preferences and contribute to these loops are shown in Fig. 2.

3.1. Do Obese People Have Different Behavior Toward Sweet Food than Lean People?

Most studies have compared lean and obese subjects for the preference or liking of sweet stimuli, usually sucrose solutions, or have compared lean and obese subjects for their intake of sweet foods in the laboratory. These studies have produced mixed results: In some studies, lean people prefer sweet food or drinks more than do obese people (76–80), and in one study the reverse was observed (36). However, the most common observation is that there is no difference in sweet preference between lean and obese people (2,31,81–87). Outside of the laboratory, when food intake is measured in situations where people choose their own meals, most studies demonstrate that lean subjects eat more of their calories as sugar compared with obese subjects (88).

Based upon these data, it would appear that there is little evidence that obese people prefer sweets or eat more sweet food and drink compared with lean people. However, there are three points that are important to consider before drawing this conclusion. First, because subjects can and do restrain their intake of foods, especially sweets, when they are dieting or trying to avoid gaining weight, food intake outside the laboratory may not correspond with sweet preference (i.e., subjects may choose to not eat their most preferred foods.) Second, food intake as reported by subjects can be biased, and when proxy measures of sweet intake such as oral bacteria associated with sucrose consumption are measured, obese women have higher indices of sweet consumption compared with lean women (89). Third, none of these studies measures people before they become obese and therefore does not directly test the hypothesis that a subject's behavior toward sweet food and drink is a factor in the development of obesity.

Once someone becomes obese, the preference for sweet may change because of a shift in the homeostatic mechanisms and feedback loops that regulate hunger and satiety (Fig. 2). To try to understand the behavior of the obese subject in the absence of obesity, investigators have studied formerly obese people who have reduced their weight and are no longer obese. These subjects demonstrate a heightened preference for sugar when it is mixed with high concentrations of fat (18). In another study, diabetic patients measured during weight loss preferred lower concentrations of sweetness compared to the preferences before weight loss (90). It is unclear what effect weight loss alone has on sweet preference, and whether changes in preference after weight loss reflect the preferences subjects had prior to

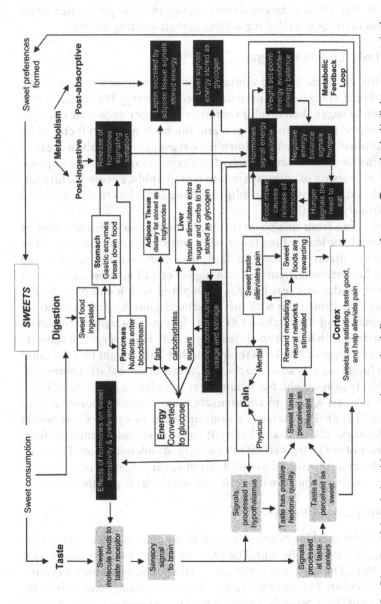

FIGURE 2 Diagram of the relationship among pathways that influence sweet preference. Boxes shaded in light gray contain the sweet-taste transduction pathway. Boxes shaded in black contain the pathway involved in the metabolic control of satiety and energy balance.

becoming obese. Lean people, who restrict their food intake, however, such as ballerinas and patients with anorexia nervosa, vary in their sweet preference (91–93). There is no consistent change in sweet preference when people restrict their food intake, regardless of their starting weight.

3.2. Metabolic Effects of Sugar

For diets with the same caloric content, the macronutrient composition affects the balance of nutrients stored or burned for energy. When excess calories are eaten as sugar, then insulin secretion and other endocrine changes convert the excess calories to glycogen and the body may also increase its overall metabolism temporarily to burn the excess calories. This process of glycogen storage and increased carbohydrate oxidation avoids the comparatively costly conversion of carbohydrate to stored lipids. Excess dietary fat, however, is stored as triglyceride in adipose tissue and is less readily oxidized compared with glycogen (94).

Extrapolating from this observation, humans who consume calories from sugar should be leaner than those who consume an equivalent number of calories from dietary fat (88). In fact, in a rodent study, substituting sucrose for other macronutrients led to a higher rate of metabolism, a lower overall caloric intake, and less body fat compared with a comparable diet without sugar (95). Consistent with this hypothesis, human patients who ate a higher proportion of their calories as sugar lost more weight after gastric surgery compared with those who ate less sugar (96). However, when subjects are asked to add sugared drinks to their diet, they gain weight (97). In other words, when liquid sugar is added to the diet, there is poor caloric compensation and subjects gain weight, but when sugar is added as a solid food (jelly beans), then subjects appear to compensate for the added calories and do not gain weight (98). The metabolic consequences of eating sugar would encourage leanness rather than obesity if sugar is replacing calories from other sources, but not if sugar is added to an already adequate diet. The composition of the calories (liquid or solid) from sugar might be important in determining whether subjects will reduce their calories from other sources.

3.3. The Pleasure of Sweet

Sugar is a fuel that provides calories, but it is also a pleasure that is rewarding in the absence of any other benefit. The pleasure of sweetness soothes crying infants (99–104). The effects of sugar are partially due to its taste because, although oral sucrose reduces pain in babies, sugar placed directly into the stomach does not (105). Sugar is soothing to adults as well as babies. Investigators examined the intake of sweet foods in women and noted a higher intake of sweets both during the menstrual cycle and in those with more incidences of psychiatric problems (12). Sweets may alleviate depression and premenstrual symptoms, and provide relief from the cravings for other drugs because sweet taste releases opiates into the blood, at least in rodents (106). Human babies exposed to the distress of cocaine withdrawal suck sweet pacifiers more than do babies without prior cocaine exposure (107). In addition to the release of opiates, the ability of sugar to bring pleasure is caused by changes in the neural circuits in specific brain areas (108,109). People may differ in their ability to perceive pleasure from sucrose because of individual differences in these neural circuits. People who derive a greater than average pleasure from sucrose and who have a greater than average amount of distress may gain weight if they eat sugar to soothe themselves and do not reduce calories from other sources.

3.4. Insulin and Leptin

Sweetness in food and drink provides a signal of the number of calories available in the form of readily digested sugar. Therefore we might expect that sweet taste sensitivity would change in the face of the metabolic need for glucose. This has proved to be the case. When metabolic changes occur that reduce glucose availability, such as increases in plasma insulin concentration, then sweet preference increases (110–112). A similar response is seen in diabetic animals with high levels of plasma glucose but limited ability to utilize it because of insulin resistance. This effect, however, may only occur during dire metabolic states, because moderate levels of hunger (and the concomitant metabolic consequences of normal food deprivation) do not influence the preference for sweet solutions (113).

In addition to hormones such as insulin that regulate immediate glucose availability, other hormones regulate long-term energy stores. Investigators have proposed that the body has a regulatory mechanism that maintains weight at or near a set point, and that obesity ensues either because people have a high set point or because the set point is overridden by increased caloric consumption (114,115). A fall below set point increases appetite and may increase the preference for energy-dense foods such as sweets and fats (116,117). One hormone proposed to provide this signal of long-term energy stores is leptin. Leptin is secreted by adipose tissue and acts as a signal to the brain to indicate high or low energy reserves. Receptors for leptin are located in the brain as well as in other peripheral tissues (118).

Mice with mutations of the leptin receptor have a higher behavioral and neural response to sugars compared with littermates without mutations, which suggests that leptin might suppress the peripheral sweet taste system (119). Evidence in support of this hypothesis comes from the observation that leptin receptors are present on taste receptor cells in mice, and the administration of exogenous leptin acts directly to suppress the neuronal activation to sweet—but not salty, sour, or bitter—stimuli (120). Obese mice that lack a functional leptin receptor (db/db) do not reduce their consumption of sweet solutions after leptin administration, but their lean littermates, which have normal leptin receptors, do reduce their consumption (121).

Although exogenous administration of leptin reduces the neural response to sweet in mice with a functioning leptin receptor, insulin resistance, inability to utilize plasma glucose, and leptin resistance induced by prolonged obesity or diabetes may override the normal ability of leptin to reduce the cellular response to sweet taste (122). In obese and diabetic animals, the increase in plasma leptin concentration does not appear to have an effect on the neural response to sweet.

To extrapolate from these studies in rodents to human behavior should be approached cautiously. The only study performed on humans to date found that the plasma leptin concentration of obese women was not correlated with sucrose preference (123). However, as demonstrated earlier, in humans, indices of the perceived intensity of a sucrose concentration do not necessarily correspond to how much that sweet concentration will be liked. Thus, future studies in humans may examine how the perceived intensity of a sucrose solution correlates with plasma leptin concentration, and if leptin is shown to have a direct effect on human taste receptor cell function, then manipulation of plasma leptin concentrations and the measure of sucrose perception would be a logical next step for human studies.

3.5. Digestion

Some people are born with an impaired ability to digest specific sugars, such as lactose or fructose. As a consequence of their inability to digest the sugar, they often do not wish to eat it and find it repugnant (124,125). Similarly, there may be cases in which sugar is more easily digested than other nutrients and therefore is more desired. One such example of this situation is the high sugar intake of patients with Crohn's disease (126). One hypothesis is that sweet preferences and aversions may be learned responses that depend upon the punishing or rewarding properties of sugar ingestion. In healthy people, the ability to digest sugars varies from person to person, and this normal variation may affect sweet preference through learning. Differences in the degree of digestive tolerance for sugars are correlated with geography and genotype. For instance, there are geographical differences in the ability to digest lactose that reflect the degree of dairy farming in a region. Therefore, differences in the efficacy of digestive enzymes by geography and traditional diet may partially account for racial differences in the preference for sugar (127).

Studies designed to assay differences in the digestion of sugar and its impact on the human sweet tooth in otherwise healthy subjects might prove useful.

4. CONCLUSIONS

Understanding human behavior toward sweetness and its influence on body weight requires further study. Longitudinal studies of people before they become obese are needed to assess the effects of sweet preference on body weight. Experimental results in mice have taught us two things: Sweet tooth is partially explained by differences in the DNA sequence of taste receptor genes, and the hormone leptin has a direct effect on taste receptor cells. Changes in sweet preference may be part of the homeostatic mechanism that regulates body weight in humans and is worthy of further study.

ACKNOWLEDGMENTS

Comments from Beverley Cowart, Alexander A. Bachmanov, Gary K. Beauchamp, Marcia Pelchat, Julie Mennella, and Yanina Pepino on this work are gratefully acknowledged. Patricia Watson provided helpful editorial suggestions.

REFERENCES

1. Moskowitz HR. The psychology of sweetness. In: Sipple H, McNutt K, eds. Sugars in Nutrition. New York: Academic Press, 1974.
2. Witherly S, Pangborn RM, Stern JS. Gustatory responses and eating duration of obese and lean adults. Appetite 1980; 1:53–63.
3. Pfaffmann C. The sensory and motivating properties of the sense of taste. In: Jones MR, ed. Nebraska Symposium on Motivation. University of Nebraska Press, 1961.
4. Drewnowski A, Henderson SA, Shore AB, Barratt-Fornell A. Sensory responses to 6-n-propylthiouracil (PROP) or sucrose solutions and food preferences in young women. Ann N Y Acad Sci 1998; 855:797–801.
5. Pangborn RM. Individual variation in affective responses to taste stimuli. Psychon Sci 1970; 21:125–126.
6. Kampov-Polevoy AB, Garbutt JC, Janowsky DS. Association between preference for sweets and excessive alcohol intake: a review of animal and human studies. Alcohol 1999; 34:386–395.
7. Greene LS, Desor JA, Maller O. Heredity and experience: their relative importance in the development of taste preference in man. J Comp Physiol Psychol 1975; 89:279–284.
8. Mattes RD, Mela DJ. Relationships between and among selected measures of sweet-taste preference and dietary intake. Chemical Senses 1986; 11:523–539.

9.   Conner MT, Haddon AV, Pickering ES, Booth DA. Sweet tooth demonstrated: individual differences in preference for both sweet foods and foods highly sweetened. J Appl Psychol 1988; 73:275–280.

10. Pelchat ML. Of human bondage: food craving, obsession, compulsion, and addiction. Physiol Behav 2002; 76:347–352.

11. Laeng B, Berridge KC, Butter CM. Pleasantness of a sweet taste during hunger and satiety: effects of gender and "sweet tooth". Appetite 1993; 21:247–254.

12. Barkeling B, Andersson I, Lindroos AK, Birkhed D, Rossner S. Intake of sweet foods and counts of cariogenic microorganisms in obese and normal-weight women. Eur J Clin Nutr 2001; 55:850–855.

13. Liem DG, Mennella JA. Sweet and sour preferences during childhood: role of early experiences. Dev Psychobiol 2002; 41:388–395.

14. Desor JA, Beauchamp GK. Longitudinal changes in sweet preferences in humans. Physiol Behav 1987; 39:639–641.

15. Thompson DA, Moskowitz HR, Campbell RG. Effects of body weight and food intake on pleasantness ratings for a sweet stimulus. J Appl Physiol 1976; 41:77–83.

16. Moskowitz HR, Dubose C. Taste intensity, pleasantness, quality of aspartame, sugars and their mixtures. J Inst Can Sci Technol Aliment 1977; 10:126–131.

17. Drewnowski A. Sensory preferences and fat consumption in obesity and eating disorders. In: Mela, ed. Dietary Fats: Determinants of Preference, Selection, and Consumption. Essex: Elsevier Science, 1992; 59–77.

18. Drewnowski A, Brunzell JD, Sande K, Iverius PH, Greenwood MR. Sweet tooth reconsidered: taste responsiveness in human obesity. Physiol Behav 1985; 35:617–622.

19. Blakeslee AF, Salmon TN. Genetics of sensory thresholds: individual taste reactions for different substances. Proceedings of the National Academy of Sciences of the United States of America 1935; 21:84–90.

20. Kahn SG. Taste perception—individual reactions to different substances. Illinois Academy of Science Transactions 1951; 44:263–269.

21. Richter C. The self-selection of diets. Essays in Biology. University of California Press, 1943; 501–505.

22. Nilsson B, Holm A-K, Sjostrom R. Taste thresholds, preferences for sweet taste and dental caries in 15-year-old children. A pilot study. Swed Dent J 1982; 6:21–27.

23. Adams D, Butterfield N. Taste thresholds and caries experience. J Dent 1979; 7:208–211.

24. Mattes RD. Gustation as a determinant of ingestion: methodological issues. Am J Clin Nutr 1985; 41:672–683.

25. Lundgren B, Jonsson B, Pangborn RM, Sontag AM. Taste discrimination vs hedonic response to sucrose in coffee beverage. An interlaboratory study. Chemical Senses and Flavour 1978; 3:249–265.

26. Bachmanov AA, Reed DR, Ninomiya Y et al. Sucrose consumption in mice: major influence of two genetic loci affecting peripheral sensory responses. Mammalian Genome 1997; 8:545–548.

27. Steiner JE, Glaser D, Hawilo ME, Berridge KC. Comparative expression of hedonic impact: affective reactions to taste by human infants and other primates. Neurosci Biobehav Rev 2001; 25:53–74.

28. Geiselman PJ, Anderson AM, Dowdy ML, West DB, Redmann SM, Smith SR. Reliability and validity of a macronutrient self-selection paradigm and a food preference questionnaire. Physiol Behav 1998; 63:919–928.

29. Beauchamp GK, Moran M, Dietary experience and sweet taste preference in human infants. Appetite 1982; 3:139–152.

30. Schiffman SS, Graham BG, Sattely-Miller EA, Peterson-Dancy M. Elevated and sustained desire for sweet taste in african-americans: a potential factor in the development of obesity. Nutrition 2000; 16:886–893.

31. Salbe AD, DelParigi A, Pratley RE, Drewnowski A, Tataranni PA. Taste preferences and body weight changes in an obesity-prone population. American Journal of Clinical Nutrition 2003.

32. Desor JA, Greene LS, Maller O. Preferences for sweet and salty in 9- to 15-year-old and adult humans. Science 1975; 190:686–687.

33. Bacon AW, Miles JS, Schiffman SS. Effect of race on perception of fat alone and in combination with sugar. Physiol Behav 1994; 55:603–606.

34. Bertino M, Beauchamp GK, Jen K-IC. Rated taste perception in two cultural groups. Chemical Senses 1983; 8:3–15.

35. Monneuse MO, Bellisle F, Louis-Sylvestre J. Impact of sex and age on sensory evaluation of sugar and fat in dairy products. Physiol Behav 1991; 50: 1111–1117.

36. Enns MP, Van Itallie TB, Grinker JA. Contributions of age, sex and degree of fatness on preferences and magnitude estimations for sucrose in humans. Physiol Behav 1979; 22:999–1003.

37. Zellner DA, Garriga-Trillo A, Rohm E, Centeno S, Parker S. Food liking and craving: A cross-cultural approach. Appetite 1999; 33:61–70.

38. Parker S, Kamel N, Zellner D. Food craving patterns in Egypt: comparisons with North America and Spain. Appetite 2003; 40:193–195.

39. Than TT, Delay ER, Maier ME. Sucrose threshold variation during the menstrual cycle. Physiol Behav 1994; 56:237–239.

40. Dippel RL, Elias JW. Preferences for sweet in relationship to use of oral contraceptives and pregnancy. Horm Behav 1980; 14:1–6.

41. Weizenbaum F, Benson B, Solomon L, Brehony K. Relationship among reproductive variables, sucrose taste reactivity and feeding behavior in humans. Physiol Behav 1980; 24:1053–1056.

42. De Graaf C, Zandstra EH. Sweetness intensity and pleasantness in children, adolescents, and adults. Physiol Behav 1999; 67:513–520.

43. James CE, Laing DG, Oram N, Hutchinson I. Perception of sweetness in simple and complex taste stimuli by adults and children. Chem Senses 1999; 24:281–287.

44. James CE, Laing DG, Oram N. A comparison of the ability of 8–9-year-old children and adults to detect taste stimuli. Physiol Behav 1997; 62: 193–197.

45. Drewnowski A, Henderson SA, Shore AB, Fischler C, Preziosi P, Hercberg S. The fat-sucrose seesaw in relation to age and dietary variety of French adults. Obes Res 1997; 5:511–518.

46. Cooke L, Wardle J, Gibson EL. Relationship between parental report of food neophobia and everyday food consumption in 2–6 year-old children. Appetite 2003; 41:205–206.

47. Pelchat ML. Food cravings in young and elderly adults. Appetite 1997; 28:103–113.

48. Cabanac M, Duclaux R. Specificity of internal signals in producing satiety for taste stimuli. Nature 1970; 227:966–967.

49. Esses VM, Herman CP. Palatability of sucrose before and after glucose ingestion in dieters and nondieters. Physiol Behav 1984; 32:711–715.

50. Kleifield EI, Lowe MR. Weight loss and sweetness preferences: the effects of recent versus past weight loss. Physiol Behav 1991; 49:1037–1042.

51. Cabanac M, Duclaux R. Obesity: absence of satiety aversion to sucrose. Science 1970; 168:496–497.

52. Pliner P, Fleming AS. Food intake, body weight, and sweetness preferences over the menstrual cycle in humans. Physiol Behav 1983; 30: 663–666.

53. Wright P, Crow RA. Menstrual cycle: effect on sweetness preferences in women. Hormones and Behavior 1973; 4:387–391.

54. Reed DR, Bachmanov AA, Beauchamp GK, Tordoff MG, Price RA. Heritable variation in food preferences and their contribution to obesity. Behavior Genetics 1997; 27:373–387.

55. Miller Jr IJ, Reedy Jr FE, Variations in human taste bud density and taste intensity perception. Physiol Behav 47:1213–1219, 1990.

56. Anderson SL, Coli R, Daly IW et al. Familial dysautonomia is caused by mutations of the IKAP gene Am J Hum Genet 68:753–758, 2001.

57. Slaugenhaupt SA, Blumenfeld A, Gill SP et al. Tissue-specific expression of a splicing mutation in the IKBKAP gene causes familial dysautonomia. Am J Hum Genet 2001; 68:598–605.

58. Gadoth N, Mass E, Gordon CR, Steiner JE. Taste and smell in familial dysautonomia. Dev Med Child Neurol 1997; 39:393–397.

59. Li X, Staszewski L, Xu H, Durick K, Zoller M, Adler E. Human receptors for sweet and umami taste. Proceedings of the National Academy of Sciences 2002; 99:4692–4696.

60. Nelson G, Hoon MA, Chandrashekar J, Zhang Y, Ryba NJ, Zuker CS. Mammalian sweet taste receptors. Cell 2001; 106:381–390.

61. Fuller JL. Single-locus control of saccharin preference in mice. J Hered 1974; 65:33–36.

62. Bachmanov AA, Li X, Reed DR et al. Positional cloning of the mouse saccharin preference (*Sac*) locus. Chem Senses 2001; 26:925–933.

63. Kitagawa M, Kusakabe Y, Miura H, Ninomiya Y, Hino A. Molecular genetic identification of a candidate receptor gene for sweet taste. Biochem Biophys Res Commun 2001; 283:236–242.

64. Max M, Shanker YG, Huang L et al. Tas1r3, encoding a new candidate taste receptor, is allelic to the sweet responsiveness locus Sac. Nat Genet 2001; 28:58–63.

65. Montmayeur JP, Liberles SD, Matsunami H, Buck LB. A candidate taste receptor gene near a sweet taste locus. Nat Neurosci 2001; 4:492–498.

66. Sainz E, Korley JN, Battey JF, Sullivan SL. Identification of a novel member of the T1R family of putative taste receptors. J Neurochem 2001; 77:896–903.

67. Reed DR, Li S, Li X et al. Polymorphisms in the taste receptor gene (*Tas1r3*) region are associated with saccharin preference in 30 mouse strains. Journal of Neuroscience, 2004; 24:938–946.

68. Damak S, Rong M, Yasumatsu K et al. Detection of sweet and umami taste in the absence of taste receptor T1r3. Science 2003; 301:850–853.

69. Zhao G, Zhang Y, Hoon MA et al. The receptors for mammalian sweet and umami taste. Cell 2003; 115:255–266.

70. Liao J, Schultz PG. Three sweet receptor genes are clustered in human chromosome 1. Mamm Genome 2003; 14:291–301.

71. Diamant H, Oakley B, Strom L, Wells C, Zotterman Y. A comparison of neural and psycohphysical responses to taste stimuli in man. Acta Physiol Scand 1965; 64:67–74.

72. Faurion A. Physiology of the sweet taste. In: Autrum H, Ottoson D, Perl E, Schmidt R, Shimazu H, Willis W, eds. Progress in Sensory Physiology. Berlin: Springer-Verlag, 1987.

73. Morabia A, Fabre J, Chee E, Zeger S, Orsat E, Robert A. Diet and opiate addiction: a quantitative assessment of the diet of non-institutionalized opiate addicts. Br J Addict 1989; 84:173–180.

74. Fantino M, Hosotte J, Apfelbaum M. An opioid antagonist, naltrexone, reduces preference for sucrose in humans. Am J Physiol 1986; 251:R91–R96.

75. Kranzler HR, Sandstrom KA, Van Kirk J. Sweet taste preference as a risk factor for alcohol dependence. Am J Psychiatry 2001; 158:813–815.

76. Warwick ZS, Schiffman SS. Sensory evaluations of fat-sucrose and fat-salt mixtures: relationship to age and weight status. Physiol Behav 1990; 48: 633–636.

77. Heymsfield SB, Allison DB, Heshka S, Pierson RN. Assessment of human body composition. Handbook of Assessment Methods for Eating Behaviors and Weight Related Problems. Beverly Hills: Sage, 1995:515–560.

78. Underwood PJ, Belton E, Hulme P. Aversion to sucrose in obesity. Proc Nutr Soc 1973; 32:93A–94A.

79. Johnson WG, Keane TM, Bonar JR, Downey C. Hedonic ratings of sucrose solutions: effects of body weight, weight loss and dietary restriction. Addict Behav 1979; 4:231–236.

80. Cox DN, van Galen M, Hedderley D, Perry L, Moore PB, Mela DJ. Sensory and hedonic judgements of common foods by lean consumers and consummers with obesity. Obesity Research 1998; 6:438–447.

81. Rodin J. Effects of obesity and set point on taste responsiveness and ingestion in humans. J Comp Physiol Psychol 1975; 89:1003–1009.

82. Malcolm R, O'Neil PM, Hirsch AA, Currey HS, Moskowitz G. Taste hedonics and thresholds in obesity. Int J Obes 1980; 4:203–212.
83. Thompson DA, Moskowitz HR, Campbell RG. Taste and olfaction in human obesity. Physiol Behav 1977; 19:335–337.
84. Drewnowski A, Grinker JA, Hirsch J. Obesity and flavor perception: multidimensional scaling of soft drinks. Appetite 1982; 3:361–368.
85. Pangborn RM, Simone M. Body size and sweetness preference. J Am Diet Assoc 1958; 34:924–928.
86. Garn SM, Solomon MA, Cole PE. Sugar-food intake of obese and lean adolescents. Ecology of Food and Nutrition 1980; 9:219–222.
87. Frijters JER, Rasmussen-Conrad EL. Sensory discrimination, intensity perception, and affective judgment of sucrose-sweetness in the overweight. The Journal of General Psychology 1982; 107:233–247.
88. Hill JO, Prentice AM. Sugar and body weight regulation. Am J Clin Nutr 62:264S–273S; discussion 1995; 273S–274S.
89. Barkeling B, Linne Y, Lindroos AK, Birkhed D, Rooth P, Rossner S. Intake of sweet foods and counts of cariogenic microorganisms in relation to body mass index and psychometric variables in women. Int J Obes Relat Metab Disord 2002; 26:1239–1244.
90. Laitinen JH, Tuorila HM, Uusitupa MI. Changes in hedonic responses to sweet and fat in recently diagnosed non-insulin-dependent diabetic patients during diet therapy. Eur J Clin Nutr 1991; 45:393–400.
91. Martin C, Bellisle F. Eating attitudes and taste responses in young ballerinas. Physiol Behav 1989; 46:223–227.
92. Sunday SR, Halmi KA. Taste perceptions and hedonics in eating disorders. Physiol Behav 1990; 48:587–594.
93. Drewnowski A, Halmi KA, Pierce B, Gibbs J, Smith GP. Taste and eating disorders. Am J Clin Nutr 1987; 46:442–450.
94. Flatt J. The biochemistry of enegy expenditure. In: Bray GA, ed. Recent advances in obesity research: II. Proceedings of the 2nd International Congress on Obesity. Westport, CT: Technomic Publishing Company, 1978:211–227.
95. Goodson S, Halford JC, Jackson HC, Blundell JE. Paradoxical effects of a high sucrose diet: high energy intake and reduced body weight gain. Appetite 2001; 37:253–254.
96. Lindroos AK, Lissner L, Sjostrom L. Weight change in relation to intake of sugar and sweet foods before and after weight reducing gastric surgery. Int J Obes Relat Metab Disord 1996; 20:634–643.
97. Tordoff MG, Alleva AM. Effect of drinking soda sweetened with aspartame or high-fructose corn syrup on food intake and body weight. Am J Clin Nutr 1990; 51:963–969.
98. DiMeglio DP, Mattes RD. Liquid versus solid carbohydrate: effects on food intake and body weight. Int J Obes Relat Metab Disord 2000; 24:794–800.
99. Blass EM, Hoffmeyer LB. Sucrose as an analgesic for newborn infants. Pediatrics 1991; 87:215–218.

100. Masters-Harte LD, Abdel-Rahman SM. Sucrose analgesia for minor procedures in newborn infants. Ann Pharmacother 2001; 35:947–952.

101. Gradin M, Eriksson M, Holmqvist G, Holstein A, Schollin J. Pain reduction at venipuncture in newborns: oral glucose compared with local anesthetic cream. Pediatrics 2002; 110:1053–1057.

102. Carbajal R, Chauvet X, Couderc S, Olivier-Martin M. Randomised trial of analgesic effects of sucrose, glucose, and pacifiers in term neonates. BMJ 1999; 319:1393–1397.

103. Barr RG, Quek VS, Cousineau D, Oberlander TF, Brian JA, Young SN. Effects of intra-oral sucrose on crying, mouthing and hand-mouth contact in newborn and six-week-old infants. Dev Med Child Neurol 1994; 36:608–618.

104. Johnston CC, Stremler R, Horton L, Friedman A. Effect of repeated doses of sucrose during heel stick procedure in preterm neonates. Biol Neonate 1999; 75:160–166.

105. Ramenghi LA, Evans DJ, Levene MI. "Sucrose analgesia": absorptive mechanism or taste perception? Arch Dis Child Fetal Neonatal Ed 1999; 80:F146–147.

106. Blass E, Fitzgerald E, Kehoe P. Interactions between sucrose, pain and isolation distress. Pharmacol Biochem Behav 1987; 26:483–489.

107. Maone TR, Mattes RD, Beauchamp GK. Cocaine-exposed newborns show an exaggerated sucking response to sucrose. Physiol Behav 1992; 51:487–491.

108. Hajnal A, Smith GP, Norgren R. Oral sucrose stimulation increases accumbens dopamine in the rat. Am J Physiol Regul Integr Comp Physiol 2004; 286:R31–37.

109. Berridge KC. Pleasures of the brain. Brain Cogn 2003; 52:106–128.

110. Mayer-Gross W, Walker JW. Taste and selection of food in hypoglycemia. British Journal of Experimental Pathology 1946; 27:297–305.

111. Rodin J, Wack J, Ferrannini E, DeFronzo RA. Effect of insulin and glucose on feeding behavior. Metabolism 1985; 34:826–831.

112. Thompson DA, Campbell RG. Hunger in humans induced by 2-deoxy-D-glucose: glucoprivic control of taste preference and food intake. Science 1977; 198:1065–1068.

113. Pangborn RM. Influence of hunger on sweetness preferences and taste thresholds. Am J Clin Nutr 1959; 7:280–287.

114. Levin BE, Keesey RE. Defense of differing body weight set points in diet-induced obese and resistant rats. Am J Physiol 1998; 274:R412–R419.

115. Keesey RE, Hirvonen MD. Body weight set-points: determination and adjustment. J Nutr 1997; 127:1875S–1883S.

116. Drewnowski A, Holden-Wiltse J. Taste responses and food preferences in obese women: effects of weight cycling. Int J Obes Relat Metab Disord 1992; 16:639–648.

117. Schiffman SS, Musante G, Conger J. Application of multidimensional scaling to ratings of foods for obese and normal weight individuals. Physiol Behav 1978; 21:417–422.

118. Friedman JM, Halaas JL. Leptin and the regulation of body weight in mammals. Nature 1998; 395:763–770.

119. Ninomiya Y, Sako N, Imai Y. Enhanced gustatory neural responses to sugars in the diabetic db/db mouse. Am J Physiol 1995; 269:R930–R937.

120. Kawai K, Sugimoto K, Nakashima K, Miura H, Ninomiya Y. Leptin as a modulator of sweet taste sensitivities in mice. Proc Natl Acad Sci USA 2000; 97:11044–11049.

121. Shigemura N, Ohta R, Kusakabe Y et al. Leptin modulates behavioral responses to sweet substances by influencing peripheral taste structures. Endocrinology 2003.

122. Shimizu Y, Yamazaki M, Nakanishi K et al. Enhanced responses of the chorda tympani nerve to sugars in the ventromedial hypothalamic obese rat. J Neurophysiol 2003; 90:128–133.

123. Karhunen L, Lappalainen R, Haffner S et al. Serum leptin, food intake and preferences for sugar and fat in obese women. Int J Obes Relat Metab Disord 1998; 22:819–821.

124. Cox TM. The genetic consequences of our sweet tooth. Nat Rev Genet 2002; 3:481–487.

125. Davidenkov S. Inherited inability to eat sugar. J Heredity 1940; 31:5–7.

126. Schutz T, Drude C, Paulisch E, Lange KP, Lochs H. Sugar intake, taste changes and dental health in Crohn's disease. Dig Dis 2003; 21:252–257.

127. Ishii H. Consumer perceptions of products containing sweeteners in Asian countries. In: Corti A, ed. Low Calorie Sweeteners: Present and Future. Basel: Karger, 1999:164–170.

4

The Adipose Renin-Angiotensin System: Genetics, Regulation, and Physiologic Function

Brynn H. Voy

Life Sciences Division, Oak Ridge National Laboratory, Oak Ridge, Tennessee, U.S.A.

Suyeon Kim, Sumithra Urs, Melissa Derfus, Young-Ran Heo, and Rashika Joshi

Department of Nutrition and Agricultural Experiment Station, University of Tennessee, Knoxville, Tennessee, U.S.A.

Florence Massiera, Michele Teboul, and Gerard Ailhaud

CNRS 6543. Centre de Biochimie, Nice, France

Annie Quignard-Boulangé

INSERM U465 - IFR58, Centre Biomédical des Cordeliers, Paris, France

Naima Moustaid-Moussa

Department of Nutrition and Agricultural Experiment Station, University of Tennessee, Knoxville, Tennessee, U.S.A.

1. INTRODUCTION

Decades of study describe the importance of the renin-angiotensin system (RAS) in pressure and volume homeostasis and its role in cardiovascular disease. In recent years, the emphasis has gradually turned to paracrine actions of angiotensin II (Ang II) due to the discovery of local RAS within such diverse organs as kidney, brain, heart, and ovary. As a result, many additional physiological roles that lie outside hemodynamic regulation have been attributed to Ang II. One of the many surprising tissues discovered to contain a local RAS is adipose tissue. In this review we will discuss the discovery and regulation of RAS in adipose tissue and summarize the evidence to date demonstrating its role in regulation of adipogenesis, adipocyte metabolism, and (possibly) obesity-related hypertension.

2. COMPONENTS OF RAS IN ADIPOSE TISSUE

The existence of the RAS in adipose tissue was first discovered in 1987 when angiotensinogen (AGT) mRNA was detected in periaortal brown adipose tissue (BAT) and in cells found within the rat aorta wall (1,2). These findings led to the subsequent detection of AGT mRNA and AGT secretion in several rat adipose tissue depots and isolated adipocytes from rat arterial vessel walls, atria, and mesentery (3,4). More recently, adipose tissue and adipocyte cell lines from rats, mice, and humans have been shown to both express components of the RAS and synthesize and secrete Ang II (5–11). Although synthesis and secretion of Ang II from adipose tissue has been established, some controversy remains regarding the cell types (preadipocytes vs. mature adipocytes) and the enzymes involved, particularly in humans. Mature adipocytes are the primary source of AGT in adipose tissue, although low levels of expression have been reported in human preadipocytes (9). With respect to biochemical pathways of Ang II formation in adipocytes, the role of renin is particularly controversial. Sharma et al. (12) described expression of renin and other RAS components in adipocytes obtained from subcutaneous adipose depots of both lean and obese patients. However, Faloia et al. (13) found no evidence for renin by RT-PCR or western blotting in adipose tissue of either lean or obese normotensive individuals. In rats, Pinterova et al. (11) found renin and other RAS components in mature adipocytes, but angiotensin-convert enzyme (ACE) was found only in the stromal-vascular fraction, consisting of blood vessels and preadipocytes. However, another group was unable to detect renin despite confirming angiotensin I (AngI) generating activity, although it may have been due to use of only Northern blot analysis instead of the more sensitive RT-PCR (14). Some of the controversy in both humans and rodents may be due to the fact that enzymes other than renin and ACE can form Ang II from AGT. Both cathepsin G, which can cleave Ang II directly from AGT, and cathepsin D, which has ACE-like activity, are expressed in human adipose tissue and thus may play a role in Ang II synthesis (8).

3. ANGIOTENSINOGEN EXPRESSION AND REGULATION IN ADIPOSE TISSUE

Levels of AGT protein are directly controlled by AGT mRNA levels, as no posttranscriptional or posttranslational mechanisms of AGT regulation have been identified. Levels of AGT in circulation (0.3–1.0 µM) are in the range of the Km for renin (0.9–1.2 µM), indicating that AGT availability is the rate-limiting factor for Ang II synthesis under most physiological conditions (15). In murine adipocyte cell lines (3T3-L1, 3T3-F442A, Ob1771), AGT mRNA levels are very low in preadipocytes but are upregulated dramatically during the course of differentiation into mature adipocytes (16–18). Like many genes necessary for lipogenesis, AGT is considered a late marker for adipocyte differentiation, with maximal levels of synthesis and secretion reached at 10–12 days after the onset of differentiation (19). Increased AGT expression occurs regardless of whether differentiation occurs with hormonal stimulation (i.e., insulin, dexamethasone, and isobutyl-methyl-xanthine) or spontaneously, indicating that it is part of the normal cellular conversion and not dependent on exogenous factors. Upregulation of AGT during adipogenesis can be explained at the molecular level by the presence of a differentiation-specific element (DSE) in the AGT promoter that is required for both sustained transcriptional expression of AGT and, along with its corresponding binding protein (DSEB), for hormonal differentiation of 3T3-L1 adipocytes (20,21). Peroxisome prolif-erator-activated receptors (PPARs) and their associated response elements may also contribute to differentiation-dependent expression of AGT. Both long-chain natural and nonmetabolized fatty acids as well as peroxisome proliferators induced AGT expression in preadipocytes with kinetics that closely resembled their effects on known fatty acid-responsive genes (19). To date, however, no known PPAR response element has been identified in the AGT gene. Furthermore, the PPARγ activator rosiglitazone had no effect on AGT expression in primary cultures of mature human adipocytes (22).

Although AGT upregulation is not dependent upon a specific differen-tiation protocol, dexamethasone and insulin, two common components of the differentiation cocktail, have been shown to regulate and enhance AGT expression. Dexamethasone increased AGT expression in Ob1771 adipo-cytes (23). Insulin effects, however, may depend on the cell line: increased AGT was reported in 3T3-L1 adipocytes (7), whereas low concentrations of insulin were shown to decrease AGT expression in 3T3-F442A and Ob 1771 adipocytes (24). In vivo, insulin deficiency induced by streptozotocin significantly reduced AGT levels in Sprague-Dawley rats; insulin administration reversed this effect (25). Very recently, the study from Harte et al. has demonstrated that insulin increased AGT protein expression of isolated human abdominal subcutaneous adipocytes in a dose-dependent manner (26). Triiodothyronine, estrogens, and Ang II, well-known regulators of hepatic AGT expression, were shown to have no effect on AGT levels in adipocytes (23).

Angiotensinogen shares nutritional control with many genes involved in lipogenesis, implicating the adipose RAS in metabolic regulation. Frederich et al. (27) reported that fasting decreased adipose but not liver AGT expression in rats, and that mRNA levels were restored by refeeding. Plasma glucose concentration may in part explain the effects of fasting/ refeeding on adipose AGT levels. Hyperglycemia in the presence of an insulin clamp (euinsulinemia) increased adipose AGT expression by three-fold in both lean and obese rats. Interestingly, however, hyperinsulinemia/ euglycemia significantly decreased adipose and hepatic AGT expression, but only in lean, insulin-sensitive animals (28). Collectively, these results suggest that insulin and glucose independently regulate AGT expression in adipose tissue. Blood pressure paralleled the changes in adipose RAS activity, suggesting that adipose tissue may contribute to blood pressure regulation by supplying Ang II.

Both bilateral nephrectomy and ACE inhibition with enalapril increased AGT expression in adipose tissue of Sprague Dawley rats (3), but sodium restriction had no effect (29). Adipose AGT expression in male rats is sensitive to testosterone levels, as castration results in a 50% reduction in AGT expression levels that are restored by testosterone administration. Increased testosterone levels are commonly observed in patients with android or upper-body obesity, and a concomitant increase in adipose Ang II production may contribute to the increased incidence of hypertension in these patients (30,31). Aging may also contribute to changes in the activity of the adipose RAS. Angiotensinogen content was shown to be to be three-fold higher in adipocytes from 8-week-old vs. 26-week-old Wistar rats (14,32), although no age-related differences were reported in obese Zucker rats (33). The effects of various factors on adipocyte AGT levels are summarized in Tables 1 and 2.

4. ANGIOTENSIN II RECEPTORS IN ADIPOSE TISSUE

The presence of both angiotensin II receptor (AT1 and AT2) subtypes has been demonstrated using molecular (Northern and western blots),

TABLE 1 Effects of Various Treatments on AGT Levels in Adipocytes in Vitro

Model	Treatment	Effect on AGT	Reference
3T3-L1, Ob1771, 3T3-F442A preadipocytes	Differentiation	Increased	16–18
Ob1771 preadipocytes	Natural and non-metabolizable long-chain fatty acids	Increased	19
Human primary adipocytes	Rosiglitazone (PPARγ agonist)	No effect	22
Ob1771 adipocytes	Dexamethasone	Increased	23
3T3-L1 adipocytes, human primary adipocytes	Insulin	Increased	7,26
3T3-F442A, Ob1771 adipocytes	Insulin	Decreased	24
Ob1771 adipocytes	Triiodothyronine	No effect	23
Ob1771 adipocytes	Estrogen	No effect	23
Ob1771 adipocytes	Ang II	No effect	23
Ob1771 adipocytes	Growth hormone	No effect	23

functional (binding assays), and pharmacological (receptor antagonists) tests. However, the identity of the predominant receptor subtype in mature adipose cells has not been established due to variable findings across models and species. The AT1 receptors were first identified in rat epididymal adipocyte membranes in 1993 (34), and the majority of subsequent studies

TABLE 2 Effects of Various Treatments on AGT Levels in Adipose Tissue In Vivo

Rat model	Treatment	Effect on AGT mRNA	Reference
Sprague-Dawley	Insulin deficiency (streptozotocin)	Reduced	25
Sprague-Dawley	Insulin replacement	Restored	25
Sprague-Dawley	Bilateral nephrectomy	Increased	3
Sprague-Dawley	ACE inhibition (enalapril)	Increased	3
Sprague-Dawley	Sodium restriction	No effect	29
Sprague-Dawley	Castration	Reduced	31
Sprague-Dawley	Testosterone replacement	Restored	31
Wistar	Aging	Reduced	14,32
Zucker obese	Aging	No effect	33

in rats point to AT1 as the primary receptor (11,35,36). However, both subtypes have been found in human preadipocytes and mature adipocytes, with AT2 being the primary type found in mature cells (37). The subtype AT2 has also been shown to predominate in adipocyte cell lines of murine origin, and Ang II functions in mice are prevented by AT2 antagonism.

5. ANGIOTENSIN II AND ADIPOCYTE GROWTH, DIFFERENTIATION, AND METABOLISM

Clinical and rodent studies in which both humans and rats lost weight in response to ACE inhibitors first suggested a possible effect of Ang II on body weight (38,39). Angiotensin II was discovered to exert trophic effects in a variety of tissues and cells (40,41), and subsequent studies have focused on a similar role in adipose tissue (42–44). One study reported that the cell cycle was accelerated by Ang II in human preadipocytes in vitro in parallel with upregulation of the cell cycle regulator cyclin D1 (44), suggesting that Ang II may increase the number of preadipocytes available for differentiation into mature, lipid-storing cells. Nonetheless, the primary mechanisms through which Ang II has been demonstrated to directly increase adipose mass include the stimulation of adipocyte differentiation and the induction of hypertrophy of existing adipocytes (42,45).

Several studies demonstrate that prostacyclin (PGI_2)—a metabolite of arachidonic acid and potent effector of preadipocyte differentiation—serves as the second messenger that links Ang II to adipocyte differentiation (46–48). PGI_2 was shown to induce differentiation of Ob1771 preadipocytes (47). Independent studies demonstrated that Ang II increased PGI_2 production in both adipocytes in vitro (48) and adipose tissue in situ (49). Darimont combined these concepts using a coculture system and demonstrated that treatment of mature adipocytes with Ang II elicited PGI_2 release, which in turn induced differentiation of cocultured preadipocytes (42). This response was blocked by simultaneous treatment with PD123177, an AT1 receptor antagonist, but not by AT2 blockade. Angiotensin II effects were also inhibited by aspirin, which inhibits the cyclooxygenase enzymes necessary for arachidonic acid conversion to PGI_2, and by antibodies against PGI_2 (42).

In further support of a link between PGI_2, Ang II, and adipocyte differentiation, (carba) prostacyclin—a stable analogue of PGI_2—was shown to upregulate the transcription rate of numerous adipogenic genes, including *Agt* (18). Finally, exposure of rat epididymal fat pads in vivo and ex vivo to Ang II increased the formation of fat cells as indexed by an increase in glycerol-3-phosphate dehydrogenase (GPDH) expression (50). Explants of epididymal fat were incubated with Ang II in the presence or absence of aspirin, which is a cyclooxygenase inhibitor, thus decreasing PGI_2 production. Stromal-vascular cells (containing the preadipocyte fraction) were then isolated, cultured for two days, and examined for spontaneous differentiation. Treatment with Ang II enhanced the appearance of cells positive for GPDH, a marker of adipocyte differentiation. This response was attenuated by the addition of aspirin or PD123177 but not by losartan, again indicating involvement of the AT2 receptor (50).

These findings markedly differ from the reported effects of Ang II on human preadipocyte differentiation. Schling and Loffler demonstrated that Ang II led to a distinct reduction in insulin-induced differentiation of primary human preadipocytes (51). Interestingly, the attenuation by Ang II was only marginal when cells were induced to differentiate with a cocktail of insulin, cortisol, and isobutyl methyl xanthine. In addition, pharmacological blockade of the AT1 receptor in the absence of exogenous Ang II significantly enhanced adipogenesis. Moreover, mature adipocytes inhibited the differentiation of cocultured preadipocytes, and this inhibition was released by addition of losartan to the media (52). On the basis of these results, it was proposed that Ang II could act as a protective factor against uncontrolled adipose expansion (51), and that blockade of the RAS could prevent diabetes by promoting adipocyte formation in adipose tissue that would counteract ectopic deposition of lipids in other tissues such as muscle, liver, and pancreas (53). Additional experiments will be necessary to validate these hypotheses.

The basis for such marked differences in the consequences of Ang II between rodents and humans is unclear. It is important to point out that the rodent studies investigated the effects of Ang II on spontaneous differentiation, presumably mediated through PGI_2 release by mature adipocytes, whereas the human experiments examined interactions of Ang II with hormonally induced differentiation. It is not known if PGI_2 affects conversion of human preadipocytes as it does in rodents. However, studies from our laboratory have shown that Ang II increases both PGI_2 and prostacyclin endoperoxide (PGE_2) levels in primary cultures of human adipocytes (54). Collectively, the data suggest that Ang II interacts differently with multiple signaling pathways that lead to adipocyte differentiation; species-specific differences in these interactions cannot be ruled out at this time. It is also important to note that in addition to Ang II differentially impacting differentiation in human vs. rodent cells, similar effects on lipogenesis and adipocyte hypertrophy have been demonstrated in both murine and human adipocytes. Figure 1 represents a schematic view of the physiological actions of adipocyte Ang II on adipocytes, preadipocytes, and extra-adipose tissues.

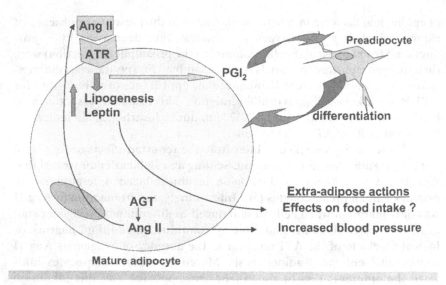

FIGURE 1 Endocrine and paracrine effects of angiotensin II in adipocytes. Adipocyte-derived angiotensin II (Ang II) regulates adipocyte metabolism, including increasing leptin secretion, gene expression, and lipogenesis. These effects are mediated via angiotensin II receptors (ATR). Within adipose tissue, adipocyte Ang II also increases prostaglandin secretion (PGI$_2$), which then promotes differentiation of the preadipocytes. Adipocyte Ang II, when secreted, exerts endocrine effects such as increasing blood pressure (as demonstrated in transgenic mice studies in Fig. 2) as well as other potential actions to be determined.

In adult humans and animals, adipose expansion occurs primarily through hypertrophy of existing adipocytes to accommodate increases in cellular lipid storage (55). Although insulin is the most well-documented stimulator of fatty acid synthesis and triglyceride storage in adipocytes, Ang II exerts similar effects in vitro and in vivo. Physiological concentrations of Ang II significantly increase fatty acid synthase and GPDH activities and gene expression, and also cellular triglyceride content, in both 3T3-L1 adipocytes and primary human adipocytes. This response is attenuated by AT2 but not AT1 receptor antagonists, indicating that Ang II acts through the type II receptor in these adipocyte models. Transcription rates of both fatty acid synthase (*Fasn*) and leptin(*Lep*) genes were also upregulated in parallel with increased fat storage in human primary adipocytes and in the 3T3-L1 cell line; this response was shown to be independent of Ang II increases in PGI$_2$ secretion, suggesting that Ang II acts through multiple second messenger pathways in adipose cells (45,54). Dissection of the *Fasn* promoter in reporter gene assays identified the E box at the proximal *Fasn* promoter as critical for this gene's regulation by Ang II. This region overlaps with a previously identified insulin response element, suggesting that Ang II and insulin share molecular mechanisms of regulation in adipocytes (56).

Potential actions of Ang II on lipolysis have been studied in humans and rats, but with conflicting results. Infusion of a pressor dose of Ang II in Sprague-Dawley rats led to a significant loss of body weight over the course of one week, and renin-AT1 (AT_1R) blockade with losartan prevented this response. However, these animals were pair-fed, and the majority of the weight loss was attributed to a reduction in food intake rather than a specific effect of Ang II on lipolysis (57). In a similar study, infusion with increasing pressor doses of Ang II for seven days resulted in a significantly reduced body weight compared to vehicle-infused controls at all Ang II doses (58). Fat pad weights were not reduced in Ang II-treated animals, except for a modest reduction in retroperitoneal fat pad weights in mice receiving the highest Ang II dosage (59). These authors reported an elevation in abdominal surface temperature with Ang II treatment and were unable to rule out a role for sympathetic activation of metabolism as the basis for the weight loss. In humans, neither subpressor nor pressor doses of infused Ang II nor ACE inhibition with enalapril impacted whole-body lipolytic rates (59). However, in a separate study microdialysis of subcutaneous adipose tissue with increasing doses of Ang II induced a dose-dependent decrease in blood flow, lipolysis, and glucose uptake that was more pronounced in the femoral than in the abdominal region (60). It was not determined whether these responses were due to direct actions of Ang II or were secondary to an accumulation of free fatty acids or a reduction in pH, both of which would result from reduced blood flow that in turn would inhibit lipolysis. In a subsequent study this group examined the specific effects of Ang II applied to the interstitial space of human adipose tissue and found a dose-dependent increase in lipolysis, in spite of a minimal effect on local blood flow (61). The latter findings suggest that the actions of Ang II on adipocyte metabolism are not exclusively due to changes in local blood flow but rather include direct effects of the hormone on adipocyte metabolism.

6. REGULATION OF ADIPOSE TISSUE RAS IN OBESITY AND HYPERTENSION

The discovery that Ang II increases adipocyte lipogenesis prompted several studies of the link between obesity and regulation of the adipose RAS.

Variable results have been reported in rodents and in humans. Angiotensinogen expression was decreased in adipose tissue of obese Zucker rats and viable yellow (A^{vy}) mice when compared by Northern blot analysis on the basis of equivalent amounts of total RNA (7). By contrast, expression levels in *ob/ob* and *db/db* mice were higher than in lean controls (27). Recently, Hainault et al. (33) examined AGT expression in isolated adipocytes from Zucker obese (*fa/fa*) rats during the onset of obesity and found that AGT protein content and secretion when expressed per cell were significantly elevated in adipocytes from obese vs. lean animals. Elevated rates of AGT secretion were not simply due to the increased adipocyte size, as this effect was still present when adipocytes were fractionated by size and comparisons were made between similar subpopulations. In fact, normalization by size enhanced the effect, with AGT secretion fivefold greater in adipocytes from obese vs. lean rats.

   Altered distribution of cholesterol in adipocyte membranes is a typical characteristic of hypertrophied adipocytes and may provide a mechanistic link to changes in AGT expression in obese adipocytes, although at least part of the increased secretion in *fa/fa* rats is cell size-independent. Le Lay et al. reported that treating adipocytes in vitro with either methyl-β-cyclodextrin or mevastatin mimicked the membrane cholesterol reduction of hypertrophied adipocytes. Expression levels of many adipocyte genes were impacted by this treatment, the most significant of which was a ninefold increase in AGT expression (62).

   Much of the interest in the adipose RAS lies in its potential role in blood pressure regulation, particularly in obesity-associated hypertension. In obese individuals, adipose tissue is potentially the most significant source of Ang II in the body. Although the specific contribution of adipose-derived Ang II to blood pressure control is certain to vary among individuals, its potential impact should not be ignored clinically. Several studies in humans suggest a positive relationship between adipose AGT levels and fat mass, although controversy remains due in part to the various study protocols that have been implemented. Van Harmelen et al. (63) measured AGT levels in subcutaneous and omental adipose depots of 20 obese men undergoing weight reduction and found that in both depots the AGT levels correlated positively with waist-to-hip-ratio, an index of central obesity. Another study found that AGT mRNA in visceral but not subcutaneous adipose tissue correlated positively and significantly with body mass index (BMI) in both lean and obese individuals (64). By contrast, Sharma et al. (12) looked only at adipocytes from subcutaneous depots of both obese normo- and hypertensive subjects and reported significantly lower AGT mRNA levels in adipocytes from obese compared to lean patients regardless of blood pressure. Interestingly, they also reported that *Ren, ACE*, and *AGTR1* genes were upregulated in cells from obese individuals that were also hypertensive. One consistent finding in both human and rodent studies is that RAS components, particularly AGT, are present at much higher levels in visceral vs. subcutaneous adipose tissue. Indeed, depot-specific regulation of synthesis of several bioactive molecules has been reported for human adipose tissue (65). Clearly, interpretation of the link between obesity and the adipose RAS is much more complex in humans than in inbred strains of rodents, largely due to heterogenous genetic background. Heritability of AGT mRNA levels is very high (estimated at 90%), and several authors have reported that AGT expression in adipose tissue varies widely from patient to patient in ways that cannot be correlated to known patient characteristics (age, gender, etc.) (7,62). Several genetic polymorphisms at the *AGT* locus have been positively associated with plasma AGT levels and risk for hypertension (66,67). One such polymorphism in the gene's promoter region, designated –20A → C, has recently been shown to be an important modifier of the relationship between body size and blood pressure in a population of African origin (68). Another promoter variant consisting of an A/G polymorphism at –217 has been shown to be much more prevalent among African-American hypertensives compared to normotensives. Interestingly, this region of the promoter binds the CAAT/enhancer-binding protein (C/EBP) family of transcription factors, which play an important role in adipocyte gene expression. Moreover, reporter constructs containing the human *AGT* gene promoter with nucleoside A at position –217 increased basal transcription activity when transiently transfected into HepG2 cells compared to constructs with nucleoside G at the same position (69). The DD allele of the *ACE* gene was also recently associated with significant increases in incidence of overweight and abdominal adiposity and elevated diastolic blood pressure, particularly with aging (70). It is therefore likely that regulation of AGT in human adipose tissue results from a complex interplay of genetic, environmental, and physiological factors that are only beginning to be unraveled. Studies suggesting a relationship between the adipose RAS, obesity, and control of blood pressure in humans are summarized in Table 3.

7. KNOCKOUT AND TRANSGENIC MOUSE MODELS OF RAS

Several transgenic mouse models have underscored the importance of AGT levels in control of blood pressure. Kim and Smithies introduced 1–4 copies of the *Agt* gene into mice and demonstrated that blood pressure was directly proportional to plasma AGT levels and number of *Agt* gene copies (71).

TABLE 3   Evidence for an Association Between the Adipose RAS, Obesity and Hypertension in Humans

Model	Association	Depot(s)	Reference
Obese males	Positive correlation between waist-to-hip ratio and AGT levels	Subcutaneous and omental	63
Lean and obese	Positive correlation between AGT levels and BMI*	Visceral but not subcutaneous	64
Normo- and hypertensive lean and obese	Lower adipocyte AGT levels in obese vs. lean; increased renin, ACE and AT1R expression in obese hypertensives	Subcutaneous	12
Lean and obese males	ACE I/D polymorphism a significant predictor of overweight and abdominal adiposity in men; homozygosity of the D allele associated with enhanced age-related increase in body weight and blood pressure and with higher incidence of overweight.	N/A	70
Normo-and hypertensive	Individuals homozygous for the 20A $\longrightarrow$ C polymorphism in the *AGT* promoter displayed positive correlation between BMI and systolic blood pressure	N/A	68

*BMI = body mass index.

Merrill et al. introduced both the human *AGT* and human *Ren* genes into mice and also found a direct effect on blood pressure (72). Transgenic models also provide the most convincing evidence that AGT plays an important role in regulation of adiposity. The most convincing evidence for a role for the adipose RAS in control of adiposity comes from recent studies with *Agt* knockout (*Agt-ko*) and transgenic mice. The knockout animals exhibited hypotrophy of adipocytes, decreased lipogenesis and increased locomotor activity (73–75). Body weight was significantly reduced in *Agt-ko* mice compared to wild-type littermates, an effect that was not due to decreased body length or fat-free mass but rather to reduced adipose mass. Epididymal fat pad weights were reduced by twofold compared to wild type controls, an effect that was surprisingly not observed in other Ang II-responsive organs (heart, kidney, etc.). Investigation of adipose cellularity revealed that the reduction in fat pad weight was due to adipocyte hypotrophy that was accompanied by a twofold reduction in fatty acid synthase activity, which is consistent with previously described effects of Ang II on lipogenesis in adipocytes in vitro (45). Interestingly, a high-fat diet did not induce significant weight gain in *Agt-ko* animals. Moreover, a high-fat diet failed to suppress FAS activity in *Agt-ko* but not in wild-type mice, suggesting that Ang II is involved in the feedback loop between dietary fat intake and control of adipocyte FAS activity. These findings were further supported by the subsequent creation of transgenic mice that overexpress *Agt* specifically in adipocytes by placing the *Agt* gene under control of the *aP2* promoter (74). Overexpression of *Agt* in adipose tissue increased plasma AGT levels by 22%, and systolic blood pressure by 16%, supporting a potential role for the adipose RAS in hypertension. These animals also exhibited dramatic increases in fat mass that were due to adipocyte hypertrophy. Interestingly, total fat cell number was lower in transgenic animals compared to wild-type controls. There was no difference in the weight of brown adipose tissue and surprisingly no difference in circulating leptin levels in *aP2-Agt* and wild-type mice, despite increased fat mass. Mean 23-hour metabolic rates were also similar across all genotypes. These transgenic mice were also bred with *Agt-ko* mice to examine the specific contribution of the adipose RAS to circulating Ang II and whole-body function. Total fat mass, epididymal fat pad weight, and adiposity, all of which were significantly reduced in *Agt-ko* mice, were restored to wild-type levels in *aP2-Agt-ko* animals, in which *Agt* expression was reintroduced specifically in adipose tissue (73–75). The effects of *Agt* deletion, and replacement and overexpression in adipose tissue, are summarized in Fig. 2.

Whereas other components of the RAS have been genetically manipulated in mice, the only other report of an effect on body weight was in animals homozygously null for both the AT1A and AT1B receptors. These mice, but not mice homozygous null for either subtype alone, displayed reduced overall body weight compared to wild-type mice (76). However, organ weights were also reduced by a comparable magnitude, leading to the conclusion that deletion of both forms of the AT1 receptor impacted overall animal growth in a general manner. Clearly, adipose tissue-specific, targeted deletions of other RAS components would provide additional insight into the mechanisms through which the adipose RAS regulates fat mass and exerts its endocrine effects.

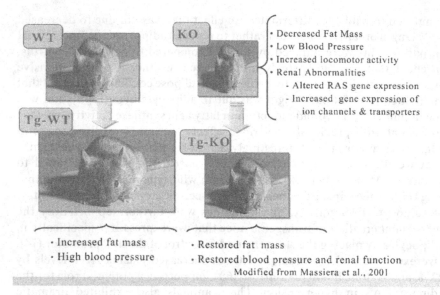

• Increased fat mass   • Restored fat mass
• High blood pressure   • Restored blood pressure and renal function
                                   Modified from Massiera et al., 2001

FIGURE 2   Endocrine and paracrine effects of targeted expression of angiotensinogen in adipose tissue. This figure summarizes the phenotypes of the angiotensinogen (*Agt*) knockout mouse (*ko*), the *ko* mouse that reexpresses AGT in adipose tissue (*Tg-ko*) and the wild-type (WT) mouse overexpressing AGT in adipose tissue (*Tg-wt*).

8. CONCLUSIONS

Angiotensin II is one of many factors secreted by adipose cells, recently found to function as endocrine cells. The discovery of a local RAS in adipose tissue laid the foundation for postulating a role of this local RAS in obesity-associated hypertension. The studies discussed in this review demonstrate both paracrine and an endocrine roles for adipose angiotensin in the regulation of adipocyte metabolism and gene expression, and for the regulation of kidney homeostasis and blood pressure, respectively. The recent studies on mice specifically overexpressing angiotensinogen in adipose tissue were especially crucial in demonstrating the latter role. Many questions, however, require further studies. For example, it is not known whether specific inactivation of angiotensinogen in adipose tissue would reduce blood pressure and alter adipocyte metabolism as might be inferred from adipose tissue-specific overexpression of AGT. Further, no studies have addressed signaling mechanisms of Ang II in adipocytes and only limited studies have addressed dietary modulation of adipocyte RAS. Such studies are only a few of those needed to further our understanding of the function and regulation of AngII in adipose tissue and further clarify the role of this system in obesity-related hypertension.

REFERENCES

1. Campbell DJ, Habener JF. Cellular localization of angiotensinogen gene expression in brown adipose tissue and mesentery: quantification of messenger ribonucleic acid abundance using hybridization in situ. Endocrinology 1987; 121:1616–1626.
2. Naftilan AJ, Zuo WM, Inglefinger J, Ryan TJ, Pratt RE, Dzau VJ. Localization of angiotensinogen messenger RNA in the rat aorta. Hypertension 1988; 11:591–596.
3. Cassis LA, Saye J, Peach MJ. Location and regulation of rat angiotensinogen messenger RNA. Hypertension 1988; 11:591–596.
4. Cassis LA, Lynch KR, Peach MJ. Localization of angiotensinogen messenger RNA in rat aorta. Circ Res 1988; 62:1259–1262.
5. Saye JA, Ragsdale NV, Carey RM, Peach MJ. Localization of angiotensin peptide-forming enzymes of 3T3-F442A adipocytes. Am J Physiol 1993; 264:C1570–C1576.
6. Jonsson JR, Game PA, Head RJ, Frewin DB. The expression and localization of ACE mRNA in human adipose tissue. Blood Press 1994; 3:72–75.
7. Jones BH, Standridge MK, Taylor JW, Moustaid-Moussa N. Angiotensinogen gene expression in adipose tissue: analysis of obese models and hormonal and nutritional control. Am J Physiol 1997; 273:R236–R242.
8. Karlsson C, Lindell K, Ottosson M, Sjostrom L, Carlsson B, Carlsson LM. Human adipose tissue expresses angiotensinogen and enzymes required for its conversion to angiotensin II. J Clin Endoclinol Metab 1998; 83:3925–3929.
9. Schling P, Mallow H, Trindl A, Loffler G. Evidence for a local rennin angiotensin system in primary cultured human preadipocytes. Int J Obes Relat Metab Disord 1999; 23:336–341.
10. Engeli S, Gorzelniak K, Kreutz R, Runkel N, Distler A, Sharma AM. Co-expression of renin-angiotensin system genes in human adipose tissue. J Hypertens 1999; 17:555–560.
11. Pinterova L, Krizanova O, Zorad S. Rat epididymal fat tissue express all components of the rennin-angiotensin system. Gen Physio Biophys 2000; 19:329–334.
12. Gorzelniak K, Engeli S, Janke J, Luft FC, Sharma AM. Hormonal regulation of the human adipose-tissue renin-angiotensin system: relationship to obesity and hypertension. J Hypertens 2002; 20:965–973.
13. Faloia E, Gatti C, Camilloni MA, Mariniello B, Sardu C, Garrapa GG, Mantero F, Giachetti G. Comparison of circulating and local adipose tissue renin-angiotensin system in normotensive and hypertensive subjects. J Endocrinol Invest 2002; 25:309–314.
14. Harp JB, DiGirolamo M. Components of the rennin-angiotensin system in adipose tissue: changes with maturation and adipose mass enlargement. J Gerontol A Biol Sci Med Sci 1995; 50:B270–B276.

15. Clauser E, Gaillard I, Wei L, Corvol P. Regulation of angiotensinogen gene. Am J Hypertens 1989; 2:403–410.
16. Saye JA, Cassis LA, Sturgill TW, Lynch KR, Peach MJ. Angiotensinogen gene expression in 3T3-L1 cells. Am J Physiol 1989; 256:C448–C451.
17. Saye JA, Lynch KR, Peach MJ. Changes in angiotensinogen messenger RNA in differentiating 3T3-F442A adipocytes. Hypertension 1990; 15:867–871.
18. Aubert J, Ailhaud G, Negrel R. Evidence for a novel regulatory pathway activated by (carba) prostacyclin in preadipose and adipose cells. FEBS Lett 1996; 397:117–121.
19. Safonova I, Aubert J, Negrel R, Ailhaud G. Regulation by fatty acids of angiotensinogen gene expression in preadipose cells. Biochem J 1997; 7: 235–239.
20. McGehee RE, Ron D, Brasier AR, Habener JF. Differentiation-specific elements: a cis-acting developmental switch required for the sustained transcriptional expression of the AGT gene during hormonal-induced differentiation of 3T3-L1 fibroblasts to adipocytes. Mol Endocrinol 1993; 7:551–560.
21. McGehee RE, Habener JF. Differentiation-specific element binding protein (DSEB) binds to a defined element in the promoter of the AGT gene required for the irreversible induction of gene expression during differentiation of 3T3-L1 adipoblasts to adipocytes. Mol Endocrinol 1995; 9:487–501.
22. Risusset J, Auwerx J, Vidal H. Regulation of gene expression by activation of the peroxisome proliferator-activated receptor gamma with rosiglitazone (BRL 49653) in human adipocytes. Biochem Biophys Res Commun 1999; 265:265–271.
23. Aubert J, Darimont C, Safonova I, Ailhaud G, Negrel R. Regulation by glucocorticoids of angiotensinogen gene expression and secretion in adipose cells. Biochem J 1997; 328:701–706.
24. Aubert J, Safonova I, Negrel R, Ailhaud G. Insulin down-regulates angiotensinogen gene expression and angiotensinogen secretion in cultured adipose cells. Biochem Biophys Res Commun 1998; 250:77–82.
25. Cassis LA. Downregulation of the renin-angiotensin system in streptozotocin-diabetic rats. Am J Physiol 1992; 262:E105–E109.
26. Harte AL, McTernan PG, McTernan CL, Crocke J, Starcynski J, Barnett AH, Matyka K, Kumar S. Insulin increases angiotensinogen expression in human abdominal subcutaneous adipocytes. Diabetes Obes Metab 2003; 5:462–467.
27. Frederich RC, Kahn BB, Peach MJ, Flier JS. Tissue-specific nutritional regulation of angiotensinogen in adipose tissue. Hypertension 1992; 19: 339–344.
28. Gabriely I, Yang XM, Cases JA, Ma XH, Rossetti L, Barzilai N. Hyperglycemia modulates angiotensinogen gene expression. Am J Physiol 2001; 281: R795–R802.
29. Naftilan AJ, Zuo WM, Inglefinger J, Ryan TJ, Pratt RE, Dzau VJ. Localization and differential regulation of angiotensinogen mRNA expression in the vessel wall. J Clin Invest 1991; 87:1300–1311.

30. Chen YF, Naftilan AJ, Oparil S. Androgen-dependent angiotensinogen and renin messenger RNA expression in hypertensive rats. Hypertension 1992; 19:456–463.

31. Serazin-Leroy V, Morot M, de Mazancourt P, Giudicelli Y. Androgen regulation and site specificity of angiotensinogen gene expression and secretion in rat adipocytes. Am J Physiol Endocrinol Metab 2000; 279:E1398–E1405.

32. Adams F, Wiedmer P, Gorzelniak K, Engeli S, Klaus S, Boschmann M. Age-related changes of Renin-Angiotensin system genes in white adipose tissue of rats. Horm Metab Res 2002; 34:716–720.

33. Hainault I, Nebout G, Turban S, Ardouin B, Ferre P, Quignard-Boulange A. Adipose tissue-specific increase in angiotensinogen expression and secretion in the obese (fa/fa) Zucker rat. Am J Physiol Endocrinol Metab 2002; 282:E59–66.

34. Crandall DL, Herzlinger HE, Saunders BD, Zolotor RC, Feliciano L, Cervoni P. Identification and characterization of angiotensin II receptors in rat epididymal adipocyte membranes. Metabolism 1993; 42:511–515.

35. Crandall DL, Herzlinger HE, Saunders BD, Armellino DG, Kral JG. Distribution of angiotensin II receptors in rat and human adipocytes. J Lipid Res 1994; 35:1378–1385.

36. Cassis LA, Fettinger MJ, Roe AL, Shenoy UR, Howard G. Characterization and regulation of angiotensin II receptors in rat adipose tissue. Angiotensin receptors in adipose tissue. Adv Exp Med Biol 1996; 396:39–47.

37. Schling P. Expression of angiotensin II receptors type 1 and type 2 in human preadipose cells during differentiation. Horm Metab Res 2002; 34:709–715.

38. McGrath BP, Matthews PG, Louis W, Howes L, Whitworth JA, Kincaid Smith PS, Fraser I, Scheinkestel C, MacDonald G, Rallings M. Double-blind study of dilevalol and captopril, both in combination with hydrochlorothiazide, in patients with moderate to severe hypertension. J Cardiovasc Pharmacol 1990; 16:831–888.

39. Campbell DJ, Duncan AM, Kladis A, Harrap SB. Converting enzyme inhibition and its withdrawal in spontaneously hypertensive rats. J Cardiovasc Pharmacol 1995; 26:426–36.

40. Meffert S, Stoll M, Steckelings UM, Bottari SP, Unger T. The angiotensin II AT2 receptor inhibits proliferation and promotes differentiation in PC12W cells. Mol Cell Endocrinol 1996; 122:59–67.

41. Nakajima M, Hutchinson HG, Fujinaga M, Hayashida W, Morishita R, Zhang L, Horiuchi M, Pratt RE, Dzau VJ. The angiotensin II type 2 (AT2) receptor antagonizes the growth effects of the AT1 receptor: gain-of-function study using gene transfer. Proc Natl Acad Sci USA 1995; 92:10663–10667.

42. Darimont C, Vassaux G, Ailhaud G, Negrel R. Differentiation of preadipose cells: paracrine role of prostacyclin upon stimulation of adipose cells by angiotensin II. Endocrinology 1994; 135:2030–2036.

43. Lyle RE, Habener JF, McGehee RE. Antisense oligonucleotides to differentiation-specific element binding protein (DSEB) mRNA inhibit adipocyte differentiation. Biochem Biophys Res Commun 1996; 228:709–715.

44. Crandall DL, Armellino DC, Busler DE, McHendry-Rinde B, Kral JG. Ang II receptors in human preadipocytes: role in cell cycle regulation. Endocrinology 1999; 140:154–158.
45. Jones BH, Standridge MK, Moustaid-Moussa N. Angiotensin II increases lipogenesis in 3T3-L1 and human adipose cells. Endocrinology 1997; 138: 1512–1519.
46. Gaillard D, Negrel R, Lagarde M, Ailhaud G. Requirement and role of arachidonic acid in the differentiation of preadipose cells. Biochem J 1989; 257:389–397.
47. Negrrel R, Ailhaud G. Metabolism of arachidonic acid and prostaglandin sysnthesis in the preadipocyte clonal line Ob 1771. Biochem Biophys Res Commun 1981; 68:768–777.
48. Axelrod L, Minnich AK, Ryan CA. Stimulation of prostacyclin production in isolated rat adipocytes by Ang II, vasopressin, and bradykinin: evidence for 2 separate mechanisms of prostaglandin synthesis. Endocrinology 1985; 116:2548–2553.
49. Darimont C, Vassaux G, Gaillard D, Ailhaud G, Negrel R. In situ microdialysis of prostaglandins in adipose tissue: stimulation of prostacyclin release by angiotensin II. Int J Obes Relat Metab Disord 1994; 18:783–788.
50. Saint-Marc P, Kozak LP, Ailhaud G, Darimont C, Negrel R. Angiotensin II as a trophic factor of white adipose tissue: stimulation of adipose cell formation. Endocrinology 2001; 142:487–492.
51. Schling P, Loffler G. Effects of angiotensin II on adipose conversion and expression of genes of the renin-angiotensin system in human preadipocytes. Horm Metab Res 2001; 33:189–95.
52. Janke J, Engeli S, Gorzelniak K, Luft FC, Sharma AM. Mature adicpoytes inhibit in vitro differentiation of human preadipocytes via angiotensin type 1 receptors. Diabetes 2002; 51:1699–1707.
53. Sharma AM, Janke J, Gorzelniak K, Engeli S, Luft FC. Angiotensin blockade prevents type 2 diabetes by formation of fat cells. Hypertension 2002; 40: 609–611.
54. Kim S, Whelan J, Claycombe K, Reath DB, Moustaid-Moussa N. Angiotensin II increases leptin secretion by 3T3-L1 and human adipocytes via a prostaglandin-independent mechanism. J Nutr 2002; 132:1135–1140.
55. Kras KM, Hausman DB, Hausman GJ, Martin RJ. Adipocyte development is dependent upon stem cell recruitment and proliferation of preadipocytes. Obes Res 1999; 7:491–497.
56. Kim S, Dugail I, Standridge M, Claycombe K, Chun J, Moustaid-Moussa N. Angiotensin II-responsive element is the insulin-responsive element in the adipocyte fatty acid synthase gene: role of adipocyte determination and differentiation factor 1/sterol-regulatory-element-binding protein 1c. Biochem J 2001; 357:899–904.
57. Brink M, Wellen J, Delafontaine P. Angiotensin II causes weight loss and decreases circulating insulin-like growth factor I in rats through a pressor-independent mechanism. J Clin Invest. 1996; 97:2509–2554.

58. Cassis LA, Marshall DE, Fettinger MJ, Rosenbluth B. RA Mechanisms contributing to angiotensin II regulation of body weight. Am J Physiol 1998; 274:E867–E876.

59. Townsend RR. The effects of angiotensin II on lipolysis in humans. Metabolism 2001; 50:468–472.

60. Boschmann M, Ringel J, Klaus S, Sharma AM. Metabolic and hemodynamic response of adipose tissue to angiotensin II. Obes Res 2001; 9:486–491.

61. Boschmann M, Jordan J, Adams F, Christensen NJ, Tank J, Franke G, Stoffels M, Sharma AM, Luft FC, Klaus S. Tissue-specific response to interstitial angiotensin II in humans. Hypertension 2003; 41:37–41.

62. Le Lay S, Krief S, Farnier C, Lefrere I, Le Liepvre X, Bazin R, Ferre P, Dugail I. Cholesterol, a cell size-dependent signal that regulates glucose metabolism and gene expression in adipocytes. J Biol Chem 2001; 276:16904–16910.

63. Van Harmelen V, Ariapart P, Hoffstedt J, Lundkvist I, Bringman S, Arner P. Increased adipose angiotensinogen gene expression in human obesity. Obes Res 2000; 8:337–341.

64. Giacchetti G, Faloia E, Sardu C, Camilloni MA, Mariniello B, Gatti C, Garrapa GG, Guerrieri M, Mantero F. Gene expression of angiotensinogen in adipose tissue of obese patients. Int J Obes Relat Metab Disord 2000; 24(Suppl 2):S142–143.

65. Dusserre E, Moulin P, Vidal H. Differences in mRNA expression of the proteins secreted by the adipocytes in human subcutaneous and visceral adipose tissues. Biochim Biophys Acta 2000; 1500:88–96.

66. Lalouel JM, Rohrwasser A, Terreros D, Morgan T, Ward K. Angiotensinogen in essential hypertension: from genetics to nephrology. J Am Soc Nephrol 2001; 12:606–615.

67. Jeunemaitre X, Gimenez-Roqueplo AP, Celerier J, Corvol P. Angiotensinogen variants and human hypertension. Curr Hypertens Rep 1999; 1:31–41.

68. Tiago AD, Samani NJ, Candy GP, Brooksbank R, Libhaber EN, Sareli P, Woodiwiss AJ, Norton GR. Angiotensinogen gene promoter region variant modifies body size-ambulatory blood pressure relations in hypertension. Circulation 2002; 106:1483–1487.

69. Jain S, Tang X, Narayanan CS, Agarwal Y, Peterson SM, Brown CD, Ott J, Kumar A. Angiotensinogen gene polymorphism at –217 affects basal promoter activity and is associated with hypertension in African Americans. J Biol Chem 2002; 277:36889–36896.

70. Strazzullo P, Iacone R, Iacoviello L, Russo O, Barba G, Russo P, D'Orazio A, Barbato A, Cappucio FP, Farinaro E, Siani A. Genetic variation in the renin-angiotensin system and regional adiposity in men: the Olivetti Prospective Heart Study. Ann Intern Med 2003; 138:17–23.

71. Kim HS, Krege JH, Kluckman KD, Hagaman JR, Hodgin JB, Best CF, Jennette JC, Coffman TM, Maeda N, Smithies O. Genetic control of blood pressure and the angiotensinogen locus. Proc Natl Acad Sci U S A 1995; 92:2735–2739.

72. Yang G, Merrill DC, Thompson MW, Robillard JE, Sigmund CD. Functional expression of the human angiotensinogen gene in transgenic mice. J Biol Chem 1994; 269:32497–32502.

73. Massiera F, Seydoux J, Geloen A, Quignard-Boulange A, Turban S, Sain-Marc P, Fukamizu A, Negrel R, Ailhaud G, Teboul M. Angiotensinogen-deficient mice exhibit impairment of diet-induced weight gain with alteration in adipose tissue development and increased locomotor activity. Endocrinology. 2001; 142:5220–5225.

74. Massiera F, Bloch-Faure M, Ceiler D, Murakami K, Fukamizu A, Gasc J, Quignard-Boulange A, Negrel R, Ailhaud G, Seydoux J, Meneton P, Teboul M. Adipose angiotensinogen is involved in adipose tissue growth and blood pressure regulation. FASEB J 2001; 15:2727–2729.

75. Kim S, Urs S, Massiera F, Wortmann P, Joshi R, Heo YR, Andersen B, Kobayashi H, Teboul M, Ailhaud G, Quignard-Boulange A, Fukamizu A, Jones BH, Kim JH, Moustaid-Moussa N. Effects of high-fat diet, angiotensinogen (agt) gene inactivation, and targeted expression to adipose tissue on lipid metabolism and renal gene expression. Horm Metab Res 2002; 34:721–725.

76. Oliverio MI, Madsen K, Best CF, Ito M, Maeda N, Smithies O, Coffman TM. Renal growth and development in mice lacking AT1A receptors for angiotensin II. Am J Physiol 1998; 274:F43–F50.

5

Regulation of Fat Synthesis and Adipogenesis

Kee-Hong Kim, Michael J. Griffin, Josep A. Villena, and Hei Sook Sul
Department of Nutritional Sciences and Toxicology,
University of California, Berkeley, California, U.S.A.

1. INTRODUCTION

Adipocytes are highly specialized cells that play a crucial role in the energy balance of most vertebrates. Adipocytes convert excess energy to triacylglycerol and deposit it during feeding in preparation for periods of food deprivation when energy intake is low. Adipocytes may become enlarged by increased fat storage. Moreover, precursor cells present in the stromal vascular fraction of adipose tissue can differentiate into adipocytes even in mature animals. These two processes, fat synthesis and adipogenesis, are under tight hormonal and nutritional control. In this review, we have summarized our work on the regulation of fat synthesis. We have focused specifically on the transcriptional activation of the fatty acid synthase (FAS) gene and on the inhibitory role of two secretory factors, preadipocyte-specific preadipocyte factor-1 (Pref-1) and adipose tissue-specific adipocyte differentiation-specific factor (ADSF), in adipose differentiation.

2. TRANSCRIPTIONAL REGULATION OF GENES ENCODING LIPOGENIC ENZYMES

2.1. Nutritional and Hormonal Regulation of Lipogenic Enzymes

Fatty acid and triacylglycerol synthesis is regulated in response to the nutritional/hormonal state in animals. Subjecting rodents to fasting causes a decrease in lipogenesis; when fasted animals are subsequently refed a diet high in carbohydrate and low in fat, there is a prompt and drastic rise in the production of fatty acids and triacylglycerol to levels well above those observed in normally fed rats. Under lipogenic conditions, excess glucose is converted to acetyl-CoA, which is used for the synthesis of long-chain fatty acids. By the action of its seven active sites, fatty and synthase (FAS) catalyzes all of the reaction steps in the conversion of acetyl-CoA and malonyl-CoA to palmitate. The fatty acids produced are then used for esterification of glycerol-3-phosphate to generate triacylglycerol. Mitochondrial glycerol-3-phosphate acyltransferase (GPAT) catalyzes the first committed step in glycerophospholipid biosynthesis by catalyzing acylation of glycerol-3-phosphate using fatty acyl-CoA to generate 1-acylglycerol-3-phosphate. The concentrations of many of the key enzymes in this pathway, including FAS and mitochondrial GPAT, are decreased during fasting and subsequently "superinduced" during the refeeding period. Induction of these enzymes is highly coordinated and these inducible genes may be regulated via common mechanisms (1).

It is generally accepted that insulin in the circulation, along with glucose, is elevated during feeding of a high carbohydrate diet and induces enzymes involved in fatty acid and triacylglycerol synthesis. Glucagon, on the other hand, is elevated during starvation and suppresses activities of enzymes in fatty acid and fat synthesis by increasing intracellular cyclic adenosine monophosphate (cAMP). In our early studies, we showed that transcription of the FAS and mitochondrial GPAT genes increased when previously fasted mice were refed a high carbohydrate diet (2,3). There was no detectable transcription of FAS or mitochondrial GPAT genes in fasted or fasted-refed streptozotocin-diabetic mice, indicating that insulin is required for transcriptional induction by fasting/refeeding. Administration of cAMP at the start of feeding in normal mice prevented an increase in the transcription of these genes by feeding. Furthermore, there was a rapid and marked increase in the transcription rates of the FAS and GPAT genes when insulin was given to diabetic mice (2,3). Overall, these genes are regulated at the transcriptional level by nutritional and hormonal stimuli. The molecular mechanisms underlying transcriptional regulation of these genes need to be elucidated.

2.2. Regulation of Fatty Acid Synthase Gene Transcription

To study the molecular mechanisms by which lipogenic enzymes such as FAS and GPAT are regulated, we employed 3T3-L1 adipocytes in culture. These cells provide a good model system for studying lipogenic gene transcription since these genes are highly induced during the differentiation process and are sensitive to hormones. In these cells the regulation of FAS and GPAT mimics regulation in vivo (2). We identified an E-box motif (5'-CATGTG-3') at position–65 that is a binding site for upstream stimulating factor (USF) (4), a ubiquitous member of the bHLH leucine zipper family of transcription factors implicated in glucose control of L-type pyruvate kinase gene transcription (5). Both USF-1 and USF-2 occupy the–65 complex (4); dominant negative mutants of USF-1 and USF-2 inhibited insulin stimulation of the FAS promoter (6), demonstrating that these proteins are required for insulin stimulation of FAS gene transcription. We also found that insulin regulation of the FAS promoter occurs via the PI3-kinase/Akt pathway (7).

Another transcription factor found to play a key role in FAS gene transcription is sterol regulatory element binding protein-1 (SREBP-1). This protein recognizes a sterol regulatory element (SRE) (5'-TCACNCCAC-3') sequence (8–12), but can also bind to E-boxes due to the presence of an atypical tyrosine residue in the DNA-binding domain (13). A major role for SREBP in transcriptional regulation of FAS was first suggested when Goldstein and Brown demonstrated that overexpression of the truncated active form of SREBP-1 in liver causes a large accumulation of triacylglycerol and the induction of a battery of lipogenic genes including FAS and mitochondrial GPAT (14). Others have shown that induction of FAS and other lipogenic enzymes by fasting/refeeding is severely impaired in SREBP-1 knockout mice (15). It has also been shown that SREBP-1c, one of two isoforms of SREBP-1, is highly induced by refeeding a carbohydrate-enriched diet and that SREBP can transactivate the FAS promoter by binding to the –65 E-box (16). However, as described earlier, our in vitro data strongly indicated to us that the critical factor functioning through the–65 E-box was USF, not SREBP (Fig. 1A). In addition, we identified a canonical SRE, at–150 (5'-ATCACCCCAC-3') in the FAS promoter suggesting a potential role of SREBP in regulation of the FAS gene (17). To determine whether SREBP functions through this site, we cotransfected a truncated active form of SREBP-1a into 3T3-L1 cells along with various FAS-LUC reporter plasmids (17) FAS-luciferase (FAS-LUC). Induction of the FAS reporter gene by SREBP in vitro was reduced when sequences between –136 and –19 were deleted (not shown), suggesting the presence of a binding site for SREBP in this region, probably the –65 E-Box. However, when binding of SREBP to the –65 E-box was prevented by mutation, deletion of sequences between –184 and –136

(A)

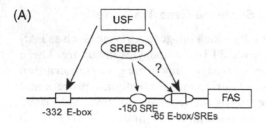

(B)

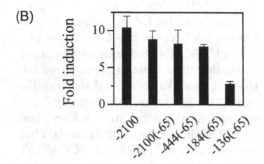

(C)

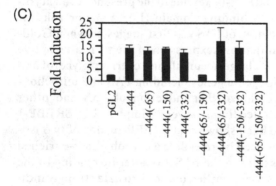

FIGURE 1 Constructs of FAS promoter and localization of the FAS promoter region mediating FAS transactivation by SREBP-1a. (A) Schematic of putative USF and SREBP binding site in the FAS promoter. The diagram represents the proximal 444 bp of the FAS promoter. (B) Localization of the FAS promoter region mediating FAS transactivation by SREBP-1a. Five micrgrams of −2100, −2100 (−65), −444 (−65), −184, and −136 (−65) FAS-LUC plasmids were cotransfected with 25 ng of an expression vector for SREBP-1a into 3T3-L1 fibroblasts. The values represent the mean ± standard deviation. (C) Role of the −150 SRE in activation of the FAS promoter by SREBP. Five micrograms of each of the indicated constructs containing −444 bp of the 5′-flanking sequence of the FAS gene bearing mutations at the indicated positions were cotransfected with 25 ng of an expression vector for SREBP-1a into 3T3-L1 fibroblasts. The values represent the mean ± standard deviation.

abolished transactivation by SREBP-1 (Fig. 1B), indicating the presence of an SREBP-responsive element in this region. In vitro, SREBP may activate and increase FAS promoter activity if any single putative SREBP binding site is present, regardless of its true physiological relevance. In support of this, only when both the −150 SRE and −65 E-box were mutated was transactivation of the FAS promoter prevented by SREBP in vitro (Fig. 1C).

To examine regulation of the FAS promoter in a physiological context in vivo, we generated transgenic mice carrying the 2.1-kb 5′-flanking promoter region of the rat FAS gene fused to the chloramphenicol acetyltransferase (CAT) reporter gene (18). The transgene was expressed strongly only in lipogenic tissues, liver, and white adipose tissue, and was drastically induced by feeding and insulin. Overall, the studies from these transgenic mice demonstrated that the first 2.1-kb 5′-flanking sequence of the FAS gene is sufficient for tissue-specific and hormonal/nutritional regulation. To further define the FAS promoter sequences required for transcriptional activation by nutrients and hormones in vivo, we generated several additional lines of transgenic mice (19), each carrying different 5′-deletion constructs: −644, −444, −278, and −131 FAS-CAT (Fig. 2A). As shown in Fig. 2B, both the −644 and −444 constructs behaved in a manner similar to that in the −2.1 kb transgenic mice, indicating that the region between −444 and −2.1 kb does not contain any sequences necessary for activation of FAS by fasting/refeeding. However, when the same experiment was conducted on −278 FAS-CAT transgenic mice, the induction of CAT, although detectable, was severely decreased. This indicates that the region between −444 and −278 contains one or more elements required for transcriptional activation of FAS. Furthermore, no CAT expression was detectable in fasted/refed −131 FAS-CAT transgenic mice, indicating that the region between −278 and −131 contains additional element(s) required for basal transcription of FAS. Similar results were obtained upon insulin administration to streptozotocin diabetic transgenic mice. We concluded that two major regions of the FAS promoter are required for transcriptional activation by refeeding/insulin: one between −278 and −131, which we showed to be required for low-level activation of the promoter, and a second region between −278 and −444, required for maximal "superinduction" of the gene. We also showed through gel shift assays that a second E-box present at position −332 is a binding site for USF-1 in vitro, and that this site may play an important role for the high-level induction of FAS that is observed with feeding/insulin (19).

To determine the regions in the FAS promoter through which SREBP-1 functions in vivo, we employed mice transgenic for the truncated, nuclear-active form of human SREBP-1a (amino acids 1–460) under the control of the PEPCK promoter (14,17). In these mice, the transgenic SREBP-1a is induced by fasting and repressed by refeeding, whereas endogenous

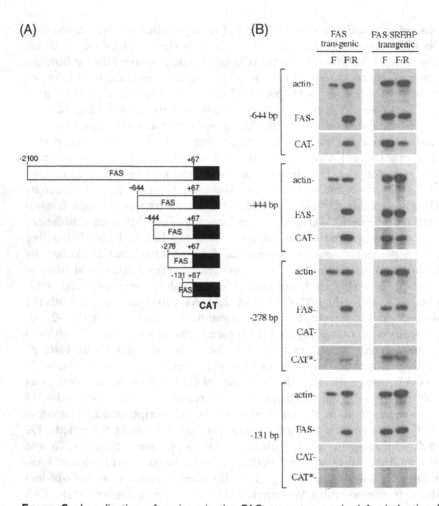

FIGURE 2 Localization of regions in the FAS promoter required for induction by fasting/refeeding and SREBP-1a in vivo. (A) Constructs used in generating the FAS promoter-CAT transgenic mice. The upper panel shows the restriction fragments inserted into pBLCAT3 to generate the various FAS promoter-CAT fusion genes indicated in the lower panel. (B) Regulation by fasting/refeeding of CAT mRNA expression driven by the various FAS promoter regions in the indicated FAS-CAT single transgenic and FAS-CAT/PEPCK-SERBP-1a double transgenic mice. Transgenic mice were either fasted for 24 h or fasted for 24 h and then refed a high-carbohydrate, fat-free diet for 16 h. Total RNA isolated from liver or adipose tissue was subjected to RNase protection assay to determine CAT, FAS, and actin mRNA levels. *Represents an autoradiogram on BioMax film (Kodak) exposed for 3 days, and F and F-R indicate fasted and fasted refed states, respectively. Essentially the same results were obtained from three independent experiments.

SREBP-1c is absent in the fasted state but induced by refeeding (17). The level of hepatic FAS mRNA was found to be high whether these animals were fasted or refed (Fig. 2B). We mated our five 5′-deletion FAS-CAT transgenic mice to SREBP transgenic mice and subjected them to fasting/refeeding. As with the single FAS-CAT transgenic mice described earlier, the −644 and −444 kb double transgenic mice both showed high CAT expression in the refed state (Fig. 2B). Notably, CAT expression was also high in the fasted state in all three constructs, indicating that the region between −2100 and −444 does not contain the putative site(s) for binding and function of SREBP-1. However, in the −278 FAS-CAT transgenic mice, CAT induction was reduced whether the animals were fasted or refed, and was completely absent in the −131 FAS-CAT double transgenic mice, suggesting that the region between −131 and −278 contains at least one element responsible for induction of FAS by SREBP. An increase in SREBP and its binding to the −150 SRE may be the major limiting mechanism for activation of FAS gene transcription by fasting/refeeding in vivo.

2.3. Fatty Acid Synthase Promoter Occupancy and Function of USF and SREBP In Vivo

Our in vitro experiments clearly established the importance of the *cis*-acting elements in the proximal FAS promoter required for insulin regulation in vitro. A critical remaining question was whether the −150 SRE and −65 E-box each were required elements in vivo; in our in vitro experiments, mutation of the −65 E-box in the context of the largest −2.1 kb promoter prevented induction by insulin (6), but had no effect on activation by cotrans-fected SREBP-1 (17). To address these questions, we introduced mutations into the −150 SRE and −65 E-box in the context of the −444 FAS-CAT transgenic construct (20). We chose the −444 promoter fragment for these experiments as it was the shortest 5′-deletion construct that conferred maximal expression of CAT. As shown in Fig. 3A, no expression of CAT was detected in three transgenic lines carrying a mutation at the −150 SRE, indicating that this element indeed is required for induction in vivo. Similar results were obtained when the −444 (−65 mut) mice were fasted and refed: no significant expression of CAT was detected, whereas the control −444 FAS-CAT mice showed a strong induction by refeeding, and endogenous FAS expression was high in all mice (Fig. 3B). Moreover, CAT expression was still not detected when −444 (−150 mut) FAS-CAT/PEPCK-SREBP-1 double transgenic mice were fasted or refed, strongly suggesting that SREBP-1 functions directly through the −150 SRE in vivo. We concluded that both the −150 SRE and −65 E-boxes are required for induction by fasting/refeeding.

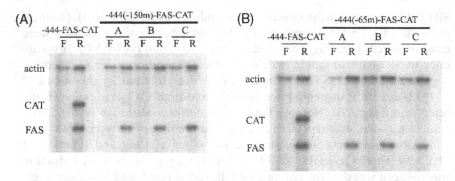

FIGURE 3 Requirement of the −150 SRE and −65 E-Box for nutritional regulation of FAS in vivo. (A) RNase protection assay was performed on mRNA extracted from the livers of fasted and refed −444 (−150 m) FAS-CAT transgenic mice to determine the mRNA levels for endogenous FAS and the reporter CAT genes. ß-Actin mRNA levels are shown as controls. A, B, and C designate the three transgenic lines used in the experiment. (B) RPA was performed with mRNA extracted from the livers of fasted and refed −444 (−65m) FAS-CAT transgenic mice to determine the mRNA levels for endogenous FAS and the reporter CAT genes. ß-Actin mRNA levels are shown as controls. Three transgenic lines were also used in this experiment.

To directly examine the occupancy of the FAS promoter by USF and SREBP in vivo, we utilized the chromatin immunoprecipitation (ChIP) technique (20). Livers from mice that were fasted or fasted/refed were chemically cross-linked and then sonicated to shear the DNA. DNA fragments that were cross-linked to USF or SREBP were then immunoprecipitated from the chromatin using polyclonal anti-USF and anti-SREBP antibodies. Captured DNA fragments were analyzed by PCR using primers complementary to the CAT transgene, as well as to endogenous FAS and low density lipoprotein receptor (LDLR) promoters used as internal controls. We found that USF binding to the endogenous promoter is unchanged by fasting/ refeeding, but a clear and strong induction of SREBP-1 binding in the refed state was detected (Fig. 4A). The patterns of USF and SREBP binding to the endogenous FAS promoter and −444 FAS-CAT transgene were indistinguishable, indicating that the proximal −444 region contains binding sites for both proteins.

A significantly different result was obtained when we performed ChIP on chromatin from fasted/refed −131 FAS-CAT transgenic mice: we consistently observed binding of USF, but not SREBP, signifying that the −65 E-box is not occupied by SREBP in vivo. The −278 construct, however, was bound by both USF and SREBP, supporting our hypothesis that the binding

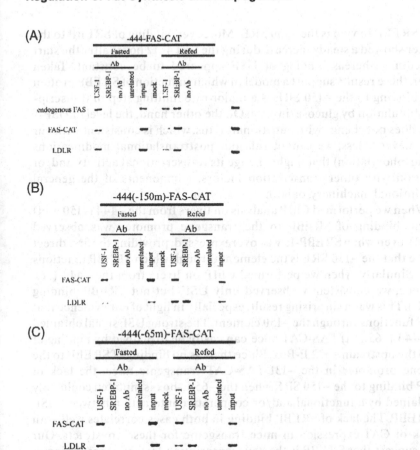

FIGURE 4 Binding of USF and SREBP to the endogenous FAS promoter and FAS-CAT transgenes in vivo. (A) Cross-linked chromatin samples from -444 FAS-CAT mice fasted or refed were immunoprecipitated with anti-USF and anti-SREBP antibodies, and the DNA was analyzed by PCR using the appropriate primers. The sizes of the PCR-generated fragments were 269 bp for endogenous FAS, 230 bp for FAS-CAT, and 242 bp for LDL receptor (LDLR). Controls included a precipitation lacking antibody (no Ab) and chromatin (mock), and immunoprecipitation with an unrelated antibody (unrelated). (B) ChIP was performed on cross-linked chromatin of livers from fasted and refed -444(-150 m) FAS-CAT transgenic mice immunoprecipitated with antibodies against USF and SREBP. DNA was analyzed by PCR with the appropriate primers. LDL receptor (LDLR) promoter is shown as a control. (C) ChIP was performed on cross-linked chromatin of livers from fasted and refed -444(-150 m) FAS-CAT transgenic mice immunoprecipitated with antibodies against USF and SREBP. DNA was analyzed by PCR with the appropriate primers. LDL receptor (LDLR) promoter is shown as a control. The results shown are representative of a minimum of three independent experiments for at least two of the founder lines.

site for SREBP in vivo is the –150 SRE. Moreover, binding of SREBP to the promoter showed a steady increase during the first 4–12 hours after the start of refeeding, whereas binding of USF appeared to be constant. Taken together, these results support a model in which induction of SREBP protein and its binding to the –150 SRE is a major-rate limiting step in transcriptional stimulation by glucose/insulin. On the other hand, the level of USF-1 protein does not change with nutritional status, which is consistent with our results; nevertheless, we cannot rule out posttranslational modifications (i.e., phosphorylation) that might change its transcriptional activity and/or interaction with other transcription factors, components of the general transcriptional machinery, or both.

When we performed ChIP analysis on livers from the –444 (–150 mut) mice, no binding of SREBP to the transgene promoter was observed (Fig. 4B), even when SREBP-1a was overexpressed, providing the first direct evidence that the –150 SRE is the element through which SREBP functions in vivo. Similarly, when we performed ChIP on livers from the –444 (–65 mut) mice, we consistently observed only USF, but not SREBP, binding (Fig. 4C). This was a surprising result, especially in light of our evidence that SREBP functions through the –150 element. The strong USF signal obtained from –444 (–65mut) FAS-CAT mice can easily be explained by binding of USF to the upstream –332 E-Box. Since there is no binding of SREBP to the transgene promoter in the –131 FAS-CAT transgenic mice, the lack of SREBP binding to the –150 SRE when the –65 E-box is mutated could only be explained by a functional and/or cooperative interaction between USF and SREBP. The lack of SREBP binding in both cases correlates well with the lack of CAT expression in mice transgenic for these constructs. Our results suggest that SREBP is the key transcription factor for regulation of FAS by fasting/refeeding and that USF probably functions as an important coregulator of SREBP binding and function. We are currently studying whether USF and SREBP form a common protein complex on the DNA, and whether these two factors directly work together to mediate transcriptional activation of FAS. Overall, our studies have clearly shown the importance of the –65 E-Box and –150 SRE for transcriptional regulation of FAS in vivo and establish a critical role of both USF and SREBP for this regulation.

3. REGULATION OF ADIPOSE DIFFERENTIATION

Gene expression studies during adipocyte differentiation have firmly established that peroxisome proliferator-activated receptor γ (PPARγ) and the CCAAT enhancer-binding protein (C/EBP) family of transcription factors play central roles in adipocyte differentiation (21,22). However, various factors in cell-cell and cell-matrix communications govern expression of these transcription factors and thereby regulate conversion of preadipocytes to adipocytes (23). Although adipose tissue serves as the major energy reservoir in higher eukaryotes, the role of adipose tissue as a secretory organ has emerged through the discovery of leptin (24,25). In addition to leptin, other factors including adiponectin and TNF-α are also secreted from adipose tissue. These factors are involved in regulating a variety of physiological functions including satiety and energy metabolism. For example, leptin levels are known to reflect adipose tissue mass and may act as a satiety signal to control food intake as well as intermediary metabolism. In addition, some of these factors may regulate adipocyte differentiation for feedforward or feedback regulation of adipogenesis. In an attempt to understand molecules that play critical roles in the conversion of preadipocytes to adipocytes, we identified two secretory factors, preadipocyte-specific Pref-1 and adipocyte-specific ADSF.

3.1. Preadipocyte Factor-1, an Epidermal Growth Factor-Repeat-Containing Protein

3.1.1. Alternative Splicing and Generation of Functionally Active Soluble Form of Pref-1

Preadipocyte factor-1 (*pref-1*) encodes a protein of 358 amino acids that contains two hydrophobic stretches: one located within the first 20 N-terminal residues that has the characteristics of a signal sequence and the other, spanning amino acids 300–322, functioning as a single membrane-spanning domain with six tandem EGF repeats in the extracellular domain (26). Sequence analysis shows that Pref-1 shares structural homology with the notch/delta/serrate family of signaling proteins that are involved in cell fate determination. However, due to the lack of a DSL domain which is conserved in all notch ligands, Pref-1 is unlikely to be a notch ligand.

The Pref-1 is found in the preadipocyte cell membrane with molecular mass ranging from 50 to 60 kDa due to alternate splicing and glycosylation (26). In addition to the full-length form, three major short forms of Pref-1 each containing in-frame deletions in the extracellular juxtamembrane region are generated by alternative splicing (Fig. 5A). Moreover, the Pref-1 protein contains two proteolytic cleavage sites at the extracellular domain; one is located near the fourth EGF repeat and the other in the juxtamembrane domain. The full-length transmembrane form of Pref-1 undergoes cleavage at those sites to release soluble forms of 24–25 kDa and 50 kDa, respectively (27). In addition to the longest alternative spliced form of Pref-1, named Pref-1A, three other shorter forms (Pref-1B, Pref-1C, and Pref-1D) containing in-frame deletions in the juxtamembrane domain and the EGF6 repeat are found to be abundant in Pref-1-expressing cells. Of the four major alternatively spliced products of Pref-1, only Pref-1A and Pref-1B can generate both large and small soluble forms. On the contrary, Pref-1C and Pref-1D generate only the small soluble form. This could be due to the membrane proximal cleavage site within a 22-amino acid juxtamembrane sequence.

3.1.2. Inhibition of Adipocyte Differentiation by Pref-1

Upon treatment with the dexamethasone/methyisobutylxanthine (Dex/MIX), 3T3-L1 preadipocytes fully differentiate into adipocytes in a period of 5–7 days. The Pref-1 is highly expressed in 3T3-L1 preadipocytes at confluence and its expression is abolished in fully differentiated adipocytes. The decrease in Pref-1 mRNA is one of the earliest responses of 3T3-L1 preadipocytes known to date upon treatment with the differentiation inducing agents, Dex/MIX (26).

To demonstrate the role of Pref-1, a *pref-1* antisense construct was stably transfected into 3T3-L1 cells. In the presence of MIX but in the absence of Dex, antisense 3T3-L1 cells can undergo a low but spontaneous differentiation into adipocytes, whereas control cells showed no sign of

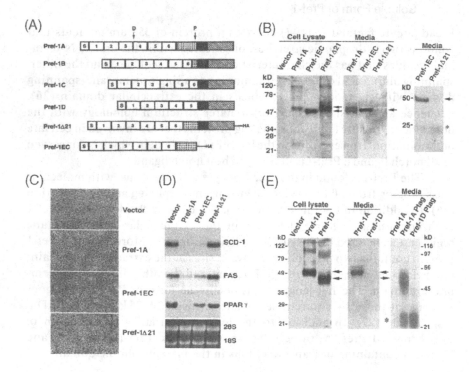

adipocyte conversion. Furthermore, preadipocytes stably transfected with *pref-1* lack the capability to differentiate or to acquire the characteristics of mature adipocytes, such as lipid accumulation and expression of genes involved in lipid metabolism. These studies convincingly demonstrate that *pref-1* expression inhibits adipose conversion of 3T3-L1 cells, and also that *pref-1* downregulation is a necessary step in adipocyte differentiation (26). The molecular mechanism of Pref-1 function in adipocyte differentiation remains unknown. However, the presence of EGF-like repeats in the protein and the demonstrated role of this motif in other molecules suggest that Pref-1 may function via interaction of its EGF-like repeats with its putative receptor, thereby maintaining a preadipocyte morphology.

The Pref-1 presents a complex expression pattern with multiples transmembrane and soluble forms. We found that the membrane form of Pref-1 is

◄ ─────────────────────────────

FIGURE 5 Pref-1 constructs and inhibition of adipocyte differentiation by soluble pref-1. (A) Pref-1A has a secretary signal sequence (S), six EGF repeats (boxes labeled 1–6), a juxtamembrane region (dotted box), a transmembrane domain (black box), and a cytoplasmic domain (hatched box). Pref-1A has proximal (P) and distal (D) proteolytic processing sites, as indicated. The major alternatively spliced forms, Pref-1B–Pref-1D, have an in-frame deletion at the juxtamembrane region and EGF repeat 6. Pref-1 Δ21 is an artificial Pref-1 with a 21-amino acid deletion corresponding to the processing site proximal to the transmembrane region (P-site). Pref-1EC includes only the extracellular domain of Pref-1, representing the large soluble form. HA-HA epitope. 3T3-L1 cells were differentiated in conditioned media from COS cells transfected with pcDNA3.1 control, Pref-1A, Pref-1EC, or Pref-1 Δ21 expression vectors. (B) Western blot analysis of Pref-1 expression in transfected COS cell lysates and conditioned media. Left-hand and middle panels: COS cells were cultured in serum-free DMEM. Right-hand panel, COS cells were cultured in DMEM with 10% FBS. The arrows indicate Pref-1 in cell lysate and the larger soluble fragment in the conditioned media. The asterisk indicates the smaller soluble fragments in conditioned media. (C) Oil Red O staining of 3T3-L1 cells differentiated in the conditioned media with 10% FBS. (D) Northern blot analysis for adipocyte marker expression in differentiated cells. PPARγ-peroxisome proliferator-activated receptor γ. (E) The top-left two panels show immunoblot analysis for Pref-1 expression in cell lysates and serum-free conditioned media from COS cells transfected with Pref-1A, Pref-1D, or empty vector. The top-right panel shows the phosphorylated secreted Pref-1 in conditioned media from COS cells transfected with the P-tagged expression vector for Pref-1A and Pref-1D, as described in the text. The arrows indicate Pref-1 in cell lysates and the larger soluble fragment in the conditioned media. The asterisk indicates the smaller soluble fragments in the conditioned media. Lower left panel: Oil Red O staining of 3T3-L1 cells differentiated in serum-containing conditioned media from COS cells transfected with Pref-1A, Pref-1D, or empty vector. Lower right panel: Northern blot analysis of adipocyte-marker expression after differentiation.

proteolytically processed at two sites in the extracellular domain, resulting in the larger (50 kDa) and smaller (25 kDa) soluble forms. Using conditioned media from COS cells transfected with various forms of Pref-1 (Pref-1A-, Pref-1D, and Pref-1Δ21), we demonstrated that only the large soluble form (50 kDa), corresponding to the full ectodomain, inhibits 3T3-L1 adipocyte differentiation, by preventing lipid accumulation and expression of different adipocyte markers such as FAS, stearyl-CoA desaturase-1 (SCD-1), and PPARγ. On the other hand, the artificial membrane form of Pref-1 without the juxtamembrane cleavage site (Pref-1Δ21) was not effective in inhibiting adipocyte differentiation. Prevention by Pref-1 of adipocyte transcription factor PPARγ expression suggests that Pref-1 acts early during the differentiation process in inhibiting adipogenesis (Fig. 5B-E). Of the four alternative-splicing products, Pref-1A and Pref-1B, which generate both large and small soluble forms, inhibit adipocyte differentiation, whereas Pref-1C, Pref-1D, and Pref-1Δ21, which lack the processing site proximal to the membrane and therefore generate only the smaller soluble form, did not show any effect. Thus, the 24–25 kDa form could be considered as a product originated by proteolysis at the membrane-distal cleavage site of the large soluble form. One can speculate that cleavage of the large soluble form may serve as a mechanism to inactivate or modulate its activity.

3.1.3. Studies of Pref-1 Knockout and Pref-1 Overexpressing Mice

In the embryonic stage of mice, Pref-1 is expressed in multiple tissues, such as liver, lung, tongue, pituitary, and developing vertabrae (26). Also, Pref-1 is found in mouse placenta, and detectable amounts of circulating Pref-1 are found in maternal serum in concentrations that correlates with the litter size (28). The increased Pref-1 concentration in maternal serum may be caused by the abundant synthesis and secretion of the protein by fetuses. However, expression of Pref-1 after birth is rapidly abolished in most tissues and becomes restricted to certain cellular types such as preadipocytes (26), pancreatic islets cells (29), thymic stromal cells (30), and adrenal gland cells (31). This expression pattern suggests that, besides its role in adipocyte differentiation, pref-1 could also be involved in embryonic development. In order to define the in vivo role of Pref-1, we have recently generated a *pref-1* knockout mouse by targeted deletion (32). The genomic *pref-1* allele was disrupted by insertion of a neomycin-resistance cassette in place of exon 2 and 3. At weaning age, both male and female *pref-1* null mice had approximately 14% lower body weight than wild-type mice. However, *pref-1* null mice rapidly started gaining more weight afterward than wild type, and no differences in body weight could be found at around 10 weeks of age (Fig. 6A). Organ analysis showed that several organs, including lungs and kidney, remained significantly smaller in null mice. In contrast, the weight of major

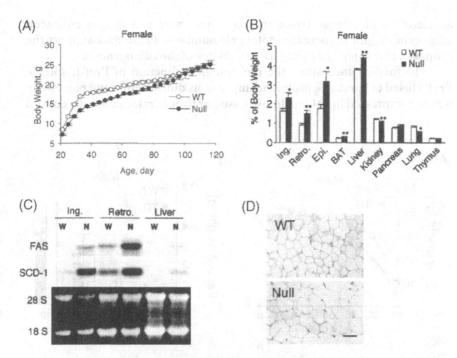

FIGURE 6 Pref-1 knockout mice exhibit accelerated adiposity. (A) Female Pref-1 null (filled circles)(Null) and wild type (open circles)(WT) mice were fed high-fat (45 kcal%) diet ad libitum from 21 days of age and were weighed at 4-day intervals. Statistics were performed with a two-tailed t test. For female mice, from day 21 to day 68, $P < 0.01$, and from day 72 and day 76, $P < 0.05$. Pref-1 null mice also display accelerated adiposity. (B) Percentage of fat pad and organ weight relative to body weight. Fat depots and organs were dissected from 16-week-old WT and Pref-1 Null mice fed a high-fat diet (n = 7 to 10 per group). Ing-inguinal fat pad; Retro-retroperitoneal fat pad; Epi-epididymal fat pad; BAT-brown adipose tissue. All values are means $\pm$ SEM. *-$P < 0.05$; **-$P < 0.01$. (C) Expression of adipocyte marker genes in adipose tissues (Ing-inguinal fat pad. Total RNA was extracted from adipose tissue of 16-week-old wild type (W) and Pref-1 null (N) mice fed a high-fat diet. (D) Paraffin section of inguinal fat pad from 16-week-old female mouse. The scale bar represents 50 μm.

fat depots (inguinal, retroperitoneal, and epididymal) was higher in null mice than in wild type, indicating that accelerated body weight gain in *pref-1* knockout mice was due to an increase in adipose tissue mass (Fig. 6B). The mRNA levels of different markers of adipocyte differentiation, such as FAS and SCD-1, were increased in adipose tissue of null mice (Fig. 6C), and also adipose cell size was increased (Fig. 6D). No changes in total DNA content between fat pads of null and wild-type mice were found, indicating that acceleration of adipose tissue in *pref-1* null mice was due to enhanced adipogenesis, not to increased total cell numbers. These data support the proposed role of *pref-1* as a negative regulator of the adipogenesis.

To further understand the physiological function of Pref-1, soluble Pref-1 fused to human Fc protein to improve its dimerization and bioactivity was overexpressed in either adipose tissue or liver in mice under the control

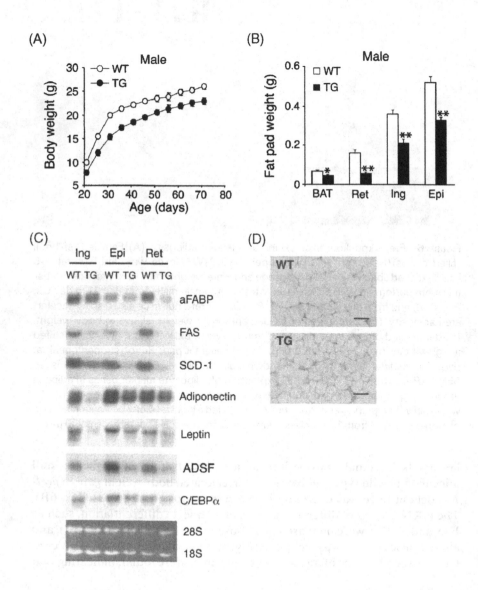

of the adipocyte fatty acid-binding protein (aP2/aFABP) or albumin promoter, respectively (33). The *aP2-pref-1/hFc* transgenic mice showed lower body weight than their wild-type littermates with a substantial decrease in total fad pad weight (Fig. 7A,B). Moreover, adipose tissue from transgenic mice showed reduced expression of adipocyte markers and adipocyte-secreted factors, including aFABP, FAS, SCD-1, C/EBPα, adiponectin, leptin, and ADSF (Fig. 7C) as well as reduced adipose cell size (Fig. 7D). This suggests that the decrease in fat mass in transgenic mice reflects an impairment of adipocyte differentiation of all adipose depots. The study clearly demonstrates that the soluble Pref-1 is sufficient to inhibit adipocyte differentiation in vivo. Furthermore, the *aP2-pref-1/hFc* transgenic mice exhibit hypertriglyceridemia, impaired glucose tolerance, and decreased insulin sensitivity. Mice expressing the *pref-1/hFc* transgene in liver also showed a decrease in adipose mass and adipocyte marker expression, suggesting an endocrine mode of Pref-1.

We also examined the role of *pref-1* in embryonic development. Our *pref-1* gene ablation in mice leads to embryonic and postnatal death. We found that the average litter size in null mice crossings was 36% lower than those of the wild types. The null newborn mice had significantly smaller body size than wild-type newborns, and approximately 50% died within two days after birth. These unexpected results indicate that lack of Pref-1 reduces embryonic survival, and that *pref-1* expression is also important for perinatal growth and survival. Moreover, the surviving *pref-1* null adult mice exhibited developmental abnormalities. These alterations included fusion between ribs, asymmetrical fusion of ribs to the sternum, and blepharophimosis. The

FIGURE 7 Transgenic mice overexpressing soluble Pref-1 in adipose tissue exhibit substantial decrease in body weight and total fat pad weight with reduced expression of adipocyte markers and adipocyte-secreted factors. (A) Growth curve for male wild type (open circles) (WT) and transgenic mice (filled circles) (TG) fed a chow diet. Body weight of mice measured at 5-day intervals is shown; each point represents mean ± SEM for 6 to 13 mice. Body weight of Pref-1/hFc transgenic mice was significantly lower ($P < 0.01$) than that of wild type mice at all ages. (B) Fat depot weights from 10-week-old mice are presented (n = 6 per group). BAT-brown adipose tissue; Ret, retroperitoneal fat pad; Ing, inguinal fat pad; Epi, epididymal fat pad. Statistically significant differences between the groups are indicated as *$P < 0.05$ and **$P < 0.01$. (C) Northern blot analysis of adipocyte marker expression in white adipose tissue of WT and Pref-1/hFcTG mice. Total RNA from three different fat pads was probed with cDNA probes for different adipocyte markers. (D) Paraffin-embedded sections of renal white adipose tissue from 10-week-old male mice were stained with hematoxylin and eosin. Scale bar-50 μm.

exact cause of *pref-1* knockout embryonic and newborn death and malformations are still unknown. However, the wide expression pattern of *pref-1* in embryos and its nearly total extinction in adult mice support the crucial role of *pref-1* during mouse development. Recently, due to differential methylation, *pref-1* has been reported to be an imprinted gene expressed from the paternal allele, but not the maternal allele (34–36). In our *pref-1* knockout mouse model, heterozygotes with maternal or paternal inheritance of the *pref-1* knockout allele showed both an expression pattern of *pref-1* and a phenotypic profile similar to those of wild-type or null mice. This clearly demonstrates expression of *pref-1* only from paternal allele. In mice, the *pref-1* gene is located on chromosome 12 in a region containing a cluster of imprinted genes that, in addition to *pref-1*, include *dat, gtl2, peg11, antipeg11*, and *meg8* (37). Maternal uniparental disomy in chromosome 12 (mUPD12) embryos are smaller than wild-type embryos, and also present diverse skeletal abnormalities (38). Furthermore, in humans, mUPD14 patients and patients with a deletion at region 14q32 encompassing the human *PREF-1/DLK1* locus show growth retardation, scoliosis, blepharophimosis, early puberty, and mild obesity (39,40). In addition, our *pref-1/hFc* transgenic mouse model exhibits a partially overlapping phenotype with mouse pUPD12. Similar to *pref-1* null mouse embryos, *pref-1/hFc* transgenic embryos showed growth retardation and skeletal abnormalities, primarily in the distal vertebra. The exact imprinted gene or genes responsible of this wide array of alterations is still unknown. However, based on the study in the *pref-1* knockout mouse model, and *pref-1* transgenic mice, we propose that characteristics of UPDs are, at least in part, due to the alteration of *pref-1* expression.

Overall, our studies on Pref-1 function have demonstrated that Pref-1 is a secreted factor that plays an important role in the regulation of adipose tissue development. Although the mechanisms and pathways through which it exerts its function still remain unknown, the secreted soluble form of Pref-1 inhibits adipocyte differentiation in an endocrine manner. Moreover, our in vivo studies using a murine model in which Pref-1 expression has been abolished or overexpressed reveal that Pref-1 is critical for normal growth and development of the organism and its perinatal survival. Our in vivo study also links Pref-1 expression with some of the pathologies associated with genetic disorders of UPDs or chromosomal deletions spanning the *pref-1/dlk1* locus.

3.2. ADSF/Resistin: a Cysteine-Rich Adipose Tissue Specific Secretory Factor

We identified ADSF as a cysteine-rich protein expressed and secreted by mature adipocytes by microarray analysis (41). This factor is synthesized as a 114-amino acid-long polypeptide containing a 20-amino acid signal peptide at the N-terminus. Two other groups have also identified ADSF independently. Holcomb et al. identified this factor as FIZZ3 (found in inflammatory zone 3), a member of protein family FIZZ by an expressed sequence tag (EST) database screen against FIZZ1, which is induced during lung inflammation (42). By subtractive cloning, Steppan et al. also identified this factor as a TZD-regulated adipocyte-derived hormone that causes insulin resistance and termed it resistin (43). ADSF/FIZZ/resistin belongs to a gene family that also includes FIZZ1/RELMα, which is highly expressed in the stromal vascular fraction of adipose tissue, heart, lung, tongue, and intestine in mouse sharing 44% amino acid homology with predicted mouse ADSF/resistin/FIZZ3 protein (44). Another member of this family is FIZZ2/RELMβ which is highly expressed in colon and intestine in mouse sharing 56% amino acid homology with predicted mouse ADSF/resistin/FIZZ3 protein (44). Recently, another member of FIZZ/RELM family, RELMγ, has been identified in mice and rats, which is highly expressed in hematopoietic tissues sharing 58% amino acid homology with predicted mouse ADSF/resistin/FIZZ3 (45). Henceforth, we refer to ADSF/resistin/FIZZ3 as ADSF in this review.

The ADSF forms a homodimer via disulfide bonding at Cys_{11}(46). FIZZ2/RELMβ also forms a homodimer, mediated by a cysteine residue that corresponds to Cys_{11} of ADSF. On the other hand, FIZZ1/RELMα which lacks this cysteine residue was reported to exist as a monomer (46). However, FIZZ1/RELMα has also been reported to form homooligomers despite the absence of Cys_{11} (47). Recent studies also demonstrated a possibility that ADSF not only forms a dimer but also forms multimeric complexes with other FIZZ/RELMs (47,48). It is likely that the presence of multiple cysteine residues in ADSF implies that intramolecular and/or intermolecular disulfide bonds might be involved in the maintenance of structural integrity of these secreted proteins. Possible heterooligomeric complex formation among FIZZ/RELM family members presents functional implication also.

The ADSF mRNA is exclusively expressed in adipose tissues including brown adipose tissue (BAT) and inguinal, epididymal, and retroperitoneal fat pad tissues (41). ADSF mRNA is not found in preadipocytes but ADSF mRNA is markedly increased at the later stage of 3T3-L1 preadipocyte differentiation into adipocytes in vitro (Fig. 8A) (41,49). Interestingly, although circulating level is controversial, ADSF mRNA expression was found to be lower in adipose tissue of obesity mouse models including ob/ob mice as compared to wild-type mice (Fig. 8B). ADSF mRNA expression may be correlated with obesity status. The human homologue of murine ADSF, having 60% amino acid identity, has been shown to be expressed in

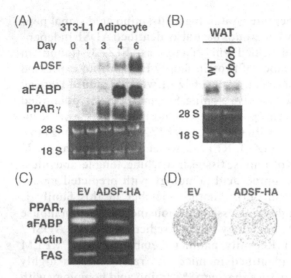

Figure 8 ADSF mRNA expression during adipocyte differentiation and conditioned media from COS cells transfected with HA-tagged ADSF expression vector inhibits 3T3-L1 adipocyte differentiation. (A) Five µg of RNA prepared from 3T3-L1 cells at the indicated time points of day 0 (at confluence) and days 1–6 were subjected to Northern blot analysis for ADSF, aFABP, SCD-1, and PPARγ. (B) Northern blot analysis of white adipose tissue (WAT) ADSF expression in male obese *ob/ob* and their wild type (WT) lean counterparts. 3T3-L1 cells were differentiated in conditioned medium from COS cells transfected with pcDNA3.1 control vector and expression vector containing HA-tagged mouse ADSF cDNA sequence. (C) 3T3-L1 cells differentiated in serum-containing conditioned medium from COS cells transfected with empty vector (EV) or HA-tagged expression vector (ADSF-HA) were analyzed by RT-PCR analysis for adipocyte markers (PPARγ, FAS, aFABP, and actin) after differentiation and (D) stained by oil red O.

abdominal adipose tissue (42,44,50,51). However, the correlation between ADSF gene expression and both insulin resistance and body mass index is controversial in humans (52).

The physiological function of ADSF remains to be determined. However, its potential role as an endocrine factor in both adipogenesis and insulin action has been implicated (43,53). Given that it is expressed only in adipose tissue and is highly induced during adipocyte differentiation, we predicted that ADSF might promote adipocyte conversion of preadipocytes. We originally thought that ADSF might be a signal to generate adipocytes for the increased capacity to store excess energy. Unexpectedly, by using conditioned media from COS cells transfected with mouse ADSF expression vector, we observed that the treatment of differentiating 3T3-L1 cells with ADSF inhibits adipocyte differentiation. Cells treated with ADSF showed a decrease in expression of adipocyte markers, PPARγ, aFABP, and FAS (Fig. 8C), as well as lipid staining (Fig. 8D), by approximately 70% as compared to control cells. These results clearly demonstrate an inhibition of adipocyte differentiation by ADSF. Since ADSF expression is induced by feeding/insulin and ADSF expression is found to be low in obesity models, we speculate that ADSF may serve as a feedback signal to restrict adipose tissue formation.

The inhibitory function of ADSF in adipocyte differentiation is likely to be mediated by its increased gene expression resulting from nutritional and hormonal changes (41). For example, ADSF mRNA level was barely detectable during fasting and dramatically increased when fasted mice were refed a high carbohydrate diet. The ADSF mRNA level was also very low in adipose tissue of streptozotocin-induced diabetic mice and highly increased upon insulin administration (Fig. 9). Although ADSF was originally identified as a TZD-downregulated gene, TZD regulation of ADSF is controversial. Another study also showed that ADSF mRNA levels are increased by TZD in adipose tissue in mice (54–56). Regulation of ADSF by TZDs in humans has not yet been examined.

Steppan et al. demonstrated that ADSF might play an important role in the regulation of insulin action (43). Administration of recombinant ADSF

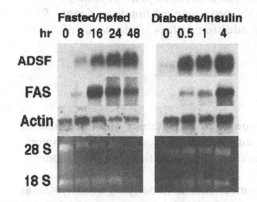

FIGURE 9 Nutritional and hormonal regulation of ADSF mRNA expression in mice. Total RNA prepared from adipose tissue of mice fasted for 48 hr or refed a high-carbohydrate diet was used for Northern blot analysis. Total RNA also isolated from the white adipose tissue of streptozotocin-diabetic mice and of streptozotocin-diabetic mice treated with insulin was used for Northern blot analysis for ADSF, FAS, and actin mRNAs. 28S and 18S ribosomal RNA from ethidium bromide-stained gels are shown.

impaired insulin action in mice and in 3T3-L1 adipocytes; immunoneutrali-zation of ADSF in serum in obese mice with the ADSF antibody revealed improved insulin sensitivity. Recently, by use of pancreatic insulin clamp techniques, Rajala et al. reported that acute administration of recombinant ADSF to rats caused impairment of insulin sensitivity. They pointed out that this effect was completely accounted for by a marked increase in the rate of glucose production by the liver (53). In addition, they also showed that RELMβ exhibited a similar effect on insulin sensitivity. Interestingly, a recent study demonstrated that RELMα is expressed in stromal vascular fraction of adipose tissue, and in several other tissues it can also inhibit 3T3-L1 adipocyte differentiation. These observations suggest that other members of *Relm/fizz* gene family might share functional properties with that of ADSF. In this regard, as mentioned earlier, various members of this family may form heteroligomers to elicit similar biological effects in vivo.

The newly identified ADSF brought wide attention from science com-munities interested in obesity and obesity-associated insulin resistance. However, the physiological function of ADSF in adipogenesis and insulin action is not yet clearly established. Evidence for the long-term physiological function of circulating ADSF in adipose tissue development as well as in glu-cose and lipid homeostasis needs to be demonstrated. Further studies on identifying ADSF target tissues and its putative receptor will make it possi-ble to investigate ADSF function at a biochemical level. Furthermore, study-ing mouse models of gain-of-function and loss-of-function will cast light on the role of ADSF in adipogenesis and its associated pathophysiology and in insulin resistance. Understanding the physiological function of other *Relm* family members will also provide additional clues to the mode of ADSF function.

4. CONCLUSIONS

We have summarized in this review our studies on transcriptional regulation of fatty acid synthase, a central enzyme in lipogenesis, and characteri-zation of Pref-1 and ADSF, secretory proteins that inhibit adipocyte differentiation.

Elucidation of the molecular mechanisms underlying transcriptional activation of the fatty acid synthase gene will help us to understand the pro-cess that leads to the increase in fat synthesis and deposition that contributes to the hypertrophy of adipose tissue. We show that both the −150 SRE and the −65 E-box are required for regulation of the FAS promoter by nutritional and hormonal regulation, although none of these elements per se is sufficient for such regulation in vivo. It is possible that common mechanisms govern regulation of fatty acid synthase, mitochondrial glycerol-3-phosphate acyltransferase, and other lipogenic enzymes.

In regard to the regulation of adipocyte differentiation, we have shown that Pref-1, via generating a biologically active soluble form, and ADSF inhibit adipocyte differentiation. The specific receptors or interacting protein(s) for Pref-1 and ADSF should be identified. Also important are the signaling pathways subsequent to pref-1 and ADSF receptor interaction that leads to the inhibition of adipocyte differentiation. Especially, the developmental role of Pref-1 needs to be examined in an in vivo context. Elucidating Pref-1 and ADSF actions will help us understand the adipocyte differentiation process that contributes to the development of obesity and its associated metabolic disorders.

REFERENCES

1. Wang D, Sul HS. Insulin stimulation of the fatty acid synthase promoter is mediated by the phosphatidylinositol 3-kinase pathway Involvement of protein kinase B/Akt. J Biol Chem 1998; 273:25420–25426.
2. Paulauskis JD, Sul HS. Hormonal regulation of mouse fatty acid synthase gene transcription in liver. J Biol Chem 1989; 264:574–577.
3. Shin DH, Paulauskis JD, Moustaid N, Sul HS. Transcriptional regulation of p90 with sequence homology to Escherichia coli glycerol-3-phosphate acyltransferase. J Biol Chem 1991; 266:23834–23839.
4. Wang D, Sul HS. Upstream stimulatory factors bind to insulin response sequence of the fatty acid synthase promoter. USF1 is regulated. J Biol Chem 1995; 270:28716–28722.
5. Lefrancois-Martinez AM, Martinez A, Antoine B, Raymondjean M, Kahn A. Upstream stimulatory factor proteins are major components of the glucose response complex of the L-type pyruvate kinase gene promoter. J Biol Chem 1995; 270:2640–2643.
6. Wang D, Sul HS. Upstream stimulatory factor binding to the E-box at -65 is required for insulin regulation of the fatty acid synthase promoter. J Biol Chem 1997; 272:26367–26374.
7. Sul HS, Smas CM, Wang D, Chen L. Regulation of fat synthesis and adipose differentiation. Prog Nucleic Acid Res Mol Biol 1998; 60:317–345.
8. Ericsson J, Jackson SM, Lee BC, Edwards PA. Sterol regulatory element binding protein binds to a cis element in the promoter of the farnesyl diphosphate synthase gene. Proc Natl Acad Sci U S A 1996; 93:945–950.
9. Guan G, Dai PH, Osborne TF, Kim JB, Shechter I. Multiple sequence elements are involved in the transcriptional regulation of the human squalene synthase gene. J Biol Chem 1997; 272:10295–10302.
10. Guan G, Dai P, Shechter I. Differential transcriptional regulation of the human squalene synthase gene by sterol regulatory element-binding proteins (SREBP)

1a and 2 and involvement of 5′ DNA sequence elements in the regulation. J Biol Chem 1998; 273:12526–12535.

11. Lopez JM, Bennett MK, Sanchez HB, Rosenfeld JM, Osborne TE. Sterol regulation of acetyl coenzyme A carboxylase: a mechanism for coordinate control of cellular lipid. Proc Natl Acad Sci U S A 1996; 93:1049–1053.

12. Yokoyama C, Wang X, Briggs MR, Admon A, Wu J, Hua X, Goldstein JL, Brown MS. SREBP-1, a basic-helix-loop-helix-leucine zipper protein that controls transcription of the low density lipoprotein receptor gene. Cell 1993; 75:187–197.

13. Kim JB, Spotts GD, Halvorsen YD, Shih HM, Ellenberger T, Towle HC, Spiegelman BM. Dual DNA binding specificity of ADD1/SREBP1 controlled by a single amino acid in the basic helix-loop-helix domain. Mol Cell Biol 1995; 15:2582–2588.

14. Shimano H, Horton JD, Hammer RE, Shimomura I, Brown MS, Goldstein JL. Overproduction of cholesterol and fatty acids causes massive liver enlargement in transgenic mice expressing truncated SREBP-1a. J Clin Invest 1996; 98:1575–1584.

15. Shimano H, Yahagi N, Amemiya-Kudo M, Hasty AH, Osuga J, Tamura Y, Shionoiri F, Iizuka Y, Ohashi K, Harada K, Gotoda T, Ishibashi S, Yamada N. Sterol regulatory element-binding protein-1 as a key transcription factor for nutritional induction of lipogenic enzyme genes. J Biol Chem 1999; 274:35832–35839.

16. Kim JB, Sarraf P, Wright M, Yao KM, Mueller E, Solanes G, Lowell BB, Spiegelman BM. Nutritional and insulin regulation of fatty acid synthetase and leptin gene expression through ADD1/SREBP1. J Clin Invest 1998; 101:1–9.

17. Latasa MJ, Moon YS, Kim KH, Sul HS. Nutritional regulation of the fatty acid synthase promoter in vivo: sterol regulatory element binding protein functions through an upstream region containing a sterol regulatory element. Proc Natl Acad Sci U S A 2000; 97:10619–10624.

18. Soncini M, Yet SF, Moon Y, Chun JY, Sul HS. Hormonal and nutritional control of the fatty acid synthase promoter in transgenic mice. J Biol Chem 1995; 270:30339–30343.

19. Moon YS, Latasa MJ, Kim KH, Wang D, Sul HS. Two 5′-regions are required for nutritional and insulin regulation of the fatty-acid synthase promoter in transgenic mice. J Biol Chem 2000; 275:10121–10127.

20. Latasa MJ, Griffin MJ, Moon YS, Kang C, Sul HS. Occupancy and function of the -150 sterol regulatory element and -65 E-box in nutritional regulation of the fatty acid synthase gene in living animals. Mol Cell Biol 2003; 23:5896–5907.

21. Gregoire FM, Smas CM, Sul HS. Understanding adipocyte differentiation. Physiol Rev 1998; 78:783–809.

22. MacDougald OA, Mandrup S. Adipogenesis: forces that tip the scales. Trends Endocrinol Metab 2002; 13:5–11.

23. Zhao L, Gregoire F, Sul HS. Transient induction of ENC-1, a Kelch-related actin-binding protein, is required for adipocyte differentiation. J Biol Chem 2000; 275:16845–16850.

24. Halaas JL, Gajiwala KS, Maffei M, Cohen SL, Chait BT, Rabinowitz D, Lallone RL, Burley SK, Friedman JM. Weight-reducing effects of the plasma protein encoded by the obese gene. Science 1995; 269:543–546.

25. Zhang Y, Proenca R, Maffei M, Barone M, Leopold L, Friedman JM. Positional cloning of the mouse obese gene and its human homologue. Nature 1994; 372:425–432.

26. Smas CM, Sul HS. Pref-1, a protein containing EGF-like repeats, inhibits adipocyte differentiation. Cell 1993; 73:725–734.

27. Smas CM, Chen L, Sul HS. Cleavage of membrane-associated pref-1 generates a soluble inhibitor of adipocyte differentiation. Mol Cell Biol 1997; 17:977–988.

28. Bachmann E, Krogh TN, Hojrup P, Skjodt K, Teisner B. Mouse fetal antigen 1 (mFA1), the circulating gene product of mdlk, pref-1 and SCP-1: isolation, characterization and biology. J Reprod Fertil 1996; 107:279–285.

29. Carlsson C, Tornehave D, Lindberg K, Galante P, Billestrup N, Michelsen B, Larsson LI, Nielsen JH. Growth hormone and prolactin stimulate the expression of rat preadipocyte factor-1/delta-like protein in pancreatic islets: molecular cloning and expression pattern during development and growth of the endocrine pancreas. Endocrinology 1997; 138:3940–3948.

30. Kaneta M, Osawa M, Sudo K, Nakauchi H, Farr AG, Takahama Y. A role for pref-1 and HES-1 in thymocyte development. J Immunol 2000; 164:256–264.

31. Halder SK, Takemori H, Hatano O, Nonaka Y, Wada A, Okamoto M. Cloning of a membrane-spanning protein with epidermal growth factor-like repeat motifs from adrenal glomerulosa cells. Endocrinology 1998; 139:3316–3328.

32. Moon YS, Smas CM, Lee K, Villena JA, Kim KH, Yun EJ, Sul HS. Mice lacking paternally expressed Pref-1/Dlk1 display growth retardation and accelerated adiposity. Mol Cell Biol 2002; 22:5585–5592.

33. Lee K, Villena JA, Moon YS, Kim KH, Lee S, Kang C, Sul HS. Inhibition of adipogenesis and development of glucose intolerance by soluble preadipocyte factor-1 (Pref-1). J Clin Invest 2003; 111:453–461.

34. Schmidt JV, Matteson PG, Jones BK, Guan XJ, Tilghman SM. The Dlk1 and Gtl2 genes are linked and reciprocally imprinted. Genes Dev 2000; 14:1997–2002.

35. Takada S, Tevendale M, Baker J, Georgiades P, Campbell E, Freeman T, Johnson MH, Paulsen M, Ferguson-Smith AC. Delta-like and gtl2 are reciprocally expressed, differentially methylated linked imprinted genes on mouse chromosome 12. Curr Biol 2000; 10:1135–1138.

36. Wylie AA, Murphy SK, Orton TC, Jirtle RL. Novel imprinted DLK1/GTL2 domain on human chromosome 14 contains motifs that mimic those implicated in IGF2/H19 regulation. Genome Res 2000; 10:1711–1718.

37. Charlier C, Segers K, Wagenaar D, Karim L, Berghmans S, Jaillon O, Shay T, Weissenbach J, Cockett N, Gyapay G, Georges M. Human-ovine comparative sequencing of a 250-kb imprinted domain encompassing the callipyge (clpg) locus and identification of six imprinted transcripts: DLK1, DAT, GTL2, PEG11, antiPEG11, and MEG8. Genome Res 2001; 11:850–862.

38. Georgiades P, Watkins M, Surani MA, Ferguson-Smith AC. Parental origin-specific developmental defects in mice with uniparental disomy for chromosome 12. Development 2000; 127:719–4728.

39. Hordijk R, Wierenga H, Scheffer H, Leegte B, Hofstra RM, Stolte-Dijkstra I. Maternal uniparental disomy for chromosome 14 in a boy with a normal karyotype. J Med Genet 1999; 36:782–785.

40. Manzoni MF, Pramparo T, Stroppolo A, Chiaino F, Bosi E, Zuffardi O, Carrozzo R. A patient with maternal chromosome 14 UPD presenting with a mild phenotype and MODY. Clin Genet 2000; 57:406–408.

41. Kim KH, Lee K, Moon YS, Sul HS. A cysteine-rich adipose tissue-specific secretory factor inhibits adipocyte differentiation. J Biol Chem 2001; 276:11252–11256.

42. Holcomb IN, Kabakoff RC, Chan B, Baker TW, Gurney A, Henzel W, Nelson C, Lowman HB, Wright BD, Skelton NJ, Frantz GD, Tumas DB, Peale FV Jr., Shelton DL, Hebert CC. FIZZ1, a novel cysteine-rich secreted protein associated with pulmonary inflammation, defines a new gene family. Embo J 2000; 19:4046–4055.

43. Steppan CM, Bailey ST, Bhat S, Brown EJ, Banerjee RR, Wright CM, Patel HR, Ahima RS, Lazar MA. The hormone resistin links obesity to diabetes. Nature 2001; 409:307–312.

44. Steppan CM, Brown EJ, Wright CM, Bhat S, Banerjee RR, Dai CY, Enders GH, Silberg DG, Wen X, Wu GD, Lazar MA. A family of tissue-specific resistin-like molecules. Proc Natl Acad Sci U S A 2001; 98:502–506.

45. Gerstmayer B, Kusters D, Gebel S, Muller T, Van Miert E, Hofmann K, Bosio A. Identification of RELMgamma, a novel resistin-like molecule with a distinct expression pattern small star, filled. Genomics 2003; 81:588–595.

46. Banerjee RR, Lazar MA. Dimerization of resistin and resistin-like molecules is determined by a single cysteine. J Biol Chem 2001; 276:25970–25973.

47. Blagoev B, Kratchmarova I, Nielsen MM, Fernandez MM, Vodby J, Andersen JS, Kristiansen K, Pandey A, Mann M. Inhibition of adipocyte differentiation by resistin-like molecule alpha. Biochemical characterization of its oligomeric nature. J Biol Chem 2002; 277:42011–42016.

48. Chen J, Wang L, Boeg YS, Xia B, Wang J. Differential dimerization and association among resistin family proteins with implications for functional specificity. J Endocrinol 2002; 175:499–504.

49. Way JM, Gorgun CZ, Tong Q, Uysal KT, Brown KK, Harrington WW, Oliver Jr WR, Willson TM, Kliewer SA, Hotamisligil GS. Adipose tissue resistin expression is severely suppressed in obesity and stimulated by peroxisome proliferator-activated receptor gamma agonists. J Biol Chem 2001; 276: 25651–25653.

50. McTernan CL, McTernan PG, Harte AL, Levick PL, Barnett AH, Kumar S. Resistin, central obesity, and type 2 diabetes. Lancet 2002; 359:46–47.

51. McTernan PG, McTernan CL, Chetty R, Jenner K, Fisher FM, Lauer MN, Crocker J, Barnett AH, Kumar S. Increased resistin gene and protein expression in human abdominal adipose tissue. J Clin Endocrinol Metab 2002; 87:2407.

52. Savage DB, Sewter CP, Klenk ES, Segal DG, Vidal-Puig A, Considine RV, O'Rahilly S. Resistin/Fizz3 expression in relation to obesity and peroxisome proliferator-activated receptor-gamma action in humans. Diabetes 2001; 50:2199–2202.

53. Rajala MW, Obici S, Scherer PE, Rossetti L. Adipose-derived resistin and gut-derived resistin-like molecule-beta selectively impair insulin action on glucose production. J Clin Invest 2003; 111:225–230.

54. Haugen F, Jorgensen A, Drevon CA, Trayhurn P. Inhibition by insulin of resistin gene expression in 3T3-L1 adipocytes. FEBS Lett 2001; 507:105–108.

55. Shojima N, Sakoda H, Ogihara T, Fujishiro M, Katagiri H, Anai M, Onishi Y, Ono H, Inukai K, Abe M, Fukushima Y, Kikuchi M, Oka Y, Asano T. Humoral regulation of resistin expression in 3T3-L1 and mouse adipose cells. Diabetes 2002; 51:1737–1744.

56. Viengchareun S, Zennaro MC, Pasual-Le Tallec L, Lombes M. Adipocytes are novel sites of expression and regulation of adiponectin and resistin. FEBS Lett 2002; 532:345–350.

6

Transcriptional Regulation of Energy Metabolism in Liver

Manabu T. Nakamura
Department of Food Science and Human Nutrition, University of Illinois at Urbana-Champaign, Urbana, Illinois, U.S.A.

1. INTRODUCTION

Liver plays a central role in the regulation of macronutrient metabolism in response to changes in physiological and dietary conditions. Classic work in biochemistry has elucidated acute regulation of enzyme activities involved in glucose metabolism by phosphorylation/dephosphorylation and by allosteric action of intermediate metabolites. In addition to this acute regulation of enzyme activity, starvation or a large change in dietary composition requires metabolic adaptation. Transcriptional regulation plays the major role in this adaptation process. Key transcription factors that regulate this adaptive response have been identified in the last decade. Those include peroxisome proliferator-activated receptors (PPARs), sterol regulatory element-binding protein (SREBP), and carbohydrate response element-binding protein (ChREBP). Furthermore, due to the advances in genomics and gene-targeting technologies, our knowledge of transcriptional regulation has been increasing dramatically. However, sorting and interpreting the flood of information is essential to utilize the abundant knowledge

119

generated by these advanced technologies. Thus, the objective of this chapter is to summarize recent findings in the transcriptional regulation of carbohydrate and lipid metabolism in liver, and place them in physiological contexts.

2. GLYCOLYSIS AND DE NOVO LIPOGENESIS

2.1. Induction by Insulin and Glucocorticoids

It has long been known that when animals are fed a high-carbohydrate diet after fasting, activities of lipogenic enzymes become much higher in the liver than in ad libitum-fed animals, and that both insulin and glucocorticoids are required for this "overshoot" response (1). Subsequent studies have shown that transcriptional induction is largely responsible for the increased activity of lipogenic enzymes. Fatty acid synthase (FAS), a key lipogenic enzyme, is primarily regulated at the transcriptional level, and has no known acute regulation (2,3). An insulin response element (IRE) was identified in the proximal promoter of the *Fas* gene (4), and subsequent studies have

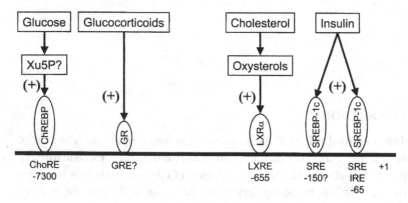

FIGURE 1 Transcriptional regulation of the rat fatty acid synthase gene in liver. Insulin, glucocorticoids, glucose, and cholesterol activate transcription of *Fas* gene via respective transcription factors. Xu5P is a likely metabolite that mediates the glucose effect (31). GRE is not identified (9). The SRE sequence at -150 is not functionally confirmed (6). Transcriptional regulation of FAS and ACC is very similar (21,33). The *SCD-1* promoter may have LXRE, but may not have ChoRE (21,61). Abbreviations: ChoRE = carbohydrate response element; ChREBP = carbohydrate response element-binding protein; FAS = fatty acid synthase; GR = glucocorticoid receptor; GRE = glucocorticoid response element; IRE = (positive) insulin response element; LXR = liver-X receptor; LXRE = LXR response element; SRE = sterol regulatory element; SREBP = sterol regulatory element binding protein; Xu5P = xylulose 5-phosphate.

shown that SREBP-1c is the transcription factor that binds the element and mediates insulin action (Fig. 1) (5,6).

Glucocorticoids are critical for metabolic adaptation during starvation (7), which is discussed in detail in the gluconeogenesis section. Moreover, glucocorticoids are required for replenishing fat storage after starvation. Adrenalectomized rats die within 10 days when they are subjected to once-a-day meal feeding because of body fat depletion (8). Insulin and glucocorticoids synergistically induce lipogenic genes after starvation or in meal feeding (1). In the rat *Fas* gene, the distal region of −4600 to −7380 is required for this glucocorticoid response (Fig. 1) (9), although the glucocorticoid response element (GRE) in this region is yet to be identified. Whereas glucocorticoids alone have little effect, glucocorticoids greatly increase the induction of lipogenic genes by insulin (9). Thus, glucocorticoids play a permissive role in the induction of lipogenic genes.

2.2. The Binding Protein SREBP-1c and its Regulation

As mentioned in the previous section, SREBP-1c (also called adipocyte differentiation and determination factor-1, ADD-1) mediates at least in part the insulin effects on lipogenic gene induction. The SREBPs are transcription factors of the basic helix-loop-helix leucine zipper (bHLHLZ) family, and were first identified as the factors that bind the sterol regulatory element (SRE) in the low-density lipoprotein receptor promoter (10,11) and independently as an E-box-binding factor in adipocytes (12). The SREBPs have three isoforms: SREBP-1a, SREBP-1c, and SREBP-2. Isoforms SREBP-1 and SREBP-2 are transcribed from different genes (11,13). The 1c and 1a isoforms are encoded from the same gene by alternative promoter usage (14). The SREBP-2 form mainly activates transcription of genes involved with cholesterol synthesis and metabolism, SREBP-1c targets genes for fatty acid synthesis, and SREBP-1a can activate both (15,16). The expression of SREBP-1a and 1c differs among tissues. Whereas SREBP-1a is the major species in all cell lines examined, as well as in intestine, thymus, spleen, and testes, SREBP-1c is highly expressed in liver, adrenal gland, adipose tissue and brain (14). The expression pattern of SREBP-1a suggests that the main role of SREBP-1a is to supply fatty acids for membrane phospholipids in proliferating cells.

The SREBP-1c differs from 1a only in that the first exon encodes shorter peptides than does the exon in 1a (17). Because exon 1 encodes part of activation domain, 1c is less potent than 1a in transcriptional activation of target genes (17). In particular, SREBP-1c is unable to activate genes for cholesterol metabolism (16). Also, unlike SREBP-1a, the SREBP-1c mRNA is not induced by cholesterol deprivation (14). The SREBP-1c form is the major

TABLE 1   SREBP-1c-Induced Genes in Liver

Enzymes for fatty acid synthesis
    Acetyl-CoA carboxylase (16,21,117)[a]
    Fatty acid synthase (5,16,21)
    Long-chain acyl-CoA elongase (16,118)
Enzymes involved with both TG and PL synthesis
    Stearoyl-CoA desaturase-1, -2 (16,42)
    Glycerol-3-phosphate acyltransferase (16)
Enzymes involved with PL synthesis
    Delta-6 desaturase (44,119)
    Delta-5 desaturase (119)
    CTP:phosphocholine cytidylyltransferase (52)
Other lipogenic enzymes
    Acetyl-CoA synthase (16,21,120)
    ATP-citrate lyase (21)
    Glucose-6-phosphate dehydrogenase (21)
    Malic enzyme (21)
    S14 (41)

---

[a]References.

one expressed in hepatocytes in vivo (14), it activates entire genes for fatty acid and glycerolipid synthesis in liver (Table 1). Accumulating evidence indicates that SREBP-1c mediates the effect of insulin on transcriptional activation of genes involved in fatty acid synthesis. The SREBP-1c expression is diminished in fasting and is rapidly increased in refeeding in an insulin-dependent manner (16,18–20). Knocking out the *Srebp-1c* gene abolished the induction, that usually occurs with refeeding after 12-hr fasting, of glucose-6-phosphate dehydrogenase (G6PDH), malic enzyme, stearoyl-CoA desaturase (SCD)-1, and glycerol-3-phosphate acyltransferase (G3PAT), but the induction of acyl-CoA carboxylase (ACC) and FAS was only partially impaired (21). This study has shown that SREBP-1c is required for maximum induction of lipogenic genes by refeeding. It is likely that the partial induction of FAS and ACC mRNA in *Srebp-1c-/-* mice upon refeeding is due to ChREBP, which is discussed in the next section.

The SREBPs are synthesized as a larger precursor protein (pSREBP), the middle part of which is inserted into the endoplasmic reticulum membrane. After proteolytic cleavage, the amino terminal domain (nSREBP) migrates to a nucleus and activates target genes (Fig. 2) (15). The SREBP activity is primarily regulated by this proteolysis. The carboxyl terminal domain of SREBP plays a regulatory role and associates with another membrane-bound protein, SREBP cleavage-activating protein (SCAP),

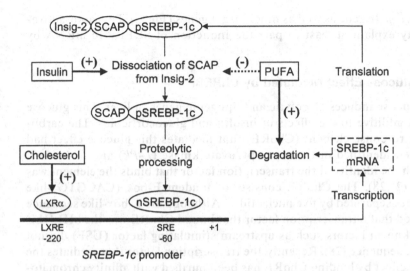

FIGURE 2 Postulated regulation of SREBP-1c activity. Insulin may increase proteolytic activation of SREBP-1c by suppressing Insig-2 expression (25). Nuclear form SREBP-1c may then activate transcription of SREBP-1c gene (26). PUFAs counteract this insulin effect by inhibiting proteolytic processing and decreasing mRNA (40,41,45,46), most likely by increasing mRNA degradation (49). Dietary cholesterol and LXR agonists induce SREBP-1c via LXRE (58). Abbreviations: Insig = insulin induced gene protein; LXR = liver-X receptor; LXRE = LXR response element; nSREBP = nuclear form SREBP; pSREBP = precursor SREBP; PUFA = polyunsaturated fatty acid; SCAP = SREBP cleavage-activating protein; SRE = sterol regulatory element; SREBP = sterol regulatory element binding protein.

which binds another protein, Insig (insulin-induced gene protein) (Fig. 2) (22,23). Dissociation between Insig and SCAP initiates sequential events that result in the proteolytic activation of SREBP. Refeeding stimulates this proteolytic processing of both pSREBP-1c and pSREBP-2 (18,24). Although refeeding increases both proteolytic activation of SREBP-1c and expression of SREBP-1c mRNA, the mechanism of this effect is not fully understood. An insulin-responsive Insig message, Insig-2a, was identified recently (25). Both Insig-2a and 2b mRNAs are encoded from the same gene with alternative promoters. Although the Insig-2a and 2b mRNAs differ in the first exon, the coding sequences are identical. The former, Insig-2a mRNA, is specifically expressed in liver, and is suppressed by insulin. Therefore, insulin may initiate the proteolytic activation of SREBP by reducing Insig-2a protein, which would then lead to dissociation of the SCAP: pSREBP complex from Insig-2 (Fig. 2) (25). In addition, a functional SRE is present in the mouse *Srebp-1c* prompter (26). Thus, nSREBP-1c is likely to activate transcription of the *Srebp-1c* gene as well as other target genes (Fig. 2). This auto-activation may explain at least in part the induction of SREBP-1c mRNA by insulin.

2.3. Glucose Effect Mediated by ChREBP

High glucose induces glycolytic and lipogenic genes in liver. This glucose effect is additive to the effects of insulin and glucocorticoids. The carbohydrate response element (ChoRE) that mediates this glucose effect had long been identified in liver-type pyruvate kinase (*L-Pk*) and *S14* genes, although the identity of the transcription factor that binds the element was elusive (27,28). The ChoRE consists of tandem E-box (CACGTG)-like sequences separated by five nucleotides. Although the E-box-like sequence suggested that a transcription factor that bound ChoRE was the bHLHLZ family, known factors such as upstream stimulating factor (USF) did not bind the sequence (28). Recently, the transcription factor that mediates the glucose effect by binding ChoRE has been purified with affinity chromatography and named carbohydrate response element-binding protein (ChREBP) (29). As predicted, ChREBP is a transcription factor of the bHLHLZ family, and is expressed exclusively in liver (29). Under a high glucose condition, ChREBP is dephosphorylated and translocated to the nucleus, resulting in the activation of target genes (30). Protein phosphatase 2A, which is activated by xylulose-5-phosphate (Xu5P, an intermediate of the pentose phosphate cycle), is capable of dephosphorylating ChREBP (31). In addition to its presence in the *L-Pk* gene, ChoRE has been identified in promoters of *Fas* (Fig. 1) (32), *Acc* (33), and *S14* (27). The presence of ChoRE in *Fas* and *Acc* is consistent with the report that these genes are still partially induced in the *Srebp-1c* -/- mice upon refeeding (21).

Because high glucose stimulates insulin secretion, the *Fas* gene induction by glucose is mediated by both ChREBP and SREBP-1c (Fig. 1). Whereas ChREBP expression is liver-specific, SREBP-1c is widely expressed in tissues including those that lack a significant capacity for triglyceride (TG) synthesis (14). Thus, ChREBP, not SREBP-1c, may play the major role in the induction of liver-specific lipogenesis for excess glucose disposal.

2.4. Role of Polyunsaturated Fatty Acids in Regulation of Lipogenesis

Polyunsaturated fatty acids (PUFAs) at 2–3 weight % of a diet suppress the induction of lipogenic genes by refeeding (34). This suppression is exerted at the transcriptional step and is unique to PUFAs. Other fatty acids such as saturated and monounsaturated fatty acids have no effects (35,36). Although

PUFA-responsive regions had been mapped in several gene promoters, the transcription factor that mediates the PUFA suppression was elusive (35,37–39). Subsequently, SREBP-1c was identified as the factor that binds the PUFA response sequence and mediates the PUFA effect in *Fas* (40), *S14* (41), *Scd* (42,43), and delta-6 desaturase (*D6d*) (44) genes. The PUFAs suppress nSREBP-1c with two mechanisms: (1) by inhibiting proteolytic processing, and (2) by decreasing the mRNA (Fig. 2) (40,41,45,46). However, the mechanism of these PUFA effects on SREBP-1c is not well understood. The SCAP contains a cholesterol-sensing sequence and dissociates from the Insig protein when membrane cholesterol is low, leading to the activation of SREBPs (22). In *Drosophila*, phosphatidylethanolamine (PE)—not cholesterol—inhibits the SREBP processing (47). Interestingly, a recent study showed that increased membrane PE resulted in downregulation of three desaturases: D6D, delta-5 desaturase (D5D), and SCD (48). Although these desaturases are targets of SREBP-1c (Table 1), it is yet to be determined whether PE regulates SREBP activation in mammalian liver. Also, it is unknown if PUFAs modulate PL species in membrane.

Polyunsaturated fatty acids also accelerate the degradation of SREBP-1c mRNA, although the mechanism has yet to be elucidated (Fig. 2) (49). In addition, PUFAs suppress SREBP-1c transcription in cell studies by antagonizing a synthetic liver-X receptor (LXR) agonist (50,51), but they reduce SREBP-1c mRNA without affecting transcription in rat liver (40). Thus, it is unclear at this moment whether the observations in the cell line are relevant to the in vivo PUFA effect in liver.

As shown in Table 1, one of the target genes of SREBP-1c is for CTP:phosphocholine cytidylyltransferase (CCTα), which is an enzyme specific for phospholipid (PL) synthesis (52). In addition, two desaturases—D6D and D5D—are also considered as enzymes for PL synthesis because the products—20:4 n-6 and 22:6 n-3—are mostly incorporated into PL, not TG (53,54). The profile of SREBP-1c target genes suggests that the physiological role of SREBP-1c in liver may be the regulation of PL synthesis. This hypothesis is further supported by the identification of PUFA as the major regulator of SREBP-1c activation. As discussed in Section 2.3, lipogenesis for excess glucose disposal may be regulated by ChREBP rather than SREBP-1c. If this hypothesis holds, SREBP-1c and its regulation by PUFA on lipogenesis may not have a significant role in fat storage from glucose.

2.5. Liver-X Receptor-α and Regulation of Lipogenic Genes by Cholesterol

The LXRs are transcription factors of the nuclear receptor family. Two isoforms have been identified, LXRα is the most abundant in liver, whereas

LXRβ is expressed ubiquitously (55). The LXRs form heterodimers with the retinoid-X receptor (RXR) and bind a direct repeat (DR)-4 element (LXR response element, LXRE) in target genes (56). Cholesterol metabolites, oxysterols, are natural ligands of LXRs and activate them. The LXRs activate a group of genes involved in reverse cholesterol transport (55,56). In liver, activated LXRα induces the messages of cytochrome P450 (CYP) 7A and ATP-binding cassette transporters (ABC) G5 and G8 (55). Because CYP7A is an enzyme that catalyzes the rate-limiting step of bile acid synthesis, and ABCG5 and ABCG8 are cholesterol transporters, an overall effect of LXRα activation in liver is an increase in the bile acid synthesis and in the secretion of cholesterol and bile acid to bile. When the LXRα gene is disrupted, the mouse becomes intolerant of dietary cholesterol and accumulates cholesterol in its liver, underscoring the physiological function of LXRα (57). In addition, synthetic LXR agonists induce SREBP-1c, and LXRE is identified in the mouse *Srebp-1c* promoter (Fig. 2) (58). Thus, LXR agonists induce the lipogenic genes targeted by SREBP-1c, and increase very-low-density protein (VLDL) production in liver (59). A physiological role of this SREBP-1c induction by LXRα in liver may be providing fatty acids for the synthesis of PLs, which are also an essential component of bile.

Recently, LXRE has been identified in the rat *Fas* promoter (Fig. 1) (60). Also, dietary cholesterol has been shown to induce the SCD-1 mRNA in an SREBP-1c-independent manner (61). These observations are consistent with the findings in a study with *Srebp-1c* null mice in which an LXR agonist failed to induce G6PDH and malic enzyme, whereas residual induction was observed in ACC, FAS, and SCD-1 (21). Therefore, like the *Fas* gene, the *Acc* and *Scd-1* promoters may also have LXRE, and LXR may induce ACC and SCD-1 both directly and indirectly via LXRE and SRE, respectively. Because cholesteryl ester acts as a reservoir when cholesterol is in excess, provision of fatty acids for cholesteryl ester synthesis may be the role for the SREBP-1c-independent induction of these three lipogenic genes by LXR. This explanation fits well to the finding that D6D and D5D were not induced by dietary cholesterol (62).

2.6. Summary

Insulin and glucocorticoids synergistically induce lipogenic enzymes in liver. The SREBP-1c is likely to mediate the insulin effect. Polyunsaturated fatty acids suppress induction of lipogenic genes by suppressing SREBP-1c processing and expression. Another transcription factor, ChREBP, is activated by high glucose. The ChoRE is identified in *L-Pk*, *S14*, and *Fas* genes. The binding protein ChREBP may play a role in converting excess glucose to TG for energy storage, whereas the physiological role of SREBP-1c may be induction of PL synthesis. The LXR, the major regulator of cholesterol reverse transport, also activates lipogenic genes in SREBP-1c-dependent and -independent manners. Fatty acid synthesis induced by LXR may be required for bile secretion as well as cholesteryl ester synthesis.

3. GLUCONEOGENESIS

Liver performs both glycolysis and gluconeogenesis. Three steps—glucokinase (GK)/glucose-6-phosphatase (G6Pase), phosphofructokinase (PFK)/fructose-1,6-bisphosphatase, and L-PK/phosphoenolpyruvate carboxykinase (PEPCK)—may form potentially futile cycles if reactions of both directions are active at the same time. Activities of GK, PFK, fructose-1, 6-bisphosphatase, and L-PK are acutely regulated by hormones such as insulin, glucagon, and catecholamines. In addition, gluconeogenic enzymes are preferentially expressed in the periportal area, whereas glycolytic enzymes are abundant in the perivenous area of liver lobules (63). This zonal difference in enzyme expression also contributes to the prevention of futile cycles. However, there is no known acute regulation of enzyme activity of either PEPCK or G6Pase, the key enzymes of gluconeogenesis. Transcription is the major regulatory mechanism of these two enzymes.

3.1. Glucagon and Glucocorticoids

Under a fasted condition, glucagon induces gluconeogenic genes, such as *Pepck* and *G6pase*, by increasing cellular cyclic adenosine monophosphate (cAMP)(Fig. 3) (64,65). The *Pepck* gene has the cAMP response element (CRE) in its promoter (Fig. 3) (65). The CRE-binding protein (CREB) binds CRE and activates gene transcription when it is phosphorylated by cAMP-dependent protein kinase A (66). This cAMP-mediated induction requires glucocorticoids although glucocorticoids alone have little effect on the induction of gluconeogenic enzymes (64,65). The *Pepck* gene has at least two GREs that are required for maximal activation of transcription (Fig. 3) (65). Blood glucocorticoid concentration increases during fasting and starts decreasing upon refeeding (67,68). As discussed in Section 2.1, glucocorticoids also synergistically induce lipogenic genes when insulin is present, although glucocorticoids alone have little effect. Thus, glucocorticoids have a permissive effect on the induction of both lipogenic and gluconeogenic genes, depending on the coexisting hormones. Therefore, the physiological role of glucocorticoids in energy metabolism is both the adaptation to fasting and the replenishing of energy stores upon refeeding.

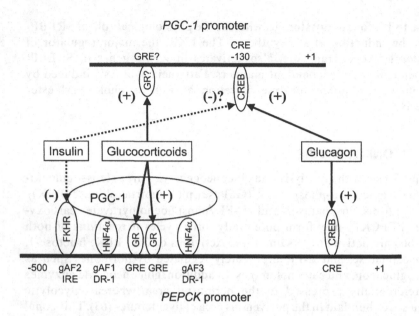

FIGURE 3 Transcriptional regulation of PEPCK gene. Glucagon and glucocorticoids induce *PEPCK* gene by increasing binding of CREB and GR, respectively, to the *PEPCK* promoter (65). In addition to GR, FKHR and HNF4α are required for recruitment of PGC-1 (69,70,78). Insulin counteracts this activation by phosphorylating FKHR, which then loses activity (75,76). Glucagon and glucocorticoids also induce PGC-1, an essential cofactor of the *PEPCK* gene activation (69). Insulin also inhibits *PGC-1* transcription possibly by inactivating CREB (79). GRE has not been identified in the *PGC-1* promoter. The *G6Pase* gene is regulated similarly (70,75,78). Abbreviations: CRE = cAMP response element; CREB = CRE-binding protein; DR = direct repeat; FKHR (=FOXO1)= Forkhead transcription factor; gAF = glucocorticoid accessory factor element; GR = glucocorticoid receptor; GRE = glucocorticoid response element; HNF4 = hepatocyte nuclear factor 4; IRE=(negative) insulin response element; PEPCK = phosphoenolpyruvate carboxykinase; PGC-1 = PPAR-gamma coactivator-1.

3.2. Hepatocyte Nuclear Factor 4α

In addition to GREs, the glucocorticoid induction of the *Pepck* gene requires other sequences that are termed glucocorticoid accessory factor elements (gAF). Recently, hepatocyte nuclear factor 4α (HNF4α) has been identified as the factor that binds DR-1 sequences in gAF 1 and 3, and is indispensable for PEPCK induction (Fig. 3) (69,70). The HNF4α is a transcription factor of the nuclear receptor family and highly expressed in liver. The primary physiological role of HNF4α is to direct liver-specific gene expression (Table 2) (70,71). Fatty acyl-CoA has been proposed as ligands of HNF4α (72).

TABLE 2   HNF4-Dependent Genes in Liver

Lipoprotein synthesis and secretion (71)[a]
Apo AII
Apo A IV
Apo C II
Apo C III
Microsomal triglyceride transfer protein
Bile acid synthesis and secretion (71)
Cytochrome P450 7A1
Organic anion transporter 1
Liver fatty acid-binding protein
Gluconeogenesis (70)
Phosphoenolpyruvate carboxykinase
Glucose-6-phosphatase

[a] References.

However, the crystal structure of HNF4α suggests that a fatty acid may constitutively bind the ligand pocket of HNF4α (73). If this is the case, HNF4α may be constitutively active. It remains unknown whether any endogenous ligands affect the activity of HNF4α and play a regulatory role in the PEPCK induction.

3.3. Insulin-Protein Kinase B-Forkhead Transcription Factor Signaling Pathway

Insulin suppresses induction of gluconeogenic genes. This effect is dominant over the induction by glucagon and glucocorticoids (65). The element required for this suppression by insulin (negative IRE) is also required for induction by glucocorticoids (gAF2 in Fig. 3). Forkhead transcription factor (FKHR/FOXO1) is identified as the protein that binds the negative IRE in the *Pepck* promoter and activates the gene (74). FKHR, an essential component for PEPCK induction, also mediates the suppression of the *Pepck* transcription by insulin (75). Protein kinase B (PKB/Akt) is a down-stream kinase in the insulin signaling pathway, and phosphorylates FKHR, which then loses binding affinity to IRE (Fig. 3) (75,76).

3.4. Peroxisome Proliferator Activated Receptor-Gamma Coactivator-1

Although the *Pepck* promoter elements and associated transcription factors had been extensively studied, the regulatory mechanism was somewhat ambiguous until recently (65,74,77). A breakthrough came when

PPAR-gamma coactivator-1 (PGC-1) was identified as a key coactivator involved in the *Pepck* gene regulation (69). This protein binds glucocorticoid receptor (GR), HNF4α, and FKHR, and acts as an essential coactivator for these transcription factors (Fig. 3) (69,78). Importantly, the expression of PGC-1 is also under the regulation of hormones. Glucocorticoids and glucagon induce PGC-1, whereas insulin suppresses it (Fig. 3) (69). Functional CRE is identified in the *Pgc-1* promoter, and may mediate suppression by insulin as well as induction by glucagon (Fig. 3) (79). Regulation of the *G6pase* transcription is very similar to *Pepck* regulation, and all components shown in Fig. 3 are involved with the *G6pase* regulation (70),(75,78).

3.5. Effects of Fructose on Lipogenesis and Gluconeogenesis

Fructose consumption in the form of high-fructose corn syrup has been increasing in the united states during the past decades (80). Liver is the primary site of fructose metabolism (81). Fructose is first phosphorylated by fructokinase to fructose-1-phosphate, which is then cleaved to 3-carbon units by aldolase B before entering the glycolytic/gluconeogenic pathway. Thus, fructose bypasses two regulatory steps of glycolysis—GK and PFK—and stimulates de novo lipogenesis (82,83) and VLDL secretion (84,85). At the same time, fructose is also converted in liver to glucose, which then can be used by liver and other organs for energy production and glycogen synthesis. Thus, high dietary fructose also stimulates gluconeogenesis in liver (86,87). The stimulation of these two pathways with opposite directions is achieved by differential gene expression between periportal and perivenous regions of the liver (63,88). When rats were fed a high-fructose diet, food intake initially decreased and then returned normal in four days (89). High-fructose feeding changes not only the activity of enzymes (82,87) but also the abundance of proteins (85) and messages (83,90,91). However, it is yet to be elucidated how liver senses high fructose intake and changes its gene expression to adapt its metabolism.

3.6. Summary

Glucagon activates transcription of gluconeogenic genes by the cAMP-mediated signaling pathway. Glucocorticoids have permissive effects on the induction of both gluconeogenic and glycolytic/lipogenic genes depending on the hormone that coexists. The HNF4α is an essential component for liver-specific gene expression including PEPCK and G6Pase, but HNF4α ligands and their regulatory role in gluconeogenesis are yet to be fully elucidated. Insulin suppresses gluconeogenic genes by inactivating FKHR.

Coactivator PGC-1 binds GR, HNF4α, and FKHR, and plays a key regulatory role in the induction of PEPCK and G6Pase. Dietary fructose induces both gluconeogenic and lipogenic genes in liver. The mechanism of the induction by fructose is unknown.

4. FATTY ACID OXIDATION AND KETOGENESIS

4.1. Peroxisome Proliferator-Activated Receptorα and Fatty Acids

Glucose and fatty acids are two major sources of energy for mammals. The choice between glucose and fatty acids as an energy source is acutely regulated by hormones and the abundance of glucose in the liver. Malonyl-CoA, a potent allosteric inhibitor of carnitine palmitoyltransferase-1, plays the central role in this fuel switching (92,93). Also, ketogenesis is acutely regulated by succinyl-CoA that inhibits mitochondrial 3-hydroxyl-3-methylglutaryl-CoA synthase, the first step of ketogenesis (94,95). Liver plays a critical role in energy metabolism during fasting by providing glucose and ketone bodies to other organs. This metabolic adaptation to fasting is largely achieved by the induction of enzymes for fatty acid oxidation and ketogenesis, not by the acute regulation of enzyme activity. A transcription factor of the nuclear receptor family, PPARα plays a central role in this process. Various long-chain fatty acids as well as hypolipidemic drugs act as ligands of PPARα (96,97). Ligand-bound PPARα forms heterodimers with RXR and binds the DR-1 element of target genes, most of which are involved with fatty acid oxidation (Table 3). The essential role of PPARα in the adaptation to fasting was demonstrated by a targeted disruption of the gene. The *Pparα*-null mice grew phenotypically normally as long as animals were fed ad libitum. However, when the animals were fasted, they showed hypoglycemia, hypoketonemia, hypothermia, and impairment in the induction of fatty acid oxidation and ketogenic enzymes in their livers (98–100). Some of the *Pparα*-null mice died within 48 hours of fasting (99). These studies demonstrated that the major role of PPARα is transcriptional adaptation of energy metabolism to fasting. Consistent with this line of the physiological role, PPARα is also required for the induction of pyruvate dehydrogenase kinase-4 that inactivates pyruvate dehydrogenase during fasting (Table 3). Moreover, PPARα may regulate amino acid metabolism during fasting (101).

A variety of long-chain fatty acids bind and activate PPARα (96,97). Fatty acids have strong binding affinity to PPARα with a reported Kd range of 5–10 nM (102). Therefore, non-esterified fatty acids released from adipose tissue are likely to act as endogenous ligands of PPARα during fasting. A group of hypolipidemic drugs such as fibrates also act as strong ligands of

TABLE 3   Genes Dependent on PPARα in Induction by Fasting or by Agonists

Gene	mRNA by fasting	mRNA by agonist	Functional PPRE
Fatty acid oxidation			
Liver carnitine palmitoyltransferase 1	+ [a](PPARα-independent) (98,99)	+(104)[b]	
Medium-chain acyl-CoA dehydrogenase	+(99)		
Acyl-CoA oxidase	+(99)	+(104,121)	Yes (122,123)
L-Bifunctional protein	+(100)	+(104,121)	Yes (123)
Cytochrome P450 4A	+(99,100,124)	+(104,121)	Yes (125)
Ketogenesis			
3-Hydroxy-3-methylglutaryl-CoA synthase (mitochondrial)	+(126)	+(121)	Yes (127)
Fatty acid transport			
Fatty acid transporter/CD36		+(121,128,129)	No (129)
Fatty acid desaturases			
Stearoyl-CoA desaturase-1	− (104,130)	+(103,104)	Yes (103)
Delta-6 desaturase	0 (104)	+(104)	Yes (105)
Delta-5 desaturase	−/0 (104)	+(104)	Unknown
Glucose oxidation			
Pyruvate dehydrogenase kinase 4	+(131)	+(121)	

[a]Compared with fed control animals: + = increased; 0 = no change; − = decreased.
[b]Reference.

PPARα (96,97). Thus, the hypolipidemic effect of fibrates may be due to the PPARα-mediated induction of the fatty acid oxidation enzymes.

As shown in Table 3, synthetic PPARα ligands, peroxisome proliferators, also strongly induce acyl-CoA desaturases (103,104), and functional PPRE has been identified in promoters of SCD-1 (103) and D6D (105). The function of desaturases is the synthesis of unsaturated fatty acids and is not related to degradation of fatty acids. Furthermore, desaturases are not induced in fasted condition (104). The physiological significance of desaturase induction by PPARα is yet to be elucidated.

4.2. Role of Polyunsaturated Fatty Acids in Fatty Acid Oxidation

Dietary fish oil rich in docosahexaenoic acid (22:6 n-3) and eicosapentaenoic acid (20:5 n-3) increases activities and mRNA of fatty acid oxidation enzymes (106–108). This induction is mediated by PPARα (109). Other PUFAs such as α-linolenic acid (18:3 n-3) (108,110) and γ-linolenic acid (18:3 n-6) (111) have similar effects, whereas linoleic acid (18:2 n-6) does not have a significant effect on saturated fats (106,108). Whether or not a fatty acid is readily stored in TG may explain the differential effects between 18:2 n-6 and other PUFAs. With the exception of 18:2 n-6, PUFAs are not incorporated into TG in large quantity (112–114). Thus, excess dietary PUFAs that are poor substrates for TG synthesis may serve as PPARα ligands in the fed state, resulting in the induction of their own oxidation pathway. In addition, dietary fish oil reduces plasma TG. The hypotriglyceridemic effect of fish oil is not due to the PPARα-mediated induction of fatty acid oxidation enzymes (109). Studies with cells suggest that fish oil may reduce VLDL secretion by stimulating insulin pathway (115,116).

4.3. Summary

The receptor PPARα plays an essential role in the metabolic adaptation to fasting by inducing genes for fatty acid oxidation and ketogenesis. Fatty acids released from adipose tissue during fasting are the likely ligands of PPARα. Dietary PUFAs except for 18:2 n-6 are likely to induce fatty acid oxidation enzymes via PPARα as a feed-forward mechanism. The hypotriglyceridemic effect of fish oil is not dependent on PPARα, and may involve the insulin-signaling pathway.

ABBREVIATIONS

ABC = ATP-binding cassette transporter; ACC = acetyl-CoA carboxylase; ADD-1 = adipocyte differentiation and determination factor-1 = SREBP-1c; bHLHLZ = basic helix-loop-helix leucine zipper; CCTα = CTP: phosphocholine cytidylyltransferase; ChoRE = carbohydrate response element; ChREBP = carbohydrate response element-binding protein; CRE = cAMP response element; CREB = CRE-binding protein; CYP = cytochrome P450; D5D = delta-5 desaturase; D6D = delta-6 desaturase; DR = direct repeat; FAS = fatty acid synthase; FKHR (=FOXO1)= Forkhead transcription factor; G3PAT =glycerol-3-phosphate acyltransferase; G6PDH = glucose-6-phosphate dehydrogenase; G6Pase = glucose-6-phosphatase; gAF =glucocorticoid accessory factor element; GK = glucokinase; GR = glucocorticoid receptor; GRE = gluco-glucocorticoid response element; HNF4 = hepatocyte nuclear factor 4; Insig = insulin-induced gene protein; IRE = insulin response element; L-PK = liver-type pyruvate kinase; LXR = liver-X receptor; LXRE = LXR response element; nSREBP = nuclear form SREBP; PE = phosphatidyletha-

sphatidylethanolamine; PEPCK = phosphoenolpyruvate carboxykinase; PFK = phosphofructokinase; PGC-1 = PPAR-gamma coactivator-1; PKB (=Akt) = protein kinase B; PL = phospholipid; PPAR = peroxisome proliferator-activated receptor; pSREBP = precursor SREBP; PUFA = polyunsaturated fatty acid; RXR = retinoid-X receptor; SCAP = SREBP cleavage-activating protein; SCD = stearoyl-CoA desaturase; SRE = sterol regulatory element; SREBP = sterol regulatory element-binding protein; TG = triglyceride; USF = upstream stimulating factor; VLDL = very-low-density lipoprotein; Xu5P = xylulose 5-phosphate.

REFERENCES

1. Berdanier CD, Shubeck D. Interaction of glucocorticoid and insulin in the responses of rats to starvation-refeeding. J Nutr 1979; 109:1766–1771.
2. Paulauskis JD, Sul HS. Hormonal regulation of mouse fatty acid synthase gene transcription in liver. J Biol Chem 1989; 264:574–577.
3. Clarke SD, Armstrong MK, Jump DB. Nutritional control of rat liver fatty acid synthase and S14 mRNA abundance. J Nutr 1990; 120:218–224.
4. Moustaid N, Beyer RS, Sul HS. Identification of an insulin response element in the fatty acid synthase promoter. J Biol Chem 1994; 269:5629–5634.
5. Magana MM, Osborne TF. Two tandem binding sites for sterol regulatory element binding proteins are required for sterol regulation of fatty-acid synthase promoter. J Biol Chem 1996; 271:32689–32694.
6. Latasa MJ, Moon YS, Kim KH, Sul HS. Nutritional regulation of the fatty acid synthase promoter in vivo: sterol regulatory element binding protein functions through an upstream region containing a sterol regulatory element. Proc Natl Acad Sci U S A 2000; 97:10619–10624.
7. Seitz HJ, Kaiser M, Krone W, Tarnowski W. Physiologic significance of glucocorticoids and insulin in the regulation of hepatic gluconeogenesis during starvation in rats. Metabolism 1976; 25:1545–1555.
8. Berdanier CD, Wurdeman R, Tobin RB. Further studies on the role of the adrenal hormones in responses of rats to meal-feeding. J Nutr 1976; 106:1791–1800.
9. Rufo C, Gasperikova D, Clarke SD, Teran-Garcia M, Nakamura MT. Identification of a novel enhancer sequence in the distal promoter of the rat fatty acid synthase gene. Biochem Biophys Res Com 1999; 261:400–405.
10. Briggs MR, Yokoyama C, Wang X, Brown MS, Goldstein JL. Nuclear protein that binds sterol regulatory element of low density lipoprotein receptor promoter. I. Identification of the protein and delineation of its target nucleotide sequence. J Biol Chem 1993; 268:14490–14496.
11. Hua X, Yokoyama C, Wu J, Briggs MR, Brown MS, Goldstein JL, Wang X. SREBP-2, a second basic-helix-loop-helix-leucine zipper protein that stimulates transcription by binding to a sterol regultory element. Proc Natl Acad Sci USA 1993; 90:11603–11607.

12. Tontonoz P, Kim JB, Graves RA, Spiegelman BM. ADD1: a novel helix-loop-helix transcription factor associated with adipocyte determination and differentiation. Mol Cell Biol 1993; 13:4753–4759.

13. Yokoyama C, Wang X, Briggs MR, Admon A, Wu J, Hua X, Goldstein JL, Brown MS. SREBP-1, a basic-helix-loop-helix-leucine zipper protein that controls transcription of the low density lipoprotein receptor gene. Cell 1993; 75:187–197.

14. Shimomura I, Shimano H, Horton JD, Goldstein JL, Brown MS. Differential expression of exons 1a and 1c in mRNAs for sterol regulatory element binding protein-1 in human and mouse organs and cultured cells. J Clin Invest 1997; 99:838–845.

15. Brown MS, Goldstein JL. The SREBP pathway: regulation of cholesterol metabolism by proteolysis of a membrane-bound transcription factor. Cell 1997; 89:331–340.

16. Horton JD, Goldstein JL, Brown MS. SREBPs: activators of the complete program of cholesterol and fatty acid synthesis in the liver. J Clin Invest 2002; 109:1125–1131.

17. Sato R, Yang J, Wang X, Evans MJ, Ho YK, Goldstein JL, Brown MS. Assignment of the membrane attachment, DNA binding, and transcriptional activation domains of sterol regulatory element-binding protein-1 (SREBP-1). J Biol Chem 1994; 269:17267–17273.

18. Horton JD, Bashmakov Y, Shimomura I, Shimano H. Regulation of sterol regulatory element binding proteins in livers of fasted and refed mice. Proc Natl Acad Sci USA 1998; 95:5987–5992.

19. Shimomura I, Bashmakov Y, Ikemoto S, Horton JD, Brown MS, Goldstein JL. Insulin selectively increases SREBP-1c mRNA in the livers of rats with streptozotocin-induced diabetes. Proc Natl Acad Sci U S A 1999; 96:13656–13661.

20. Foretz M, Pacot C, Dugail I, Lemarchand P, Guichard C, Le Liepvre X, Berthelier-Lubrano C, Spiegelman B, Kim JB, Ferre P, et al. ADD1/SREBP-1c is required in the activation of hepatic lipogenic gene expression by glucose. Mol Cell Biol 1999; 19:3760–3768.

21. Liang G, Yang J, Horton JD, Hammer RE, Goldstein JL, Brown MS. Diminished hepatic response to fasting/refeeding and liver X receptor agonists in mice with selective deficiency of sterol regulatory element- binding protein-1c. J Biol Chem 2002; 277:9520–9528.

22. Yang T, Espenshade PJ, Wright ME, Yabe D, Gong Y, Aebersold R, Goldstein JL, Brown MS. Crucial step in cholesterol homeostasis: sterols promote binding of SCAP to INSIG-1, a membrane protein that facilitates retention of SREBPs in ER. Cell 2002; 110:489–500.

23. Yabe D, Brown MS, Goldstein JL. Insig-2, a second endoplasmic reticulum protein that binds SCAP and blocks export of sterol regulatory element-binding proteins. Proc Natl Acad Sci U S A 2002; 99:12753–12758.

24. Xu J, Cho H, O'Malley S, Park JH, Clarke SD. Dietary polyunsaturated fats regulate rat liver sterol regulatory element binding proteins-1 and -2 in three distinct stages and by different mechanisms. J Nutr 2002; 132:3333–3339.

25. Yabe D, Komuro R, Liang G, Goldstein JL, Brown MS. Liver-specific mRNA for Insig-2 down-regulated by insulin: implications for fatty acid synthesis. Proc Natl Acad Sci U S A 2003; 100:3155–3160.

26. Amemiya-Kudo M, Shimano H, Yoshikawa T, Yahagi N, Hasty AH, Okazaki H, Tamura Y, Shionoiri F, Iizuka Y, Ohashi K, et al. Promoter analysis of the mouse sterol regulatory element-binding protein-1c gene. J Biol Chem 2000; 275:31078–31085.

27. Shih HM, Towle HC. Definition of the carbohydrate response element of the rat S14 gene. Evidence for a common factor required for carbohydrate regulation of hepatic genes. J Biol Chem 1992; 267:13222–13228.

28. Kaytor EN, Shih H, Towle HC. Carbohydrate regulation of hepatic gene expression. Evidence against a role for the upstream stimulatory factor. J Biol Chem 1997; 272:7525–7531.

29. Yamashita H, Takenoshita M, Sakurai M, Bruick RK, Henzel WJ, Shillinglaw W, Arnot D, Uyeda K. A glucose-responsive transcription factor that regulates carbohydrate metabolism in the liver. Proc Natl Acad Sci U S A 2001; 98:9116–9121.

30. Kawaguchi T, Takenoshita M, Kabashima T, Uyeda K. Glucose and cAMP regulate the L-type pyruvate kinase gene by phosphorylation/dephosphorylation of the carbohydrate response element binding protein. Proc Natl Acad Sci U S A 2001; 98:13710–13715.

31. Kabashima T, Kawaguchi T, Wadzinski BE, Uyeda K. Xylulose 5-phosphate mediates glucose-induced lipogenesis by xylulose 5- phosphate-activated protein phosphatase in rat liver. Proc Natl Acad Sci U S A 2003; 100:5107–5112.

32. Rufo C, Teran-Garcia M, Nakamura MT, Koo SH, Towle HC, Clarke SD. Involvement of a unique carbohydrate-responsive factor in the glucose regulation of rat liver fatty-acid synthase gene transcription. J Biol Chem 2001; 276:21969–21975.

33. O'Callaghan BL, Koo SH, Wu Y, Freake HC, Towle HC. Glucose regulation of the acetyl-CoA carboxylase promoter PI in rat hepatocytes. J Biol Chem 2001; 276:16033–16039.

34. Clarke SD, Romsos DR, Leveille GA. Specific inhibition of hepatic fatty acid synthesis exerted by dietary linoleate and linolenate in essential fatty acid adequate rats. Lipids 1976; 11:485–490.

35. Jump DB, Clarke SD, MacDougald O, Thelen A. Polyunsaturated fatty acids inhibit S14 gene transcription in rat liver and cultured hepatocytes. Proc Natl Acad Sci USA 1993; 90:8454–8458.

36. Clarke SD, Jump DB. Dietary polyunsaturated fatty acid regulation of gene transcription. Annu Rev Nutr 1994; 14:83–98.

37. Soncini M, Yet S-F, Moon Y, Chun J-Y, Sul HS. Hormonal and nutritional control of the fatty acid synthase promoter in transgenic mice. J Biol Chem 1995; 270:30339–30343.

38. Fukuda H, Iritani N, Katsurada A, Noguchi T. Insulin/glucose-, pyruvate- and polyunsaturated fatty acid-responsive region(s) of rat fatty acid synthase gene promoter. Biochem Mol Biol Int 1996; 38:987–996.

39. Waters KM, Miller CW, Ntambi JM. Localization of a polyunsaturated fatty acid response region in stearoyl-CoA desaturase gene 1. Biochim Biophys Acta 1997; 1349:33–42.

40. Xu J, Nakamura MT, Cho HP, Clarke SD. Sterol regulatory element binding protein-1 expression is suppressed by dietary polyunsaturated fatty acids: a mechanism for the coordinate suppression of lipogenic genes by polyunsaturated fats. J Biol Chem 1999; 274:23577–23583.

41. Mater MK, Thelen AP, Pan DA, Jump DB. Sterol Response Element-binding Protein 1c (SREBP1c) Is Involved in the Polyunsaturated Fatty Acid Suppression of Hepatic S14 Gene Transcription. J Biol Chem 1999; 274:32725–32732.

42. Tabor DE, Kim JB, Spiegelman BM, Edwards PA. Identification of conserved cis-elements and transcription factors required for sterol-regulated transcription of stearoyl-CoA desaturase 1 and 2. J Biol Chem 1999; 274:20603–20610.

43. Ntambi JM. Regulation of stearoyl-CoA desaturase by polyunsaturated fatty acids and cholesterol. J Lipid Res 1999; 40:1549–1558.

44. Nara TY, He WS, Tang C, Clarke SD, Nakamura MT. The E-box like sterol regulatory element mediates the suppression of human delta-6 desaturase gene by highly unsaturated fatty acids. Biochem Biophys Res Commun 2002; 296:111–117.

45. Kim HJ, Takahashi M, Ezaki O. Fish oil feeding decreases mature sterol regulatory element-binding protein 1 (SREBP-1) by down-regulation of SREBP-1c mRNA in mouse liver. A possible mechanism for down-regulation of lipogenic enzyme mRNAs. J Biol Chem 1999; 274:25892–25898.

46. Yahagi N, Shimano H, Hasty AH, Amemiya-Kudo M, Okazaki H, Tamura Y, Iizuka Y, Shionoiri F, Ohashi K, Osuga J, et al. A crucial role of sterol regulatory element-binding protein-1 in the regulation of lipogenic gene expression by polyunsaturated fatty acids. J Biol Chem 1999; 274:35840–35844.

47. Dobrosotskaya IY, Seegmiller AC, Brown MS, Goldstein JL, Rawson RB. Regulation of SREBP processing and membrane lipid production by phospholipids in Drosophila. Science 2002; 296:879–883.

48. Shimada Y, Morita T, Sugiyama K. Dietary eritadenine and ethanolamine depress fatty acid desaturase activities by increasing liver microsomal phosphatidylethanolamine in rats. J Nutr 2003; 133:758–765.

49. Xu J, Teran-Garcia M, Park JH, Nakamura MT, Clarke SD. Polyunsaturated fatty acids suppress hepatic sterol regulatory element- binding protein-1 expression by accelerating transcript decay. J Biol Chem 2001; 276: 9800–9807.

50. Ou J, Tu H, Shan B, Luk A, DeBose-Boyd RA, Bashmakov Y, Goldstein JL, Brown MS. Unsaturated fatty acids inhibit transcription of the sterol regulatory element-binding protein-1c (SREBP-1c) gene by antagonizing ligand- dependent activation of the LXR. Proc Natl Acad Sci U S A 2001; 98:6027–6032.

51. Yoshikawa T, Shimano H, Yahagi N, Ide T, Amemiya-Kudo M, Matsuzaka T, Nakakuki M, Tomita S, Okazaki H, Tamura Y, et al. Polyunsaturated fatty acids suppress sterol regulatory element-binding protein 1c promoter activity by

inhibition of liver X receptor (LXR) binding to LXR response elements. J Biol Chem 2002; 277:1705–1711.

52. Kast HR, Nguyen CM, Anisfeld AM, Ericsson J, Edwards PA. CTP:phospho-choline cytidylyltransferase, a new sterol- and SREBP- responsive gene. J Lipid Res 2001; 42:1266–1272.

53. Nakamura MT, Nara TY. Gene regulation of mammalian desaturases. Biochem Soc Trans 2002; 30:1076–1079.

54. Nakamura MT, Nara TY. Essential fatty acid synthesis and its regulation in mammals. Prostaglandins Leukot Essent Fatty Acids 2003; 68:145–150.

55. Repa JJ, Mangelsdorf DJ. The liver X receptor gene team: potential new players in atherosclerosis. Nat Med 2002; 8:1243–1248.

56. Edwards PA, Kast HR, Anisfeld AM. BAREing it all: the adoption of LXR and FXR and their roles in lipid homeostasis. J Lipid Res 2002; 43:2–12.

57. Peet DJ, Turley SD, Ma W, Janowski BA, Lobaccaro J-MA, Hammer RE, Mangelsdorf DJ. Cholesterol and bile acid metabolism are impared in mice lacking the nuclear oxysteriol receptor LXR alpha. Cell 1998; 93:693–704.

58. Repa JJ, Liang G, Ou J, Bashmakov Y, Lobaccaro JM, Shimomura I, Shan B, Brown MS, Goldstein JL, Mangelsdorf DJ. Regulation of mouse sterol regulatory element-binding protein-1c gene (SREBP-1c) by oxysterol receptors, LXRalpha and LXRbeta. Genes Dev 2000; 14:2819–2830.

59. Grefhorst A, Elzinga BM, Voshol PJ, Plosch T, Kok T, Bloks VW, van der Sluijs FH, Havekes LM, Romijn JA, Verkade HJ, et al. Stimulation of lipogenesis by pharmacological activation of the liver X receptor leads to production of large, triglyceride-rich very low density lipoprotein particles. J Biol Chem 2002; 277:34182–34190.

60. Joseph SB, Laffitte BA, Patel PH, Watson MA, Matsukuma KE, Walczak R, Collins JL, Osborne TF, Tontonoz P. Direct and indirect mechanisms for regulation of fatty acid synthase gene expression by liver X receptors. J Biol Chem 2002; 277:11019–11025.

61. Kim HJ, Miyazaki M, Ntambi JM. Dietary cholesterol opposes PUFA-mediated repression of the stearoyl- CoA desaturase-1 gene by SREBP-1 independent mechanism. J Lipid Res 2002; 43:1750–1757.

62. Leikin AI, Brenner RR. Cholesterol-induced microsomal changes modulate desaturase activities. Biochim Biophys Acta 1987; 922:294–303.

63. Jungermann K, Kietzmann T. Zonation of parenchymal and nonparenchymal metabolism in liver. Annu Rev Nutr 1996; 16:179–203.

64. Krone W, Huttner WB, Seitz HJ, Tarnowski W. Induction of rat liver phosphoenolpyruvate carboxykinase (GTP) by cyclic AMP during starvation. The permissive action of glucocorticoids. Biochim Biophys Acta 1976; 437:62–70.

65. Hanson RW, Reshef L. Regulation of phosphoenolpyruvate carboxykinase (GTP) gene expression. Annu Rev Biochem 1997; 66:581–611.

66. Gonzalez GA, Montminy MR. Cyclic AMP stimulates somatostatin gene transcription by phosphorylation of CREB at serine 133. Cell 1989; 59:675–680.

67. De Boer SF, Van der Gugten J. Daily variations in plasma noradrenaline, adrenaline and corticosterone concentrations in rats. Physiol Behav 1987; 40:323–328.

68. Bergendahl M, Vance ML, Iranmanesh A, Thorner MO, Veldhuis JD. Fasting as a metabolic stress paradigm selectively amplifies cortisol secretory burst mass and delays the time of maximal nyctohemeral cortisol concentrations in healthy men. J Clin Endocrinol Metab 1996; 81:692–699.

69. Yoon JC, Puigserver P, Chen G, Donovan J, Wu Z, Rhee J, Adelmant G, Stafford J, Kahn CR, Granner DK and others. Control of hepatic gluconeogenesis through the transcriptional coactivator PGC-1. Nature 2001; 413:131–138.

70. Rhee J, Inoue Y, Yoon JC, Puigserver P, Fan M, Gonzalez FJ, Spiegelman BM. Regulation of hepatic fasting response by PPARgamma coactivator-1alpha (PGC-1): Requirement for hepatocyte nuclear factor 4alpha in gluconeogenesis. Proc Natl Acad Sci U S A 2003; 100:4012–4017.

71. Hayhurst GP, Lee YH, Lambert G, Ward JM, Gonzalez FJ. Hepatocyte nuclear factor 4alpha (nuclear receptor 2A1) is essential for maintenance of hepatic gene expression and lipid homeostasis. Mol Cell Biol 2001; 21:1393–1403.

72. Hertz R, Magenheim J, Berman I, Bar-Tana J. Fatty acyl-CoA thioesters are ligands of hepatic nuclear factor-4 alpha. Nature 1998; 392:512–516.

73. Dhe-Paganon S, Duda K, Iwamoto M, Chi YI, Shoelson SE. Crystal structure of the HNF4 alpha ligand binding domain in complex with endogenous fatty acid ligand. J Biol Chem 2002; 277:37973–37976.

74. Hall RK, Yamasaki T, Kucera T, Waltner-Law M, O'Brien R, Granner DK. Regulation of phosphoenolpyruvate carboxykinase and insulin-like growth factor-binding protein-1 gene expression by insulin. The role of winged helix/forkhead proteins. J Biol Chem 2000; 275:30169–30175.

75. Schmoll D, Walker KS, Alessi DR, Grempler R, Burchell A, Guo S, Walther R, Unterman TG. Regulation of glucose-6-phosphatase gene expression by protein kinase Balpha and the forkhead transcription factor FKHR. Evidence for insulin response unit-dependent and -independent effects of insulin on promoter activity. J Biol Chem 2000; 275:36324–36333.

76. Nakae J, Park BC, Accili D. Insulin stimulates phosphorylation of the forkhead transcription factor FKHR on serine 253 through a Wortmannin-sensitive pathway. J Biol Chem 1999; 274:15982–15985.

77. Yeagley D, Guo S, Unterman T, Quinn PG. Gene- and activation-specific mechanisms for insulin inhibition of basal and glucocorticoid-induced insulin-like growth factor binding protein-1 and phosphoenolpyruvate carboxykinase transcription. Roles of forkhead and insulin response sequences. J Biol Chem 2001; 276:33705–33710.

78. Puigserver P, Rhee J, Donovan J, Walkey CJ, Yoon JC, Oriente F, Kitamura Y, Altomonte J, Dong H, Accili D, et al. Insulin-regulated hepatic gluconeogenesis through FOXO1-PGC-1alpha interaction. Nature 2003; 423:550–555.

79. Herzig S, Long F, Jhala US, Hedrick S, Quinn R, Bauer A, Rudolph D, Schutz G, Yoon C, Puigserver P, et al. CREB regulates hepatic gluconeogenesis through the coactivator PGC-1. Nature 2001; 413:179–183.

80. Elliott SS, Keim NL, Stern JS, Teff K, Havel PJ. Fructose, weight gain, and the insulin resistance syndrome. Am J Clin Nutr 2002; 76:911–922.
81. Bonthron DT, Brady N, Donaldson IA, Steinmann B. Molecular basis of essential fructosuria: molecular cloning and mutational analysis of human ketohexokinase (fructokinase). Hum Mol Genet 1994; 3:1627–1631.
82. Park OJ, Cesar D, Faix D, Wu K, Shackleton CH, Hellerstein MK. Mechanisms of fructose-induced hypertriglyceridaemia in the rat. Activation of hepatic pyruvate dehydrogenase through inhibition of pyruvate dehydrogenase kinase. Biochem J 1992; 282:753–757.
83. Fiebig R, Griffiths MA, Gore MT, Baker DH, Oscai L, Ney DM, Ji LL. Exercise training down-regulates hepatic lipogenic enzymes in meal-fed rats: fructose versus complex-carbohydrate diets. J Nutr 1998; 128:810–817.
84. Zavaroni I, Chen YD, Reaven GM. Studies of the mechanism of fructose-induced hypertriglyceridemia in the rat. Metabolism 1982; 31:1077–1083.
85. Taghibiglou C, Rashid-Kolvear F, Van Iderstine SC, Le-Tien H, Fantus IG, Lewis GF, Adeli K. Hepatic very low density lipoprotein-ApoB overproduction is associated with attenuated hepatic insulin signaling and overexpression of protein-tyrosine phosphatase 1B in a fructose-fed hamster model of insulin resistance. J Biol Chem 2002; 277:793–803.
86. Tobey TA, Mondon CE, Zavaroni I, Reaven GM. Mechanism of insulin resistance in fructose-fed rats. Metabolism 1982; 31:608–612.
87. Korieh A, Crouzoulon G. Dietary regulation of fructose metabolism in the intestine and in the liver of the rat. Duration of the effects of a high fructose diet after the return to the standard diet. Arch Int Physiol Biochim Biophys 1991; 99:455–460.
88. Anundi I, Kauffman FC, Thurman RG. Gluconeogenesis from fructose predominates in periportal regions of the liver lobule. J Biol Chem 1987; 262:9529–9534.
89. Zavaroni I, Sander S, Scott S, Reaven GM. Effect of fructose feeding on insulin secretion and insulin action in the rat. Metabolism 1980; 29:970–973.
90. Nagai Y, Nishio Y, Nakamura T, Maegawa H, Kikkawa R, Kashiwagi A. Amelioration of high fructose-induced metabolic derangements by activation of PPARalpha. Am J Physiol Endocrinol Metab 2002; 282:E1180–1190.
91. Pagliassotti MJ, Wei Y, Bizeau ME. Glucose-6-phosphatase activity is not suppressed but the mRNA level is increased by a sucrose-enriched meal in rats. J Nutr 2003; 133:32–37.
92. Saggerson D, Ghadiminejad I, Awan M. Regulation of mitochondrial carnitine palmitoyl transferases from liver and extrahepatic tissues. Adv Enzyme Regul 1992; 32:285–306.
93. Park EA, Cook GA. Differential regulation in the heart of mitochondrial carnitine palmitoyltransferase-I muscle and liver isoforms. Mol Cell Biochem 1998; 180:27–32.
94. Lowe DM, Tubbs PK. Succinylation and inactivation of 3-hydroxy-3-methylglutaryl-CoA synthase by succinyl-CoA and its possible relevance to the control of ketogenesis. Biochem J 1985; 232:37–42.

95. Hegardt FG. Mitochondrial 3-hydroxy-3-methylglutaryl-CoA synthase: a control enzyme in ketogenesis. Biochem J 1999; 338:569–582.

96. Forman BM, Chen J, Evans RM. Hypolipidemic drugs, polyunsaturated fatty acids, and eicosanoids are ligands for peroxisome proliferator-activated receptors alpha and delta. Proc Natl Acad Sci USA 1997; 94: 4312–4317.

97. Kliewer SA, Sundseth SS, Jones SA, Brown PJ, Wisely GB, Koble CS, Devchand P, Wahli W, Wilson TM, Lenhard JM, et al. Fatty acids and eicosanoids regulate gene expression through direct interaction with peroxisome proliferator-activated receptors alpha and gamma. Proc Natl Acad Sci USA 1997; 94:4318–4323.

98. Kersten S, Seydoux J, Peters JM, Gonzalez FJ, Desvergne B, Wahli W. Peroxisome proliferator-activated receptor alpha mediates the adaptive response to fasting. J Clin Invest 1999; 103:1489–1498.

99. Leone TC, Weinheimer CJ, Kelly DP. A critical role for the peroxisome proliferator-activated receptor alpha (PPARalpha) in the cellular fasting response: the PPARalpha-null mouse as a model of fatty acid oxidation disorders. Proc Natl Acad Sci U S A 1999; 96:7473–7478.

100. Hashimoto T, Cook WS, Qi C, Yeldandi AV, Reddy JK, Rao MS. Defect in peroxisome proliferator-activated receptor alpha-inducible fatty acid oxidation determines the severity of hepatic steatosis in response to fasting. J Biol Chem 2000; 275:28918–28928.

101. Kersten S, Mandard S, Escher P, Gonzalez FJ, Tafuri S, Desvergne B, Wahli W. The peroxisome proliferator-activated receptor alpha regulates amino acid metabolism. FASEB J 2001; 15:1971–1978.

102. Lin Q, Ruuska SE, Shaw NS, Dong D, Noy N. Ligand selectivity of the peroxisome proliferator-activated receptor alpha. Biochemistry 1999; 38:185–190.

103. Miller CW, Ntambi JM. Peroxisome proliferators induce mouse liver stearoyl-CoA desaturase 1 gene expression. Proc Natl Acad Sci USA 1996; 93:9443–9448.

104. He WS, Nara TY, Nakamura MT. Delayed induction of delta-6 and delta-5 desaturases by a peroxisome proliferator. Biochem Biophys Res Commun 2002; 299:832–838.

105. Tang C, Cho HP, Nakamura MT, Clarke SD. Regulation of human delta-6 desaturase gene transcription: identification of a functional direct repeat-1 element. J Lipid Res 2003; 44:686–695.

106. Halminski MA, Marsh JB, Harrison EH. Differential effects of fish oil, safflower oil and palm oil on fatty acid oxidation and glycerolipid synthesis in rat liver. J Nutr 1991; 121:1554–1561.

107. Baillie RA, Takada R, Nakamura M, Clarke SD. Coordinate induction of peroxisomal acyl-CoA oxidase and UCP-3 by dietary fish oil: a mechanism for decreased body fat deposition. Prostaglandins Leukot Essent Fatty Acids 1999; 60:351–356.

108. Ide T, Kobayashi H, Ashakumary L, Rouyer IA, Takahashi Y, Aoyama T, Hashimoto T, Mizugaki M. Comparative effects of perilla and fish oils on the

activity and gene expression of fatty acid oxidation enzymes in rat liver. Biochim Biophys Acta 2000; 1485:23–35.

109. Dallongeville J, Bauge E, Tailleux A, Peters JM, Gonzalez FJ, Fruchart JC, Staels B. Peroxisome proliferator-activated receptor alpha is not rate-limiting for the lipoprotein-lowering action of fish oil. J Biol Chem 2001; 276:4634–4639.

110. Kabir Y, Ide T. Activity of hepatic fatty acid oxidation enzymes in rats fed alpha- linolenic acid. Biochim Biophys Acta 1996; 1304:105–119.

111. Takada R, Saitoh M, Mori T. Dietary γ-linolenic acid-enriched oil reduces body fat content and induces liver enzyme activities relating to fatty acid β-oxidation in rats. J Nutr 1994; 124:469–474.

112. Nakamura MT, Tang AB, Villanueva J, Halsted CH, Phinney SD. The body composition and lipid metabolic effects of long-term ethanol feeding during a high omega 6 polyunsaturated fatty acid diet in micropigs. Metabolism 1993; 42:1340–1350.

113. Nakamura MT, Phinney SD, Tang AB, Oberbauer AM, German JB, Murray JD. Increased hepatic delta 6-desaturase activity with growth hormone expression in the MG101 transgenic mouse. Lipids 1996; 31:139–143.

114. Zhou L, Nilsson A. Sources of eicosanoid precursor fatty acid pools in tissues. J Lipid Res 2001; 42:1521–1542.

115. Fisher EA, Pan M, Chen X, Wu X, Wang H, Jamil H, Sparks JD, Williams KJ. The triple threat to nascent apolipoprotein B. Evidence for multiple, distinct degradative pathways. J Biol Chem 2001; 276:27855–27863.

116. Murata M, Kaji H, Iida K, Okimura Y, Chihara K. Dual action of eicosapentaenoic acid in hepatoma cells: up-regulation of metabolic action of insulin and inhibition of cell proliferation. J Biol Chem 2001; 276:31422–31428.

117. Magana MM, Lin SS, Dooley KA, Osborne TF. Sterol regulation of acetyl coenzyme A carboxylase promoter requires two interdependent binding sites for sterol regulatory element binding proteins. J Lipid Res 1997; 38:1630–1638.

118. Moon YA, Shah NA, Mohapatra S, Warrington JA, Horton JD. Identification of a mammalian long chain fatty acyl elongase regulated by sterol regulatory element-binding proteins. J Biol Chem 2001; 276:45358–45366.

119. Matsuzaka T, Shimano H, Yahagi N, Amemiya-Kudo M, Yoshikawa T, Hasty AH, Tamura Y, J-i Osuga, Okazaki H, Iizuka Y, et al. Dual regulation of mouse {Delta}5- and {Delta}6-desaturase gene expression by SREBP-1 and PPAR{alpha}. J Lipid Res 2002; 43:107–114.

120. Luong A, Hannah VC, Brown MS, Goldstein JL. Molecular characterization of human acetyl-CoA synthetase, an enzyme regulated by sterol regulatory element-binding proteins. J Biol Chem 2000; 275:26458–26466.

121. Cherkaoui-Malki M, Meyer K, Cao WQ, Latruffe N, Yeldandi AV, Rao MS, Bradfield CA, Reddy JK. Identification of novel peroxisome proliferator-activated receptor alpha (PPARalpha) target genes in mouse liver using cDNA microarray analysis. Gene Expr 2001; 9:291–304.

122. Tugwood JD, Issemann I, Anderson RG, Bundell KR, McPheat WL, Green S. The mouse peroxisome proliferator activated receptor recognizes a response

element in the 5′ flanking sequence of the rat acyl CoA oxidase gene. EMBO J 1992; 11:433–439.

123. Marcus SL, Miyata KS, Zhang B, Subramani S, Rachubinski RA, Capone JP. Diverse peroxisome proliferator-activated receptors bind to the peroxisome proliferator-responsive elements of the rat hydratase/dehydrogenase and fatty acyl-CoA oxidase genes but differentially induce expression. Proc Natl Acad Sci U S A 1993; 90:5723–5727.

124. Kroetz DL, Yook P, Costet P, Bianchi P, Pineau T. Peroxisome proliferator-activated receptor alpha controls the hepatic CYP4A induction adaptive response to starvation and diabetes. J Biol Chem 1998; 273:31581–31589.

125. Aldridge TC, Tugwood JD, Green S. Identification and characterization of DNA elements implicated in the regulation of CYP4A1 transcription. Biochem J 1995; 306:473–479.

126. Le May C, Pineau T, Bigot K, Kohl C, Girard J, Pegorier JP. Reduced hepatic fatty acid oxidation in fasting PPARalpha null mice is due to impaired mitochondrial hydroxymethylglutaryl-CoA synthase gene expression. FEBS Lett 2000; 475:163–166.

127. Rodriguez JC, Gil-Gomez G, Hegardt FG, Haro D. Peroxisome proliferator-activated receptor mediates induction of the mitochondrial 3-hydroxy-3-methylglutaryl-CoA synthase gene by fatty acids. J Biol Chem 1994; 269:18767–18772.

128. Motojima K, Passilly P, Peters JM, Gonzalez FJ, Latruffe N. Expression of putative fatty acid transporter genes are regulated by peroxisome proliferator-activated receptor alpha and gamma activators in a tissue- and inducer-specific manner. J Biol Chem 1998; 273:16710–16714.

129. Sato O, Kuriki C, Fukui Y, Motojima K. Dual promoter structure of mouse and human fatty acid translocase/CD36 genes and unique transcriptional activation by peroxisome proliferator- activated receptor alpha and gamma ligands. J Biol Chem 2002; 277:15703–15711.

130. Thiede MA, Strittmatter P. The induction and characterization of rat liver stearyl-CoA desaturase mRNA. J Biol Chem 1985; 260:14459–14463.

131. Wu P, Sato J, Zhao Y, Jaskiewicz J, Popov KM, Harris RA. Starvation and diabetes increase the amount of pyruvate dehydrogenase kinase isoenzyme 4 in rat heart. Biochem J 1998; 329:197–201.

7

Design and Analysis of Microarray Studies for Obesity Research

**Marie-Pierre St-Onge and
Steven B. Heymsfield**

New York Obesity Research Center, Columbia University College of
Physicians & Surgeons, New York, New York, U.S.A.

**Grier P. Page, Kui Zhang,
Kyoungmi Kim, and David B. Allison**

Section on Statistical Genetics, Department of Biostatistics,
University of Alabama at Birmingham, Birmingham, Alabama, U.S.A.

Maria DeLuca

Department of Environmental Health Sciences,
University of Alabama at Birmingham, Birmingham, Alabama, U.S.A.

1. INTRODUCTION

Over the last 5 years, microarrays for measuring mRNA expression have emerged as one of the most exciting new technologies available to help us understand basic biological underpinnings of complex biological phenomena. Microarrays offer the opportunity of measuring the transit level of thousands of genes simultaneously. This opens up new levels of questions that can be addressed at a more genomic than simply genetic level, that is, we can begin asking questions about coordinated gene activity, gene pathways, multivariate gene expression predictors of response or outcome, and the development of complex multivariate biomasses. Just as microarrays offer many opportunities, they also offer many challenges. Many questions remain unanswered about the measurement quality of existing microarray methods and how to maximize that measurement of quality. Statistically rigorous approaches to experimental design, data analysis, and interpretation are just now being articulated. A plethora of papers have appeared in the literature describing a wide variety of methods. It is not clear that all of the new methods offered are equally valid, and many may not be valid at all. Massive multiple testing brings both philosophical and statistical hurdles and at the same time offers novel statistical opportunities in the area of inference and estimation and inference. Perhaps one of the best outcomes of the recent focus on microarray research is that in trying to adapt new methodologies to these complex data structures, biologists, statisticians, and computer scientists are talking with each other more than ever in trying to develop methodologic approaches that build on the prior knowledge that biologists bring to the table. Herein, we try to describe some of these advances, offer caveats about the use of microarray research and several commonly utilized methods, and offer guidelines on some methods that may be especially useful to both. Further research in this area is needed and we encourage investigators to join in helping to build understanding and bring clarity to this new and complex area of research.

1.1. What is a Microarray?

There are several classes of microarrays (1) utilized for such purposes as genotyping (2), proteomic analysis (3), and quantification of the amount of a given type of mRNA in a biological specimen (4). For the remainder of this chapter, when we use the term microarray, we refer only to this last class of microarrays.

A microarray consists of a flat surface on which one can place a substance derived from a biological specimen, subsequently process the microarray through an appropriate reader, and thereby estimate the quantity of mRNA present for multiple genes or expressed sequence tags in the original biological specimen from which the substance was extracted. There are multiple types of microarrays as will be briefly reviewed later in this chapter. The critical point that makes microarrays an interesting advance is that they can be used to measure the amount of mRNA in a specimen for thousands of genes simultaneously. This advance not only allows traditional questions about the expression levels of many genes, each considered in isolation, to be addressed with markedly increased efficiency, but also allows new genomic-level questions to be addressed in which the expression levels of multiple genes are considered simultaneously to understand gene networks, identify gene families, develop multivariate gene expression profiles that characterize organisms belonging to specific classes or to predict what class an organism belongs to, and address other such questions that bring us squarely into the realm of modern high dimensional biology (5).

The power of microarrays has not been lost on the biomedical research community. Figure 1 displays the number of citations in the *Web of Science* (the Institute for Scientific Information's *Science Citation Index*, *Social Science Citation Index*, and *Arts & Humanities Citation Index* combined) database that include the term microarrays or microarray and the number that include either term plus one of several obesity-oriented terms (any of the following: obese, obesity, adipose, adiposity, fat, adipocyte, adipocytes). As can be seen, since 1995, both have increased in a rapidly accelerating and roughly exponential and parallel fashion. Microarrays have taken the scientific community by storm, but their full potential has, in our opinion, yet to be realized. This appears to be due to several factors, including (but not limited to) the relatively high cost of some types of arrays and the need for more training in the conduct of array research, particularly with respect to design and analysis. We hope that this chapter can help to address the latter issue.

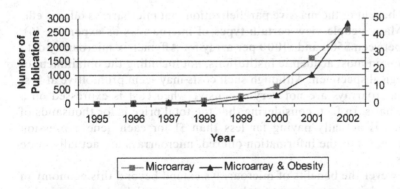

Figure 1 Growth of microarray research in Web of Science database. The line demarcated with triangles indicates the number of citations in Web of Science in each year containing the word microarray or microarrays. The line demarcated with squares indicates the number among these citations that include phrases that imply the paper is about statistical methods.

2. WHY DO MICROARRAY STUDIES IN OBESITY RESEARCH?

2.1. Gene Expression in General

At one level, microarrays are just another method of measuring gene expression levels. The study of gene expression has many purposes. By detecting genes expressed during key phases of development, we may be able to discern the underlying molecular and physiological mechanisms that control development. By detecting genes that have their expression up or down regulated by toxins, nutrients, and drugs, we may increase our understanding of the pathways through which the substances affect organisms. By detecting genes that are differentially expressed with body composition changes during aging, we may begin to understand the underlying process of these age-related changes at a fundamental level. By detecting genes that are differentially expressed in experimental mutant models, we may discover new candidate genes for traits as diverse as muscle growth, intramuscular adipose tissue, adipogenesis, appetite regulation and any other traits affected by these genetic mutations. By knowing in which tissues and to what degree a gene is expressed, we can increase our understanding of the anatomic structures and pathways involved in the process in question. To some extent, all of these questions can be addressed one gene at a time and, therefore, do not require microarrays. However, by allowing the measurement of many genes simultaneously, microarrays offer many additional benefits.

2.2. Benefits of Massive Parallelization

The first benefit of the massive parallelization that microarrays offer is efficiency. Many people view certain types of microarrays as expensive (6), costing between $500 and $1000 per array for Affymetrix microarray data acquisition in most academic institutions, not including the obtainment of the biological specimen. Although such costs may seem prohibitive at first glance, microarrays are not very expensive when cost is expressed on a per gene basis. In fact, considering that the total price is for thousands of genes, one is actually paying far less than $1 for each gene expression measurement. For the information offered, microarrays are actually quite inexpensive.

However the benefits of microarrays extend beyond this economy of scale. By studying many genes together, one can potentially detect genes that are coexpressed (i.e., genes whose expression levels are correlated). In turn, by detecting genetic sequences that are coexpressed with other genes as part of a functional family, we may be able to assign functions to previously unknown genes or genes of unknown function. By offering us such highly multivariate data, microarrays potentially allow us to uncover gene networks (7,8), predict outcomes (e.g., Ref. 9), classify objects (10), and develop indices or biomarkers against which to judge the effects of various interventions (11).

3. WHAT IS THE CURRENT STATE OF ART IN THE OBESITY FIELD?

Before proceeding further, we offer an overview of the use of microarrays to date as indicated in the published literature. This offers some perspective on where the field is and where it can benefit from expansion or modification.

3.1. What Questions Are Being Addressed?

Extant studies applying microarray methodology to the study of obesity and obesity-related metabolic abnormalities have examined the role of dietary fat intake (12–14), food restriction (15–17), and soy protein intake (18) on gene expression. There is much ongoing research examining differences in gene expression between leptin-deficient and wild-type animals (15,16,19–21), lean versus obese animal models (22–24), and in insulin-resistant states (25,26). Furthermore, the identification of genes that are differentially expressed in various adipose tissue compartments is being extensively explored (27–29). In addition, the expression of genes involved in preadipocyte differentiation has been the topic of some research (30–32).

Other goals of microarray studies, in the context of obesity-related research, have been to identify genes involved in pre-eclampsia (33) and in muscle preferential energy source during space flight (34), as well as those genes activated by sterol regulatory element binding proteins (SREBP) (35). There also seems to be much interest in investigating the genes associated with adipocyte differentiation and maintenance of adiposity as those genes could be involved in the development of metabolic disorders associated with obesity. There are data to suggest that some such genes are mutated or aberrant in animal models of obesity and metabolic disorders (36). The characterization of those genes may ultimately make possible the identification of new therapeutic targets for the treatment of obesity and other metabolic disorders (30).

In fact, most studies seem to conclude, from the results of their microarray analyses, that further studies of the identified genes and their biochemical pathways are necessary to increase our understanding of the mechanisms involved in the various hypotheses tested (17,19,26,27,30,31).

In addition, several authors suggest that the results obtained through their research provide new genetic targets for the treatment of obesity or related disorders (17,19,21,24,30,35) or that results suggest the implication of certain genes in depot- and gender-related differences in the metabolic complications of obesity (28). Conclusions from several studies also propose that microarray analyses provide tools for searching for novel genes related to obesity and insulin resistance (29) and are useful to gain more insight into organ-specific gene expression in complex diseases (37).

3.2. What Species Are Being Used?

Studies to date have been mostly done using animal models with fewer studies using human subjects (26–29,33,37,38). The most frequently employed species have been mice and rats, either lean or obese, with and without genetic mutations. The ob/ob mouse is the most studied animal, being involved in studies of leptin research and of genetic obesity (15,16,20,21,35). Other transgenic animals have included bombesin receptor substrate-deficient mice for the study of body weight and energy metabolism in adipose tissue (22), angiotensinogen knock-out mice to study angiotensinogen expression in relation to adiposity (14), insulin receptor substrate 1 and 2 knockout mice to determine the regulation of sterol SREBP-1c expression (25), and SREBP-1a transgenic mice to investigate the expression profile and nutritional regulation of acetyl-CoA synthetase cDNA and gene promoter (17). Lean and obese Zucker rats have also been used to measure hypothalamic mRNA signals for melanin concentrating hormone and its receptor (24), while obese Zucker rats were studied to examine the role of soy protein on liver mRNA expression (18). The only obesity-related study found that did not use a rodent animal model has used lean and fat chickens to analyze genes that are differentially expressed in the livers of these animals and how they may play a regulatory role in adiposity (23).

Animals have ranged in age from 4 weeks (16) to 8 months (22) but most studies have used animals ranging from 6–10 weeks (17,20–22,32). However, not all studies report the age of the animals examined (12,13,15,23,24,34), and animal age is not always obvious from the description of the methodologies in the various reports.

Other studies have used cell lines to examine gene expression (19,23,30–32). Mouse 3T3–L1 cells have those that are the most frequently used in microarray studies of obesity (30–32), but other cell lines have included chicken hepatoma LMH cells (23), GT1-7 mouse hypothalamic neuronal cells (19), mouse 3T3-C2 fibroblasts (31) and human embryonic kidney 293T cells (32).

3.3. What Sample Sizes Are Being Used?

Sample size varies with different studies and has not been reported in all publications (15,17,23,29,35). In animal experiments, analyses have been made on samples ranging from three (12,16,20,25) to 20 (24) animals per group. However, in most studies, sample sizes have been of four–six animals/group (14,18,22,27,32,34).

Sample size is also small in human studies. Of the studies reviewed here, one has included five patients and four controls (37), while others examined six patients/group (33), 10 (38) or 15 subjects (28), and 17 and 18 subjects of varying background (26).

3.4. What Tissues Are Being Studied?

Table 1 shows a list of the different tissues that have been analyzed for gene expression in studies linked to obesity research. The liver is the most frequently examined organ (close to 50% of studies) followed by epididymal fat pads (approximately 25% of studies). Peri-epididymal fat pads and intrascapular brown adipose tissue are other adipose tissue depots that have been analyzed, along with subcutaneous and visceral adipose tissue in human studies. Skeletal muscles have also been studied but somewhat less frequently. Kidney, brain, hypothalamus specifically, and placenta are

TABLE 1   Tissues Extracted and Examined in Microarray Studies of Obesity

Tissue	Reference
Abdominal subcutaneous adipose tissue	28,35,37
Abdominal visceral adipose tissue	29,35,38
Brain	22
Epididymal fat pad	12–14,21
Hypothalamus	24
Interscapular brown adipose tissue	22
Kidney	14
Liver	14,16–18,20,21,23,25,35
Peri-epididymal fat pad	22
Placenta	33
Skeletal muscle	21
Soleus muscle	34
Soleus muscle	34
Subcutaneous brown adipose tissue	27
Subcutaneous white adipose tissue	27
Vastus lateralis muscle	26

other organs that have been extracted and whose gene expressions have
been studied.

3.5. What Platforms (Types of Arrays) Are Being Used?

Most studies reviewed here have used Affymetrix microarrays and software
for analyses. mRNA are usually hybridized to the species-specific Affymetrix GeneChip microarray. The most often used genome array is the murine
U74v2 for studies involving a mouse model or the U34A rat genome for
studies employing a rat model. Other platforms have included Hy Bond N^+
nylon membranes and polylysine-coated glass slides. Arrays are then
scanned using a confocal scanner, usually Affymetrix, when an Affymetrix
platform is utilized, and appropriate analysis software. Table 2 shows a list
of the different types of microarrays used in the studies examined in this
chapter.

   Also, a large proportion of the studies examined have combined the
microarray methodology with Northern blot analyses (12,13,16,17,20,23,
25,30,32,35) and, less frequently, with Western blotting (19,21) and immuno-
blotting (32). In addition, many studies have also included PCR analyses,
either reverse transcriptase (12,19,22,29,38), real-time (21,24,26), or real-
time reverse transcriptase PCR (28,33). Reverse-transcriptase PCR has a
higher dynamic range than microarrays and is often used to validate
observed trends with chip experiments (33).

3.6. What Data Analytic Strategies Are Being Employed?

The statistical methods used to determine differential gene expression varied
widely between the experiments reviewed here. Furthermore, there was
great variability in the extent to which statistical tools and tests were utilized. In addition, several reports did not describe any analytic strategy to
statistically manipulate the data obtained (19,21–23,35,37).

TABLE 2 Types of Microarray Assays Used in Obseity Research

Type of array	Reference
Affymetrix array	13–15,19,21,25,29,30,32–34
Dye terminator sequencing	17,18
ExpressHyb solution	37
Hybond N^+ nylon membranes	12,29,35,38
Polysine-coated plates	Ferrante et al., 2002, (16)

Most studies report fold-change (FC)* (12–14,17,25,30,32,33,37), but few have applied statistical tests to these results. Of those studies which have conducted statistical analyses, one has done pair-wise comparisons of mRNA levels between groups (14), while another used analysis of variance on log-transformed normalized intensities (32). In addition, the criterion used to assess whether genes were differentially expressed between groups differed between studies, with no apparent consensus. Fold changes of ≥ 2.0 (14,37), ≥ 3.0 (33), and >5 or <-5 (30) have been reported. A few studies have reported K-means (15,20) and hierarchical means (20). As has been noted elsewhere, declaring genes to be differentially expressed or not differentially expressed solely on the basis of whether they exceed some FC cutoff without incorporating an assessment of variability has no theoretically sound basis and should be abandoned (26,39,40).

Many authors also normalize their data. Ferrante et al. (16) reported mean normalized expression ratios and used 30% greater increase or decrease in signal intensity to establish differential expression. Ramis et al. (28) normalized their data for the amount of spotted PCR product, local hybridization conditions within a slide, and hybridization conditions between slides. They also used a ratio of ≥ 2.0 in normalized fluorescence intensity to determine differential expression and applied Student's paired t-test statistic to their data. In the study by Boeuf et al. (27), data with signal intensity <2.0 times the background level were excluded and the mean of the Cy5/Cy3 ratio calculated for each remaining clone was normalized with the median Cy5/Cy3 ratio of all spots. Data obtained by Iqbal et al. (18) were normalized with reference to the intensity of β-actin gene after subtraction of the background intensity. Orthogonal contrasts were then used to compare and contrast changes between treatments. Finally, Yang et al. (38), in 2003, normalized their data among arrays based on the sum of background-subtracted signals from all genes on the membrane. Signal to noise ratios of ≥ 2.0 were considered as positive signals and Chi-squared tests with one degree of freedom were computed.

Other statistical tests have been employed by different groups. These statistical methods include cluster analysis (20), Wilcoxon Rank Sum Test (26), and Fisher's t-test (24).

---

*Fold-change (FC) is a measure of effect size. It is not defined perfectly consistently across studies, but is generally taken to be: $FC \equiv [I(\bar{X}_1 > \bar{X}_2) - I(\bar{X}_1 \leq \bar{X}_2)]^{\frac{\max(\bar{X}_1, \bar{X}_2) - \min(\bar{X}_1, \bar{X}_2)}{\min(\bar{X}_1, \bar{X}_2)}}$, where $I(\bullet)$ is the indicator function, $\bar{X}_1$ is the mean of group 1 and $\bar{X}_2$ is the mean of group 2.

4. ASSOCIATION OF GENOME-WIDE TRANSCRIPTIONAL PROFILES WITH QUANTITATIVE TRAIT LOCI

Recent studies have attested the power of using microarray analysis in combination with quantitative trait loci (QTL) mapping to identify candidate loci affecting variation in obesity-related phenotypes (41). The latter approach allows identification of the chromosomal regions containing loci responsible for producing variation in quantitative traits. In general, QTL mapping is a statistical analysis of the association between a complex phenotype and the occurrence of specific marker alleles in the genome of selected individuals. If a QTL is linked to or at the marker locus, there will be a statistical association between the marker genotypes at that locus and the mean values of the trait. Over the past few years, QTL mapping has been used to identify genomic regions associated with obesity-related phenotypes in humans (42,43) and model animals (reviewed in Ref. 44). However, the genomic regions identified are usually very broad and encompass anywhere from tens to hundreds of genes. On the other hand, carrying out genome-wide genetic analyses of gene-expression data in the same experimental population and treating the expression level of each gene as a quantitative trait will give the highest probability of successfully dissecting the genetic basis of the trait. The approach consists of using standard QTL mapping techniques to identify those genetic regions that can account for variation in the levels of gene expression. Then, if the differentially expressed genes are located within a QTL, more confidence can be assigned to the possibility that a differentially expressed gene is a predisposing factor for the expression of the selected phenotype. Schadt and colleagues have successfully used microarray analysis coupled with QTL mapping to identify candidate loci affecting variation in fat pad mass (FPM) in mice (41). They utilized a microarray chip representing 23,574 mouse genes to analyze the expression levels of the genes in the liver tissues from 111 F2 mice, and found that 7861 are significantly differentially expressed in the two parental strains. Measuring, as quantitative traits, the expression of each of the 7861 genes in the population, they were able to identify two patterns of expression that characterize two distinct groups, high FPM and low FPM. Interestingly, the high FPM subtype was clearly subdivided in two further expression patterns. This points out the problem of heterogeneity in the phenotype that affects the power of detecting associations to the causative loci using the population as a whole in QTL mapping studies. Indeed, after classifying, on the basis of clinical and gene expression data, each of the 111 F2 animals into distinct phenotypes, genetic analysis of these subtypes allowed Schadt et al. (41) to identify chromosome 2 and 19 QTL regions that affect one subgroup but not the other. The final step in the study was the identification of positional candidate genes simplified, as mentioned above, by the correlation between DNA variations and gene expression levels. Two main positional candidate genes encoding for a dolichyl-diphospho-oligosaccharide-protein glycosyltransferase and a cation-transporting ATPase were identified at the chromosome 2 locus. Notably, both genes have human orthologues mapping on the chromosome 20q12–q13.12 region, previously linked to human obesity-related phenotypes (45,46).

4.1. Summary of State of the Art

Given the above, microarray-based research in the obesity field can be characterized as being at a very early and perhaps somewhat primitive state of development. This is not markedly different from the situation for most other applied fields in their early phase of implementation. In the overwhelming majority of studies, sample sizes are almost certainly far too small to yield studies with sufficient power and precision. Statistical analytic techniques used are generally very unsophisticated. They do not utilize newer methods available, often do not correct for multiple testing, and rarely capitalize on the added information available by having so many genes studied at once.

5. DESIGN

Here we discuss the design phase of the microarray experiment. Perhaps the most important thing we can state is that applied investigators are strongly encouraged to contact statisticians for consultation on the conduct of microarray experiments *before* the data are collected. All too often, investigators only approach the statistician after the data are collected and then receive the frustrating news that there has been some fatal design flaw that makes it difficult or impossible to extract meaningful insights from the data. Here we address some of the design issues that investigators may consider.

5.1. Choice of Platform

One of the first questions to consider in planning a microarray experiment is which platform (type of array) to use. There are multiple choices including prefabricated, commercially available oligonucleotide arrays (Affymetrix being the most popular), custom cDNA arrays, custom spotted oligonucleotide arrays, nylon membranes, and others (for reviews, see Refs. 47 and 48). Four primary considerations are generally brought to bear when choosing among platforms: availability, applicability, cost, and measurement quality.

5.2. Availability

Availability was more of an issue in past years than it is today. Most major research universities have one or more core facilities that offer investigators access to microarray technology. Moreover, obesity researchers can benefit from the investment that the National Institute of Diabetes, Digestive, and Kidney Diseases (NIDDK) has made in national biotechnology centers. For example, at Yale University, through an NIDDK Microarray Biotechnology Center Grant, investigators, who need not be at Yale, can access array technology for many platforms and many species at moderate cost (see http://info.med.yale.edu/wmkeck/dna_arrays.htm). Thus, having access to the technology is no longer a major issue.

5.3. Applicability

Nevertheless, having access to the broad technology is not the same as having access to an applicable microarray. For species such as human, mouse, rat, yeast, and a few other commonly used model organisms, this is not an issue. However, for most agricultural species, marine species, plants, and the occasional odd model such as lemmings, prefabricated microarrays are less likely to be available. For such situations, custom-made, glass cDNA arrays (49) may be the best option, although using an array from another species seems to also yield useful information in some circumstances (e.g., Refs. 50 and 51). For example, Affymetrix human chips have been successfully used with rhesus macaques (e.g., Ref. 52).

5.4. Cost

Currently, in-house custom arrays, once the system is set up, are far less (at least an order of magnitude) expensive than are Affymetrix microarrays. Other systems tend to lie in the middle with respect to cost. Given our comments in other sections of this chapter about the strong desirability of radically increasing sample sizes, cost is an important issue. Thus, at present, cost strongly favors custom glass cDNA arrays but, as with most technologies, prices tend to fall rapidly and this gap may close in the future.

5.5. Measurement Quality

Given the enormous use of microarrays, one might think that many thorough studies of the measurement properties of each microarray technique would have been done with sufficient sample size to yield interpretable results; that those studies would have been analyzed with well-established methods in the field of measurement theory (53); and that results would be presented in a clear and interpretable manner. Unfortunately, this is not the case.

Although a number of studies have been reported that yield some information about measurement quality, these tend to be modestly sized, idiosyncratically conducted studies of one or two microarray platforms and account for only one or two sources of variability. The field of microarrays could do well to take a page from the psychometricians who, through dealing with different instruments and different potential sources of bias, have delineated careful approaches to assessing measurement quality in an organized fashion (for a dense but superlative example, see Ref. 54). Larger and better-designed studies are underway, but at present, it is difficult to find data that convincingly favor one platform over another or are thoroughly informative about any platform. Nevertheless, we review some of the available information here.

On a positive note, Nimgaonkar et al. (55) compared expression levels obtained across old and new generation Affymetrix chips. This can be seen as a form of parallel forms reliability. Nimgaonkar et al. (55) found that, as long as a high proportion of common probes were used and the degree of expression was relatively high, reliability was also high. Cheung et al. (56) investigated the heritability of gene expression levels in humans using cDNA microarrays. They found significant heritability for some gene expression levels which, ipso facto, implies some reliability of the measurements (or correlated errors across family members which seems unlikely). Moreover, in their Fig. 1, they plotted the variance between replicate measurements against the variance within replicate measurements for all genes examined. Simply eye-balling the figure indicates that, on average, the variance between is at least twice the variance within replicates, suggesting some reasonable degree of reproducibility.

In contrast, Kothapalli et al. (57) compared two different commercial microarray systems and reported:

Our analysis revealed several inconsistencies in the data obtained from the two different microarrays. Problems encountered included inconsistent sequence fidelity of the spotted microarrays, variability of differential expression, low specificity of cDNA microarray probes, discrepancy in fold-change calculation and lack of probe specificity for different isoforms of a gene.

Kuo et al. (58) compared "mRNA measurements of 2895 sequence-matched genes in 56 cell lines from the standard panel of 60 cancer cell lines from the National Cancer Institute (NCI 60)."

They reported: By calculating the correlation between matched measurements and calculating concordance between cluster from two high-throughput DNA microarray technologies, Stanford type cDNA microarrays and Affymetrix oligonucleotide microarrays, corresponding measurements from the two platforms showed poor correlation. Clusters of genes and cell lines were discordant between the two technologies, suggesting that relative intra-technology relationships were not preserved. GC-content, sequence length, average signal intensity, and an estimator of cross-hybridization were found to be associated with the degree of correlation. This suggests gene-specific or, more correctly, probe-specific factors influencing measurements differently in the two platforms, implying a poor prognosis for a broad utilization of gene expression measurements across platforms.

Many other papers include some information about measurement quality of various microarray systems (e.g., Refs. 12 and 59–64). Perhaps one of the most important points concerning measurement quality is illustrated by Jenssen et al. (65). These workers assessed the repeatability of results using replicate cDNA spots on six cDNA microarray data sets. Their results "indicate a high degree of variation in data quality, both across the data sets and between arrays within data sets." This high degree of variability in data quality across data sets implies that one should not be comforted by reports of good measurement quality in the literature nor should one necessarily be disturbed by reports of poor measurement quality in some studies. Rather, the ideal information is the reliability and validity of the microarray measurements taken in the laboratory, under the conditions of the study, and using the microarray platform used in the study. Without such information, which is lacking in the overwhelming majority of studies, some skepticism about any particular dataset is in order.

5.6. Choice of Controls

One of the key questions in designing a microarray study in obesity is the definition of the control group, if there is to be one. This obviously depends critically on the question being addressed (66). Although the choice of the appropriate control is critical, we will not belabor it here because it is not obvious that choosing appropriate controls is any different for a microarray study than any other study.

5.7. Randomization

As in any experimental design, the ability to infer causation is critically dependent on how cases (mice, people, etc.) are assigned to conditions and randomization is clearly the preferred technique (67). It is important to realize that random assignment is not equivalent to haphazard assignment, alternating assignment, or any other predictable process. If, for example, mice are to be assigned to either of two different diets before the tissue of interest is extracted for analysis, they should be assigned randomly.

5.8. Counterbalancing Technical Factors

In addition, it has been observed that factors such as technician handling the samples, the particular day on which samples were analyzed, lot of chips, dyes used in cDNA studies, and other such factors can have a major impact on measured expression levels. Therefore, it is essential to avoid confounding such factors with the independent variable(s) under study. That is, one should not use, for example, red dye on all of the control specimens and green dye on all of the experimental specimens or vice versa (68,69). Similarly, one should not analyze all of the control specimens on one day and all of the experimental specimens on another. Although this may seem trivial, in our experience, these factors have nontrivial effects. Alternatively, although the merit of counterbalancing such factors may seem obvious to some readers, in our experience, this is often overlooked in practice and adversely affects investigator's ability to draw meaningful conclusions.

5.9. Sample Size and Power

Determining sample size and power for microarray experiments is challenging because of the newness of the endeavor. Our field has only just begun to generate the methodology and shared experience that allows for sound planning. Certainly statisticians are unified on at least two points. First, *replication* is critical, by replication we mean including more than one case in each condition in a microarray study (69,70). Replication is necessary to allow measures of within condition variance to be constructed, without which, formal statistical inference is all but impossible. Again, this may seem obvious to investigators working in other areas, but this has apparently not been obvious to early microarray investigators. Second, most statisticians would strongly agree that the extremely small sample sizes commonly used in microarray studies (e.g., ≤three cases per condition) is an extreme limitation and severely impairs power and precision. Although statisticians are lining up to creatively attempt to make sense of data involving such small samples (e.g., Ref. 71), even our best statistical creativity cannot fully compensate for small sample sizes.

When advising our collaborators on sample size, the first rule of thumb is to use no less than five cases per group when dealing with designs involving two or more groups. Although there are, of course, exceptions such as when each group represents a point on a continuum (e.g., age), many groups will be used, or linear effects of the continuum will be assessed, in general, we strongly adhere to this rule for the following reason. When sample sizes are small, parametric statistical tests of the differences between the mean levels of gene expression for each of the genes will be more sensitive to distributional forms of the expression data. Deviations from assumed distributions or other assumption violations may increase type I or type II error rates with small sample sizes (72). This can be resolved by use of an appropriate nonparametric test when the differences between the mean levels of gene expression for each of the genes are tested. We often recommend the use of a bootstrap test (with pivoting) for such situations (72). If one chooses the bootstrap as a method to nonparametrically produce p-values, the maximum number of different bootstrap samples is only

$$W_{max} = \left[\frac{(2n-1)!}{n!(n-1)!} \right]^2$$

where n is the number of cases per group in a two group comparison. When $n = 2$, $W_{max} = 9$. When $n = 3$, $W_{max} = 100$. When $n = 4$, $W_{max} = 1225$. When $n = 5$, $W_{max} = 15,876$. Most statisticians believe that *at least* 1000 bootstrap resamples should be used to offer valid inference and the number should be even larger if one uses very small alpha levels that may be required for multiple testing corrections. For the above-stated reasons, we advocate that no less than five cases per group be used. Interestingly, a recent report from Pavlidis et al. (73) also recommends no less than five replicates per group based on empirical observations.

Although we encourage that *no less* than five cases per group be used, this does not mean that we support the notion that *no more* than five cases per group be used. We (5) have developed a novel approach to study design and sample size determination for microarray research that is decidedly different than the classic approach that might be taken by, for example, a clinical researcher testing a single hypothesis, the outcome of which may lead immediately to a clinical recommendation. Such scientists want to be sure that their family-wise type I error rate (FWER) across the whole study is less than alpha (e.g., 0.05). However, in our experience, basic scientists conducting microarray experiments are more interested in three questions: (1) Of the genes declared significant at a particular threshold, what is the expected proportion that these are indeed genes that are differentially expressed? (2) Of the genes not declared significant, what is the expected proportion that these are indeed genes that are not differentially expressed (we refer to these two proportions as the true positive (TP) and true negative (TN) rates, respectively)? (3) Of the genes that are truly differentially expressed, what is the expected proportion that these genes will be detected in the data at a particular threshold? This last question is akin to the notion of power; however, we explore it from a slightly different angle and thus, to avoid confusion, will refer to this quantity as the expected discovery rate (EDR). These proportions are related to the quantities indicated in Table 3 below.

The proportions that are of interest are that of a true positive

$$TP = \frac{D}{C + D}$$

that of a true negative,

$$TN = \frac{A}{A + B} \text{ and } EDR = \frac{D}{B + D}$$

and where each proportion is defined to be zero if its denominator is zero. This latter proportion, EDR, represents the probability that a gene will be detected in the data at a particular threshold given that it is a gene that is differentially expressed. Ideally, we would like each of these quantities to be equal to 1.0. That would mean that we detect everything that there is to be detected and never have a false positive. This can only be achieved (asymptotically) with an infinite sample size. With finite sample sizes, we can trade these quantities against each other and try to find a combination of a threshold for declaring results to be significant and sample size that yields expected proportions we find acceptable.

Using a mixture model (74) that we will describe in slightly greater detail below, and extended in Gadbury et al. (5), we are able to use a parametric bootstrap approach to estimate the quantities TP, TN, and EDR based upon observed pilot data. Figure 2 contains an example plot of such estimates. These values are plotted at a threshold (alpha) level of 0.01.

TABLE 3 Truth Versus Decision Table for Microarray Inference

	Genes for which there is not a real effect	Genes for which there is a real effect
Genes not declared significant at designated threshold level	A	B
Genes declared significant at designated threshold level	C	D

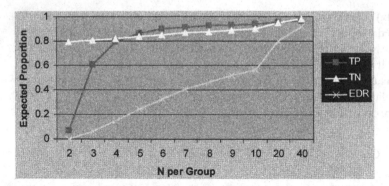

FIGURE 2   Power and sample size calculations for Dr. X (threshold is alpha = .01). Example of plot facilitating determination of desirable sample size for a planned microarray study.

Based on these data, we would estimate that with $n = 10$ per group, over 90% of those genes we declare significant will be true positives, and over 90% of those genes we declare nonsignificant will be true negatives, and we will detect approximately 60% of the genes that are truly differentially expressed. Given that we often estimate that there are literally thousands of genes differentially expressed in the experiments we analyze, this would provide a very ample list with which to conduct follow-up research. Other statisticians have proposed similar approaches (75,76). Methods for determining sample size in classification problems have now also begun to be elaborated (77,78).

5.10. Pooling

A common technique in microarray research involves pooling mRNA from multiple tissue specimens. This may be done for either of two reasons. First, sometimes the tissue of interest (e.g., hypothalamus in mice) is too small to permit extraction of enough mRNA for a full microarray analysis, thereby prompting investigators to pool multiple specimens (e.g., Ref. 79). Although pooling is one reasonable response to small tissue size, alternatives are available, such as various forms of amplification (80,81).

Second, microarrays can be expensive. Some investigators therefore ask the statistician "If you tell me that I need to have N cases in each condition, can I save money by pooling the biological samples from, for example, $\frac{1}{2}N$ cases on each of two chips in each condition?" Several investigators have addressed this issue (82–84). A critical assumption in addressing this question is that the expected value of an expression measurement obtained from a pool of N subjects on a single chip equals the expected value that will be obtained from the arithmetic mean of those same N subjects measured on N separate chips. For convenience, although it is not strictly necessary, let us assume that the gene expression scores are normally distributed (within condition for each gene), that the number of cases per group studied is constant across groups and is an even number, and that there is homogeneity of variance across groups. An investigator could conduct a nonpooled analysis comparing the two groups under study. If the cases (subjects) were independent, one could test for group differences in the expression level for each gene using Student's t-test (Eq. 1).

$$\frac{\overline{X}_1 - \overline{X}_2}{\sqrt{\frac{S_1^2}{N} + \frac{S_2^2}{N}}} \sim t_{2N-2} \tag{1}$$

If one adopted a pooling approach, one could divide the N cases from each group into two pools of $\frac{1}{2}N$ cases. Then, one could test whether the two pools from the first group differed from the two pools from the second group. The test statistics to use would also be a Student's t-test with a slight alteration as in Eq. 2.

$$\frac{\overline{X}_1 - \overline{X}_2}{\sqrt{\frac{S_1^2}{2} + \frac{S_2^2}{2}}} \sim t_2 \tag{2}$$

In Eq. 1, the variances are the variances among the individual cases, whereas in Eq. 2 the variances are the variance among chips containing mRNA from pools of $\frac{1}{2}N$ subjects. These two t-distributions have noncentrality parameters that are given by Eqs. 3 and 4.

$$\delta = \frac{\mu_1 - \mu_2}{\sqrt{\frac{2\sigma_y^2}{N} + \frac{2\sigma_e^2}{N}}} \tag{3}$$

$$\delta = \frac{\mu_1 - \mu_2}{\sqrt{\frac{2\sigma_y^2}{N} + \sigma_e^2}} \tag{4}$$

These noncentrality parameters differ only in the term on the right under the square root. Here, σ_y^2 represents the biological variability in gene expression scores and σ_e^2 represents the variability due to measurement error. Unlike Eq. 4, in Eq. 3, σ_e^2 will be reduced by a factor of $2/N$. This illustrates that the noncentrality parameters, which largely drive power, will be identical when there is no measurement error.

When there is much measurement error, the noncentrality parameter for the nonpooled design will be much greater and is therefore likely to offer greater power. In contrast, when measurement error is 0, barring the differences in the degrees of freedom, virtually identical power can be obtained with the pooled as with the nonpooled design at much lower cost because fewer chips will be needed. This indicates that the careful assessment of the degree of measurement error will be critical to helping us design better studies in the future, and ascertaining whether, and under what circumstances, pooling is an effective strategy for microarray research. This discussion highlights the importance of investigators obtaining sound information about the measurement reliability of their microarray systems.

6. CDNA SPECIFIC DESIGNS

Multichannel Specific Design Issues and Opportunities

When using Affymetrix GeneChips a single sample or pool of samples is hybridized to each chip that does not provide many options for complicated experimental design; however, spotted cDNA and oligo chips allow for two or more samples, each labeled with a different fluorochrome, to be hybridized to a single chip. This adds complexity and opportunities to the design of microarray experiments using this technology.

There are three general designs for a spotted microarray experiment: (1) reference design, (2) balanced block, and (3) loop design. Over all of these is the need for dye swapping. The varieties of fluorochromes incorporate into RNA and DNA at different rates due to a variety of biochemical factors. While normalization procedures can remove average dye biases but cannot remove gene specific differences in dye incorporation, each sample or set of samples should be labeled with each dye and run on different chips. This is called reverse labeling or dye swapping (1,2).

The common reference design uses an aliquot of a single sample (the reference) on all chips in the experiment (Fig. 3 illustrates the design), with an experimental sample being the other sample of the chip. This will allow for all the samples in the experiment to be compared spot for spot to a single reference sample and thus all the experimental conditions readily compared. A single reference sample can be used across many different experiments so that the experiments can be directly compared. The reference design is also relatively straight-forward to implement, but requires more chips than other designs: $2n$ chips for n samples with dye swap. The dye swap may or may not be used in a reference design for any dye bias not removed by normalization will affect all samples equally.

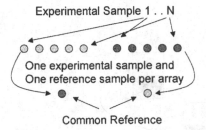

FIGURE 3 Common reference with dye swap.

The balanced block design (Fig. 4) has an advantage over the reference design since half of all the chips are not wasted on a reference sample. In the balanced block design each array is hybridized with a sample from each experimental group with half of each group labeled with one dye and half with the other dye. The analysis for this experiment is based on ANOVA. The balanced block design is very efficient since n samples can be tested on n chips as opposed to $2n$ in a reference design. There are also some limitations with this type of design, including difficulty to account for variation in the size and shape of the spots. In addition, it is difficult to implement an experiment with multiple classes, thus it is most efficient for two class comparisons. It is also inefficient if there is large inter-sample variability and when samples rather than arrays are the limiting factor (68,85,86).

The loop design (Fig. 5) has received much attention lately (87–89). The loop design has several advantages over the balanced block design. In the loop design, each sample is labeled with each dye. In its simplest form, the green (cy3) of sample 1 is hybridized with the red (cy5) of sample 2. The green of sample 2 is hybridized with red of sample 3, etc. until the green of

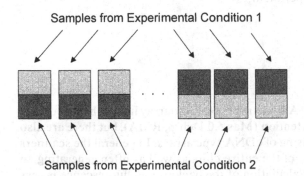

FIGURE 4 Balanced block design with dye swap.

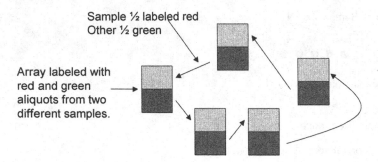

Sample ½ labeled red
Other ½ green

Array labeled with
red and green
aliquots from two
different samples.

FIGURE 5   Loop deisgn with dye swap.

sample *n* is hybridized with the red of sample 1. This allows *n* samples to be assayed on *n* arrays with dye swap. This design is very sample efficient and allows for all experimental groups to be compared, but there can be quite a bit of error in the comparisons if the loop is large, especially if there is large spot to spot variability (89). In addition, if a chip is lost, the loop is broken and much of the information is lost. Advanced multiple-cyclic loops have been developed to reduce some of the issues while only requiring a few more than *n* chips for *n* samples. The statistical analysis of these types of experiments can be quite complicated. Loops are more efficient than a reference design, but they can be very complicated to implement in the lab, and the incorrect hybridization of a single sample can destroy an experiment. Thus care must be taken when conducting a loop design.

The choice of the design to be used in a spotted array experiment will have a large impact upon the results and interpretability of a microarray experiment. Investigators must be aware of the strengths and limitations as well as of the complexities of the analysis of some of these designs before initiating a microarray experiment.

7.   ANALYSIS

7.1   Image Processing

7.1.1.   Image Processing

Issues in the processing of Affymetrix type microarray images have received an extensive amount of attention (Mas 5.0 Dchip, RMA), but there are also many issues in the processing of cDNA type arrays. In general the scanners all generate a TIFF image of the entire chip, as well as often generating an image for each channel. Calculation of the background fluorescence is very important, but a variety of different methods have been developed including the calculation of local background using a variety of techniques as well as global background intensities. Spot finding is a very difficult issue and ideally all spotted cDNA will be of the exact same size and shape as each other and in the exactly correct position, but this is not what happens in reality. Likely, the tips of some of the spotting needles may be sharper than others, the springs may be slightly less springy, etc. making the spots slightly nonhomogeneous. Manual adjustment is usually still required, which can be difficult for 40,000 spots.

Segmentation involves deciding on the size and shape of spots as well as identifying the foreground and background spots. Some methods allow for fixed-diameter round spots while others are more flexible in shape and diameter.

A variety of image-processing software tools are available and one should be aware of which tools are used in an experiment as well as the techniques that each one uses.

7.1.2. Preprocessing

The first step in the analysis of microarray data is sometimes referred to as preprocessing. The idea of preprocessing is, to the greatest extent possible, to remove extraneous sources of variance or bias in the measurements obtained and to transform the data in ways that hopefully lead to more interpretable and valid inferences and more stable and precise estimates.

7.1.3. Normalization

In the microarray field, normalization does not necessarily refer to transforming data to make it normally distributed as it would in the field of statistics (90). Rather, normalization refers to a process by which the original data observed on a microarray is adjusted in some way to account for artifacts that can affect the absolute numbers observed. There are many ways to normalize microarray data (91–99) and no one is universally accepted as the best. The simplest form of normalization is chip mean normalization in which one simply multiplies all the chips by a constant (usually target mean expression level/chip mean expression level) to take each of the chips to the same mean expression level. Quantile–quantile normalization converts the expression levels to percentile ranks (across genes, within arrays) which are then transformed to standard normal deviates via the inverse cumulative density function for the standard normal distribution. Unlike mean normalization, this not only forces the mean expression level to be constant across microarrays, but it also constrains the variance and shape of the distributions to be constant across microarrays. Whether this is a good thing or not is unclear. Other forms of normalization try to achieve linearity in the within-array across-gene regression of the measured levels for one channel (dye) on the measured levels for the other channel. The biochemical properties of each of the dyes on a two-color array cause the dyes to be incorporated at different rates, which results in an apparent dye bias on a single array. This is often correct using a dye swap, where the same sample is divided and labeled with both dyes and run on different chips. This will allow the biases to be reduced. Alternatively, lowess (or Lowess) normalization (locally weighted regression) (100,101) can be used to correct these differences, based upon the assumption that the same number of genes are changing up as down, a biased move in expression will bias the results. Lowess normalization involves estimating the mean of one variable conditional on another, usually made robust by using a moving window. Other methods of normalization use a common set of housekeeping genes that are, in theory, expressed at about the same level in all tissues (102). The housekeeping genes can then either be used for chip mean normalization or lowess normalization.

Microarrays, both single and multichannel, can have spatial variation in hybridization, which can affect results. A variation on lowess normalization is used to correct for these artifacts. A polynomial is fit to the two-dimensional space of measured fluorescence across a chip, areas of high or low fluorescence may then be detected and smoothed (92,103).

7.1.4. Transformation

Transformation is generally done after the normalization step and is intended to improve the fit of the observed distributions of data (within-gene, across-microarrays) to some desired distribution, usually the normal distribution. A log transformation is by far the most commonly utilized and experience suggests that this is often a good choice, though no formal proof exists to state that it is the best choice. Another alternative would be to use the Box-Cox (104) family of transformation defined as:

$$Y' \equiv \frac{(Y+p)^{\lambda} - 1}{\lambda}.$$

where Y is the original expression score, Y' is the transformed score, p is an additive constant set to ensure that $Y+p > 0$ for all Y, and λ is a constant. If $\lambda = 0$, then Y' is simply taken to be $LN(Y+p)$. A value of λ can be selected to insure that the residuals of Y (in a study of two or more groups or conditions, the residuals would be the values of Y after subtracting the group means) follow a normal distribution as closely as possible. An open question is whether a single value of λ should be selected, whether one common value should be chosen for all genes, or some middle ground selected. To our knowledge, this has not been addressed to date, but our intuition is that given sample sizes commonly used (e.g., < 20 per group), estimating a common λ for all genes is probably wise. Still other transformation approaches exist (105–107) although to date they have not been widely employed.

7.1.5. PM–MM

An aspect of preprocessing that is unique to Affymetrix oligonucleotide arrays involves subtracting values of the amount of hybridization to a mismatch (MM) sequence corresponding to each perfect match (PM) sequence on the chip. The MM sequences are identical to their corresponding PM sequences with the exception that they differ by a single nucleotide. The idea behind the subtraction of MM from PM is that MM will assess background hybridization [noise or error (e)] whereas PM should asses both true signal (t) plus noise (i.e., $t + e$). Therefore, subtracting MM from PM should yield $(t + e) - e = t$; that is a more accurate estimate of true signal intensity. However, this thinking is simplistic because: (1) both t and e are random variables; (2) e actually represent two separate random variables (one for PM and one for MM) that may or may not be correlated; and (3) MM scores may also be influenced by t, though perhaps to a lesser degree. Because of these points, the simple subtraction is probably not best. Alternatives that are now gaining favor include ignoring MM altogether or including MM as a covariate in a linear model that has PM as the dependent variable (107,108). Although yet to be widely used, this last approach may be the most sound.

7.1.6. Inference

In statistical analysis when we refer to inference, we refer to the process of drawing a conclusion about some aspect of a population (i.e., about some aspect of the state of nature) on the basis of some sample data. Most often, inference is conducted from the context of classical frequentist null hypothesis significance testing. Typically we ask "Is some effect or association zero in the population?" In the context of microarray research the most common inferential question is usually "Does the level of gene expression for this gene differ between the two groups (or conditions) I have studied?" There is no one method for making such inferences. Many methods exist and several factors may be considered when choosing among them including use of instrumentation, availability of software, and, in our opinion most importantly, the properties of the testing procedure and how well those properties are established. When we refer to the properties of the testing procedure we are generally referring to its error rate. Looking back to Table 3, decisions landing in cells C or B represent errors whereas decisions landing in cells A or D represent correct decisions. We want inferential testing procedures that maximize the number of correct decisions and minimize the number of errors. Among correct decisions (A and D) investigators may choose to place different value on correct decisions of type A and correct decisions of type D. Similarly, among incorrect decisions, investigators may choose to place different value on incorrect decisions of type C and incorrect decisions of type D. There is no right answer as to how one should weight these correct and incorrect decisions—it is simply a matter of choice, values, and resources. Traditionally, frequentist statistical testing has demanded procedures that fix the ratio of

$$\frac{c}{a+c}$$

to some level, most often 0.05. This is the classical type I error rate. Thus, in considering which statistical procedures one wants to use, the answer may be the one which has the best error rate when errors are weighted according to the investigator's values. To make such a judgment, investigators need to know the error rate of their procedures. This is challenging because there have been very few (if any) thorough simulation studies or mathematical proofs demonstrating that some of the preferred methods necessarily have better error rates than others across a broad variety of circumstances. In fact, for many methods that have been offered in the literature we do not even have information about what their properties are. This is because many methodologists who are rapidly offering methods in this rapidly evolving area seem to have forgotten the traditional need to validate methods and demonstrate their properties by simulation and/or analytic proof when offering a new method. Thus, the microarray researcher is at a significant disadvantage in being able to intelligently choose among the currently available methods. We try to give some guidance below.

Perhaps the most commonly used method for drawing inference about differential expressions is the use of FC. FC is defined in greater detail in the section on estimation found later in this chapter. The commonly used approach is to define any two groups as having differential gene expression if the FC for the gene in question is greater than or equal to some value FC_T. FC_T can be thought of as the threshold for declaring genes to be differentially expressed. The most commonly used thresholds appear to be 2.0 and 3.0 although some authors use more complex rules. Currently, there is broad consensus among statisticians that simply using fixed cutoffs of FC to determine differential gene expression regardless of sample size, variability in expression levels for the gene and study under consideration, and other factors is without sound basis. This is so for many reasons. First, the distribution of FC is not known. Therefore, there is no way of knowing that, under the null hypothesis of differential gene expression, FC values greater than FC_T will only occur less than or equal to $100 \times \alpha\%$ of the time. Without such knowledge, we have no way of describing the type 1 error rate of our procedures. Second, FC does not explicitly incorporate a measure of variability in expression. Thus, for example, an FC value on a gene with a much greater inter-subject variance is treated the same as an FC value with a smaller inter-subject variance. Because of this, FC is not what is referred to as a pivotal statistic, even asymptotically. Therefore, its distribution will depend on unknown factors again rendering it useless for formal inference. Third, the use of such cut offs for FC stands in contradiction to the law of large numbers which implies that (under virtually all circumstances) a sample estimate of a population parameter converges in probability to the population parameter as the sample size upon which the estimate is based approaches infinity. In other words, as our samples become larger, the estimates that we derive from them become increasingly more accurate to the point that, when sample sizes are very large, our estimates should be so accurate that even very small deviations from a value expected under the null hypothesis will be significant. Contrarily, in very small samples our estimates may vary so widely from the parameters that we are trying to estimate that even apparently huge deviations may be meaningless. For example, in a sample of only two cases, an FC of 2 may be quite unimpressive whereas in a sample of two million cases, even an FC value of 1.01 might be highly statistically significant. Fourth, it is sometimes stated that investigators are less interested in statistical significance than biological significance. We believe that such statements represent an inappropriate either/or point of view. That is, investigators should be concerned about both statistical and biological significance. Statistical significance can be thought of as telling one whether something is really there. Biological significance can be thought of as telling one, if it is really there, how big and important is it. FC may help us to determine how big potentially important things are, but not whether they are really there. Even for determining how big effects are, FC value may have some limitations as described in the section on estimation found later in this chapter. Finally, some investigators have taken the approach of taking a single biological specimen and aliquoting into two subsamples. These two subsamples are then treated as though they were different samples and run on microarrays. The FC value for each gene is then calculated and for the k genes on the microarray, a distribution of k FC values is presented. FC cutoff points corresponding to some percentile values of interest (e.g., the 1st and 99th) are then defined and subsequently used as cutoff points for determining statistical significance in subsequent research involving samples from two or more groups or conditions. Although this may seem sound, it makes at least two mistakes. First, it implicitly assumes that the distribution of FC, under the null hypothesis, is the same for all genes. Because, as we have stated above, FC is not a pivotal statistic, we have no basis for making this assumption. Second, this whole endeavor confuses the standard error of the mean with the standard error of the measurement. Therefore, even if our first criticism of this approach were irrelevant, this approach would at best be offering correct answers to incorrect questions. That is, it would tell us whether two measurements were different not whether the means of two populations were different. For all of these reasons, we and many other statisticians recommend that the simplistic use of FC and fixed cutoff points for determining statistical significance be abandoned.

If FC won't work for determining whether genes are differentially expressed, what methods can be used? Although it may seem less than staggeringly dramatic to say so, the good old-fashioned Student's t-test works extraordinarily well. When examining a single gene at a time, the Student's t-test may offer the best combination of valid inference and statistical power available. In 1908, Student's showed that when the data are normally distributed this test is valid and, moreover, it is known that when data (residuals actually) are normally distributed, there can be no more powerful test of whether or not any two means are equal. [This statement is not necessarily true when one considers testing many pairs of genes simultaneously, see Ref. 110). Therefore, we recommend strong consideration of the use of Student's t-test when looking at each gene separately. However, we still need to note that in using Student's t-test we are assuming that the data are normally distributed within each group or condition and that the variances are equal within each condition (i.e., homoscedasticity). To the extent that these assumptions are not met, Student's t-test is not necessarily valid. We recently explored this question in detail in an extensive simulation study (72). We compared Student's t-test to a large number of nonparametrical alternatives including permutation and bootstrap tests to evaluate the relative power and validity (by validity we refer to a per test control over the type I error rate) of each method under a variety of distributions, sample sizes, effect sizes, and degrees of heterogeneity of variance (i.e., heteroscedasticity). We found that when samples sizes are equal across the groups or conditions being studied and homoscedasticity holds, Student's t-test performs remarkably well regardless of the normality of the distribution. When variances were unequal, Welch's (111) adjusted t-test performed quite well. However, when samples sizes were unequal, we found that only when the groups have equal variance and the data is normally distributed can one use Student's t-test. In other cases, only the Chebby Checker, and not permutation tests, was found to be valid. Thus one should strive for equal sample size per group or transform the data to be normally distributed with equal variance.

7.1.7. Multiple Testing

Of course, in microarray research we are not simply testing a single gene at a time but are testing for differences in many genes. This brings up two questions. First, should we take this multiple testing into account in the inferential process to correct for the increased probability of making one or more type I errors (decisions of type C in Table 3) as the total number of tests, and presumably the total number of true null hypotheses, increases? Second, is there some way we can take advantage of the fact that we are conducting so many tests and have so much more information and thereby conduct better tests? With respect to the first question, this is ultimately a philosophical issue that has troubled statisticians and investigators for decades. Saville (112) stated that each investigator must ultimately "cut the Gordian knot" and decide whether they simply wish to have their per test type I error rate held at some level or whether they wish to have the type I error rate for an entire set of tests held at that level. We will not go into all of the philosophical questions that this issue raises (for review, see Refs. 112–115) but will simply state here that the zeitgeist of the field seems to be that some correction for multiple testing is in order. The question then becomes how to do it. By far the easiest and most robust way of doing so is the Bonferonni correction. The Bonferonni correction entails conducting each individual test at level α/k where α is some FWER one wishes to maintain and k is the number of tests conducted. This ensures that probability of making one or more type 1 error remains less than or equal to alpha [i.e., $P(C > 0) \leq \alpha$]. Unfortunately, it can also be extremely stringent. That is, when there are certain types of correlation structures among the tests conducted, types that are considered plausible and common, the Bonferonni correction will be too conservative. In addition, when many of the null hypotheses are not true, the Bonferonni correction will also be too conservative. Therefore, people are seeking alternative methods. One of the best outcomes of microarray studies may be the fact that it has stimulated so much new and good work in the area of multiple testing. With respect to these new methods, they can generally be divided into two types. One type tries to control the family-wise error rate. The other tries to control the false discovery rate. The false discovery rate (FDR) is equivalent to $1 - TP$ where TP is defined as it was using the notation in Table 3. Although the concept of FDR goes back longer, it was formally introduced by Benjamini and Hochberg (116). Although neither controlling the FWER nor the FDR can be said to be intrinsically better than the other, there is currently a groundswell of enthusiasm for methods that control the FDR because controlling the FDR is both more powerful than controlling the FWER and, more importantly, seems to be more in line with the way applied investigators think. That is, investigators seem to be more interested in asking "What proportion of those things that I decide to follow up on will be good leads rather than ghost chases?" In contrast, they seem less interested in asking "If all the null hypotheses are true, what is the probability that I have incorrectly rejected one or more?"

For both FWER and FDR control approaches there are multiple methods available. Some assume independence of all the tests conducted, others do not. Clearly those that do not are more robust and generally applicable than those that do. Those that do not assume independence tend to be more computationally demanding because they often use computer intensive methods to simulate the effects of the correlation structure among genes on the distribution of the test statistics involved. Those that assume independence can often be implemented with very simple, back-of-the-envelope calculations whereas those that do not assume independence often require more sophisticated computer programs. Among known FDR methods, some are adaptive in that they try to use all tests simultaneously to get some sense of how many null hypotheses may be false and then build that information into the FDR control procedure for greater statistical power (e.g., see Refs. 74, 116 and 117). Thus, with respect to multiplicity control we do recommend that some method of taking multiple testing into account be utilized. We favor methods that utilize the FDR as opposed to controlling the FWER, though both have merit. For both types of methods we advocate methods that take into account the dependency in the data and are adaptive to the extent that such methods can be utilized.

The second issue with multitesting is more positive than negative. That is, it asks whether we somehow utilize the potential additional information available in multiple other tests to strengthen the inferences we conduct with each individual test. The answer seems to be yes. A variety of methods that based on Bayesian, empirical Bayes, and frequentist testing are becoming available. In a different context, Berry and Berry (118) considered the same issue in deciding how to make inferences about multiple side effects in a clinical trial. Berry's Bayesian method unfortunately ignores potential correlations among the observations. Other methods are based upon empirical Bayes methodology and involve shrinking estimates of effect toward a grand mean of effect estimates across all genes (e.g., (119)). These shrunken effect size estimates tend to be more precise, that is have less variance, than ordinary estimates. Because they are estimated with greater precision, they may be statistically significant where ordinary estimators are not. That is, by taking into account the multiple testing, in some cases one may actually be able to increase power. Recently, Brand et al. (120) developed an adaptive alpha-spending function that takes into account the tendency among genes induced by a presumed underlying biological cascades to be correlated in their expression. This method can improve the power of both methods that control the FWER and methods that control FDR. Finally, there are procedures that are becoming available that simultaneously consider all genes and try to draw valid inferences by simultaneously considering information from the ensemble (see Refs. 74, 87, 88, and 121–124). We cannot review all of these methods in great detail here and certainly cannot review many of the other methods that have been proposed.

The mixed-model-based methods (87,88,123,124) are some of the more statistically sound and well-thought-out approaches. These methods are highly adaptable to a broad range of circumstances, can often be run in widely available software (e.g., SAS) and enjoy a stronger statistical foundation than most methods presented in the literature for microarray analysis. A key difference among these ANOVA-based methods is the extent to which they constrain the residual variances for each gene (that is the variance in gene expression that is not explained by a model fitted to the data) to be equal across all genes. Kerr and Churchill's (87,88) method constrained all variances to be identical across all genes. Wolfinger et al. (124) allowed the variances to be different for every gene. Several investigators have now noted that the optimal answer may lie somewhere in between methods to either shrink variances towards a common mean variance or to cluster genes together and allow for common variances within clusters but different variances across clusters are being developed (125,126). While these methods do enjoy a generally sound theoretical basis, this basis is founded in asymptotics. That is, we know that test statistics utilized in these models are asymptotically correct. They will give the right answer as the sample size approaches infinity. Unfortunately, as we have discussed above, the sample sizes used in microarray research have not begun to approach infinity. Therefore, it remains an open question whether these methods yield valid type I error rates in the sample sizes typically used in microarray research. This is an important question for methodologic research. The results of Catellier and Muller (127) give one grounds for doubt that these asymptotic methods perform well in small samples.

An alternative approach to analyzing data is a two-stage approach in which one first conducts a significance test for each gene and then uses the distribution of p-values that result to fit a higher level model. Allison et al. (74) have developed a method based upon applying finite mixture models to these p-values. We have publicly available software that implements this method (see www.soph.uab.edu/ssg_content.asp?id = 1164) and refer to the software algorithm that implements this as the Mix-o-matic. The Mix-o-matic takes the distribution of p-values and fits it to several components including a uniform distribution (the distribution that will be expected if all null hypotheses were true) and one or more beta distributions. Once the model is estimated, we can use it to calculate an FDR level for every gene.

We are thereby able to provide investigators with statistics that they seem to find meaningful. Specifically, we are able to offer statements such as "for this gene our best estimate is that there is a 95% chance that it is a gene that is truly differentially expressed." It is important to recognize that the preceding statement must be interpreted in a Bayesian framework. Examples of the use of this methodology with real data can be found in Ref. 11. Although our method certainly can benefit from further development, we find it useful because it allows us to communicate with investigators in ways that answer the questions that are on their minds and allow us to offer statements about our error rate that we have validated. Thus, investigators can have some confidence in inferences we offer them and, importantly, know exactly how confident they should or should not be.

While we certainly favor the approach we have developed, it is not the only approach we utilize when analyzing microarray data and it is certainly not the only approach that we think has any merit. As of yet, there is no clear consensus as to which of all the available analytic techniques is best or even if one is best. That is, future research will need to tell us which of the current and future methods are worth using at all and among those that are worth using at all whether any one stands out as best or whether some are better for some purposes and under some circumstances and others better for different purposes and under different circumstances.

Finally, after the formal inference using the microarray data, some investigators choose to take additional steps to validate their inferences. A variety of approaches exist for such validation. For an overview, see Ref. 128.

7.1.8. Estimation

In addition to simply attempting to determine whether a gene is differentially expressed across two or more conditions, investigators will probably want to know the magnitude of the difference. A statistic that estimates the magnitude of an effect is generally referred to as an *effect size*. The concept of magnitude of effect is very closely related to biological significance though they are hardly identical. There are many metrics of effect size (129–131). In choosing among effect sizes and estimation approaches, two key issues are interpretability and statistical properties.

With respect to interpretability, metrics that are scale free seem to be best. A raw mean difference would generally not be interpretable. For example, how would one meaningfully interpret the statement that the mean difference between some wild type and some knockout mice for WNT expression was 9346? In contrast, the FC metric is likely so popular, in part, because it is seemingly easy to interpret. Whether this seeming ease of interpretation for FC is warranted is open to question however. FC is a ratio-based metric. Such metrics are usually only justified when the variable under consideration is measured on a ratio scale (132). A ratio scale is a scale of measurement for which equal ratios of two pairs of measurements correspond to equal ratios of the corresponding underlying trait, for any two pairs of measurements. This can only occur when the measurement scale has a true, fixed, and known zero point. Body weight, for example, is a ratio-scale measurement whereas temperature, in Fahrenheit or Celsius, is not. Thus, we would get the same FC if weight were our outcome in a study regardless of whether we recorded weight in pounds or kilograms. In contrast, we would not get the same FC if body temperature were our outcome in a study regardless of whether we recorded the data in Fahrenheit or Celsius. In the process of normalization and the use of PM—MM measurements, we are often adding arbitrary constants to gene-expression measurements such that they are not necessarily made on a ratio scale. Therefore, the FC results can be tremendously dependent on these arbitrary factors. Second, there is some linguistic confusion regarding the term FC. First, it is often confused with relative amount such that if the mean of group A is 1.0 and the mean of group B is 2.0, some investigators label this as a two-fold change when in fact it is a one-fold change. That is, the difference (change) between the two measurements is one times the smaller measurement. Second, even though FC can be defined as we did earlier to allow numbers to theoretically range from negative to positive infinity, a linguistic purist would not allow that for a quantity that cannot go below zero (i.e., the amount of mRNA molecules in a specimen). It is nonsensical to state that an FC is less than −1. That is, it is impossible to have more than a one-fold reduction because to do so would require having less than zero mRNA.

Another scale-free metric is the standardized mean difference (δ) defined as the difference between the two means in question divided by their pooled within-group standard deviation (131). This statistic, though perhaps a bit unfamiliar to basic biologists, is very familiar to meta-analysts who use it routinely to summarize the strength of effect when two groups are compared on a quantitative outcome variable. In essence, this statistic standardizes the gene expression scores under consideration to have a standard deviation of 1.0 such that differences can then be described in standard deviation units. There is a 1:1 relationship between the δ and the proportion of variance in the gene expression score explained by the independent variable which we will denote τ. This proportion of variance represents yet another potential effect-size metric that is scale free.

Finally, Wolfinger et al. (124) and Chu et al. (133) suggest estimating differences between means after transforming the data with a log base 2 transformation which we will denote ϖ. The rationale is that, on this scale, mean differences have an FC-like interpretation. However, it should be noted that this is more analogous than strictly correct and really refers to expression ratios rather than true FCs.

Regarding statistical properties, the properties of δ, τ, and ϖ are all well known under normality and, even without assuming normality, asymptotically. They are consistent estimators of their corresponding parameters, and formulae for placing confidence intervals around sample estimates are available. The same can not be said of FC. Moreover, such sample estimates of effect size expressed as δ, τ, and ϖ when accompanied by estimates of their standard error, can be useful for subsequent power calculations. For example, if one wished to know how many subjects one should use to try and replicate a finding in a microarray study using a fresh sample of subjects and a method or assaying a single gene expression level (e.g., rtPCR), then these metrics would be useful in standard power analysis formulae. Again, this is not true of FC. Finally, it should be noted that for all of these effect-size estimators, if one only focuses on the estimates of effect size for which some degree of statistical significance has been achieved and/or that are the largest among the entire set, then these estimates of FC will almost assuredly be highly biased upward. That is, the effects will appear, on average, much greater than they actually are. This has been well- established in other contexts (e.g., (74)) and there is no reason to believe it will be otherwise in the microarray context. Because of this, if one is interested in accurate estimation among a subset of genes with apparently large effects and/or among a subset of genes that appear to have statistically significant differential expression, then either of two things should be considered.

First, it is wise to consider estimation procedures that are essentially ensemble estimation procedures and shrink estimated values toward the grand mean of the entire ensemble or toward some predicted value based upon a prediction model. Such approaches are often referred to as empirical Bayes approaches. We use an empirical Bayes approach developed by Morris (134) to obtain estimates of δ that are shrunken in the microarray context (119) and find that these are substantially more accurate (i.e., have smaller means for error) than ordinary estimates. A similar approach to estimating correlation coefficients in microarray analysis prior to clustering data has been provided (135). Second, one should consider replicating the finding in an independent data set. It is important to note that replicating an independent data set is distinctly different from showing that one can reproduce the results obtained from the same specimens when those specimens are under alternative measurement procedure such as rtPCR. We are referring here to obtaining fresh measurements, for example, if one is studying the adipose tissue of mice and wanted to confirm or reestimate the effect of a gene detected in a differentially expressed microarray, one might get a fresh set of mice, excise the adipose tissue of interest, extract the RNA, and then run both specimens through rtPCR. The effect-size estimates obtained in these second studies would then be unbiased.

7.1.9. Prediction

We use the term prediction here broadly to refer to a construction of scoring procedures involving multiple gene expression levels to classify objects into preexisting known categories or to develop indexes that can serve as some form of biomarker. In the field of cancer research, this first type of class prediction is very common. For example, investigators are interested in using microarrays to predict whether the biopsy of a tumor represents a benign or a malignant tumor, or a tumor that will progress rapidly and aggressively, or a tumor that will progress slowly and innocuously. There may be situations where this type of approach will be useful in the field of obesity research. For example, one could envision conducting microarray analysis on muscle tissue biopsies from obese individuals who currently show no signs of insulin resistance or type II diabetes. One might be able to follow these individuals and predict via a microarray profile who subsequently goes on to become insulin resistant or diabetic and who does not. Were such a method possible, this might have great utility. If one could predict which obese individuals would go on to have subsequent complications and which would not, one could more effectively target for treatment those individuals that are expected to go on to complications. However, at this time, this is a less common use of microarrays in the obesity field. For readers interested in techniques for such prediction, see Ref. 136 and 137. We have used this general type of predictive approach to develop biomarkers using microarray measurements. For example, in Lee et al. (176), we constructed an index that was a linear multivariate composite of all the expression scores on a microarray. We constructed the index such that it provides a very good separation between old and young animals. We were able to show through simulation that the degree of separation provided was statistically significant. That is, the degree of separation was greater than that which would be seen if there were no differences between old and young animals. Subsequently we were able to compare calorically restricted animals to both old and young animals with respect to this index. We found that caloric restriction tended to lower an animal's apparent age at the transcription level by approximately 19%. This is consistent with the observation that caloric restriction prolongs life and seems to decrease the fundamental process of aging. When we develop such an index, we can then use it for future research to evaluate the effects of compounds or treatments that are hypothesized to alter the rate of aging. To the extent that these compounds or treatments reduce the value of this overall gene expression aging index, we can state that they have lowered a biomarker of aging. It then provides greater rationale for taking these compounds or treatments fully for use in further research in actual longevity testing in animals, a time consuming and expensive procedure. Work using this index in this way is currently under way.

There are many methods for constructing such indexes. The classic case is Fisher's discriminant function analysis. Under situations in which the number of cases exceeds the number of variables and the data are normally distributed, then Fisher's method is the optimal method. However, neither of these two conditions necessarily are met in microarray analysis. In particular, the number of measurements available generally exceeds the number of cases available by several orders of magnitude. In such cases, alternative approaches can be utilized. These include machine-learning approaches such as support vector machines, least-squares approaches, and related techniques (138). Which of these techniques is best is not currently known. Essentially, all are trying to do the same thing. Specifically they are trying to find the best compromise between complexity and simplicity. That is, as one makes one's predicted model more and more complex utilizing more and more of the sample information available, the predicted ability in the sample at hand will steadily increase. Life is complex and choosing ever more complex models will always be a little bit better if they are correct. When the model is based upon the sample data, it will always perform in predicting in that sample data because it is optimized for that purpose. However, the sample data contain not only information about the true compilation structure among the variables but it also contains noise due to sample variation. As one begins to fit more and more complex models, it becomes increasingly likely that one is simply adapting the model to take into account the chance variation due to sampling in the sample data set— that is, one is likely to be fitting the model to the noise. When one then goes to use this model to predict in the population at large or in a fresh sample, the prediction may be poor. The more one overfits the data in this manner, the poorer the subsequent prediction (sometimes referred to as cross-validation) is likely to be. Therefore, one does not want to make the model too complex. The question is how does one choose the optimal compromise between complexity and simplicity. The answer is unknown. At present, the best we can offer is the statement that whatever model is chosen it should be rigorously cross-validated if at all possible. At minimum, its statistical significance should be evaluated via simulation or analytically if possible. Cross-validating requires that one have sufficient data to hold some out of the estimation process so that one can subsequently check how good the prediction is using the data that is held out from the model estimation process. It is important to note that in order for the correct validation to be a legitimate exercise, the data used in the cross-validation must have had absolutely no way in the selection of the structure of the model used for prediction, what variables go into that model, and what parameter estimates go into that model. It has not been uncommon for investigators to mistakenly allow their cross-validation data to be used in some aspects of the estimation or model-development process thereby invalidating the whole cross-validation. Those interested in further details should see Ambroise and McLachlan (139).

7.1.10. Class Construction

Cluster analysis has provided a set of methods that has been very useful for exploring gene expression patterns from microarray data. The goal of such analysis is to construct classes of genes or classes of samples such that observations within a class are more similar to each other than they are to observations in different classes according to their expression levels. There are several reasons for interest in cluster analysis of microarray data. First, there is evidence that many functionally related genes have similar expression patterns (140,141). By grouping genes in a coordinated manner according to their expression under multiple conditions, we may be able to reveal the function of those genes which were previously unknown. Second, a class of genes with similar expression pattern may reveal much about regulatory mechanisms. The common regulatory elements (e.g., motifs) identified in a class of genes would greatly facilitate our understanding of genetic networks (142,143). Third, it provides a more reliable and precise way to distinguish different subtypes of tumors (e.g., breast cancers), which are not achievable by standard microscopic or molecular approaches, by classifying the samples on the basis of their gene expression levels (144–147). The new subtype of tumors can also be identified. Eventually, such classifications can lead the advancement for successful prognosis, diagnosis, and therapeutics of diseases. Fourth, given that a microarray can potentially contain expression of tens of thousand of genes over several to hundreds of samples, by grouping either genes or samples, or both simultaneously (140,148), clustering analysis potentially provides an effective way to reduce the complexity of data for easy organization, visualization, and interpretation.

A variety of clustering algorithms, have been applied to analyze microarray data to partition genes or samples into mutually exclusive classes. These methods include hierarchical clustering algorithms (140,145), K-means (149,150), self-organizing maps (151–154), the support vector machine (155,156), model-based clustering (157,158) and other algorithms (159–162). The detailed description of each algorithm is beyond the scope of this chapter. Thus, we use the hierarchical algorithm as an example to illustrate the common features existing in many clustering algorithms. Researchers can refer to the corresponding literatures for other algorithms they are interested in.

In the context of clustering analysis, the raw data matrix, X, is obtained by microarray experiments and is used as input. For a study with m genes for n samples, X is a matrix with m rows and n columns and can be represented as:

	Gene	Chip 1	Chip 2	...	Chip n
	1	$x_{1,1}$	$x_{1,2}$	...	$x_{1,m}$
X	2	$x_{2,1}$	$x_{2,1}$	...	$x_{2,n}$
	...	...	...	...	...
	m	$x_{m,1}$	$x_{m,2}$	...	$x_{m,n}$

The matrix element $x_{i,j}$ corresponds to the expression level of gene i (row) for sample j (column) and the clustering algorithm can be applied either to rows (grouping genes) or to columns (grouping samples) or to both rows and columns (grouping genes and samples simultaneously).

The first step is the normalization (or standardization) and transformation of the raw data. Although the techniques previously described can be virtually applied in this step, they have a different purpose than when used to meet requirements of statistical tests (e.g., to improve the fit of the observed distributions of data to some desired distributions). Hierarchical cluster analysis depends on a distance measure requiring commensurable variables. This means that the variables must have equal scales, which is not always the case in microarray data due to large variability between chips and makes the normalization and transformation more critical in clustering analysis. The second step is to find a distance measure between two observations and calculate it for each pair of observations. For example, the Euclidean distance can be used and computed between two genes, i and j by following formula based on the elements of row data X:

$$d_{ij} = \sqrt{(x_{i,1} - x_{j,1})^2 + (x_{i,2} - x_{j,2})^2 + \cdots + (x_{i,n} - x_{j,n})^2}$$

There are many other distance measures available, including Pearson correlation coefficient, Manhattan distance, and Spearman Rank-Order correlation (163), etc. The actual choice should reflect the nature of the biological question and the technology that was used to obtain the data. The third step is to seek clusters of observations on the basis of the pair-wise distance matrix previously calculated. Generally, there are two different methods available: (1) the agglomerative method and (2) the divisive method. The agglomerative method starts with the assumption that each object is considered as a cluster. The algorithm merges two most similar clusters iteratively until there is only one cluster left. The divisive method starts with one single cluster containing all objects. The algorithm splits it step by step until each single object is a separate cluster. Suppose we are cluster genes with the Euclidian distance by the agglomerative method. Two genes with least distance are merged to a cluster and the distance between this cluster and the other genes are recalculated. Three common options that can be used to calculate the distance between two clusters are single linkage, average linkage, and complete linkage. At the last step, a dendrogram is generated to represent the results of hierarchical clustering analysis and the clusters are obtained by cutting the dendrogram at a specific height, which can be determined by a specified number of clusters or some external criteria.

Cluster analysis is a powerful tool for microarray data exploration. However, there are several limitations for such analysis (82). First, clustering algorithms always generate clusters for any data set but most procedures do not have a probabilistic foundation. Because of this, there is no universally accepted method to assess whether a gene cluster is significant or even a clear understanding of what it would mean to be significant (164). Thus, although cluster analysis can be used for data reduction and hypothesis generation, caution is required in interpreting its results. Second, in most published studies of clustering analysis, the variation of gene expression levels is greatly ignored. Obviously, this variation will influence clusters derived from gene expression data (165). The third important consideration is how many clusters to actually make when performing these clustering methods. Currently, this number is determined on the basis of external criteria or visual inspection of the dendrogram, which is often artificial and arbitrary. Many methods have been developed to overcome the weaknesses of cluster analysis, including the bootstrap and consensus method (165), the average silhouette width (166), the gap statistic (148), the Model-based methods (158), resampling methods (160), etc.

Many clustering algorithms are available in general statistical software, such as SAS and S-Plus. Many researchers have also implemented their own software for cluster analysis to adapt to the high dimensionality and nuances of microarray data. However, given the availability of a large number of clustering algorithms and substantial differences between them, it is far from trivial to select an appropriate algorithm. One approach is to use several clustering algorithms for the same data and combine the results by consensus method (167).

7.2. Software and Tools for the Analysis of Microarray Data

This section provides a cursory overview of software for the analysis of microarray data categorized by their purposes and characteristics. Readers

TABLE 4 Annotated, abridged list of Software for Microarray Research

1. Statistics software and technical programming languages

Product	Company/ institute	Interface/ server operating system	Features	License	URL
Clementine	SPSS	Windows, Mac	Able to apply practices for analyzing microarray gene expression data and leads through all steps of gene expression analysis: data preparation, feature, selection, classification, clustering, visualization and deployment.	$499 Plus	http://www.spss.com/ spssbi/clementine/
SAS® Microarray Solution	SAS	Windows, Mac	Provides both the forum and technology to implement, use, and share powerful analysis methods with a strong statistical foundation.		http://www.sas.com/ industry/pharma/mas/

1.1. Matlab Mfiles/tools for microarray analysis (in alphabetical order)

M-file/tools	Author	Interface/ operating system	Features	License	URL
MA-ANOVA	Gary Churchill's statistical Genetics Group, The Jackson Laboratory	Windows, Linux	MA-ANOVA is a set of functions written in Matlab for the analysis of variance on microarray data.	Free for academic use	http://www.jax.org/staff/ churchill/labsite/software/ anova/rmaanova/ index.html

	Author	Interface/operating system	Features	License	URL
MatArray toolbox	David Venet	Windows, Unix running in the Matlab environment	Its main features are advanced normalization schemes; very efficient implementation of hierarchical clustering, with leaf ordering and export to TreeView; efficient implementation of K-means clustering.	Free.	http://www.ulb.ac.be/medecine/iribhm/microarray/toolbox/

1.2. General packages useful for microarray analysis (in alphabetical order)

Package	Author	Interface/operating system	Features	License	URL
cclust (convex clustering methods and clustering indexes)	Evgenia Dimitriadou	Windows, Unix	Convex clustering methods, including Kmeans algorithm, on-line update algorithm (hard competitive learning) and neural gas algorithm (soft competitive learning) and calculation of several indexes for finding the number of clusters in a data set.	GNU GPL (version 2 or later).	http://cran.r-project.org/src/contrib/packages.html#cclust
Cluster	S original by Peter Rousseeuw, Anja Struyf, Mia Hubert. R port by Kurt Hornik and Martin Maechler.	Windows, Unix	Functions for cluster analysis.	GNU GPL (version 2 or later).	http://cran.r-project.org/src/contrib/packages.html#cluster

(continued)

Package	Author	Interface/ operating system	Features	License	URL
	Edwards, Page, Gadbury, Trivedi, Wang, Brand, Allison	Windows, Mac, Unix	Easy to use Java-based package that incorporates statistically validated methods for the quality control, estimation, and inferential testing of microarray data. Automatically generates reports.	Free for Academic and government investigators.	http://www.soph.uab.edu/ssg.content.asp?id=1164
HdbsTAT!	C. Fraley and A. E. Raftery. R port by Ron Wehrens	Windows, Unix	Model-based cluster analysis.	Permission granted for noncommercial use only.	http://cran.r-project.org/src/contrib/packages.html#mclust
mclust					
multiv (multivariate data analysis routines)	S original by F. Murtagh. R port by Kurt Hornik, Friedrich Leisch, and Achim Zeileis	Windows, Unix	Multivariate data analysis routines including hierarchical clustering, PCA, Sammon mapping, and correspondence analysis.	Free for noncommercial purposes.	http://cran.r-project.org/src/contrib/packages.html#multiv

1.3. R packages for microarray analysis (in alphabetical order)

Package	Author	Interface/ operating system	Features	License	URL
affy (methods for affymetrix oligonucleotide de arrays)	Rafael A. Irizarry, Laurent Gauiter	Windows, Unix	The package contains some methods for analyses of Affymetrix oligonucleotide array data.	GNU GPL (version 2 or later).	http://biosun01.biostat.jhsph.edu/~ririzarr/Raffy

BioConductor	Many	Windows, Unix	An open source software project with several goals. Main goals: providing infrastructure in terms of design and software for analyzing genomic data, some form of graphical user interface for selected libraries, and a mechanism for linking together different groups with common goal.	GNU GPL (version 2 or later).	http://www.bioconductor.org
CyberT	Tony Long, Harry Mangalam	Windows, Unix	t-test for statistically significant differences between sample sets for arrays; Bayesian probabilistic framework to estimate the variance among replicates.	GNU GPL (version 2 or latter).	http://genomics.biochem.uci.edu/ genex/cybet/
GeneClust	Kim-Anh Do	Windows, Unix	GeneClust is a piece of computer software that can be used as a tool for exploratory analysis of gene expression microarray data; hierarchical and gene shaving; simulation to assess the clustering performance.	No pricing information is available.	http://www.ulb.ac.be/medecine/ iribhm/microarray/toolbox/
GeneSOM	Jun Yan	Windows, Unix	Clustering gene using self-organizing map.	GNU GPL (version 2 or later).	http://lib.stat.cmu.edu/R/CRAN/ src/contrib/PACKAGES. HTML#GeneSOM
Mixture modeling	Debashis Ghosh	Windows, Unix	Mixture modeling of gene expression data from microarray experiments.	Open source.	http://www.sph.umich.edu/ ~ghoshd/COMPBIO/mixture1/ inde0x.html

(continued)

Package	Author	Interface/ operating system	Features	License	URL
PAM (prediction analysis for microarrays)	Tibshirani Lab	Windows, Unix	Performs sample classification from gene expression data, estimates prediction error via cross-validation, provides a list of significant genes whose expression characterizes each diagnostic class.	GPL2.0.	http://www.stat.standord.edu/%7Etibs/PAM/Rdist/index.html
POE (probability of expression)	Elizabeth Garrett, Jiang Hu, Giovanni Parmigiani, Rob Scharpf	Windows, Unix	Statistical approaches to molecular classification that emphasize simple molecular profiles based on latent categories signifying under-, over-, and baseline-expression.	GNU GPL 2.	http://astor.som.jhmi.edu/poe/
OOMAL (object-oriented microarray analysis library) (*require S-PLUS!)	MD Anderson Cancer Center, The University of Texas	Windows, Unix	Object-oriented library for analyzing microarray data in S-PLUS, flexible tools for loading raw quantification data from a variety of microarray formats, normalization, identified differentially expressed genes, classification and discrimination between samples.		http://www3.mdanderson.org/depts./cancergenomics/oomal.html

Product	Company/institute	Interface/server operating system	Features	License	URL
R/maanova	Gary Churchill's Statistical Genetics Group, the Jackson Laboratory	Windows, Unix	R/maanova is an extensible, interactive environment for the analysis of variance on microarray data. The package contains some simple function for exploratory microarray analysis, M-A plots, lowess curve fitting, and handles replicate array data by Bayesian methods.	Free for academic use.	http://www.jax.org/research/churchill/software/anova/Rmaanova/index.html
SMA (statistics for microarray analysis)	Sandrine Dudoit, Yee Hwa (Jean) Yang, Benjamin Milo BOLSTAD (UC Berkeley)	Windows, Unix		GNU GPL (version 2 or later).	http://www.stat.berkeley.edu/users/terry/zarray/Software/smocode.html
SMA extension (com. braju. sma)	Henrik Bengtsson	Windows, Unix	Extensions of SMA.	GNU GPL.	http://www.maths.lth.se/help/R/com.braju.sma/
YASMA (yet another statistical microarray analysis)	Lorenz Wernisch and others	Windows, Unix	Correlation between array replicates, ANOVA analysis, p-values for ANOVA analysis, standard t-tests.	GNU GPL.	http://www.cryst.bbk.ac.uk/~wernisch/yasma.html

2. Data annotation software (in alphabetical order)

Product	Company/institute	Interface/server operating system	Features	License	URL
DRAGON			Upload microarray data, annotate and visualize the position and expression level of genes in cellular pathways.	No pricing information is available.	http://207.123.190.10/dragon.htm

(continued)

Product	Company/ institute	Interface/server operating system	Features	License	URL
EASE	NCBI	Windows	Customizable, stand alone software application that facilitates the biological interpretation of gene lists derived from the results of microarray, proteomic, and SAGE experiments.	Free.	http://david.niaid.nih.gov/david/ease.htm
RESOURCERER	TIGR	Windows	Provides annotation based on the TIGR Gene Indices (TGI) for commonly available microarray resources.	Public source.	http://pga.tigr.org/tigr-scripts/magic/r1.pl
3. Network/pathway reconstruction and analysis software					
GenMAPP	Conklin lab at the Gladstone Institutes and the University of California at San Francisco.	Windows	A computer application designed to visualize gene expression data on maps representing biological pathways and grouping of genes.	Free.	http://www.genemapp.org
Pathway Processor	Paul Grosu	Windows	A statistical package that organizes profiles of gene expression according to the metabolic pathways that are affected and features a unique graphical output.	Free for academic use.	http://cg.harvard.edu/cavalieri/pp.html
PubGene	PubGene Inc.	Windows	Analyze gene expression data with literature network information	Public source.	http://www.pubgene.org

Source: Refs. 68 and 85.

are referred to a more comprehensive description of the online list which contains further details on the particular software systems features, platform requirements, license prices, and other information.

Table 4 shows some commonly used academic and commercial statistical software and technical programming languages, including our own, HDBStat! (www.soph.uab.edu/ssg_content.asp?id = 1164). Of course, there are many other software programs with good qualities not included in these tables. Each software has its own uses, advantages, and disadvantages. Many have a feature that can extend functionality by programming new tools as extensions. This allows software to incorporate new analyses promptly. However, it requires users to have an understanding of the structure of their data and the statistical background of a particular analysis in order to perform an operation properly and conduct the analysis correctly.

There are two broad categories of software. One is a comprehensive software that incorporates many different analyses simultaneously like data preprocessing, dimensionality reduction, normalization, clustering, and visualization in a single package. Another is software that performs only one or a few specific analyses. Certain commercial software packages are quite expensive, make fancy graphic displays, and may appear very comprehensive at first glance. However, our experience is that these generally do not have much true analytic capability under the hood. In this respect, they are generally far outstripped by freely available software from academic and government based investigators. On the other hand software packages from academic sources are often harder to use, are not as well documented, and are not as thoroughly debugged, so greater effort and sophistication may be needed to get them to fully utilize them.

8. INTERPRETATION

The final step in the analysis involves interpreting what has been found. This is as much an endeavor in biology and bioinformatics as it is in statistics. When ultimately used to tie the statistical information obtained to the former analysis to other biological information to draw conclusions about what is going on with respect to the phenomena under study, this can be an extremely time-consuming and demanding process. It would be beyond the scope of this chapter to address it fully, but other references may be useful in this regard (168–173).

9. CONCLUSION

In conclusion, we believe that microarray research has opened up new and exciting opportunities for investigators. Many challenges remain to be addressed and investigators must stay on their toes to distinguish between the hope of gutsy analyses and procedures with backing of validity. Nevertheless, in our opinion, the ratio of fun and discovery to confusion is a good one and we encourage investigators to join in.

ACKNOWLEDGMENTS

Supported in part by NIH grants R01DK56366, P30DK56336, U24DK058776, and R01ES09912, NSF grant 0090286, and a grant from the UAB-HSF-GEF.

REFERENCES

1. Kallioniemi OP. Biochip technologies in cancer research. Ann Med 2001; 33:142–147.
2. Shi MM. Technologies for individual genotyping: detection of genetic polymorphisms in drug targets and disease genes. Am J Pharmacogenomics. 2002; 2:197–205.
3. Templin MF, Stoll D, Schrenk M, Traub PC, Vohringer CF, Joos TO. Protein microarray technology. Drug Discov Today 2002; 7:815–822.
4. Schena M. Genome analysis with gene expression microarrays. Bioessays. 1996; 18:427–431.
5. Gadbury GL, Page GP, Edwards J, Kayo T, Prolla TA, Weindruch R, Permana PA, Mountz J, Allison DB. Power and sample size estimation in high dimensional biology. Stat Methods Med Res. In press.
6. Gerhold D, Lu M, Xu J, Austin C, Caskey CT, Rushmore T. Monitoring expression of genes involved in drug metabolism and toxicology using DNA microarrays. Physiol Genomics 2001; 5:161–170.
7. Thieffry D. From global expression data to gene networks. Bioessays 1999; 21:895–899.
8. Yang D, Zakharkin SO, Page G, Brand JPL, Edwards JW, Bartolucci AA, Allison DB. Applications of Bayesian statistical methods in microarray data analysis. American Journal of Pharmaco Genomics. In press.
9. Ohno R, Nakamura Y. Prediction of response to imatinib by cDNA microarray analysis. Semin Hematol 2003; 40:42–49.
10. Zhang H, Yu CY. Tree-based analysis of microarray data for classifying breast cancer. Front Biosci 2002; 7:c63–c67.
11. Lee ML, Whitmore GA. Power and sample size for DNA microarray studies. Stat Med 2002; 21:3543–3570.
12. Li J, Yu X, Pan W, Unger RH. Gene expression profile of rat adipose tissue at the onset of high-fat-diet obesity. Am J Physiol Endocrinol Metab 2002; 282:E1334–E1341.
13. Lopez IC, Marti A, Milagro FI, de los Angeles Zulet M, Moreno-Aliaga MJ, Martinez JA, De Miguel C. DNA Microarray analysis of genes differentially expressed in diet-induced (cafeteria) obese rats. Obes Res 2003; 11:188–194.

14. Kim S, Urs S, Massiera F, Wortman P, Joshi R, Heo Y-R, Andersen B, Kobayashi H, Teboul M, Ailhaud G, Quignard-Boulangé A, Fukamiza A, Jones BH, Kim JH, Moustaid-Moussa N. Effects of high-fat diet, angiotensinogen (agt) gene inactivation, and targeted gene expression to adipose tissue on lipid metabolism and renal gene expression. Horm Metab Res 2002; 34:721–725.

15. Cohen P, Miyazaki M, Socci ND, Hagge-Greenberg A, Liedtke W, Soukas AA, Sharma R, Hudgins LC, Ntambi JM, Friedman JM. Role for stearoyl-CoA desaturase-1 in leptin-mediated weight loss. Science 2002; 297:240–243.

16. Ferrante AW Jr, Thearle M, Liao T, Leibel RL. Effects of leptin deficiency and short-term repletion on hepatic gene expression in genetically obese mice. Diabetes 2001; 50:2268–2278.

17. Sone H, Shimano H, Sakakura Y, Inoue N, Amemiya-Kudo M, Yahagi N, Osawa M, Suzuki H, Yokoo T, Takahashi A, Iida K, Toyoshima H, Iwama A, Yamada N. Acetyl-coenzyme A synthetase is a lipogenic enzyme controlled by SREBP-1 and energy status. Am J Physiol Endocrinol Metab 2002; 282: E222–E230.

18. Iqbal MJ, Yaegashi S, Ahsan R, Lightfoot DA, Banz WJ. Differentially abundant mRNAs in rat liver in response to diets containing soy protein isolate. Physiol Genomics 2002; 11:219–226.

19. Kaszubska W, Douglas Falls H, Schaefer VG, Haasch D, Frost L, Hessler P, Kroeger PE, White DW, Jirousek MR, Trevillyan JM. Protein tyrosine phosphatase 1B negatively regulates leptin signaling in a hypothalamic cell line. Mol Cell Endo 2002; 195:109–118.

20. Liang C-P, Tall AR. Transcriptional profiling reveals global defects in energy metabolism, lipoprotein, and bile acid synthesis and transport with reversal by leptin treatment in Ob/ob mouse liver. J Biol Chem 2001; 276: 49066–49076.

21. Waring JF, Ciurlionis R, Clampit JE, Morgan S, Gum RJ, Jolly RA, Kroeger P, Frost L, Trevillyan J, Zinker BA, Jirousek M, Ulrich RG, Rondinone CM. PTP1B antisense-treated mice show regulation of genes involved in lipogenesis in liver and fat. Mol Cell Endocrinol 2003; 203:155–168.

22. Aoki K, Sun Y-J, Aoki S, Wada K, Wada E. Cloning, expression, and mapping of a gene that is upregulated in adipose tissue of mice deficient in bombesin receptor subtype-3. Biochem Biophys Res Comm 2002; 290: 1282–1288.

23. Carré W, Bourneuf E, Douaire M, Diot C. Differential expression and genetic variation of hepatic messenger RNAs from genetically lean and fat chickens. Gene 2002; 299:235–245.

24. Stricker-Krongrad A, Dimitrov T, Beck B. Central and peripheral dysregulation of melanin-concentrating hormone in obese Zucker rats. Mol Brain Res 2001; 92:43–48.

25. Tobe K, Suzuki R, Aoyama M, Yamauchi T, Kamon J, Kubota N, Terauchi Y, Matsui J, Akanuma Y, Kimura S, Tanaka J, Abe M, Ohsumi J, Nagai R, Kadowaki T. Increased expression of the sterol regulatory element-binding

protein-1 gene in insulin receptor substrate-2^{-1-} mouse liver. J Biol Chem 2001; 276:38337–38340.

26. Yang X, Pratley RE, Tokraks S, Bogardus C, Permana PA. Microarray profiling of skeletal muscle tissues from equally obese, non-diabetic insulin-sensitive and insulin-resistant Pima Indians. Diabetologia 2002a; 45:1584–1593.

27. Boeuf S, Keijer J, Franssen-van Hal NLW, Klaus S. Individual variation of adipose gene expression and identification of covariated genes by cDNA microarrays. Physiol Genomics 2002; 11:31–36.

28. Ramis JM, Franssen-van Hal NLW, Kramer E, Llado I, Bouillard F, Palou A, Keijer J. Carboxypeptidase E and thrombospondin-1 are differently expressed in subcutaneous and visceral fat of obese subjects. Cell Mol Life Sci 2002; 59:1960–1971.

29. Yang Y-S, Song H-D, Shi W-J, Hu R-M, Han Z-G, Chen J-L. Chromosome localization analysis of genes strongly expressed in human visceral adipose tissue. Endocrine 2002; 18:57–66.

30. Jessen BA, Stevens GJ. Expression profiling during adipocyte differentiation of 3T3-L1 fibroblasts. Gene 2002; 299:95–100.

31. Moldes M, Lasnier F, Gauthereau X, Klein C, Pairault J, Fève B, Chambault-Guérin AM. Tumor necrosis factor-alpha-induced adipose-related protein (TIARP), a cell surface protein that is highly induced by tumor necrosis factor-alpha and adipose conversion. J Biol Chem 2001; 276:33938–33946.

32. Ross SE, Erickson RL, Gerin I, DeRose PM, Bajnok L, Longo KA, Misek DE, Kuick R, Hanash SM, Atkins KB, Andresen SM, Nebb HI, Madsen L, Kristiansen K, MacDougald OA. Microarray analyses during adipogenesis: understanding the effects of WNT signaling on adipogenesis and the roles of liver X receptor α in adipocyte metabolism. Mol Cell Biol 2002; 22:5989–5999.

33. Reimer T, Koczan D, Gerber B, Richter D, Thiesen HJ, Friese K. Microarray analysis of differentially expressed genes in placental tissue of pre-eclampsia: up-regulation of obesity-related genes. Mol Hum Reprod 2002; 8:674–680.

34. Stein TP, Schluter MD, Galante AT, Soteropoulos P, Tolias PP, Grindeland RE, Moran MM, Wang TJ, Polansky M, Wake DE. Energy metabolism pathways in rat muscle under conditions of simulated microgravity. J Nutr Biochem 2002; 13:471–478.

35. Matsuzaka T, Shimano H, Yahagi N, Yoshikawa T, Amemiya-Kudo M, Hasty AH, Okazaki H, Tamura Y, Iizuka Y, Ohashi K, Osuga J-I, Takahashi A, Yato S, Sone H, Ishibashi S, Yamada N. Cloning and characterization of a mammalian fatty acyl-CoA elongase as a lipogenic enzyme regulated by SREBPs. J Lipid Res 2002; 43:911–920.

36. Robinson SW, Dinulescu DM, Cone RD. Genetic models of obesity and energy balance in the mouse. Annu Rev Genet 2000; 3:687–745.

37. Eurlings PMH, van der Kallen CJH, Geurts JMW, Kouwenberg P, Boecks WD, de Bruin TWA. Identification of differentially expressed genes in subcutaneous adipose tissue from subjects with familial combined hyperlipidemia. J Lipid Res 2002; 43:930–935.

38. Yang Y-S, Song H-D, Li R-Y, Zhou L-B, Zhu Z-D, Hu R-M, Han Z-G, Chen J-L. The gene expression profiling of human visceral adipose tissue and its secretory functions. Biochem Biophys Res Comm 2003; 300:839–846.

39. Mariani TJ, Budhraja V, Mecham BH, Gu CC, Watson MA, Sadovsky Y. A variable fold change threshold determines significance for expression microarrays. FASEB 2003; J17:321–323.

40. Townsend JP, Hartl DL. Bayesian analysis of gene expression levels: statistical quantification of relative mRNA level across multiple strains or treatments. Genome Biol 2002; 3:RESEARCH0071.

41. Schadt EE, Monks SA, Drake TA, Lusis AJ, Che N, Colinayo V, Ruff TG, Milligan SB, Lamb JR, Cavet G, Linsley PS, Mao M, Stoughton RB, Friend SH. Genetics of gene expression surveyed in maize, mouse and man. Nature 2003; 422:297–302.

42. Deng HW, Deng H, Liu YJ, Liu YZ, Xu FH, Shen H, Conway T, Li JL, Huang QY, Davies KM, Recker RR. A genomewide linkage scan for quantitative-trait loci for obesity phenotypes. Am J Hum Genet 2002; 70:1138–1151.

43. Zhu X, Cooper RS, Luke A, Chen G, Wu X, Kan D, Chakravarti A, Weder A. A genome-wide scan for obesity in African-Americans. Diabetes 2002; 51:541–544.

44. Brockmann GA, Bevova MR. Using mouse models to dissect the genetics of obesity. Trends Genet 2002; 18:367–376.

45. Borecki IB, Rice T, Pérusse L, Bouchard C, Rao DC. An exploratory investigation of genetic linkage with body composition and fatness phenotypes: the Quebec Family Study. Obesity Res 1994; 2:213–219.

46. Lembertas AV, Pérusse L, Chagnon YC, Fisler JS, Warden CH, Purcell-Huynh DA, Dionne FT, Gagnon J, Nadeau A, Lusis AJ, Bouchard C. Identification of an obesity quantitative trait locus on mouse chromosome 2 and evidence of linkage to body fat and insulin on the human homologous region 20q. J Clin Invest 1997; 100:1240–1247.

47. Barrett JC, Kawasaki ES. Microarrays: the use of oligonucleotides and cDNA for the analysis of gene expression. Drug Discov Today 2003; 8:134–141.

48. Heller MJ. DNA microarray technology: devices, systems, and applications. Annu Rev Biomed Eng 2002; 4:129–153.

49. Lorenz MG, Cortes LM, Lorenz JJ, Liu ET. Strategy for the design of custom cDNA microarrays. Biotechniques 2003; 34:1264–1270.

50. Chismar JD, Mondala T, Fox HS, Roberts E, Langford D, Masliah E, Salomon DR, Head SR. Analysis of result variability from high-density oligonucleotide arrays comparing same-species and cross-species hybridizations. Biotechniques 2002; 33:516–518,520,522.

51. Moody D, Zou Z, McIntyre L. Cross-species hybridisation of pig RNA to human nylon microarrays. BMC Genomics 2002; 3:27.

52. Kayo T, Allison DB, Weindruch R, Prolla TA. Influences of aging and caloric restriction on the transcriptional profile of skeletal muscle from rhesus monkeys. Proc Natl Acad Sci USA 2001; 98:5093–5098.

53. Allen MJ, Yen WM. Introduction to measurement theory. Monterey, CA: Brooks/Cole, 1979:72–94.

54. Jackson DN. A sequential system for personality scale development. Spielberger CD, ed. Current Topics in Clinical and Community Psychology. New York: Academic Press1970:2:61–96.

55. Nimgaonkar A, Sanoudou D, Butte AJ, Haslett JN, Kunkel LM, Beggs AH, Kohane IS. Reproducibility of gene expression across generations of Affymetrix microarrays. BMC Bioinformatics 2003; 4:27.

56. Cheung VG, Conlin LK, Weber TM, Arcaro M, Jen KY, Morley M, Spielman RS. Natural variation in human gene expression assessed in lymphoblastoid cells. Nat Genet 2003; 33:422–425.

57. Kothapalli R, Yoder SJ, Mane S, Loughran TP Jr. Microarray results: how accurate are they? BMC Bioinformatics 2002; 3:22.

58. Kuo WP, Jenssen TK, Butte AJ, Ohno-Machado L, Kohane IS. Analysis of matched mRNA measurements from two different microarray technologies. Bioinformatics 2002; 18:405–412.

59. Costigan M, Befort K, Karchewski L, Griffin RS, D'Urso D, Allchorne A, Sitarski J, Mannion JW, Pratt RE, Woolf CJ. Replicate high-density rat genome oligonucleotide microarrays reveal hundreds of regulated genes in the dorsal root ganglion after peripheral nerve injury. BMC Neurosci 2002; 3:16.

60. Novak JP, Sladek R, Hudson TJ. Characterization of variability in large-scale gene expression data: implications for study design. Genomics 2002; 79:104–113.

61. Spruill SE, Lu J, Hardy S, Weir B. Assessing sources of variability in microarray gene expression data. Biotechniques 2002; 33:916–920,922–923.

62. Wang HY, Malek RL, Kwitek AE, Greene AS, Luu TV, Behbahani B, Frank B, Quackenbush J, Lee NH. Assessing unmodified 70-mer oligonucleotide probe performance on glass-slide microarrays. Genome Biol 2003; 4:R5.

63. Yuen T, Wurmbach E, Pfeffer RL, Ebersole BJ, Sealfon SC. Accuracy and calibration of commercial oligonucleotide and custom cDNA microarrays. Nucleic Acids Res 2002; 30:e48.

64. Yue H, Eastman PS, Wang BB, Minor J, Doctolero MH, Nuttall RL, Stack R, Becker JW, Montgomery JR, Vainer M, Johnston R. An evaluation of the performance of cDNA microarrays for detecting changes in global mRNA expression. Nucleic Acids Res 2001; 29:E41.

65. Jenssen TK, Langaas M, Kuo WP, B Smith-Sorensen, Myklebost O, Hovig E. Analysis of repeatability in spotted cDNA microarrays. Nucleic Acids Res 2002; 30:3235–3344.

66. Page GP, Edwards JW, Barnes S, Weindruch R, Allison DB. A design & statistical perspective on microarray gene expression studies in nutrition: the need for playful creativity and scientific hard-mindedness. Nutrition. In press.

67. Rubin DB. Practical implications of modes of statistical-inference for causal effects and the critical role of the assignment mechanism. Biometrics 1991; 47:1213–1234.

68. Dobbin K, Shih JH, Simon R. Statistical design of reverse dye microarrays. Bioinformatics 2003; 19:803–810.

69. Liang M, Briggs AG, Rute E, Greene AS, Cowley Jr AW. Quantitative assessment of the importance of dye switching and biological replication in cDNA microarray studies. Physiol Genomics 2003; 14:199–207.

70. Lee ML, Kuo FC, Whitmore GA, Sklar J. Importance of replication in microarray gene expression studies: statistical methods and evidence from repetitive cDNA hybridizations. Proc Natl Acad Sci USA 2000; 97:9834–9839.

71. Beasley TM, Page GP, Brand JPL, Gadbury GL, Mountz JD, Allison DB. Chebyshev's inequality for non-parametric testing with small N and a in microarray research. J R Stat Soc Ser C Appl Stat. In press.

72. Brand JPL, Gadbury GL, Beasley TM, Page GP, Long JD, Edwards JW, Allison DB. Non-parametric alternatives for inferential testing in microarray research. Submitted.

73. Pavlidis P, Li Q, Noble WS. The effect of replication on gene expression microarray experiments. Bioinformatics 2003; 19:1620–1627.

74. Allison DB, Beasley TM, Fernandez J, Heo M, Zhu S, Etzel C, Amos CI. Bias in estimates of quantitative-trait-locus effect in genome scans: Demonstration of the phenomenon and a method-of-moments procedure for reducing bias. Am J Hum Genet 2002; 70:575–585.

75. Lee ML, Whitmore GA. Power and sample size for DNA microarray studies. Stat Med 2002; 21:3543–3570.

76. Pan W, Lin J, Le CT. How many replicates of arrays are required to detect gene expression changes in microarray experiments? A mixture model approach. Genome Biol 2002; 3:RESEARCH0022.

77. Hwang D, Schmitt WA, Stephanopoulos G, Stephanopoulos G. Determination of minimum sample size and discriminatory expression patterns in microarray data. Bioinformatics 2002; 18:1184–1193.

78. Mukherjee S, Tamayo P, Rogers S, Rifkin R, Engle A, Campbell C, Golub TR, Mesirov JP. Estimating dataset size requirements for classifying DNA microarray data. J Comput Biol 2003; 10:119–142.

79. Jiang CH, Tsien JZ, Schultz PG, Hu Y. The effects of aging on gene expression in the hypothalamus and cortex of mice. Proc Natl Acad Sci USA 2001; 98:1930–1934.

80. Cantz T, Jochheim A, Cieslak A, Hillemann T, Scharf J, Manns MP, Ott M. PCR-based quantification of amplified RNA from laser microdissected mouse liver samples. Exp Mol Pathol 2003; 75:53–57.

81. Saghizadeh M, Brown DJ, Tajbakhsh J, Chen Z, Kenney MC, Farber DB, Nelson SF. Evaluation of techniques using amplified nucleic acid probes for gene expression profiling. Biomol Eng 2003; 20:97–106.

82. Allison DB. Statistical methods for microarray research for drug target identification. Proceedings of the American Statistical Association, Biopharmaceutical Section [CD-ROM], Alexandria, VA, 2002, http://www.soph.uab.edu/statgenetics/Research/Allison-JSM-2002-paper-002a.pdf.

83. Kendziorski CM, Zhang Y, Lan H, Attie AD. The efficiency of mRNA pooling in microarray experiments. Biostatistics 2003; 4:465–477.

84. Peng X, Wood CL, Blalock EM, Chen KC, Landfield PW, Stromberg AJ. Statistical implications of pooling NA samples for microarray experiments. BMC Bioinformatics 2003; 4:26.
85. Simon R, Radmacher MD, Dobbin K. Design of studies using DNA microarrays. Genet Epidemiol 2002; 23:21–36.
86. Wrobel G, Schlingemann J, Hummerich L, Kramer H, Lichter P, Hahn M. Optimization of high-density cDNA-microarray protocols by 'design of experiments'. Nucleic Acids Res 2003; 31:e67.
87. Kerr MK, Churchill GA. Experimental design for gene expression microarrays. Biostatistics 2001; 2:183–201.
88. Kerr MK, Churchill GA. Statistical design and the analysis of gene expression microarray data. Genet Res 2001; 77:123–128.
89. Kerr MK. Experimental design to make the most of microarray studies. Methods Mol Biol 2003; 224:137–147.
90. Tukey J. On the comparative anatomy of transformation. Annals of Mathematical Statistics 1964; 28:602–632.
91. Finkelstein D, Gollub J, Ewing R, Sterky F, Cherry J.M., Somerville S. Microarray data quality analysis: lessons from the AFGC project. Plant Molecular Biology 2002; 48:119–131.
92. Yang YH, Dudoit S, Luu P, Lin DM, Peng V, Ngai J, Speed TP. Normalization for cDNA microarray data: a robust composite method addressing single and multiple slide systematic variation. Nucleic Acids Res 2002; 30:e15.
93. Hoffmann R, Seidl T, Dugas M. Profound effect of normalization on detection of differentially expressed genes in oligonucleotide microarray data analysis. Genome Biol 2002; 3:research0033.1–research0033.11.
94. Bolstad B, Irizarry R, Astrand M, Speed T. A comparison of normalization methods for high density oligonucleotide array based on variance and bias. Bioinformatics 2003; 19:185–193.
95. Workman C, Jensen L, Jarmer H, Berka R, Gautier L, Nielser H, Saxild HH, Nielsen C, Brunak S, Knudsen S. A new non-linear normalization method for reducing variability in DNA microarray experiments. Genome Biol 2002; 3:RESEARCH0048.
96. Quackenbush J. Microarray data normalization and transformation. Nat Genet 2002; 32:496–501.
97. Astrand M. Contrast normalization of oligonucleotide arrays. J Comput Biol 2003; 10:95–102.
98. Edwards D. Non-linear normalization and background correction in one-channel cDNA microarray studies. Bioinformatics 2003; 19:825–833.
99. Wilson DL, Buckley MJ, Helliwell CA, Wilson IW. New normalization methods for cDNA microarray data. Bioinformatics 2003; 19:1325–1332.
100. Cleveland WS. Robust Locally Weighted Regression and Smoothing Scatterplots. Journal of the American Statistical Association 1979; 74:829–836.
101. Hastie T, Tibshirani R. Exploring the nature of covariate effects in the proportional hazards model. Biometrics 1990; 46:1005–1016.

102. Hsiao LL, Dangond F, Yoshida T, Hong R, Jensen RV, Misra J, Dillon W, Lee KF, Clark KE, Haverty P, Weng Z, Mutter GL, Frosch MP, Macdonald ME, Milford EL, Crum CP, Bueno R, Pratt RE, Mahadevappa M, Warrington JA, Stephanopoulos G, Stephanopoulos G, Gullans SR. A compendium of gene expression in normal human tissues. Physiol Genomics. Dec 21 2001; 7(2):97–104.

103. Yang IV, Chen E, Hasseman JP, Liang W, Frank BC, Wang S, Sharov V, Saeed AI, White J, Li J, Lee NH, Yeatman TJ, Quackenbush J. Within the fold: assessing differential expression measures and reproducibility in microarray assays. Genome Biol 2002; 3:research0062.

104. Box GEP, Cox DR. An analysis of transformations. J R Stat Soc Ser B 1964; 26:211–252.

105. Durbin BP, Hardin JS, Hawkins DM, Rocke DM. A variance-stabilizing transformation for gene-expression microarray data. Bioinformatics 2002; 18:S105–S110.

106. Durbin B, Rocke DM. Estimation of transformation parameters for microarray data. Bioinformatics 2003; 19:1360–1367.

107. Rocke DM, Durbin B. Approximate variance-stabilizing transformations for gene-expression microarray data. Bioinformatics 2003; 19:966–972.

108. Lemon WJ, Palatini JJ, Krahe R, Wright FA. Theoretical and experimental comparisons of gene expression indexes for oligonucleotide arrays. Bioinformatics 2002; 18:1470–1476.

109. Li C, Hung Wong W. Model-based analysis of oligonucleotide arrays: model validation, design issues and standard error application. Genome Biol 2001; 2:RESEARCH0032.

110. Efron B, Tibshirani R. Empirical Bayes methods and false discovery rates for microarrays. Genet Epidemiol 2002; 23:70–86.

111. Welch BL. The significance of the difference between two means when the population variances are unequal. Biometrika 1938; 25:350–362.

112. Saville DJ. Multiple comparison procedures: The practical solution. The American Statistician 1990; 44:174–180.

113. Curran-Everett D. Multiple comparisons: philosophies and illustrations. Am J Physiol Regul Integr Comp Physiol 2000; 279:R1–R8.

114. Greenland S, Robins JM. Empirical-Bayes adjustments for multiple comparisons are sometimes useful. Epidemiology 1991; 2:244–251.

115. Rothman KJ. No adjustments are needed for multiple comparisons. Epidemiology 1990; 1:43–46.

116. Benjamini Y, Hochberg Y. On adaptive control of the false discovery rate in multiple testing with independent statistics. Journal of Educational and Behavioral Statistics 2000; 25:60–83.

117. Reiner A, Yekutieli D, Benjamini Y. Identifying differentially expressed genes using false discovery rate controlling procedures. Bioinformatics 2003; 19:368–375.

118. Berry SM, Berry DA. Accounting for multiplicities in assessing drug safety. Biometrics. In press.

119. Edwards JW, Page GP, Gadbury G, Heo M, Kayo T, Weindruch R, Allison DB. Empirical Bayes (EB) estimation of gene-specific effects in microarray research. Submitted.

120. Brand JPL, Page GP, Gadbury GL, Edwards JW, Kyoungmi K, Beasley TM, Barnes S, Allison DB. Alpha sepending algorithm for more powerful multiplicity control in high dimensional biology: an application to inferential testing in microarray data analysis. Proceedings of Plant and Animal Genome XII, 2004.

121. Baldi P, Long AD. A Bayesian framework for the analysis of microarray expression data: regularized t -test and statistical inferences of gene changes. Bioinformatics 2001; 17:509–519.

122. Cui X, Churchill GA. Statistical tests for differential expression in cDNA microarray experiments. Genome Biol 2003; 4:210.

123. Kerr MK, Martin M, Churchill GA. Analysis of variance for gene expression microarray data. J Comput Biol 2000; 7:819–837.

124. Wolfinger RD, Gibson G, Wolfinger ED, Bennett L, Hamadeh H, Bushel P, Afshari C, Paules RS. Assessing gene significance from cDNA microarray expression data via mixed models. J Comput Biol 2001; 8:625–637.

125. Cofield SS. Mixed-effects models with variance groups in R and the trace information criteria (TRIC). Ph.D. dissertation, Virginia Commonwealth University, Richmond, VA, 2003.

126. Cui X, Hwang JTG, Qiu J, Blades NJ, Churchill GA. Improved statistical tests for differential gene expression by shrinking variance components estimates. Submitted.

127. Catellier DJ, Muller KE. Tests for Gaussian repeated measures with missing data in small samples. Stat Med 2000; 19:1101–1114.

128. Chuaqui RF, Bonner RF, Best CJ, Gillespie JW, Flaig MJ, Hewitt SM, Phillips JL, Krizman DB, Tangrea MA, Ahram M, Linehan WM, Knezevic V, Emmert-Buck MR. Post-analysis follow-up and validation of microarray experiments. Nat Genet 2002; 32:509–514.

129. Choi JK, Yu U, Kim S, Yoo OJ. Combining multiple microarray studies and modeling interstudy variation. Bioinformatics 2003; 19:184–190.

130. Feinstein AR. Indexes of contrast and quantitative significance for comparisons of two groups. Stat Med 1999; 18:2557–2581.

131. Rosenthal R. Meta-analysis procedures for social science research. Beverly Hills: Sage Publications, 1984:110–126.

132. Stevens SS. On the theory of scales of measurement. Science 1946; 103: 677–680.

133. Chu TM, Weir B, Wolfinger R. A systematic statistical linear modeling approach to oligonucleotide array experiments. Math Biosci 2002; 176:35–51.

134. Morris CN. Parametric Empirical Bayes Inference: Theory and Applications. J Amer Statist Assoc 1983; 78:47–65.

135. Cherepinsky V, Feng J, Rejali M, Mishra B. Shrinkage-based similarity metric for cluster analysis of microarray data. Proc Natl Acad Sci USA 2003; 100:9668–9673.

136. Somorjai RL, Dolenko B, Baumgartner R. Class prediction and discovery using gene microarray and proteomics mass spectroscopy data: curses, caveats, cautions. Bioinformatics 2003; 19:1484–1491.

137. Radmacher MD, McShane LM, Simon R. A paradigm for class prediction using gene expression profiles. J Comput Biol 2002; 9:505–511.

138. Ringner M, Peterson C, Khan J. Analyzing array data using supervised methods. Pharmacogenomics 2002; 3:403–415.

139. Ambroise C, McLachlan GJ. Selection bias in gene extraction on the basis of microarray gene-expression data. Proc Natl Acad Sci USA 2002; 99:6562–6566.

140. Eisen MB, Spellman PT, Brown PO, Botstein D. Cluster analysis and display of genome-wide expression patterns. Proc Natl Acad Sci. USA 1998; 95:14863–14868.

141. Spellman PT, Sherlock G, Zhang MQ, Iyer VR, Anders K, Eisen MB, Brown PO, Botstein D, Futcher B. Comprehensive identification of cell cycle-regulated genes of the yeast Saccharomyces cerevisiae by microarray hybridization. Mol Biol Cell 1998; 9:3273–3297.

142. Bussemaker HJ, Li H, Siggia ED. Regulatory element detection using correlation with expression. Nat Genet 2001; 27:167–171.

143. Pilpel Y, Sudarsanam P, Church GM. Identifying regulatory networks by combinatorial analysis of promoter elements. Nat Genet 2001; 29:153–169.

144. Alizadeh AA, Eisen MB, Davis RE, Ma IS Lossos C, Rosenwald A, Boldrick JC, Sabet H, Tran T, Yu X, Powell JL, Yang L, Marti GE, Moore T, Hudson J Jr, Lu L, Lewis DB, Tibshirani R, Sherlock G, Chan WC, Greiner TC, Weisenburger DD, Armitage JO, Warnke R, Levy R, Wilson W, Grever MR, Byrd JC, Botstein D, Brown PO, Staudtl LM. Distinct types of diffuse large B-cell lymphoma identified by gene expression profiling. Nature 2000; 403:503–511.

145. Alon U, Barkai N, Notterman DA, Gish K, Ybarra S, Mack D, Levine AJ. Broad patterns of gene expression revealed by clustering analysis of tumor and normal colon tissues probed by oligonucleotide arrays. Proc Natl Acad Sci USA 1999; 96:6745–6750.

146. Bittner M, Meltzer P, Chen Y, Jiang Y, Seftor E, Hendrix M, Radmacher M, Simon R, Yakhini Z, Ben-Dor A, Sampas N, Dougherty E, Wang E, Marincola F, Gooden C, Lueders J, Glatfelter A, Pollock P, Carpten J, Gillanders E, Leja D, Dietrich K, Beaudry C, Berens M, Alberts D, Sondak V. Molecular classification of cutaneous malignant melanoma by gene expression profiling. Nature 2000; 406:536–540.

147. Golub TR, Slonim DK, Tamayo P, Huard C, Gaasenbeek M, Mesirov JP, Coller H, Loh ML, Downing JR, Caligiuri MA, Bloomfield CD, Lander ES. Molecular classification of cancer: class discovery and class prediction by gene expression monitoring. Science 1999; 286:531–537.

148. Hastie T, Tibshirani R, Eisen MB, Alizadeh A, Levy R, Staudt L, Chan WC, Botstein D, Brown P. 'Gene shaving' as a method for identifying distinct sets of genes with similar expression patterns. Genome Biol 2000; 1:1–21.

149. Dembele D, Kastner P. Fuzzy C-means method for clustering microarray data. Bioinformatics 2003; 19:973–980.

150. Tavazoie S, Hughes JD, Campbell MJ, Cho RJ, Church GM. Systematic determination of genetic network architecture. Nat Genet 1999; 22:281–285.

151. Herrero J, Dopazo J. Combining hierarchical clustering and self-organizing maps for exploratory analysis of gene expression patterns. J Proteome Res 2002; 1:467–470.

152. Tamayo P, Slonim D, Mesirov J, Zhu Q, Kitareewan S, Dmitrovsky E, Lander ES, Golub TR. Interpreting patterns of gene expression with self-organizing maps: methods and application to hematopoietic differentiation. Proc Natl Acad Sci USA 1999; 96:2907–2912.

153. Torkkola K, Gardner RM, Kaysser-Kranich T, Ma C. Self-organizing maps in mining gene expression data. Inform Sciences 2001; 139:79–96.

154. Toronen P, Kolehmainen M, Wong C, Castren E. Analysis of gene expression data using self-organizing maps. Febs Lett 1999; 451:142–146.

155. Brown MP, Grundy WN, Lin D, Cristianini N, Sugnet CW, Furey TS, Ares M Jr, Haussler D. Knowledge based analysis of microarray gene expression data by using support vector machines. Proc Natl Acad Sci USA 2000; 97:262–267.

156. Lee Y, Lee CK. Classification of multiple cancer types by tip multicategory support vector machines using gene expression data. Bioinformatics 2003; 19:1132–1139.

157. Pan W, Lin J, Le CT. Model-based cluster analysis of microarray gene-expression data. Genome Biol 2002; 3:RESEARCH009.

158. Yeung KY, Fraley C, Murua A, Raftery AE, Ruzzo WL. Model-based clustering and data transformations for gene expression data. Bioinformatics 2001; 17:977–987.

159. Ben-Dor A, Yakhini Z. Clustering gene expression patterns. J Comput Biol 1999; 6:281–297.

160. Dudoit S, Fridlyand J. Bagging to improve the accuracy of a clustering procedure. Bioinformatics 2003; 19:1090–1099.

161. Yoskioka Y, Kurei S, Machida Y. Identification of a monofunctional aspartate kinase gene of Arabidopsis thaliana with spatially and temporally regulated expression. Genes Genet Syst. 2001 Jun; 76:189–198.

162. Zhang HP, Yu CY, Singer B, Xiong MM. Recursive partitioning for tumor classification with gene expression microarray data. Proc Natl Acad Sci USA 2001; 98:6730–6735.

163. Claverie JM. Computational methods for the identification of differential and coordinated gene expression. Hum Mol Genet 1999; 8:1821–1832.

164. Datta S, Datta S. Comparisons and validation of statistical clustering techniques for microarray gene expression data. Bioinformatics 2003; 19:459–466.

165. Zhang K, Zhao H. Assessing reliability of gene clusters from gene expression data. Funct Integr Genomics 2000; 1:156–173.

166. Kaufman L, Rousseeuw PJ. Finding groups in data: an introduction to cluster analysis. New York: Wiley, 1990.

167. Gordon AD. Classification. London-New York, 1999.

168. Hosack DA, Dennis G Jr, Sherman BT, Lane HC, Lempicki RA. Identifying biological themes within lists of genes with EASE. Genome Biol 2003; 4:R70.
169. Dennis G Jr, Sherman BT, Hosack DA, Yang J, Gao W, Lane HC, Lempicki RA. DAVID: Database for annotation, visualization, and integrated discovery. Genome Biol :R602003.
170. Zeeberg BR, Feng W, Wang G, Wang MD, Fojo AT, Sunshine M, Narasimhan S, Kane DW, Reinhold WC, Lababidi S, Bussey KJ, Riss J, Barrett JC, Weinstein JN. GoMiner: a resource for biological interpretation of genomic and proteomic data. Genome Biol 2003; 4:R28.
171. Diehn M, Sherlock G, Binkley G, Jin H, Matese JC, Hernandez-Boussard T, Rees CA, Cherry JM, Botstein D, Brown PO, Alizadeh AA. SOURCE: a unified genomic resource of functional annotations, ontologies, and gene expression data. Nucleic Acids Res 2003; 31:219–223.
172. Masys DR, Welsh JB, Lynn Fink J, Gribskov M, Klacansky I, Corbeil J. Use of keyword hierarchies to interpret gene expression patterns. Bioinformatics 2001; 17:319–326.
173. Draghici S, Khatri P, Bhavsar P, Shah A, Krawetz SA, Tainsky MA. Onto-Tools, the toolkit of the modern biologist: Onto-Express, Onto-Compare, Onto-Design and Onto-Translate. Nucleic Acids Res 2003; 31:3775–3781.

8

Solving Clinical Problems in Nutrition Research with Microarrays

Steven R. Smith, Andrey Ptitsyn, Dawn M. Graunke, Hui Xie, and Robert A. Koza
Pennington Biomedical Research Center, Louisiana State University, Baton Rouge, Louisiana, U.S.A.

1. INTRODUCTION

The sequencing of the human genome provides a fantastic opportunity to advance our understanding of nutritional diseases. Researchers have struggled through the collection of tissue samples from volunteers enrolled in clinical studies and measured gene expression one gene at a time with limited success. The advent of usable, reliable microarray technology promises to revolutionize how we view—and investigate—the functioning of the cells and tissue that make up our bodies. The purpose of this chapter is to introduce the clinical investigator to the day-to-day problems encountered in the design, implementation, and analysis of microarray analysis of clinical samples and to provide the background necessary to develop solutions to these problems. We acknowledge a bias toward samples that are pertinent to nutrition/obesity and accessible outside the surgery suite (e.g., skeletal muscle and adipose tissue). It should be kept in mind that other easily accessible cells, such as immortalized lymphoblasts, may serve as surrogates for the tissue of primary interest (1).

1.2. "Having More Data Is Good." (2)

How can we best use microarrays in clinical nutrition research? This question has several very obvious answers and some that are not so obvious. First, microarrays can be used to identify single genes, or groups of genes that are upregulated or downregulated in a particular disease or nutritional condition. This is certainly the most common usage of the technology, and has provided important insight into physiology and pathophysiology of the human condition.

A second and important use of the microarray is to better understand the results of an intervention on the transcriptosome (a global view of the mRNAs in a particular tissue). The statistical techniques and experimental design issues for these uses of microarray data are covered briefly in the subsequent section. The details of the statistical analysis of before-and-after intervention studies or across-tissue comparisons will not be covered herein to prevent redundancies within this book (Allison chapter).

Another use of the microarray data is to separate a common clinical phenotype (disease, nutritional deficiency, etc.) into subtypes based on the pattern of mRNAs present in an individuals RNA sample. The pattern of mRNAs in a sample is referred to as an expression profile or a molecular fingerprint, which reflects the underlying pathophysiology of the tumor or diseased tissue relative to the normal state. For example, when reduced in complexity, microarray data can be used to predict the prognosis of cancer patients and their response to chemotherapy (3). Environment and genetics converge to regulate cellular function through coordinated regulation of large sets of genes. In other words, "...data sets produced in this way have emergent properties...patterns and systematic features become apparent and we begin to build an integrated picture of the whole system." (2)

Importantly, the design and analysis techniques identify molecular pathways underlying specific clinical/tissue subtypes and can identify single genes whose expression level serves as a surrogate marker for the complex RNA signature. These patterns and representative individual genes can be used to predict the prognosis of cancer patients and their response to chemotherapy. For example, Alizadeh et al. (3) demonstrated that patients with B-cell lymphomas whose tumors clustered into the activated B-cell category had a higher mortality rate and responded poorly to chemotherapy. The patients whose tumours clustered into the germinal center cluster had a much better response to chemotherapy and survived longer (3). An identical approach was applied to prostate and breast cancers with great success.

We propose this approach as a means to diagnose and subtype nutritional diseases.

Two important points must be considered regarding this experimental paradigm. First, this technique does not rely on a complete understanding of the underlying biology to advance our diagnostic and therapeutic acumen. In other words, we do not necessarily need to understand the function of any of the genes on the microarray to identify patterns of genes that make up a particular biological condition. For example, we did not need to know the genetic basis of blood types in order to identify and separate individuals and use this information to guide transfusion therapy. Similarly, we can subtype individuals using microarray data combined with clinical phenotypes to advance our understanding of the underlying pathophysiological state and potentially use this information to our advantage from a therapeutic standpoint. The latter use of microarrays is a branch of the field of pharmacogenomics that is only now beginning to emerge.

Second, the application of microarray technology to clinical practice will be difficult. As such, the identification of a single gene, or small set of genes, that are representative of the overall expression pattern can be identified using discriminant analysis. Dhanasekaran et al. demonstrated that the novel gene hepsin is related to survival of prostate cancer patients (17). This gene had not been previously linked to prostate cancer and was discovered as a result of microarray analysis. A similar approach can be applied to nutritional diseases.

Following this pathway, the discussions in this chapter revolve around four key areas:

1. The design of the microarray analysis,
2. The sample requirements and techniques for dealing with small RNA samples,
3. The basics of clustering algorithms for the classification of individuals into diagnostic categories, and
4. Discriminant analysis to select a single gene as a classifier or discriminant of diagnostic categories previously identified by clustering algorithms and clinical phenotype.

We will address each of these issues from the perspective of a clinical research study and point the reader toward more detailed information and resources when necessary.

2. DESIGN OF THE CLINICAL MICROARRAY STUDY

Once the clinician has determined the type of array to employ in the study (eg., Affymetrix GeneChip® arrays, cDNA microarrays, or oligonucleotide (oligo) arrays) the amount of RNA required to perform microarray experiments can be determined. In many instances of clinical applications, one concern is significantly small sample size and potentially heterogeneous sample composition. By optimizing the experimental design, researchers will make the most of the available sample. This section discusses the potential sources of variation, options for decreasing variability, and the applicable statistical tests with relation to generating the most efficient experimental design using clinical samples.

2.1. Experimental Variability

There are four main sources of experimental variability: (1) measurement error, (2) technical variation, (3) biological variation, and (4) treatment-dependent variation. Although the sources of experimental variability are universal, potential sources of error for microarray experiments include:

1. Dust or scratches on the microarray slide or chip,
2. Variations due to the laser scanner used to measure the fluorescence of probes hybridized to the target genes spotted, or immobilized on the microarray slide,
3. User-generated errors in spot location by the quantification software
4. gene × dye-specific incorporation biases.

The scanning process may involve gene spots that have either a very low intensity or a very high, overexposed (outside of the linear range of the laser) intensity. Traditionally, the operator rescans the slide at either a higher or lower laser power or PMT/gain (photo-multiplier tube) setting. To date, there is no standard way of integrating the data from numerous scans of the same slide into a single quantification of gene expression. Although this problem is of special interest to our group, it is also an area in microarray software that needs attention and a practical solution.

Next, technical variation may occur during the extraction of RNA for probe generation, reverse or in vitro transcription of the same into cDNA, hybridization of the labeled cDNA onto spotted targets, washing of the slides following hybridization, and in some cases (depending upon the technology used for detection), the development and detection of the labeled cDNA hybridized to targets. Measurement error and technical variation may be controlled and perhaps minimized by the consistent use of a well-tested and proven protocol by trained workers. Lee et al. (4) have proposed a statistical model to pool the output from a number of microarray hybridizations and suggest that the optimal number of replications of a single hybridization is three.

2.2. Interindividual Variability: Noise or Signal?

Genetically diverse populations such as humans are likely to show even greater variability in gene expression than what we have observed among inbred mice. In addition, environmental conditions cannot be carefully controlled in humans. These factors present challenges for microarray-based studies of human gene expression in vivo (5).

Biological variation is intrinsic to all organisms. Biological variation remains an important variable in clinical research. Subjects, although sometimes related, are not genetically identical. Biological variation can be due to genetic differences between organisms or changes in gene expression in response to exposure to environmental factors as demonstrated in studies of phenotypes and gene expression patterns of monozygotic twins.

Some view biological variation as noise. However, it is possible that biological variation represents the integration of genetic and environmental influences (e.g., nutrition, social influences, etc.) and as such is a signal representative of the overall genetic and environmental influences. Obviously noise is "bad" and signal is "good." Separating the two is a difficult task and the observed inter-individual variability can be viewed in either light. This will be discussed in further detail in the section on clustering algorithms.

The variability in which the scientist or clinician is most interested is the treatment-dependent or disease-dependent source of variation. By reducing the amount of the other sources of variation, the scientist or clinician greatly enhances the probability of finding genuine treatment-dependent or disease-dependent differences in gene expression using microarray technology.

2.3. Experimental Design

In order to effectively deal with the problem of variability—also called system noise—an efficient, appropriate experimental design is imperative. When designing gene expression experiments using clinical samples, it is important to consider four main factors: (1) the type of microarray to be used, (2) how the sample will be labeled for hybridization, (3) proper randomization of the study, and (4) the relevant statistical models and conclusions that can be drawn from these models. How samples are collected and processed, and how the resulting microarrays are analyzed depends upon the technology of the chip. Affymetrix data have their own caveats and will not be discussed at length in this chapter [see reviews published by the Speed laboratory (6–8) for a review of Affymetrix-specific normalization procedures]. Affymetrix chips, by design, do not require as many replications as do the cDNA or oligo microarray technologies, and a number of different human GeneChips are commercially available. The labeling technology for the cDNA and oligo microarrays is discussed later in this chapter. The choice of microarray type and the labeling method determine the first factor in experimental design, the amount of RNA required for single microarray hybridizations. The quantity of RNA available for an experiment may not be trivial when considering the size and difficulty in obtaining clinical samples.

The next factor considered is the order in which RNA samples will be labeled and compared to each other for the determination of gene expression patterns. Traditionally, an RNA sample is labeled with Cyanine-5 (Cy5) or Cyanine-3 (Cy3) and hybridized to the microarray slide. A single RNA sample may be used and the gene expression quantified. Although this process yields information regarding relative gene expression, it is not indicative of absolute gene expression in the given sample. If two samples are used, one is labeled with Cy5 and the other with Cy3. The two samples are then combined and hybridized to the gene spots on the microarray slide and later quantified using a laser scanner capable of detecting both Cy5 and Cy3.

2.4. Randomization

The concept of randomization—that the selected sample is truly representative of the population due to the randomness of the selection process—is also critical in experimental design. For microarray experiments, as well as for all well-designed experiments, if a treatment is linked to some unit (e.g., sex or genetic background), the changes in gene expression can only be discussed as associations and not true causal effects. Only careful randomization in a study will lead to a causal inference. Randomization has always been integral to well-designed clinical studies and the process of random selection is not discussed in this review.

2.5. Pooling

By creating pools of related units, i.e., samples from the same tissue from individuals in the same treatment group, biological variance is decreased. However, a gene that is grossly overexpressed or underexpressed in an individual may be missed due to the dilution effects of pooling. The most ideal situation would be to create a number of different pools using different combinations of the same RNA samples of interest and performing separate analyses. In this case, the assumption is that the variability between different pools of similar subjects would be negligible. However, when pools are composed of RNA from only a few subjects of interest, as may occur in the clinical setting, the resulting pools are not necessarily homogenous. In order to address this issue, one may use multiple assays of a single pool of RNAs to have sufficient replication of experiments (9). In general, pooling is not an optimal design but may be necessary for limiting quantities of RNA.

2.6. Experimental Plan

The first design (see Fig. 1) is a simple dye-swap experiment in which both sample A and sample B are labeled with Cy5 and Cy3 in four separate labeling reactions. Two hybridizations are represented in panel *a*, one with sample A labeled with Cy5 and sample B labeled with Cy3 (top) and the

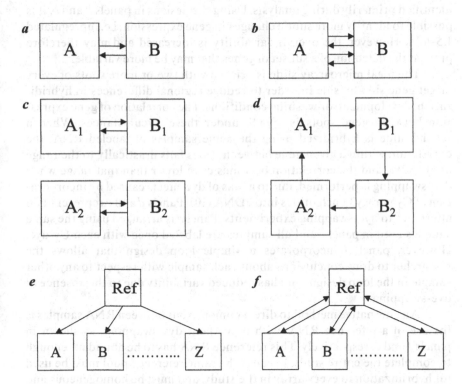

FIGURE 1 Six basic hybridization designs for microarray experiments. In all panels, the tails of arrows represent the Cy5 (red)-labeled sample and the heads of the arrows represent the Cy3 (green)-labeled samples. Panel *a* represents a single, balanced dye-swap experiment, whereas panel *b* represents multiple dye-swap experiments (technical replication). Panel *c* depicts an experimental replication of a single dye-swap experiment. Panels *d* and *e* show two alternative methods of dye-swapping experiments, with panel *d* depicting a circular design and panel *e* representing the use of a reference RNA sample. The final panel (*f*) represents a reference design incorporating dye-swapping. (Figure adapted from Ref. 12).

second hybridization has sample A labeled with Cy3 and sample B labeled with Cy5 (bottom). The dye-swap technique can be taken a step further, and multiple slides may be hybridized in order to assess technical variability, as shown in panel *b*. It is important to note, however, that although using the repeated dye-swap design, it is possible to assess the variability: The more slides used for hybridizations, the greater the degree of variability seen for a single gene. This variability is further increased by the use of the repeated dye-swap in which the experiment is reproduced with a second cohort of subjects, as shown in panel *c*. The selection criteria for up- or downregulated genes, therefore, have to become less stringent; the more robust changes are identified primarily during analysis. Using the designs in panels *b* and *c*, it is possible to identify more subtle changes in gene expression, i.e., upregulated 1.5-fold. However, the overall variability is increased and may therefore prevent the detection of a subset of genes that may be more variable.

The ideal microarray slide is printed with two or more spots of every target gene side-by-side in order to reduce regional differences in hybridization, development, or washing conditions. The correlation of gene expression between gene spots is $\geq 95\%$ under these circumstances. When a second slide is hybridized using the same samples of labeled RNA, the correlation within a given gene between slides falls drastically to the range of 60–80%, and the correlation becomes even lower than that range when dye-swapping is performed, due to a bias of dye effects caused by incorporation of Cy5 or Cy3 nucleotides into cDNA (10). Panels *d* and *e* represent two alternates to dye-swapping experiments. Panel *d* is arranged using the same samples used in panel *c* and all samples are labeled once with each Cy-dye. However, panel *d* incorporates a simple loop design that allows the researcher to draw conclusions about each sample with respect to any other sample in the loop design, and has reduced variability due to the absence of dye-swapping.

An alternative method to direct comparisons between RNA samples is the use of a reference RNA, with or without dye-swapping, as shown in panels *f* and *e*, respectively. This reference RNA has to be abundant enough to complete the entire study, because the same reference pool must be used for hybridization to every array in the study, and must be homogeneous and stable because studies may take years to complete. The reference RNA sample may be generated from specific tissues, created by mixing together RNA samples in order to have as many gene spots as possible with a positive signal, or pooled from all of the RNA samples to be used in the study. The main pitfall of the reference sample design is that half of all resources from labeling to analysis are devoted to a sample that is not of interest, and that dye effects are confounded with treatment effects unless dye-swapping is employed (11).

Given a small sample size (i.e., amount of tissue obtainable or allocated to RNA isolation), the two most likely candidates for experimental microarray design to diagnose or identify subtypes are the designs shown in panels *d* and *e* in Fig. 1. Each of these has unique challenges. One weakness of the design shown in panel *e* is that half of the experiment is dedicated to a noninformative sample, the reference RNA. Dye-incorporation differences across genes may lead to under-representation or overrepresentation of gene expression. A strong benefit of using the design shown in panel *e* is that all the microarray hybridizations are performed using the reference sample, and all the experimental samples are labeled using the same dye, which simplifies data analysis and interpretation. Given this information, the design depicted in panel *d* may be the best choice. All samples in this scenario are labeled with both dyes, eliminating dye-biases, and all gene expression data from the arrays can be expressed relative to each other. The main hindrance in using this approach is that the designs, analysis, and conclusions rapidly become very complex when more than a few samples or pools are used in an experiment, as depicted in Fig. 2. The interested reader is referred to the excellent review by Churchill (12) for additional information regarding complex experimental design.

In summary, the design of microarray studies is fairly complex. Consideration of each source of variability within the samples and experimental

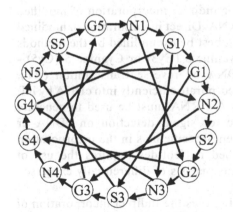

FIGURE 2 A complex microarray hybridization design incorporating three experimental groups that were sampled five times. In this design, direct comparisons between samples are performed. Note that two forward and two reverse labelings are performed for each sample, maximizing the amount of data collected for each experimental group. By setting up the hybridizations in such a loop design, it is possible to test within treatment group differences. (Figure redrawn from Ref. 12 and adapted from Ref. 8.)

design will lead the investigator to the optimal experimental design. As we will see in the next section, newer techniques for amplification and labeling of RNA may reduce the need for large samples by increasing the number of replicates possible from each sample and increasing statistical power through increased use of replicate hybridizations.

3. DEALING WITH SMALL SAMPLES IN THE CLINICAL MICROARRAY STUDY

Standard methods using direct incorporation of fluorescent-labeled nucleotides into first-strand cDNA products typically require more than 20 and up to 200 μg of total RNA per reaction. These methods become impractical when tissue sample size is limited, as is often the case in clinical studies. In recent years, methodologies have been developed enabling investigators to analyze minute amounts of starting material using high-density oligonucleotide or cDNA microarray analysis. Following an overview of labeling strategies, several recent advances in the use of cDNA and oligonucleotide microarray analysis for small tissue samples and limited RNA amounts are discussed.

3.1. Labeling Strategies

Methods for labeling samples for microarray experiments can be divided into two general categories: direct or indirect incorporation of modified nucleotides into cDNA or antisense RNA. Direct incorporation of modified or fluorescent-tagged nucleotides can best be exemplified by the methods traditionally used to incorporate Cyanine 3 (Cy3) or Cyanine 5 (Cy5) - labeled dNTPs into first-strand cDNA by reverse transcription (RT). Because Cy-labeled dNTPs do not incorporate efficiently into cDNA by RT, substantial amounts of total or poly A+ RNA must be used to generate sufficient amounts of labeled probe for signal detection on cDNA or oligonucleotide microarrays. Also, gene-specific bias in the incorporation efficiencies of Cy3- and Cy5-modified nucleotides requires the use of dye-swapping techniques for accurate analysis of differentially expressed genes.

An alternative method of labeling uses the indirect incorporation of 5-(3-aminoallyl)-dNTPs (aa-dNTPs) into first-strand cDNA synthesis followed by postlabeling coupling to N-hydroxysuccinimide-activated fluorescent dyes. The aa-dNTPs are incorporated into cDNA more efficiently than Cy-linked dNTPs. Postlabeling coupling of fluorescent dyes to aminoallyl moieties eliminates much of the bias observed with direct incorporation methods (13,14). Sensitivity comparable to direct incorporation can be achieved with as little as 10–20 μg of total RNA. Xiang et al.

2002 (15) used a modification of this method where aa-dUTP was incorporated into random hexamers used for priming for first-strand cDNA synthesis and probe generation. This increased sensitivity dramatically and allowed the use of as little as 1–5 μg total RNA for labeling.

3.2. RNA Amplification

A number of strategies have been developed to amplify starting material for use as a probe in microarray applications. These methods can be divided into two categories: (1) linear amplification and (2) exponential polymerase chain reaction (PCR) based amplification. Linear RNA amplification methods have been used to amplify as little as 10 nanograms (ng) of RNA while still conserving abundance relationships in gene expression (16–18). Briefly, first-strand cDNA is synthesized using an oligo-dT primer with an attached T7 RNA polymerase oligonucleotide-binding site. Following second-strand cDNA synthesis, the cDNA product is then subjected to in vitro transcription (IVT) using T7 RNA polymerase to produce a 1000- to 5000-fold amplification of antisense RNA (18–20). Modified nucleotides can be directly incorporated during the IVT process, or after conversion of antisense RNA to cDNA. A second round of linear amplification can also be applied to achieve $> 10^6$-fold amplification of original starting material and has been used to study gene expression in a single cell containing only 0.1 to 1 picograms (pg) of mRNA (20,21). The Affymetrix system utilizes a combination of both IVT and antibody-based signal amplification for probe labeling and signal detection (Affymetrix Inc., Santa Clara, CA). Because IVT methods tend to be quite laborious and time-consuming, several PCR-based methods have also been employed to amplify limited starting material for analysis of gene expression. Some of the disadvantages of PCR-based techniques include the possibility of disproportionate amplification of cDNAs because of size or abundance (17,22), gene-specific differences in efficiency due to sequence or secondary structure (23), and error rate of nucleotide incorporation by TAQ polymerase (20). Advantages of PCR-based amplification methods include its ease of use and ability to produce amplification ranges of up to 3×10^{11}-fold from as little as 10 pg of RNA (24). A number of commercially available products are presently available for both linear and exponential amplification of RNA for microarray analysis.

3.3. Detection Amplification

Several amplification methods have been developed to enhance the signal obtained in a microarray experiment. Signal amplification using tyramide chemistry (tyramide signal amplification; TSA), originally introduced to improve the sensitivity of immunohistochemistry (25), has been used by a number of laboratories for microarray analysis (13,26). This commercially available method (MICROMAX TSA; PerkinElmer Life Sciences, Inc., Boston, MA) is based on the enzymatic deposition of Cy3- or Cy5-tyramide adjacent to immobilized horseradish peroxidase. Signal detection can be achieved with 20–100 times less RNA than direct cDNA-labeling methods, enabling investigators to use as little as 1 μg total RNA. We routinely use this method to amplify 2–3 μg of total RNA for an 18,000+ gene oligonucleotide microarray with excellent signal-to-noise ratios. A second commercially available signal amplification method with sensitivity similar to the TSA method is based on a fluorescent oligonucleotide dendrimeric signal amplification system [(27); Genisphere, Philadelphia, PA]. Unlike the TSA method, the dendrimer detection process does not utilize the incorporation of modified dNTPs, but rather uses primers containing capture sequences for first-strand cDNA synthesis. The probes with modified 5′ ends are hybridized to the microarray followed by a secondary detection step in which specific Cy3- or Cy5-bound dendrimers are hybridized to the capture sequences. Thus far, the outlined strategies all use fluorescent detection methods for measuring changes in gene expression. A nonfluorescent ultrasensitive detection method based on resonance light scattering (RLS) by nano-sized gold and silver particles after illumination with white light (28) has been recently demonstrated for use in microarray analysis [(29); Genicon Sciences Corporation, San Diego, CA]. The signal obtained from individual RLS particles can be more than 10,000 times greater than the most sensitive fluorescent molecules and, unlike fluorescence, the RLS signal is stable and does not photo-bleach, quench, or decay. These characteristics make it possible to integrate signal strength by varying exposure time and to allow multiple measurements without risk of signal degradation.

3.4. Tissue Heterogeneity

Another important consideration with small samples, such as those obtained from clinical biopsies, is lack of tissue homogeneity. Heterogeneous tissue samples having infiltrations of contaminating tissue or cells can easily affect gene expression patterns, especially as sample sizes become smaller. Methods for amplification of RNA in gene expression analysis have become increasingly advantageous for use with small samples containing only hundreds of cells such as those isolated using laser capture microdissection or laser pressure catapulting (30–38). These methods enable investigators to select specific cell types for gene expression analysis with a minimum amount of contaminating RNA from surrounding tissue.

3.5. Degraded RNA

If not properly stored immediately following excision, RNA can rapidly become degraded in tissue samples. This can be problematic in a clinical sense because samples removed during surgical procedures cannot always be rapidly frozen or fixed in RNA preserving media (e.g., RNALater, Ambion, Inc, Austin, TX). Methods have recently been developed that enable the use of partially degraded RNA samples for gene expression studies. Generally these methods are similar to the IVT amplification methods described previously, except that an RNA polymerase-binding oligonucleotide sequence is attached to random 9-mer oligonucleotides instead of to oligo-dT to enable priming within the RNA molecule and not exclusively at the 3'-end (39).

In summary, many advances in recent years have improved both the sensitivity and the specificity of microarray analysis. These improvements have made it possible for clinical investigators to utilize a minimum amount of tissue sample and RNA while maintaining minimal ratio bias in gene expression studies.

4. CLUSTERING OF THE CLINICAL MICROARRAY DATA

4.1. General Definition and Purpose of Cluster Analysis

Cluster analysis encompasses a number of algorithms serving the purpose of assigning objects to certain categories according to their measured or counted features, so that similar objects would fall in the same category and dissimilar objects would fall – in different ones. As a statistical discipline, cluster analysis has been applied to a wide variety of scientific problems. For example, in medicine, clustering diseases, cures for diseases, or symptoms of diseases can be very useful. The correct diagnosis of clusters of symptoms such as paranoia, schizophrenia, etc. is essential for successful therapy. In general, cluster analysis is an indispensable tool in every situation when a huge amount of data has to be reduced to smaller, more manageable and meaningful chunks. Naturally, cluster analysis has been part of microarray technology ever since it was introduced. The fundamental issue of any classification is the definition of what is similar and dissimilar, in other words, the distance metric.

4.2. Choice of Distance Metric

The distance metric can be based on a single dimension or multiple dimensions. For example, if we were to cluster fast foods, we could take into account the calories and other nutritional values they contain, their price, taste rating, convenience, etc. The most straightforward way of computing distances between objects in a multidimensional space is to compute Euclidean distances. If we had a two- or three-dimensional space this measure is the actual geometric distance between objects in the space (i.e., as if measured with a ruler). However, for the purpose of cluster analysis it is irrelevant whether the distances are actual real distances, or some other derived measure of similarity that is more meaningful to the researcher; and it is up to the researcher to select the right method for the specific application.

4.2.1. Euclidean Distance

A Euclidean distance is probably the most commonly chosen type of distance. It is simply the geometric distance in multidimensional space. It is computed as:

$$\text{distance}(x, y) = \sqrt{\sum_i (x_i - y_i)^2}.$$

A common variation is squared Euclidean distance, which can be used to place progressively greater weights on objects situated further apart. Both Euclidean and squared Euclidean distances are usually computed from raw data, and not from standardized data. This method has certain advantages (e.g., the distance between any two objects is not affected by the addition of new objects to the analysis, which may be outliers). However, the distances can be greatly affected by differences in scale among the dimensions from which the distances are computed. For example, if one of the genes is expressed at a low level and then scaled up to the mean expression of the genes represented on the microarray, the resulting Euclidean or squared Euclidean distances (computed from multiple dimensions) can be greatly affected, and consequently, the results of cluster analyses may be very different. Using Euclidean distance requires a very careful choice of normalization schema.

4.2.2. Correlation Distance

This distance is probably the second most widely used in microarray data. Correlation distance is particularly useful in time-series experiments (e.g., changes in gene expression over time after the application of a treatment). It is reasonable to suppose that if two genes are coexpressed, their expression profiles (i.e., expression values throughout a series of experiments) are correlated. This distance is insensitive to the direction of change in gene expression; both up- and downregulated genes can be placed in the same cluster. The distance is computed as:

$$\text{distance}(x, y) = \frac{\sum_i (x_i - \bar{x})(y_i - \bar{y})}{\sqrt{\sum_i (x_i - \bar{x})^2 \sum_i (y_i - \bar{y})^2}}.$$

4.2.3. City-Block (Manhattan) Distance

This distance is simply the average difference across dimensions. In most cases, this distance measure yields results similar to the simple Euclidean distance. However, note that in this measure, the effect of single large differences (outliers) is dampened (because they are not squared). The city-block distance is computed as:

$$\text{distance}(\mathbf{x}, \mathbf{y}) = \sum_i |x_i - y_i|.$$

Binary or Hemming distance is a case of Manhattan distance, when all measurements are not quantitative, but binary, [i.e., only indicating presence or absence of a particular property (gene)]. In this case the distance can also be computed as the exclusive OR (XOR) of two binary vectors.

4.2.4. Chebychev Distance

This distance measure may be appropriate in cases when one wants to define two objects as different if they are different on any one of the dimensions. The Chebychev distance is computed as:

$$\text{distance}(\mathbf{x}, \mathbf{y}) = Max|x_i - y_i|.$$

4.2.5. Power Distance

Sometimes one may want to increase or decrease the progressive weight that is placed on dimensions on which the respective objects are very different. This can be accomplished via generalization of the Euclidean distance, the power distance, also called Minkowki distance. The power distance is computed as:

$$\text{distance}(\mathbf{x}, \mathbf{y}) = \left(\sum_i |x_i - y_i|^p\right)^{1/r},$$

where r and p are user-defined parameters. Parameter p controls the progressive weight that is placed on differences on individual dimensions, parameter r controls the progressive weight that is placed on larger differences between objects. If r and p are equal to 2, then this distance is equal to the Euclidean distance.

4.2.6. Percent Disagreement

This measure is particularly useful if the data for the dimensions included in the analysis are categorical in nature. This distance is computed as:

$$\text{distance}(\mathbf{x}, \mathbf{y}) = (\#x_i \neq y_i)/i.$$

4.2.7. Tahimoto Distance

This metric is most often used in taxonomy, in cases where two patterns of features (the elements in the set) are either same or different, but there is no natural notion of graded similarity:

$$\text{distance}(x, y) = \frac{n_1 + n_2 - 2n_{12}}{n_1 + n_2 - n_{12}},$$

where n_1 and n_2 are the number of elements in the first and the second sets, respectively, and n_{12} is the number of elements shared by both sets.

4.3. Types of Cluster Analysis

Once the distance metric is established, the clustering procedure can be conducted in a few radically different ways. There are virtually hundreds of published algorithms and variations, but all of them can in turn be classified into a few strategic approaches. First, clustering can be either agglomerative or disruptive. Agglomerative clustering starts with each object assigned to some cluster. This can be a cluster, containing only one element (singleton) or a group of elements, based on preexisting assumption—for example, a group sample from normal tissue introduced into cluster analysis of cancer samples. During the clustering process the objects are joined together based on their proximity to each other, measured by selected distance metric. As a result, the initial clusters grow in size, absorbing singletons and other clusters until the process stops. The great majority of algorithms currently used in microarray cluster analysis exploit the agglomerative approach. The opposite—so-called disruptive clustering—starts with all objects placed in one cluster. During the clustering process this cluster is partitioned into smaller clusters, which in turn can be subpartitioned into yet smaller clusters and so on. The process naturally stops when all clusters are broken into singletons. However, the sensible results are achieved before the final stage. On each step of subdivision a statistical metric can be devised to estimate the quality of classification. This metric can be based on local properties of the clusters (for example, density) or global property of classification, calculated for the whole data set. Further subdivision can be stopped when this classification quality metric reaches its maximum. Although agglomerative approach is not currently implemented in the major software packages for microarray analysis, there are several ongoing research projects attempting to prove the advantages of this approach for class discovery (40,41).

4.3.1. Supervised vs. Unsupervised Clustering

Second, all cluster-analysis algorithms can be classified as supervised or unsupervised. Supervised clustering utilizes some kind of preexisting information on the cluster structure in the dataset. For examples, all clusters can be predefined by known typical representatives. Otherwise, clusters can be predefined by artificially created expression profiles. For example, in a series of experiments researcher may be interested only in the genes, demonstrated one of the expected models of behavior – up- or downregulated or upregulated and returning to the initial state, etc. after a treatment.

Unsupervised algorithms utilize no predefined information about cluster structure. Some authors suggest that only application of unsupervised algorithms should be called clustering. However, there is no strict line dividing supervised and unsupervised approaches. The process of classification may include sequential application of two different algorithms or, in some algorithms, strict supervision can be substituted by assignment of priorities or weights to the clustered objects according to the preexisting information. In extreme cases, if some "model" objects are assigned priority I and all other objects assigned priority 0, clustering becomes supervised. Assigning all objects equal priority makes the algorithm unsupervised.

4.3.2. Commonly Used Clustering Algorithms

Hierarchical Tree. A hierarchical tree is the most commonly used algorithm and, arguably, the most intuitively understandable. The clustering process starts with computation of a distance matrix, $n \times n$ elements, where n is the number of objects in the data set. Each element of this matrix contains a distance measured between two objects. Then the algorithm assigns a leaf of the tree to each object (gene). On each step of the algorithm:

1. Two most similar objects of a current matrix are computed and a node joining these objects is created.
2. An expression profile for the joining node is created by averaging the expression profiles of the nodes it joins.
3. A new, smaller distance matrix is computed with a new "joining" node substituting two joined elements.

The process repeats until a single node remains. There are many possible variations of this algorithm, for example, using the group-weighted average of the distances to compute a new distance between centers. The hierarchical tree algorithms are widely used in biological sequence analysis, phylogenetic analysis, etc. The output of a hierarchical tree algorithm is not a set of clusters, but a nested tree of all possible clusters. Clusters can be isolated by cutting the branches of the tree at more or less arbitrary points.

Selecting the meaningful cutting points and verifying the resulting clusters presents a significant computational and statistical challenge.

K-Means Clustering.   K-means clustering is very different from the hierarchical tree. In a typical implementation, the number of clusters is fixed to some value and has to be defined by the researcher before the start of computation. This user-defined parameter (K) can be a reasonable guess of the number of expected regulatory patterns of gene expression profiles or a number of expected different phenotypes, known from the histological analysis or clinical data. The algorithm will assign the objects in the data set to exactly K different clusters. The computation starts with placing K seeds, or starting points. The seeding can be done differently in different implementations. Starting points can be placed randomly, or in K most distant points of the whole data set, or in the vicinity of predefined "model" objects, etc. During the clustering process, the seeds—or starting points—become the centroids of the created clusters. On each step:

1.   Each object is assigned to the cluster associated with the nearest centroid.
2.   New centroids are computed by averaging or taking the center of gravity of all objects, including the newly associated.

The process is iterated until the centroids stop, or change between iterations falls below the significance cutoff.

Self-Organizing Maps (SOMs).   Self-organizing maps (SOMs) have a set of nodes with a simple topology (a two-dimensional grid) and a distance function d on the nodes. The topology has to be defined by the researcher, for example as a 2×3 grid. Nodes are iteratively mapped into the k-dimensional space (of gene expression profiles, for example) at first by random, then iteratively adjusted. On each iteration, an object is randomly selected from the pool of data and the nodes are moved toward the selected data point proportional to the distance to that object in the initial geometry. The nearest node is moved the most, while the other nodes are moved a shorter distance. The process is iterated until a satisfactory classification is produced. The number of iterations is usually a free parameter and can be set to anything from 12,500 (default value in Spotfire Decisionsite) to 20,000–50,000 (42). The SOM algorithm is closely related to the K-means algorithm. SOMs impose structure on the data, whereas neighboring nodes tend to define related clusters. This approach tends to make the algorithm more robust in dealing with real-life data, which may contain noisy, ill-defined items with irrelevant variables and outliers. On the other hand, SOMs share the same inherent problems as K-means clustering. Particularly, the number of clusters has to be arbitrarily fixed from the beginning. An incorrect guess about the expected number of clusters in the data may lead to misleading results. For example, it is very common that most of the gene expression profiles in microarray experiments are irrelevant to the subject of the research, which may be a treatment or a particular physiological condition. In this case such irrelevant data may populate most or all of the allowed clusters, while a minority of the most interesting expression profiles would remain undiscovered.

4.3.3. Other Clustering Algorithms

Most other clustering algorithms currently used in microarray analysis can be viewed as a more or less radical modification of either hierarchical tree or K-mean-clustering algorithms, aiming to improve its applicability in a particular type of research or mitigating the problems of these algorithms. The list includes (but is not limited to) the Expectation Maximization algorithm, Bayesian Clustering, Quality clustering, and many others.

4.4. Which Clustering Algorithm Is "Best"?

Clustering algorithms cannot be categorized as good, better, or best. The utility of different algorithms can only be estimated by their applicability to particular data and their purpose of study. Selection, modification, or ab initio development of a clustering procedure suitable for a particular research project is one of the most challenging tasks in expression analysis.

4.5. Clinical Data and Microarray Data Can Be Merged Before or After Clustering

The initial data for any cluster analysis is a set of objects. Each object is defined by a set of measured or counted features. The objects can be viewed as points in n-dimensional space, where n is a number of different measurements or properties of each object. In this space, points situated close to each other reflect similar objects; clusters of similar objects can be imagined as clouds of points. The measurements, or properties, of the objects can be of various natures. Most familiar to the practicing physician would be patient's body temperature, blood pressure, erythrocyte count, etc., used for diagnostic and prognostic purpose. Clustering or classification of observations into similar groups has been practiced since prehistoric times. With the advent of molecular methods it became possible to quickly measure the specific amount of antibodies or hormones to supplement the traditional diagnosis. Introduction of microarrays does not change the principles—it only adds a huge amount of measurable things. The expression levels of thousands of genes, estimated by probe intensities of a single microarray, can be used to diagnose and separate otherwise indistinguishable cases of disease or health condition. The microarray gene-expression measurements can be used for classification separately (as is done in most contemporary studies) or mixed with the "classic" clinical and biometric data. However, the very number of measurable properties for each patient or tissue sample creates a revolution in biomedical research (and potentially in patient diagnosis and care) and facilitates application of computer (and even supercomputer) analysis and the development of new, more effective algorithms.

4.6. Examples of Clustering in Clinical Studies

All applications of cluster analysis in microarray research can be divided in two categories: clustering of genes by the similarity of their expression patterns across a number of different conditions, or clustering of phenotypes (observations, conditions) by the similarity in gene-expression pattern. Gene-expression data can be presented in the form of a matrix $n \times m$, where n is a number of observations (microarrays) and m is a number of genes interrogated by the microarray in the experiment. One can choose either rows n or columns m to be the objects of classification. Some algorithms allow simultaneous classification of a data matrix by rows and columns. A classic example of clustering genes by the similarity of their expression profiles is given in Spellman et al. as an identification of S. cerevisae genes implicated in the cell-cycle regulation (43). In this work, 6125 genes were the data objects, each of them defined by a vector of 77 expression measurements derived from the normalized microarray experiments. An example of clustering observations of phenotypes can be found in any of a few dozen works, attempting to build a molecular classification of cancer types based on the microarray data. Bhattacharjee et al. (44) investigated a set of 203 samples, containing 125 cases of adenocarcinoma, 21 squamous cell lung carcinomas, 20 pulmonary carcinoids, 6 SCLC, and 17 normal lung samples. In this research 203 objects (samples) were classified using 3312 most variably expressed transcripts, i.e., each object was represented by a vector of 3312 measurements.

These results are interesting to compare with another molecular classification of lung cancers from Garber et al. (45). The data in this research includes 67 lung tumors (41 adenocarcinomas, 16 squamous cell carcinomas, 5 LCLC, and 5SCLC, along with five normal lung samples and one fetal lung tissue). In this research, a different set of genes has been used. Only 918 cDNA clones representing 835 unique genes has been selected for classification, out of 23,100 original cDNAs, representing 17,108 genes. The number of objects for classification in this case was 67, while dimensionality of the space in which these objects were analyzed was 835 (i.e., each object

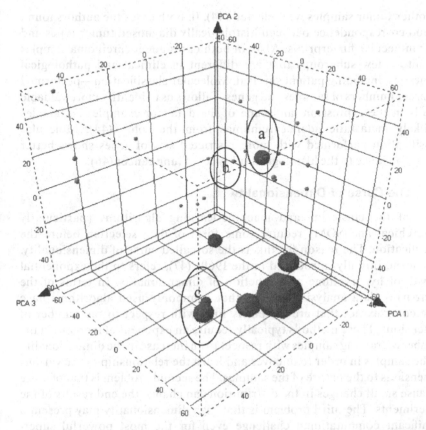

FIGURE 3 FOREL version of the cluster structure of lung cancer samples [Bhattacharjee et al. (38)]. This figure shows a 3-D plot of the first three principal components of the Euclidean intercentroid distances for all FOREL clusters. Size of the spheres is proportional to the number of elements in the cluster. The color reflects the cluster fitness metric, derived from the variance inside the luster. Singletons have zero fitness. The blue cluster (a) in the center that is apart from most other samples—clusters and singletons alike—includes the control noncancer lung samples. The only closely situated singleton (b) marked cyan belongs to a fetal lung tissue. Two closely situated midsize clusters (c) on the very bottom of the plot are made of Squamous Cell Carcinoma samples (SCC). FOREL clustering algorithm is developed at the Pennington Biomedical Research Center (Ptitsyn, A., paper in preparation).

defined by the vector of 835 measurements.) In both cases, there was some preliminary selection of the genes involved—either genes showing general variability in expression (45) were selected, or a subset of genes whose expression is very similar between the tumor pairs but varied widely among the other tumor samples was selected (44). In both cases the authors found a good correspondence between histologically diagnosed tumor types and their molecular fingerprints. Also, in both cases adenocarcinoma samples fall into a few subgroups that are different in clinical and pathological properties, including patient survival. Molecular classification—performed on larger numbers of samples and genes—allows us to identify more valuable details in the expression landscape of the data; for example, 12 samples of likely metastatic adenocarcinomas from the colon (44). Molecular classification performed with more restricted sets of genes shows better correspondence to the histological subtypes of lung cancer (46).

4.7. The Curse of Dimensionality

Most of the clustering and machine-learning algorithms (particularly hierarchical and SOM) require some form of data selection before the classification. The reason for this is the so-called curse of dimensionality. This term, initially introduced in the 1960s (47), refers to the exponential growth of hypervolume as a function of dimensionality. In terms of the microarray data analysis it means that commonly used algorithms have time complexities that are extremely high with respect to the number of dimensions. These methods typically require an exponential increase in the number of training samples with respect to an increase in the dimensionality of the samples in order to uncover and learn the relationship of the various dimensions to the nature of the samples. The second problem is that of noise because small changes in the distribution can change the end results of the experiments. The third problem is that high dimensionality may present a significant computational challenge even for the most powerful super-computers. The selection of genes adequate to the purpose of the particular research can have a dramatic effect on the results (48,49).

4.8. Goals of Clustering

The gene selection procedure should always be considered with respect to the purpose of the cluster analysis. There are three possible research goals, to which cluster analysis can be applied (50): (1) class discovery, (2) class comparison, and (3) class prediction.

4.8.1. Class Discovery

Class discovery is probably the most appropriate application for cluster analysis. In class discovery no classes are predefined and the classification is derived entirely from the statistical properties of the data. However, the cluster structure can be seriously affected by the choice of the genes selected for clustering. Selecting genes by their relevance to certain clinical data would inadvertently impose a predefined cluster structure. Selection based only on the variation in gene expression may result in classification that has no relevance to the practical goals of research—diagnostic, prognostic, or drug development. Examples of the class discovery application of cluster analysis can be found in multiple attempts at molecular classification of cancer.

4.8.2. Class Comparison

Class comparison deals with comparison of groups, defined outside the microarray experiment, i.e. independently from the gene expression profiles or "molecular fingerprints," depending on what is an object for classification. The specific purpose of such analysis is to figure out if the predefined classes are different by the pattern of gene expression and, if so, to identify the genes that make the difference. An example of this approach can be found in the analysis of expression fingerprints in breast cancer patients with and without germline BRCA1 mutations (51–53).

4.8.3. Class Prediction

Class prediction is closely related to class comparison with emphasis on the development of a multivariate function (often referred to as a classificatory or predictor) able to predict the class membership of a new sample on the basis of the expression level of the key genes. Class prediction and class comparison are often combined in one study.

Because of the great variety of algorithms and distance, metrics cluster analysis is extremely subjective by its nature. The subjective choice of the clustering strategy, algorithms, statistical metric, and parameters leaves a considerable number of traps open for research planning and results interpretation (49). Because it is likely that gene expression profiles will provide information that will affect clinical decision making, such profiling studies must be performed with statistical rigor and be reported clearly and with unbiased statistics.

5. DISCRIMINANT ANALYSIS TO IDENTIFY SINGLE, REPRESENTATIVE GENES

5.1. Discriminative Gene Selection

Like the clustering analysis, classification analysis is another important strategy to explore the microarray data. Classification, also called the "supervised" method, is a discriminant technique used to develop a predictor or classification rule using a learning set of samples with known classification to distinguish unknown samples (54). For many classification tasks based on microarray data, it is not necessary to consider many genes simultaneously. In many cases it has been shown that a few genes are sufficient for classifying two groups of samples, and in some cases as few as one or two genes are sufficient for a perfect classification (55). The process of how to identify the small amount of the feature genes is called gene selection. The goals of this process are to retain the relative feature gene and cut off irrelevant genes that may obscure the useful signatures.

The most commonly used methods for selecting discriminative genes are the standard two-sample t-test or its variants (54). This test computes sample means for each of two groups of observations and tests the hypothesis that the two means were different in two conditions. Genes out of the rejection region for the null hypothesis could be selected as discriminative genes. There are several versions of the two-sample t-test, depending on whether the two gene expression levels have equal variance (48,56). The application of this approach is often restricted by the assumption of normal distribution of genes.

There are many alternative statistical methods, including univariate and multivariate techniques. Most techniques allow for the selection of genes from only two levels, e.g., disease or health, tumor or normal. However, in many cases, biological and clinical researchers may want to identify more than two types of patients based on the selected genes. For instance, Ramaswamy et al. (57) diagnosed 14 different tumor classes from 218 random samples and Su et al. (58) classified 13 types of human carcinomas from 100 primary samples.

In this section, we mainly address logistical discriminant analysis (LDA) with stepwise selection, which can select feature genes for classifying more than two classes of unknown samples. We then briefly introduce canonical discriminant analysis (CDA) with stepwise selection that also can be used to select genes to allocate two and more types of unknown samples as a supplemental method for Fisher's linear discriminant analysis and stepwise optimization process introduced by Li (55).

However, no method can guarantee that the selected subset is best. At the very least, the selected feature gene should be reevaluated with a set of known samples. Murray (59) suggested a better idea may be to split the known samples into a number of batches and determine the best subset for each batch. The genes that appear the most frequently in the "best" subset can be used for future classification.

5.2. Logistical Discriminant Analysis (LDA)

The advantage of LDA is that there is no requirement for normal distribution of samples (compared to the two-sample t-test and Fisher's Discriminant

Analysis); nor is there a requirement for common variance–covariance matrices (compared to Fisher's Discriminant Analysis). This approach is often considered the first choice if the response variables (e.g., types of tumors or disease states) are categorical.

As the simplest way to generalize, we use three classification groups (e.g., normal, disease A, disease B) as to illustrate how LDA with stepwise selection procedure selects the discriminative genes based on the label samples. Let x be a randomly selected gene from population and let $y = 0$ if x comes from the normal group and $y = 1$ if x comes from disease group A, and $y = 2$ if x comes from disease group B. One possible analysis strategy is to create a dichotomous response variable by combining two of the response categories, i.e., either Pr (normal) vs. Pr (disease A or disease B) or Pr (normal or disease A) vs. Pr (disease B), which simultaneously compares all types of samples. Refer to Stokes et al. for details (60). The probability that $y = 1$ is:

$$\Pr(y = 0) = \frac{\exp\left(\beta_0 + \sum_{j=1}^{n} \beta_j x_j\right)}{1 + \exp\left(\beta_0 + \sum_{j=1}^{n} \beta_j x_j\right)}$$

and the probability that $y = (0 + 1)$ combined is:

$$\Pr(y = 0 + 1) = \frac{\exp(\beta_{01} + \beta' x)}{1 + \exp(\beta_{01} + \beta' x)}$$

or

$$\Pr(y = 1) = \frac{\exp\left(\beta_{01} + \sum_{j=1}^{n} \beta_j x_j\right)}{1 + \exp\left(\beta_{01} + \sum_{j=1}^{n} \beta_j x_j\right)} - \frac{\exp\left(\beta_0 + \sum_{j=1}^{n} \beta_j x_j\right)}{1 + \exp\left(\beta_0 + \sum_{j=1}^{n} \beta_j x_j\right)}$$

and the probability that $y = 2$ is:

$$\Pr(y = 2) = 1 - \Pr(y = 0 + 1) = \frac{1}{1 + \exp(\beta_{01} + \beta' x)}$$

where $\beta_0, \beta_{01}, \beta_j$ are the parameters of a logistical model that can be estimated by the maximum likelihood method and $j = (1, 2, \ldots, n)$ references discriminative genes. Based on the above notation, the stepwise selection option is involved to select the feature genes. At first, only one random gene is forced into the model to fit the parameters, and then the Chi-square of each gene not in model is computed and examines the largest of these statistics. If it is significant at the required level, say $p = 0.05$, then the corresponding gene is selected into the model. The process is repeated until none of the remaining genes meets the specified level ($p = 0.05$). PROC

LOGISTIC (SAS V8.2) with stepwise options is a suitable tool to complete the above analysis.

5.3. Canonical Discriminant Analysis (CDA)

Another statistical method addressed here is canonical discriminant analysis (CDA). Li (55) presented a program for gene selection called Tclass. The main statistical method incorporated into the program is Fisher's linear discriminant analysis. The drawback of the program is that Fisher's linear discriminant function can select feature genes from only two levels of populations. The CDA method can discriminate between two classes of known samples. In other words, CDA can select and identify two or more groups of discriminative genes from the samples.

Similar to the principal components analysis (PCA), the canonical discriminant analysis first creates latent variables by taking the discriminant linear function of the original variables. It then translates information from the original variables into new ones (eigenvector). Suppose n_j size of genes are selected from population Π_i with the assumption of normal distribution $N_p\,(m,\Sigma)$ and common variance–covariance matrices, for $i=1, 2, 3, \ldots, m$. This means there are m categories of samples taken into account. Let

$$B = \sum_{i=1}^{m} n_i(\hat{\mu}_i - \hat{\mu}.)'$$

where

$$\hat{\mu}. = \frac{1}{n.} \sum_{i=1}^{m} n_i\hat{\mu} \quad \text{and} \quad n. = \sum_{i=1}^{m} n_i$$

and let

$$W = \sum_{i=1}^{m} \sum_{r=1}^{ni} (x_{ri} - \hat{\mu}i)(x_{ri} - \hat{\mu})$$

where B and W are the usual between-group and within-group sum-of-square and cross-product matrices from one-way MANOVA. And the ratio:

$$R = \frac{a'Ba}{a'(B + W)a}$$

measures the variability between the groups relative to the common variability within groups, and a' is called the canonical discriminant function.

By selection of a', we can maximize this ratio and get the eigenvalue l_i of $(B + W)^{-1}B$. In order to classify more than two populations, we need two mutually orthogonal canonical discriminant functions, $y_1 = a_1'x$ and $y_2 = a_2'x$, which is different from the linear discriminant analysis with only one discriminant function. With these two functions, we can then calculate the distance

$$d^2 = \left[(a_1'x - a_1'\hat{\mu}_i)^2 + (a_2'x - a_2'\hat{\mu}i)^2 \right]$$

and assign the gene x to one of the groups of labeled samples with the minimum value of d^2. According to Li, each gene can be considered as an explanatory variable to enter the canonical discriminant functions, and compute the classification accuracy for each of the features, and select the feature with the best value. This process repeats until reaching the prespecified dimension of the eigenvalues.

ACKNOWLEDGMENTS

USDA 2003-34323-14010, Takeda Pharmaceuticals North America, Community Foundation of SE Michigan, NIH [Healthy transitions] R01DK50736-04, The Health Excellence Fund of Louisiana.

REFERENCES

1. Kakiuchi C, Iwamoto K, Ishiwata M, Bundo M, Kasahara T, Kusumi I, Tsujita T, Okazaki Y, Nanko S, Kunugi H, Sasaki T, Kato T. Impaired feedback regulation of XBP1 as a genetic risk factor for bipolar disorder. Nat Genet 2003; 35:171–175.

2. Brown PO, Botstein D. Exploring the new world of the genome with DNA microarrays. Nat Genet 1999; 21:33–37.

3. Alizadeh AA, Eisen MB, Davis RE, Ma C, Lossos IS, Rosenwald A, Boldrick JC, Sabet H, Tran T, Yu X, Powell JI, Yang L, Marti GE, Moore T, Hudson J, Jr., Lu L, Lewis DB, Tibshirani R, Sherlock G, Chan WC, Greiner TC, Weisenburger DD, Armitage JO, Warnke R, Staudt et al. LM. Distinct types of diffuse large B-cell lymphoma identified by gene expression profiling [see comments]. Nature 2000; 403:503–511.

4. Lee ML, Kuo FC, Whitmore GA, Sklar J. Importance of replication in microarray gene expression studies. Statistical methods and evidence from repetitive cDNA hybridizations. Proc Natl Acad Sci U S A 2000; 97:9834–9839.

5. Pritchard CC, Hsu L, Delrow J, Nelson PS. Project normal. Defining normal variance in mouse gene expression. Proc Natl Acad Sci U S A 2001; 98: 13266–13271.

6. Bolstad BM, Irizarry RA, Astrand M, Speed TP. A comparison of normalization methods for high density oligonucleotide array data based on variance and bias. Bioinformatics 2003; 19:185–193.

7. Irizarry RA, Bolstad BM, Collin F, Cope LM, Hobbs B, Speed TP. Summaries of Affymetrix GeneChip probe level data. Nucleic Acids Res 2003; 31:e15.

8. Irizarry RA, Hobbs B, Collin F, Beazer-Barclay YD, Antonellis KJ, Scherf U, Speed TP. Exploration, normalization, and summaries of high density oligonucleotide array probe level data. Biostatistics 2003; 4:249–264.

9. Churchill GA, Oliver B. Sex, flies and microarrays. Nat Genet 2001; 29:355–356.

10. Jin W, Riley RM, Wolfinger RD, White KP, Passador-Gurgel G, Gibson G. The contributions of sex, genotype and age to transcriptional variance in Drosophila melanogaster. Nat Genet 2001; 29:389–395.

11. Kerr MK, Churchill GA. Statistical design and the analysis of gene expression microarray data. Genet Res 2001; 77:123–128.

12. Churchill GA. Fundamentals of experimental design for cDNA microarrays. Nat Genet 2002; 32(Suppl):490–495.

13. Richter A, Schwager C, Hentze S, Ansorge W, Hentze MW, Muckenthaler M. Comparison of fluorescent tag DNA labeling methods used for expression analysis by DNA microarrays. Biotechniques 2002; 33:620–628, 630.

14. Hoen PA t, de Kort F, van Ommen GJ, den Dunnen JT. Fluorescent labelling of cRNA for microarray applications. Nucleic Acids Res 2003; 31:e20.

15. Xiang CC, Kozhich OA, Chen M, Inman JM, Phan QN, Chen Y, Brownstein MJ. Amine-modified random primers to label probes for DNA microarrays. Nat Biotechnol 2002; 20:738–742.

16. Wang E, Miller LD, Ohnmacht GA, Liu ET, Marincola FM. High-fidelity mRNA amplification for gene profiling. Nat Biotechnol 2000; 18:457–459.

17. Baugh LR, Hill AA, Brown EL, Hunter CP. Quantitative analysis of mRNA amplification by in vitro transcription. Nucleic Acids Res 2001; 29:E29.

18. Polacek DC, Passerini AG, Shi C, Francesco NM, Manduchi E, Grant GR, Powell S, Bischof H, Winkler H, Stoeckert CJ Jr, Davies PF. Fidelity and enhanced sensitivity of differential transcription profiles following linear amplification of nanogram amounts of endothelial mRNA. Physiol Genomics 2003; 13:147–156.

19. Van Gelder RN, von Zastrow ME, Yool A, Dement WC, Barchas JD, Eberwine JH. Amplified RNA synthesized from limited quantities of heterogeneous cDNA. Proc Natl Acad Sci U S A 1990; 87:1663–1667.

20. Eberwine J, Yeh H, Miyashiro K, Cao Y, Nair S, Finnell R, Zettel M, Coleman P. Analysis of gene expression in single live neurons. Proc Natl Acad Sci U S A 1992; 89:3010–3014.

21. Phillips J, Eberwine JH. Antisense RNA Amplification. A linear amplification method for analyzing the mRNA population from single living cells. Methods 1996; 10:283–288.

22. Phillips JK, Lipski J. Single-cell RT-PCR as a tool to study gene expression in central and peripheral autonomic neurones. Auton Neurosci 2000; 86:1–12.

23. Jeffreys AJ, Wilson V, Neumann R, Keyte J. Amplification of human mini-satellites by the polymerase chain reaction towards DNA fingerprinting of single cells. Nucleic Acids Res 1988; 16:10953–10971.

24. Iscove NN, Barbara M, Gu M, Gibson M, Modi C, Winegarden N. Representation is faithfully preserved in global cDNA amplified exponentially from sub-picogram quantities of mRNA. Nat Biotechnol 2002; 20:940–943.

25. Speel EJ, Hopman AH, Komminoth P. Amplification methods to increase the sensitivity of in situ hybridization. play card(s). J Histochem Cytochem 1999; 47:281–288.

26. Karsten SL, Van Deerlin VM, Sabatti C, Gill LH, Geschwind DH. An evaluation of tyramide signal amplification and archived fixed and frozen tissue in micro-array gene expression analysis. Nucleic Acids Res 2002; 30:E4.

27. Stears RL, Getts RC, Gullans SR. A novel, sensitive detection system for high-density microarrays using dendrimer technology. Physiol Genomics 2000; 3:93–99.

28. Yguerabide J, Yguerabide EE. Light-scattering submicroscopic particles as highly fluorescent analogs and their use as tracer labels in clinical and biological applications. Anal Biochem 1998; 262:137–156.

29. Bao P, Frutos AG, Greef C, Lahiri J, Muller U, Peterson TC, Warden L, Xie X. High-sensitivity detection of DNA hybridization on microarrays using resonance light scattering. Anal Chem 2002; 74:1792–1797.

30. Luzzi V, Holtschlag V, Watson MA. Expression profiling of ductal carcinoma in situ by laser capture microdissection and high-density oligonucleotide arrays. Am J Pathol 2001; 158:2005–2010.

31. Luzzi V, Mahadevappa M, Raja R, Warrington JA, Watson MA. Accurate and reproducible gene expression profiles from laser capture microdissection, transcript amplification, and high density oligonucleotide microarray analysis. J Mol Diagn 2003; 5:9–14.

32. Mori M, Mimori K, Yoshikawa Y, Shibuta K, Utsunomiya T, Sadanaga N, Tanaka F, Matsuyama A, Inoue H, Sugimachi K. Analysis of the gene-expression profile regarding the progression of human gastric carcinoma. Surgery 2002; 131:S39–S47.

33. Miura K, Bowman ED, Simon R, Peng AC, Robles AI, Jones RT, Katagiri T, He P, Mizukami H, Charboneau L, Kikuchi T, Liotta LA, Nakamura Y, Harris CC. Laser capture microdissection and microarray expression analysis of lung adenocarcinoma reveals tobacco smoking- and prognosis-related molecular profiles. Cancer Res 2002; 62:3244–3250.

34. Kamme F, Salunga R, Yu J, Tran DT, Zhu J, Luo L, Bittner A, Guo HQ, Miller N, Wan J, Erlander M. Single-cell microarray analysis in hippocampus CA1. demonstration and validation of cellular heterogeneity. J Neurosci 2003; 23:3607–3615.

35. Kitahara O, Furukawa Y, Tanaka T, Kihara C, Ono K, Yanagawa R, Nita ME, Takagi T, Nakamura Y, Tsunoda T. Alterations of gene expression during colo-rectal carcinogenesis revealed by cDNA microarrays after laser-capture micro-dissection of tumor tissues and normal epithelia. Cancer Res 2001; 61:3544–3549.

36. Luo L, Salunga RC, Guo H, Bittner A, Joy KC, Galindo JE, Xiao H, Rogers KE, Wan JS, Jackson MR, Erlander MG. Gene expression profiles of laser-captured adjacent neuronal subtypes. Nat Med 1999; 5:117–122.

37. Scheidl SJ, Nilsson S, Kalen M, Hellstrom M, Takemoto M, Hakansson J, Lindahl P. mRNA expression profiling of laser microbeam microdissected cells from slender embryonic structures. Am J Pathol 2002; 160:801–813.

38. Ohyama H, Zhang X, Kohno Y, Alevizos I, Posner M, Wong DT, Todd R. Laser capture microdissection-generated target sample for high-density oligonucleotide array hybridization. Biotechniques 2000; 29:530–536.

39. Xiang CC, Chen M, Ma L, Phan QN, Inman JM, Kozhich OA, Brownstein MJ. A new strategy to amplify degraded RNA from small tissue samples for microarray studies. Nucleic Acids Res 2003; 31:e53.

40. Zhang H, Yu CY, Singer B, Xiong M. Recursive partitioning for tumor classification with gene expression microarray data. Proc Natl Acad Sci U S A 2001; 98:6730–6735.

41. von Heydebreck A, Huber W, Poustka A, Vingron M. Identifying splits with clear separation. a new class discovery method for gene expression data. Bioinformatics 2001; 17(Suppl 1):S107–S114.

42. Tamayo P, Slonim D, Mesirov J, Zhu Q, Kitareewan S, Dmitrovsky E, Lander ES, Golub TR. Interpreting patterns of gene expression with self-organizing maps. methods and application to hematopoietic differentiation. Proc Natl Acad Sci U S A 1999; 96:2907–2912.

43. Spellman PT, Sherlock G, Zhang MQ, Iyer VR, Anders K, Eisen MB, Brown PO, Botstein D, Futcher B. Comprehensive identification of cell cycle-regulated genes of the yeast Saccharomyces cerevisiae by microarray hybridization. Mol Biol Cell 1998; 9:3273–3297.

44. Bhattacharjee A, Richards WG, Staunton J, Li C, Monti S, Vasa P, Ladd C, Beheshti J, Bueno R, Gillette M, Loda M, Weber G, Mark EJ, Lander ES, Wong W, Johnson BE, Golub TR, Sugarbaker DJ, Meyerson M. Classification of human lung carcinomas by mRNA expression profiling reveals distinct adenocarcinoma subclasses. Proc Natl Acad Sci U S A 2001; 98: 13790–13795.

45. Garber ME, Troyanskaya OG, Schluens K, Petersen S, Thaesler Z, Pacyna-Gengelbach M, van de Rijn M, Rosen GD, Perou CM, Whyte RI, Altman RB, Brown PO, Botstein D, Petersen I. Diversity of gene expression in adenocarcinoma of the lung. Proc Natl Acad Sci U S A 2001; 98:13784–13789.

46. Beer DG, Kardia SL, Huang CC, Giordano TJ, Levin AM, Misek DE, Lin L, Chen G, Gharib TG, Thomas DG, Lizyness ML, Kuick R, Hayasaka S, Taylor JM, Iannettoni MD, Orringer MB, Hanash S. Gene-expression profiles predict survival of patients with lung adenocarcinoma. Nat Med 2002; 8: 816–824.

47. Bellman R. Adaptive Control Processes. A Guided Tour. Princeton University Press, 1961.

48. Stolovitzky G. Gene selection in microarray data. the elephant, the blind men and our algorithms. Curr Opin Struct Biol 2003; 13:370–376.

49. Simon R, Radmacher MD, Dobbin K, McShane LM. Pitfalls in the use of DNA microarray data for diagnostic and prognostic classification. J Natl Cancer Inst 2003; 95:14–18.

50. Golub TR, Slonim DK, Tamayo P, Huard C, Gaasenbeek M, Mesirov JP, Coller H, Loh ML, Downing JR, Caligiuri MA, Bloomfield CD, Lander ES. Molecular classification of cancer. Class discovery and class prediction by gene expression monitoring. Science 1999; 286:531–537.

51. Hedenfalk I, Ringner M, Ben-Dor A, Yakhini Z, Chen Y, Chebil G, Ach R, Loman N, Olsson H, Meltzer P, Borg A, Trent J. Molecular classification of familial non-BRCA1/BRCA2 breast cancer. Proc Natl Acad Sci U S A 2003; 100:2532–2537.

52. Hedenfalk IA, Ringner M, Trent JM, Borg A. Gene expression in inherited breast cancer. Adv Cancer Res 2002; 84:1–34.

53. Hedenfalk I, Duggan D, Chen Y, Radmacher M, Bittner M, Simon R, Meltzer P, Gusterson B, Esteller M, Kallioniemi OP, Wilfond B, Borg A, Trent J. Gene-expression profiles in hereditary breast cancer. N Engl J Med 2001; 344:539–548.

54. Berrar DP, Downes CS, Dubitzky W. Multiclass cancer classification using gene expression profiling and probabilistic neural networks. Pac Symp Biocomput 2003; 5–16.

55. Li X, Rao S, Elston RC, Olson JM, Moser KL, Zhang T, Guo Z. Locating the genes underlying a simulated complex disease by discriminant analysis. Genet Epidemiol 2001; 21(Suppl 1):S516–S521.

56. Pan W. A comparative review of statistical methods for discovering differentially expressed genes in replicated microarray experiments. Bioinformatics 2002; 18:546–554.

57. Ramaswamy S, Tamayo P, Rifkin R, Mukherjee S, Yeang CH, Angelo M, Ladd C, Reich M, Latulippe E, Mesirov JP, Poggio T, Gerald W, Loda M, Lander ES, Golub TR. Multiclass cancer diagnosis using tumor gene expression signatures. Proc Natl Acad Sci U S A 2001; 98:15149–15154.

58. Su AI, Welsh JB, Sapinoso LM, Kern SG, Dimitrov P, Lapp H, Schultz PG, Powell SM, Moskaluk CA, Frierson HF Jr, Hampton GM. Molecular classification of human carcinomas by use of gene expression signatures. Cancer Res 2001; 61:7388–7393.

59. Murray G. A cautionary note on selection of variables in discriminant analysis. Applied Statistics 1977; 26:246–250.

60. Stokes M, Davis C, Koch G. Categorical Data Analysis Using the SAS System. NC: SAS Institute Inc., 1995.

9

Genomic Approaches to Understanding Vitamin D Action

James C. Fleet

Purdue University, West Lafayette, Indiana, U.S.A.

1. INTRODUCTION

The molecular actions of vitamin D metabolites have been studied extensively over the past 30 years. This has led researchers to recognize roles for vitamin D nutriture and vitamin D metabolite action in a variety of physiological systems, e.g., calcium homeostasis, immune function, and the control of cell proliferation, differentiation, and apoptosis resulting in the prevention of various cancers. The following review is intended to summarize our understanding of the molecular actions of vitamin D, to review the limited approaches taken to date using genomic approaches to study vitamin D action, and to identify issues that may benefit from a genomic approach to vitamin D action.

2. OVERVIEW OF VITAMIN D AND HEALTH

Vitamin D is a conditionally required nutrient. UV light–stimulated skin conversion of 7-dehydrocholesterol to vitamin D can meet the physiological needs of most individuals. However, low vitamin D status is a common condition during the winter months in people who live in the Northern United

States, Northern Europe, and in Canada, in people who limit their sun exposure by wearing protective clothing and sunscreen, and in the elderly (1). High vitamin D status has been associated with protection from osteoporosis, through its traditional effects on calcium homeostasis (2), and protection from cancer, due to its ability to suppress cellular proliferation, promote differentiation, and activate apoptosis (3). These later features of vitamin D biology may also account for the anti-inflammatory and immunoregulatory actions of vitamin D (4). Recent studies suggest that current recommendations for vitamin D intake (400–600 IU per day) are not sufficient to protect bone health, a classic role for vitamin D in the optimization of human health (1,5,6).

2.1. Metabolism of Vitamin D

Vitamin D, whether from the diet or produced in skin, is hydroxylated in the liver to form 25 hydroxyvitamin D_3 (25-OH D) (7), a marker of vitamin D status (8,9). The biological actions of vitamin D require further activation of 25-OH D to $1\alpha,25$ dihydroxyvitamin D_3 ($1,25(OH)_2$ D) by a 1α hydroxylase before the hormone is biologically active (10). Alterations in renal 1α hydroxylase activity are responsible for changes in circulating $1,25(OH)_2$ D levels associated with variations in dietary calcium intake (i.e., low calcium intake increases renal 1α hydroxylase activity through elevated parathyroid hormone production). However, extra-renal 1α hydroxlase has been documented in a variety of tissues, including skin, prostate epithelial cells, colonocytes, and mammary epithelial cells. Thus, $1,25(OH)_2D$, which has traditionally been viewed as an endocrine hormone, may also function as an autocrine- or paracrine-signaling molecule.

Vitamin D compounds can also be modified by the actions of cytochrome P-450 family member, 24-hydroxylase (CYP24). When 25-OH D is the substrate, $24,25(OH)_2$ D results. This vitamin D metabolite has been implicated in chondrocyte biology and in bone-fracture repair (11,12) Twenty-four hydroxylation of $1,25(OH)_2$ D is the first step in the metabolic degradation of the active hormone. CYP24 gene transcription and activation is strongly activated by $1,25(OH)_2$ D (13). Thus, CYP24 induction can be viewed as a feedback mechanism to control the biological actions of $1,25(OH)_2$ D.

2.2. Vitamin D Mediated Gene Transcription

Classically, $1,25(OH)_2D$ alters cell biology by activating the nuclear vitamin D receptor (nVDR), a member of the steroid hormone receptor superfamily, leading to the induction of gene transcription (10). The nVDR is expressed in a wide variety of cell types, from those that are involved in whole body calcium metabolism, i.e. enterocytes, renal tubule epithelial cells, and osteoblasts, to nontraditional vitamin D target tissues, e.g. immune cells, epithelial cells (mammary, prostate, colon, lung), pancreatic β cells, and adipocytes (14). The biological actions of 1,25(OH)₂ D depend upon the presence and level of the nVDR. For example, vitamin D–mediated calcium absorption is increased in nVDR-overexpressing Caco-2 cells (15) and lower in nVDR null mice (16,17) while nVDR level is an important determinant of the growth inhibition in response to 1,25(OH)₂D in prostate cancer cells (18–21).

The steps leading to vitamin D–mediated gene transcription are summarized in Fig. 1. Ligand binding promotes heterodimerization of the nVDR with the retinoid X receptor (RXR) and is required for migration of the RXR-nVDR-ligand complex from the cytoplasm to the nucleus (22–25) where it then regulates gene transcription by interacting with specific vitamin D response elements (VDRE) in the promoters of vitamin D–responsive genes (14). Although the consensus is that only a direct repeat with a 3 base spacing (DR3)-type VDRE is functional in vivo (14,26), Makishima et al., (27) recently found that both 1,25(OH)₂ D and lithocholic acid bind to the nVDR and induces CYP3A gene transcription through a nontraditional ER6 (everted repeat with a 6 base spacing) element. This suggests that the promoter elements conferring molecular regulation of gene expression by 1,25 (OH)₂ D may be more diverse than researchers have traditionally considered.

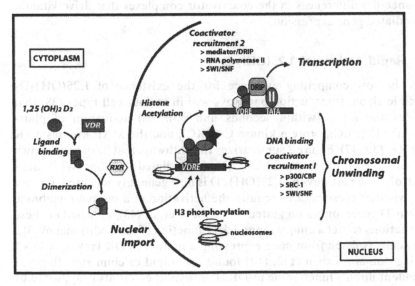

FIGURE 1   Steps required for activation of gene transcription by 1,25(OH)₂ D.

Access to VDREs in their chromosomal context may be limited (28) and may require the release of constraints imposed by chromosomal structure through phosphorylation of histone H3, acetylation of histones H3 and H4, and SWI/SNF complex-mediated phosphorylation events (29–31). Protein–protein interactions mediated by the nVDR are critical for chromosomal unwinding. The nVDR-RXR dimer recruits a complex with histone acetyl transferase (HAT) activity (e.g., CBP/p300, SRC-1 (32,33)) and the BAF57 subunit of mammalian SWI/SNF directly interacts with p160 family members like SRC-1 as well as steroid hormone receptors (34). After chromosomal unwinding, the nVDR-RXR dimer recruits the mediator D complex (DRIP) and utilizes it to recruit and activate the basal transcription unit containing RNA polymerase II (35,36). It is known that the composition of the mediator complex can vary depending upon the anchoring transcription factor (31). Thus, mediator D complex contains 16 proteins, only 14 of which are a part of the 18 protein mediator T/S complex involved in thyroid hormone receptor gene transcription. Further examination of coactivator complexes associated with nVDR-mediated gene transcription may be warranted. Preliminary evidence from kerotinocytes indicates that the major anchoring protein in the mediator complex, DRIP205, is replaced by the the steroid receptor coactivator (SRC) family members SRC-2 and SRC-3 in differentiated cells (37). Several smaller members of the mediator complex were still present in the complex. This suggests that there may be cell stage-specific (and perhaps cell type-specific) differences in the coactivator complexes that drive vitamin D–mediated gene expression.

2.3. Rapid Actions of 1,25(OH)$_2$ D

There is now compelling evidence for the existence of 1,25(OH)$_2$D-inducible signal transduction pathways within various cell types (38) that includes the rapid (within seconds and minutes) activation of phospholipase C (PLC), protein kinase C (PKC), and the MAP kinases JNK and ERK (39–42). Figure 2 summarizes the pathways that have been shown to be activated through rapid 1,25(OH)$_2$ D-mediated signaling. While activation of these pathways by 1,25(OH)$_2$ D is now generally accepted, it is not clear whether these actions require the activation of a unique membrane vitamin D receptor, as suggested by Nemere et al. (43), or whether these rapid actions reflect a unique, nonnuclear function of the traditional nVDR. Using cells isolated from mice expressing a mutant nVDR lacking a DNA binding domain, Erben et al. (44) found that rapid calcium signaling was dependent upon a functioning nVDR. This hypothesis is also supported by recent work in myocytes, where nVDR binds to, and is a target for src kinase

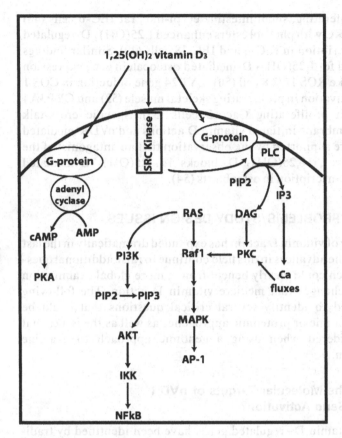

FIGURE 2 A summary of signaling pathways shown to be activated by treatment of various cell types with 1,25(OH)$_2$ D.

(45) and in the enterocyte-like cell line, Caco-2, where 1,25(OH)$_2$ D binding to nVDR induces an interaction between a ser/thr phosphatase that results in cell cycle arrest (46). In contrast, Wali et al. (47) found that in osteoblasts from nVDR null mice, rapid increases in calcium fluxes and PKC translocation did not require the presence of the nVDR.

2.4. Do Rapid and Nuclear Signaling Pathways Interact?

Several studies support the hypothesis that signal transduction pathways are important regulators of nVDR-mediated gene expression. For example, suppression of PKC activity with staurosporine or H7 inhibited 1,25(OH)$_2$ D-regulated 25-hydroxyvitamin D 24-hydroxylase (CYP24) gene expression in proliferating, small intestine crypt-like, rat IEC-6 cells (48) and activation of PKC with phorbol esters enhanced $1,25(OH)_2$ D-regulated CYP24 gene transcription in IEC-6 and IEC-18 cells (49). Similar findings have been observed for $1,25(OH)_2$ D-mediated osteocalcin gene expression in the osteoblast-like ROS 17/2.8 cell (50), CYP24 gene induction in COS-1 cells (51), c-myc activation in proliferating skeletal muscle (52) and CYP3A4 gene regulation in proliferating Caco-2 cells (53). Specific cross-talk between rapid, membrane initiated vitamin D actions and nVDR-mediated genomic actions are supported by the observation that an antagonist of the nongenomic pathway, $1\beta,25(OH)_2$ D, blocks 1α, $25(OH)_2$ D-mediated osteocalcin gene transcription in osteoblasts (54).

3. QUESTIONS/PROBLEMS/STUDY DESIGN ISSUES

Our understanding of vitamin D action has expanded dramatically in the last decade. However, the advances in this field continue to raise additional questions, some of which could clearly benefit from a more global examination of the molecular changes that mediate vitamin D action. The following section is intended to identify several critical questions that could be addressed with genomic or proteomic approaches as well as the issues that need to be considered when using a genomic approach to examine $1,25(OH)_2$ D action.

3.1. What Are the Molecular Targets of nVDR Mediated Gene Activation?

Although many vitamin D–regulated genes have been identified by traditional means, the recent identification of CYP3A as a vitamin D target gene, especially in light of its nontraditional ER6 motif (27), suggests that there are other, less obvious, molecular targets of vitamin D action that remain to be identified (CYP3A is involved in bile acid metabolism). Such discovery projects have traditionally been done with laborious methods like subtraction cloning or differential display hybridization. However, global gene expression profiling has the potential advantage of identifying all of the regulated transcripts simultaneously. Regardless of the research question being asked, the following sections present issues to consider when designing a global gene expression profiling experiment.

3.1.1. Timing

When one treats a cell with $1,25(OH)_2$ D, transcript levels may change quickly or more gradually. Figure 3A shows that after a single injection of $1,25(OH)_2$ D into mice, duodenal CaT1 mRNA levels increase quickly

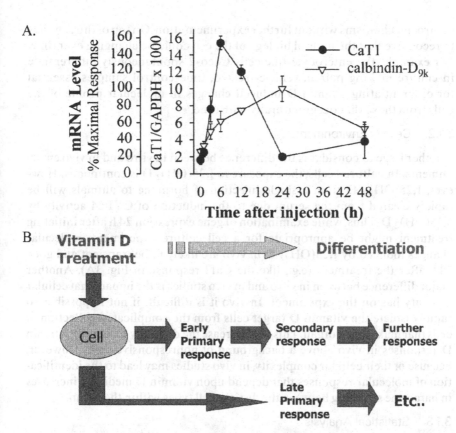

FIGURE 3 Time after 1,25 (OH)$_2$ D dose significantly influences the biological response in vivo. (A) Induction of CaT1 and calbindin D9k transcript level in mouse duodenum by a single injection of 1,25 (OH)$_2$ D (200 ng/100 g BW) (From Ref. (72)), (B) A theoretical bioresponse to 1,25 (OH)$_2$ D treatment demonstrating the relationship between primary responses, down-stream events, and natural changes in the biology of a cell.

(8-fold by 3 h, 15-fold by 6 h) while the effect of 1,25(OH)$_2$ D injection on calbindin D$_{9k}$ mRNA levels is gradual, leading to a peak expression of three-fold greater than control levels after 24 h. Thus, examination of a single time point may miss a subset of biological responses to vitamin D. In addition, although the earliest changes are likely to be direct, the later changes could be a combination of slow, direct transcriptional changes, slow direct, mRNA stabilization effects, or secondary responses caused by the first wave of direct changes (e.g. the first response is to increase the level of a transcription factor that subsequently increases transcription of another gene, see Fig. 3B). It will likely be impossible to differentiate between these control mechanisms without further experimentation. On top of this, one has to recognize that the natural biology of the cell could be changing over time. For example, the enterocyte-like cells Caco-2 spontaneously differentiate in culture making paired, vehicle-treated, time-control samples essential for differentiating vitamin D-induced changes in molecular profile of the cells from those that occur naturally with time.

3.1.2. Cellular Environment

Another issue to consider is the difference between in vitro and in vivo environments. In cultured cells, the exposure to $1,25(OH)_2$ D is continuous. However, $1,25(OH)_2$ D from a single injection of hormone to animals will be rapidly cleared from the serum due to the induction of CYP24 activity by $1,25(OH)_2$ D. Thus, while examination of gene expression 24 h after initiating treatment might be appropriate for a cell culture experiment, molecular changes induced by $1,25(OH)_2$ D in vivo are likely to be transient and gone 24 h after the treatment (e.g., like the CaT1 response in Fig. 3A). Another major difference between in vitro and in vivo studies is the impact that cellular diversity has on the experiment. In vivo it is difficult, if not impossible, to rapidly isolate the vitamin D target cells from the complicated collection of cells that make up an organ. This will increase the difficulty to identify vitamin D responses in vivo above a background of nonresponsive cells. However, because of their cellular complexity, in vivo studies may lead to the identification of molecular responses that depend upon vitamin D mediated increases in paracrine signaling between the distinct cell types within the organ.

3.1.3. Statistical Analysis

Although the proper statistical analysis of microarray data is currently under debate (55), all of the design and analysis issues that we consider in our traditional experiments also apply to the global analysis of gene expression. Thus, sample size estimates and power calculations should be conducted prior to conducting experiments. Pooling of individual samples can also be done to reduce sample variability and microarray analysis costs simultaneously (56). After collection of data, investigators should utilize various approaches (e.g., log transformation of data, Bootstapping adjustments to p-values) to ensure that the central assumption of statistics (i.e., that the data are normally distributed) has not been violated. Finally, due to the large number of comparisons conducted in gene expression profiling experiments, researchers need to be careful to avoid type I errors (i.e., false positives) but not so careful that they lose control of the type II error rate (i.e., false negatives). Excessive type I error rate is a common characteristic of using t-tests; a conservative correction for type I error rate that leads to a high type II error rate is the Bonferroni correction. More balanced approaches to controlling the type I and type II error rates are available (e.g., the false detection rate, or FDR procedure).

We have previously applied these concepts to the analysis of gene expression that occurs during the spontaneous differentiation of the BBe subclone of the intestinal cell line, Caco-2 (57).

3.2. Which Vitamin D–Regulated Transcripts are Modulated by a Common Mechanism?

After identifying the proper times and doses for $1,25(OH)_2$ D–mediated molecular action, additional studies must be conducted to clarify the mechanisms used to modulate specific molecular events or families of molecular events. In yeast, gene knockout strains that have been created for almost the entire 6000-gene genome have been tremendously valuable for elucidating molecular regulatory systems or pathways (58,59). A complete panel of gene knockout cells is not available for mammals. However, some alternative approaches that could prove fruitful are

3.2.1. Use of Pharmacological Inhibitors of Transcription and Translation

Identification of transcripts that are modulated by direct transcriptional activation versus those that require prior protein synthesis is traditionally done using specific inhibitors of transcription (e.g., actinomycin D) or translation (e.g., cycloheximide). In addition, once actinomycin D-sensitive regulation of specific genes can be confirmed, a bioinformatics approach can be attempted to assess whether changes in mRNA levels are associated with the presence of classical or nonclassical VDREs. Global, RNA half-life studies may also be fruitful in this area. Although activation of gene transcription has dominated the field of vitamin D biology, there is evidence for stimulation of posttranscriptional regulatory mechanisms by $1,25(OH)_2$ D. For example, Mosavin and Mellon (60) found that although osteocalcin gene transcription was increased 2-fold by $1,25(OH)_2$ D treatment, osteocalcin mRNA half life was increased four-fold by the treatment.

3.2.2. Use of Vitamin D Analogs nVDR-Mediated or Membrane-Initiated Pathways

As summarized in Figs 2 and 4, both classical, nVDR-mediated transcriptional and membrane-initiated rapid events are stimulated by $1,25(OH)_2$ D treatment. Creative use of pharmacological and molecular inhibitors of various kinases and signaling pathways, coupled to global examination of gene expression, would inform us whether nVDR-mediated nuclear, membrane-initiated rapid, or cross-talk between these pathways is needed for induction

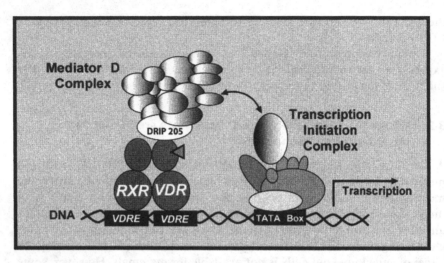

FIGURE 4 The Mediator D complex is required for 1,25(OH)$_2$ D-mediated gene transcription.

of transcript levels by vitamin D. For example, Norman et al. (61) have synthesized a cis locked analog of 1,25(OH)$_2$ D that preferentially activates the membrane-initiated pathway of vitamin D action and 1β, 25(OH)$_2$ D is an antagonist of that pathway (62).

3.2.3. Use of Genetically Modified Cells

While a full spectrum of gene modifications is not available for mammalian cells, some genetically modified animal models do exist and others could be developed for directed examination of the role of specific proteins in the control of vitamin D–mediated events. As such, cells with a lower expression of genes isolated from gene knockout mice (e.g., nVDR or 1α hydroxylase knockout mice) or due to treatment with antisense oligonucleotides or small interfering RNA (RNAi), or cells over-expressing specific genes isolated from transgenic mice or transfected cells, could help tease out the molecular roles of specific proteins in vitamin D–mediated action or the role of specific transcripts induced by 1,25(OH)$_2$ D treatment.

4. GENOMIC EVALUATION OF VITAMIN D ACTION

The application of genomic technology to the study of vitamin D action has been relatively limited to date. Like most of the work that has been conducted using arrays, the full power of the technology has not been applied. This is because until very recently, arrays capable of profiling the entire transcriptome of 30,000–40,000 transcripts did not exist. The studies that do exist have examined gene expression profiles in both classical (e.g., bone, kidney, intestine, Caco-2, ROS/17/2.8) cells and nonclassical (e.g., HL-60, squamous cell carcinoma, B) cells and used a variety of platforms (e.g., filters, spotted cDNA arrays, Affymetrix Genechips), sometimes with a limited number of highly focused transcripts (e.g., 406 transcripts related to human hematology), and rarely with a significant transcript target overlap with other platforms. This lack of consistency makes it very hard to compare the results of one experiment to the next. However, even with this caveat, the few studies available have been very promising. This section will review the available array studies on vitamin D action.

4.1. Preliminary Reports in Classical Vitamin D Target Tissues

Surprisingly, a genomic examination of classical vitamin D target tissues is not yet available as a peer reviewed report. However, several preliminary reports are available, although caution should be used when interpreting these reports due to the lack of experimental description (e.g., replicates, validation, statistical analysis, number of genes represented on array that are present). Henry et al. (63) used the Affymetrix U74B Genechip (12,000 targets; 6000 named genes) to compare the response of nVDR null mice and wild-type mice to a single i.p. injection with $1,25(OH)_2D$ (250 ng, 8 h). Using a two-fold cut off, they identified 43 bone transcripts, 20 intestinal transcripts, and 98 kidney transcripts as $1,25 (OH)_2D$ regulated in wild-type, but not in nVDR null, mice. Peng et al. (64) injected vitamin D depleted mice with $1,25(OH)_2D$ three times over 48 h (30 ng per injection at 48, 24, and 6 h prior to the end of the experiment) and examined the induction of renal transcripts using the U74B chip. They found only 57 genes increased by 50% or greater and they confirmed vitamin D regulation of two of them, C/EBP β and FK506. C/EBP β was subsequently shown to be involved in the regulation of another $1,25(OH)_2D$-inducible gene, CYP24.

Megalin is a protein involved in the renal reabsorption of fat soluble vitamins like vitamin D; as such, the megalin null mouse is somewhat equivalent to a vitamin D depleted animal (plasma $1,25(OH)_2D$ and 25-OH D are 60% lower in these mice). When Hilpert et al. (65) examined renal gene expression in megalin knockout mice using the Affymetrix Mu11K B chip (6,000 known transcripts), they found that the level of only six transcripts fell and 13 transcripts increased. Finally, Wood et al. (66) examined gene expression in the enterocyte-like Caco-2 cells after treatment with 100 nM $1,25(OH)_2$ D for 24 h using the Affymetrix U95A chip

(12,000 targets). Using a two-fold cut off, 25 genes were upregulated (including amphiregulin, alkaline phosphatase, carbonic anhydrase XII, and CYP 24) and five genes were downregulated (including dihydrofolate reductase and a Ras-like protein). While these preliminary reports are interesting, the genomic analysis of classical vitamin D target tissues clearly requires additional examination.

4.2. Nonclassical Cells

$1,25(OH)_2D$ action has been studied in a number of nonclassical cell systems due to its ability to initiate growth arrest and differentiation—characteristics that may be useful for the prevention and treatment of cancer. A short report by Savli et al. (67) used the Atlas hematology spotted filter array (406 genes) to examined the impact of $1,25(OH)_2$ treatment (5nM, 24 or 72 h) on HL-60 leukemia cell gene expression. At 24 h 7 transcript levels were upregulated and 25 transcript levels were downregulated. Twelve of these transcripts were also downregulated at 72 h, including c-myc and 3 other oncogenes, providing a glimpse into the mechanisms of chemoprevention by $1,25(OH)_2D$.

The most extensive genomic profiling of vitamin D action has been reported in squamous cell carcinoma cell lines (68,69). Akutsu et al. (68) found that 24 h of treatment with 100 nM EB 1089 (a $1,25(OH)_2D$ analog that is resistant to 24-hydroxylation) increased 38 transcript levels (1.5-fold cut off) based on a combined screening with an Atlas spotted cDNA filter array (588 genes) and a Research Genetics GF211 spotted cDNA filter array (4000 named genes). This is likely a conservative estimate due to problems the authors encountered with filter-to-filter, and hybridization variability (a common problem with spotted filter arrays). Still, this analysis identified up-regulation of several interesting transcripts that were validated by Northern blot analysis: gadd45α, a p53 target gene that is involved in DNA repair, components of various signal transduction pathways like the growth factor amphiregulin and transcription factors AP-4, STAT3, and fra-1, and cell adhesion proteins like integrin α7B. Six transcripts continued to be regulated in subsequent experiments even in the presence of cycloheximide (e.g., p21, amphiregulin, VEGF, fra-1, gadd45α, and integrin α7B). The mode of vitamin D regulation was not explored.

Lin et al. (69) conducted a time course of response to 100 nM $1,25(OH)_2D$ and EB1089 in squamous cell carcinoma cells (SCC25). Using the Affymetrix FL array and a 2.5-fold cut-off, 152 genes were identified as vitamin D regulated (89 up, 63 down). Where overlap occurred, the results from Akutsu et al. (68) were validated and a number of expected changes in transcripts previously shown to be vitamin D regulated were also seen (e.g., CYP24, osteopontin, TGF β, PTHrp, CD14, VDUP1, carbonic anhydrase II). Clustering was done based upon the pattern of expression or the functional classification of the transcripts. Figure 5 shows the diversity of the vitamin D responses in these cells. Even within the category of genes with documented, functional VDREs, there was heterogeneity in the response. For example, the CYP24 transcript level was rapidly increased (significantly increased in 1 h) and was placed in group 1 (U1) while osteopontin transcript levels increase more slowly (maximum expression at 12 h). This suggests that similar DR3-type VDREs are differentially regulated depending upon the promoter context, a finding that is consistent with studies by Toell (26). Another interesting finding from this study is that the vitamin D–induced responses were much more diverse than one might have previously predicted. For example, a number of transcripts encoding for proteins involved in the protection from oxidative stress were gradually up-regulated by EB1089 (falling into class U4 and U5); these include glucose 6 phosphate dehydrogenase (NADPH generation), glutathione peroxidase, and selenoprotein P. In addition, the thioredoxin reductase transcript was increased by 1 h after treatment with a peak induction by 6 h. Rapid suppression of a transcripts for a variety of signaling peptides (e.g., PTHrp, galanin) and induction of intracellular cell signaling proteins (e.g., cox-2, PI3K p85 subunit) was also seen after treatment. It is not clear which of these responses is primary; none of these genes has previously been shown to be vitamin D regulated or contain a functional VDRE. However, since $1,25(OH)_2$ D promotes cellular differentiation, the up regulation of some transcripts may represent a vitamin D–induced shift to a more differentiated phenotype. Regardless, these data suggest that the traditional approach of examining cell cycle proteins alone provides only a limited

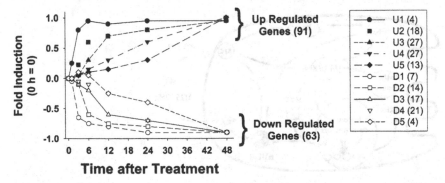

FIGURE 5 Treatment of squamous cell carcinoma cells with the vitamin D analog EB1089 results in a heterogeneous regulatory response. (Adapted from Ref. 69).

picture regarding the biological mechanisms of 1,25(OH)$_2$D action on proliferating cancer cells.

4.3. Preliminary Evidence for a Broader Mode of Molecular Regulation by 1,25(OH)$_2$D

Two additional preliminary reports warrant mention in this review. Both of them address the potential that membrane-initiated actions of 1,25(OH)$_2$D may account for a significant proportion of the alteration in transcript levels that result from treatment of cells with 1,25(OH)$_2$D. In the first report, Norman et al. (70) examined whether 1,25(OH)$_2$D–induced expression of transcripts (10 nM, 3 or 8 h) could be attenuated by either a MAP kinase inhibitor (PD98059, 10 uM, 3 h) or an agonist of the membrane-initiated actions of 1,25(OH)$_2$D, 1β,25(OH)$_2$D. Of the 44 transcripts upregulated > two-fold by 1,25(OH)$_2$D at 3 h, the MAP kinase inhibitor blocked the increase in 20 of these genes. Similarly, while 8 h of 1,25(OH)$_2$D treatment increased expression of 128 transcripts > two-fold, the membrane antagonist 1β,25(OH)$_2$ D blocked the increase in 79 of the genes. These data are consistent with the vitamin D–signal pathway cross-talk model shown in Fig. 6, i.e., activation of MAP kinase activity through membrane interactions is required for a subset the 1,25(OH)$_2$D-mediated increases in transcript levels. Farach-Carson and Xu (71) examined gene expression in proliferating osteoblast-like ROS 17/2.8 cells induced by either 1 nM 1,25(OH)$_2$D (0, 3, 24 h) or a vitamin D analog that can activate calcium influx but which has minimal ability to bind the nVDR (25(OH)-16ene-23yne D) using the Research Genetics rat gene filter array (GF300; 5000 genes). Using a cut off of two-fold increased or 50% decreased, they found that 1,25(OH)$_2$D increased the expression of 224 genes and reduced

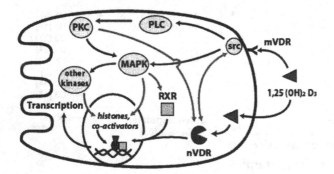

FIGURE 6 Potential points of interaction between membrane initiated and nuclear 1,25(OH)$_2$ D signaling pathways within an enterocyte.

the expression of 280 transcripts. Since they found the same pattern of regulation at 3 h when cells were treated with either $1,25(OH)_2D$ or the analog, they concluded that most of the genomic responses resulting from $1,25(OH)_2D$ treatment are independent of binding to the nVDR. These findings are provocative and if confirmed in subsequent experiments would dramatically alter our view of the relative importance of the nVDR-mediated versus membrane-initiated pathways of $1,25(OH)_2D$ action.

5. CONCLUSIONS

Our understanding of the molecular mechanisms by which $1,25(OH)_2$ D alters cell biology has diversified over the last decade. It is now clear that $1,25(OH)_2$ D action is not limited to the nVDR-mediated transcriptional response but may also include posttranscriptional stabilization of messages and rapid activation of kinase and intracellular calcium-mediated signaling events that could have nVDR-independent effects on gene transcription and message stabilization. The diversity of these nonclassical responses, as well as the breadth of the molecular response through direct nVDR-mediated transcriptional activation is just now being addressed. As a result, the analysis of vitamin D action in classical and nonclassical target tissues is ripe for a global genomic analysis. Preliminary data using this technology suggest that $1,25(OH)_2$ D influences a broader array of cellular processes and that it is doing so by a mechanism that is more complex than we have traditionally believed. This suggests that a more extensive examination of $1,25(OH)$ 2 D-mediated regulation of transcript levels is warranted.

REFERENCES

1. Heaney RP, Davies KM, Chen TC, Holick MF, Barger-Lux MJ. Human serum 25-hydroxycholecalciferol response to extended oral dosing with cholecalciferol. Am J Clin Nutr 2003; 77:204–210.
2. Lips P. Vitamin D deficiency and secondary hyperparathyroidism in the elderly: consequences for bone loss and fractures and therapeutic implications. Endocr Rev 2001; 22:477–501.
3. Grant WB. An estimate of premature cancer mortality in the U.S. due to inadequate doses of solar ultraviolet-B radiation. Cancer 2002; 94:1867–1875.
4. DeLuca HF, Cantorna MT. Vitamin D: its role and uses in immunology. FASEB J 2001; 15:2579–2585.
5. Glerup H, Mikkelsen K, Poulsen L, Hass E, Overbeck S, Thomsen J, Charles P, Eriksen EF. Commonly recommended daily intake of vitamin D is not sufficient if sunlight exposure is limited. J Intern Med 2000; 247:260–268.
6. Vieth R, Ladak Y, Walfish PG. Age-related changes in the 25-hydroxyvitamin D versus parathyroid hormone relationship suggest a different reason why older adults require more vitamin D. J Clin Endocrinol Metab 2003; 88:185–191.

7. Ponchon G, DeLuca HF. The role of the liver in the metabolism of vitamin D. J Clin Invest 1969; 48:1273–1279.
8. Dawson-Hughes B, Harris SS, Dallal GE. Plasma calcidiol, season, and serum parathyroid hormone concentrations in healthy elderly men and women. Am J Clin Nutr 1997; 65:67–71.
9. Jacques PF, Felson DT, Tucker KL, Mahnken B, Wilson PW, Rosenberg IH, Rush D. Plasma 25-hydroxyvitamin D and its determinants in an elderly population sample. Am J Clin Nutr 1997; 66:929–936.
10. Haussler MR, Whitfield GK, Haussler CA, Hsieh JC, Thompson PD, Selznick SH, Dominguez CE, Jurutka PW. The nuclear vitamin D receptor: Biological and molecular regulatory properties revealed. J Bone Min Res 1998; 13: 325–349.
11. Boyan BD, Sylvia VL, Dean DD, Schwartz Z. 24,25-(OH)(2)D(3) regulates cartilage and bone via autocrine and endocrine mechanisms. Steroids 2001; 66:363–374.
12. Seo EG, Norman AW. Three-fold induction of renal 25-hydroxyvitamin D3-24-hydroxylase activity and increased serum 24,25-dihydroxyvitamin D3 levels are correlated with the healing process after chick tibial fracture. J Bone Min Res 1997; 12:598–606.
13. Omdahl JL, Morris HA, May BK. Hydroxylase enzymes of the vitamin D pathway: expression, function, and regulation. Annu Rev Nutr 2002; 22:139–166.
14. Haussler MR, Whitfield GK, Haussler CA, Hsieh JC, Thompson PD, Selznick SH, Dominguez CE, Jurutka PW. The nuclear vitamin D receptor: biological and molecular regulatory properties revealed. J Bone Miner Res 1998; 13: 325–349.
15. Shao A, Wood RJ, Fleet JC. Increased vitamin D receptor level enhances 1,25-dihydroxyvitamin D3-mediated gene expression and calcium transport in Caco-2 cells. J Bone Miner Res 2001; 16:615–624.
16. Van Cromphaut SJ, Dewerchin M, Hoenderop JG, Stockmans I, Van Herck E, Kato S, Bindels RJ, Collen D, Carmeliet P, Bouillon R, Carmeliet G. Duodenal calcium absorption in vitamin D receptor-knockout mice: functional and molecular aspects. Proc Natl Acad Sci U S A 2001; 98:13324–13329.
17. Song Y, Kato S, Fleet JC. Vitamin D Receptor (VDR) knockout mice reveal VDR-independent regulation of intestinal calcium absorption and ECaC2 and calbindin D(9k) mRNA. J Nutr 2003; 133:374–380.
18. Miller GJ, Stapleton GE, Hedlund TE, Moffat KA. Vitamin D receptor expression, 24-hydroxylase activity, and inhibition of growth by 1 alpha, 25-dihydroxyvitamin D3 in seven human prostatic carcinoma cell lines. Clin Cancer Res 1995; 1:997–1003.
19. Hedlund TE, Moffatt KA, Miller GJ. Vitamin D receptor expression is required for growth modulation by 1 alpha,25-dihydroxyvitamin D3 in the human prostatic carcinoma cell line ALVA-31. J Steroid Biochem Mol Biol 1996; 58:277–288.
20. Zhuang SH, Schwartz GG, Cameron D, Burnstein KL. Vitamin D receptor content and transcriptional activity do not fully predict antiproliferative effects

of vitamin D in human prostate cancer cell lines. Mol Cell Endocrinol 1997; 126:83–90.

21. Hedlund TE, Moffatt KA, Miller GJ. Stable expression of the nuclear vitamin D receptor in the human prostatic carcinoma cell line JCA-1: Evidence that the antiproliferative effects of 1 alpha,25-dihydroxyvitamin D3 are mediated exclusively through the genomic signaling pathway. Endocrinology 1996; 137:1554–1561.

22. Barsony J, Pike JW, DeLuca HF, Marx SJ. Immunocytology with microwave-fixed fibroblasts shows 1 alpha,25-dihydroxyvitamin D3-dependent rapid and estrogen-dependent slow reorganization of vitamin D receptors. J Cell Biol 1990; 111:2385–2395.

23. Prufer K, Barsony J. Retinoid X receptor dominates the nuclear import and export of the unliganded vitamin D receptor. Mol Endocrinol 2002; 16:1738–1751.

24. Barsony J, Renyi I, McKoy W. Subcellular distribution of normal and mutant vitamin D receptors in living cells. J Biol Chem 1997; 272:5774–5782.

25. Prufer K, Racz A, Lin GC, Barsony J. Dimerization with retinoid X receptors promotes nuclear localization and subnuclear targeting of vitamin D receptors. J Biol Chem 2000; 275:41114–41123.

26. Toell A, Polly P, Carlberg C. All natural DR3-type vitamin response D elements show a similar functionality in vitro. Biochem J 352 Pt 2000; 2: 301–309.

27. Makishima M, Lu TT, Xie W, Whitfield GK, Domoto H, Evans RM, Haussler MR, Mangelsdorf DJ. Vitamin D receptor as an intestinal bile acid sensor. Science 2002; 296:1313–1316.

28. Paredes R, Gutierrez J, Gutierrez S, Allison L, Puchi M, Imschenetzky M, van Wijnen A, Lian J, Stein G, Stein J, Montecino M. Interaction of the 1alpha, 25-dihydroxyvitamin D3 receptor at the distal promoter region of the bone-specific osteocalcin gene requires nucleosomal remodelling. Biochem J 2002; 363:667–676.

29. Berger SL. Histone modifications in transcriptional regulation. Curr Opin Genet Dev 2002; 12:142–148.

30. Dilworth FJ, Chambon P. Nuclear receptors coordinate the activities of chromatin remodeling complexes and coactivators to facilitate initiation of transcription. Oncogene 2001; 20:3047–3054.

31. Rachez C, Freedman LP. Mediator complexes and transcription. Current Opinion in Cell Biology 2001; 13:274–280.

32. Freedman LP. Increasing the complexity of coactivation in nuclear receptor signaling. Cell 1999; 97:5–8.

33. Chen H, Lin RJ, Xie W, Wilpitz D, Evans RM. Regulation of hormone-induced histone hyperacetylation and gene activation via acetylation of an acetylase. Cell 1999; 98:675–686.

34. Belandia B, Orford RL, Hurst HC, Parker MG. Targeting of SWI/SNF chromatin remodelling complexes to estrogen-responsive genes. EMBO J 2002; 21:4094–4103.

35. Rachez C, Lemon BD, Suldan Z, Bromleigh V, Gamble M, Naar AM, Erdjument-Bromage H, Tempst P, Freedman LP. Ligand-dependent transcription activation by nuclear receptors requires the DRIP complex. Nature 1999; 398:824–828.

36. Chiba N, Suldan Z, Freedman L, Parvin J. Binding of liganded vitamin D receptor to the vitamin D receptor interacting protein coactivator complex induces interaction with RNA polymerase II holoenzyme. J Biol Chem 2000; 275:10719–10722.

37. Bikle DD, Tu CL, Xie Z, Oda Y. Vitamin D regulated keratinocyte differentiation: Role of coactivators. J Cell Biochem 2003; 88:290–295.

38. Nemere I, Farach-Carson MC. Membrane receptors for steroid hormones: a case for specific cell surface binding sites for vitamin D metabolites and estrogens. Biochem Biophys Res Comm 1998; 248:443–449.

39. Chen A, Davis BH, Bissonnette M, Scaglione-Sewell B, Brasitus TA. 1,25-Dihydroxyvitamin D(3) stimulates activator protein-1-dependent caco-2 cell differentiation. J Biol Chem 1999; 274:35505–35513.

40. Khare S, Bolt MJ, Wali RK, Skarosi SF, Roy HK, Niedziela S, Scaglione-Sewell B, Aquino B, Abraham C, Sitrin MD, Brasitus TA, Bissonnette M. 1,25 dihydroxyvitamin D3 stimulates phospholipase C-gamma in rat colonocytes: role of c-Src in PLC-gamma activation. J Clin Invest 1997; 99:1831–1841.

41. Wali RK, Baum CL, Sitrin MD, Brasitus TA. 1,25(OH)2 vitamin D3 stimulates membrane phosphoinositide turnover, activates protein kinase C, and increases cytosolic calcium in rat colonic epithelium. J Clin Invest 1990; 85:1296–1303.

42. Tien XY, Brasitus TA, Qasawa BM, Norman AW, Sitrin MD. Effect of 1,25 (OH)2 D3 and its analogues on membrane phosphoinositide turnover and [Ca21]i in Caco-2 cells. Am J Physiol 1993; 265:G143–G148.

43. Nemere I, Ray R, McManus W. Immunochemical studies on the putative plasmalemmal receptor for 1, 25(OH)(2)D(3). I. Chick intestine. Am J Physiol Endocrinol Metab 2000; 278:E1104–E1114.

44. Erben RG, Soegiarto DW, Weber K, Zeitz U, Lieberherr M, Gniadecki R, Moller G, Adamski J, Balling R. Deletion of deoxyribonucleic acid binding domain of the vitamin D receptor abrogates genomic and nongenomic functions of vitamin D. Molecular Endocrinology 2002; 16:1524–1537.

45. Capiati D, Benassati S, Boland RL. 1,25(OH)2-vitamin D3 induces translocation of the vitamin D receptor (VDR) to the plasma membrane in skeletal muscle cells. J Cell Biochem 2002; 86:128–135.

46. Bettoun DJ, Buck DW, Lu J, Khalifa B, Chin WW, Nagpal S. A vitamin D receptor-Ser/Thr phosphatase-p70 S6 kinase complex and modulation of its enzymatic activities by the ligand. J Biol Chem 2002; 277:24847–24850.

47. Wali RK, Kong J, Sitrin MD, Bissonnette M, Li YC. Vitamin D receptor is not required for the rapid actions of 1,25-dihydroxyvitamin D3 to increase intracellular calcium and activate protein kinase C in mouse osteoblasts. J Cell Biochem 2003; 88:794–801.

48. Koyama H, Inaba M, Nishizawa Y, Ohno S, Morii H. Protein kinase C is involved in 24-hydroxylase gene expression induced by 1,25 (OH)$_2$ D$_3$ in rat intestinal epithelial cells. J Cell Biochem 1994; 55:230–240.

49. Armbrecht HJ, Boltz MA, Hodam TL, Kumar VB. Differential responsiveness of intestinal epithelial cells to 1,25- dihydroxyvitamin D3–role of protein kinase C. J Endocrinol 2001; 169:145–151.

50. Desai RK, van Wijnen AJ, Stein JL, Stein GL, Lian JB. Control of 1,25-dihydroxyvitamin D3 receptor-mediated enhancement of osteoclacin gene transcription: effects of perturbing phosphorylation pathwasy by okadaic acid and staurosporine. Endocrinology 1995; 136:5685–5693.

51. Dwivedi PP, Hii CS, Ferrante A, Tan J, Der CJ, Omdahl JL, Morris HA, May BK. Role of MAP kinases in the 1,25-dihydroxyvitamin D3-induced transactivation of the rat cytochrome P450C24 (CYP24) promoter. Specific functions for ERK1/ERK2 and ERK5. J Biol Chem 2002; 277:29643–29653.

52. Buitrago C, Boland R, de Boland AR. The tyrosine kinase c-Src is required for 1,25(OH)2-vitamin D3 signalling to the nucleus in muscle cells. Biochim Biophys Acta 2001; 1541:179–187.

53. Hara H, Yasunami Y, Adachi T. Alteration of cellular phosphorylation state affects vitamin D receptor-mediated CYP3A4 mRNA induction in Caco-2 cells. Biochem Biophys Res Commun 2002; 296:182–188.

54. Baran DT, Sorensen AM, Shalhoub V, Owen T, Stein G, Lian J. The rapid nongenomic actions of 1 alpha,25-dihydroxyvitamin D3 modulate the hormone-induced increments in osteocalcin gene transcription in osteoblast-like cells. J Cell Biochem 1992; 50:124–129.

55. Nadon R, Shoemaker J. Statistical issue with microarrays: processing and analysis. Trends in Genetics 2002; 18:265–271.

56. Bakay M, Chen YW, Borup R, Zhao P, Nagaraju K, Hoffman EP. Sources of variability and effect of experimental approach on expression profiling data interpretation. BMC Bioinformatics 2002; 3:4.

57. Fleet JC, Wang L, Vitek O, Craig BA, Edenberg HJ. Gene expression profiling of Caco-2 BBe cells suggests a role for specific signaling pathways during intestinal differentiation. Physiol Genomics 2003; 13:57–68.

58. Bochner BR. New technologies to assess genotype-phenotype relationships. Nat Rev Genet 2003; 4:309–314.

59. Eide DJ. Functional genomics and metal metabolism. Genome Biol 2001; 2:1028.

60. Mosavin R, Mellon WS. Mechanism of osteocalcin mRNA stabilization in clonal osteobalst cells by 1,25-dihydroxyvitamin D3. FASEB J 1996; 10:A753.

61. Norman AW, Okamura WH, Farach-Carson MC, Allewaert K, Branisteanu D, Nemere II, Muralidharan KR, Bouillon R. Structure-function studies of 1,25-dihydroxyvitamin D3 and the vitamin D endocrine system. 1,25-di-hydroxy-pentadeuterio-previtamin D3 (as a 6-s-cis analog) stimulates nongenomic but not genomic biological responses. J Biol Chem 1993; 268:13811–13819.

62. Norman AW, Bishop JE, Bula CM, Olivera CJ, Mizwicki MT, Zanello LP, Ishida H, Okamura WH. Molecular tools for study of genomic and rapid signal transduction responses initiated by 1 alpha,25(OH)(2)-vitamin D(3). Steroids 2002; 67:457–466.

63. Henry HL, Heim M, Bishop JE, Zielinski R, Shah HD, Kwan C, Bendik I, Hunziker W, Norman AW. Microarray analysis of RNA from tissues of vitamin D receptor knockout and wild type mice treated with $1\alpha,25(OH)_2$ D3or 24R,25(OH)2D3. J Bone Min Res 2002; 17:S394.

64. Peng X, Aris VM, Galante A, Ghanny S, Soteropoulos P, Tolias P, Christakos S. Identification of genes induced by 1,25 dihydroxyvitamin D3 in mouse kidney using gene chip arrays. J Bone Min Res 2001; 16:S553.

65. Hilpert J, Wogensen L, Thykjaer T, Wellner M, Schlichting U, Orntoft TF, Bachmann S, Nykjaer A, Willnow TE. Expression profiling confirms the role of endocytic receptor megalin in renal vitamin D3 metabolism. Kidney Int 2002; 62:1672–1681.

66. Wood RJ, Tchack L, Angelo G, Pratt RE, Sonna LA. DNA microarray analysis of vitamin D-induced gene expression in a human colon carcinoma cell line. Physiol Genomics 2004; 17:122–129.

67. Savli H, Aalto Y, Nagy B, Knuutila S, Pakkala S. Gene expression analysis of 1,25(OH)2D3-dependent differentiation of HL-60 cells: a cDNA array study. Br J Haematol 2002; 118:1065–1070.

68. Akutsu N, Lin R, Bastien Y, Bestawros A, Enepekides DJ, Black MJ, White JH. Regulation of gene expression by 1alpha,25-dihydroxyvitamin D3 and its analog EB1089 under growth-inhibitory conditions in squamous carcinoma cells. Mol Endocrinol 2001; 15:1127–1139.

69. Lin R, Nagai Y, Sladek R, Bastien Y, Ho J, Petrecca K, Sotiropoulou G, Diamandis EP, Hudson TJ, White JH. Expression profiling in squamous carcinoma cells reveals pleiotropic effects of vitamin D(3) analog EB1089 signaling on cell proliferation, differentiation, and immune system regulation. Mol Endocrinol 2002; 16:1243–1256.

70. Norman AW, Bishop JE, Ishizuka S, Henry HL. $1\alpha,25(OH)_2$-vitamin D_3-mediated cross-talk from rapid to genomic responses in human leukemic NB4 cells as studied by microarray analysis and transfected reporter genes. J Bone Min Res 2002; 17:S394.

71. Farach-Carson MC, Xu Y. Microarray detection of gene expression changes induced by 1,25(OH)(2)D(3) and a Ca(2+) influx-activating analog in osteoblastic ROS 17/2.8 cells. Steroids 2002; 67:467–470.

72. Song Y, Peng X, Porta A, Takanaga H, Peng JB, Hediger MA, Fleet JC, Christakos S. Calcium transporter 1 and epithelial calcium channel messenger ribonucleic acid are differentially regulated by 1,25 dihydroxyvitamin D_3 in the intestine and kidney of mice. Endocrinology 2003; 144:3885–3894.

10

Gene Expression Profiling in Adipose Tissue

**Naima Moustaid-Moussa, Sumithra Urs,
Brett Campbell, Hyoung Yon Kim,
Suyeon Kim, and Patrick Wortman**

Department of Nutrition and Agricultural Experiment Station,
University of Tennessee, Knoxville,
Tennessee, U.S.A.

**Richard Giannone, Bing Zhang, and
Jay Snody**

University of Tennessee–Oak Ridge National Laboratory Graduate School of
Genome Science and Technology,
Oak Ridge, Tennessee, U.S.A.

Brynn H. Voy

Oak Ridge National Laboratory,
Oak Ridge, Tennessee, U.S.A.

1. INTRODUCTION

Obesity is a worldwide public health problem and has reached epidemic proportions, with 64.5% of adult Americans being overweight or obese Overweight is defined as a body mass index (BMI) between 25 and 30 and obesity as a BMI over 30 [BMI = Weight (kg)/Height (m^2)]. Obesity prevalence has increased in the past two decades at an alarming rate across gender, race, ethnicity, and age. Obesity is also well recognized now as a significant risk factor for many disorders such as sleep apnea, respiratory problems, dyslipidemia, hypertension, and coronary heart disease, the number one killer in the United States. Higher BMIs are also associated with an increase in all-cause mortality (1).

While there is strong evidence linking obesity to several degenerative diseases and evidence for the role of both genetic and environmental factors such as high fat diets and lack of physical activity in the pathogenesis of obesity (2,3), the cellular and molecular mechanisms of this relationship are far from being understood. With the recent advances in analytical and genetic methodologies, obesity studies can now be further advanced.

Given the emerging importance of adipose tissue, especially because of its recently discovered endocrine function linking adipocyte protein secretion to disease states (4,5), it is crucial to dissect adipose tissue development to better understand its function and role in metabolic diseases.

The drive to understand basic biological mechanisms of disease and gene-environment interactions has led to two distinct, yet related, approaches in the study of molecular biology: genomics and proteomics (6). This chapter will focus on the genomic approach using microarray analysis

---

Abbreviations used: AD, adipocytes; ADD1/SREBP, ADD1 sterol regulatory element binding protein; ADFP, adipose differentiation related protein; AGT, angiotensinogen; ALDH, aldehyde dehydrogenase; APM, adipose most abundant transcript; BMI, Body mass index; C/EBP, EBP, CCAAT/enhancer binding protein; COL, Collagen, DCN, decorin; DPT, dermatopontin; ECM, extracellular matrix protein; FABP, fatty acid binding protein; FACL, fatty acid coenzyme A ligase; FAS, fatty acid synthase; FN, fibronectin; GO TM: Gene Ontology Tree Machine; GPD1, glycerol 3-phosphate dehydrogenase; LOX, lysyl oxidase; LPL, lipoprotein lipase; MMP, matrix metalloprotein; NIDDM, Non Insulin Dependent Diabetes Melitus, PA, preadipocytes; PLIN, perilipin; PPAR, peroxisome proliferator-activated receptor; RT-PCR, reverse transcription-polymerase chain reaction; RXR retinoid X receptor; SMARC, SWI/SNF-related, matrix associated, actin-dependent regulator of chromatin; SMD, Stanford Microarray database; SOM, Self Organizing maps; SPARC (Secreted Protein that is Acidic and Rich in Cysteine) or osteonectin; SVF, Stromal Vascular Fraction; THBS, thrombospondin; VTN, Vitronectin.

Note: Gene nomenclature used is based on gene identity given by the University Health Network Microarray Center (UHNMC), Toronto, Canada for the 19 k human slides.

to understand adipogenesis-dependent gene expression. Such an approach may identify novel genes with important adipogenic functions and may help us understand the mechanism linking adipose tissue function and gene regulation to human obesity and comorbid conditions.

2. ADIPOSE TISSUE FUNCTION

Adipose tissue is composed of mature adipocytes (AD), the stromal vascular fraction (SVF), which contains the preadipocytes (PA), blood vessels, lymph nodes, and nerves. However, the main cellular components of adipose tissue are adipocytes, which compose more than one third of the fat tissue (7–9). The primary function of the adipocyte is to store energy as triacylglycerols during periods of positive energy balance and release the stored energy, in the form of fatty acids, during starvation. Fat deposition results from the absorption of circulating fatty acids through the action of lipoprotein lipase (LPL), de novo lipogenesis from glucose and acetate in excess of the lipolysis stimulated by hormone-sensitive lipase and in situ lipid oxidation (10–16). Adipose tissue thus plays an essential role in the regulation of the energy balance of vertebrates.

Although the traditional role attributed to white adipose tissue is energy storage and release of fatty acids when fuel is required, the metabolic role of white fat is more complex. Until the 1980s, adipose tissue was viewed almost exclusively as a depot for energy storage. However, modern molecular biological approaches and a better understanding of adipocyte biology have radically altered this view. A significant step in this change in perspective followed the discovery of leptin as an adipocyte secreted hormone (17) that is critical in energy balance regulation. Another obesity gene, agouti, whose product is also secreted by human fat cells, was reported to act at multiple sites to regulate energy homeostasis (18). Adipocyte secreted proteins include angiotensinogen, adipsin, acylation-stimulating protein, adiponectin, tumor necrosis factor α, interleukin 6, plasminogen activator inhibitor-1, and others (4,5,16–21). These proteins regulate inflammation, lipid metabolism, vascular hemostasis, complement system and immunity, cardiovascular function, and other systems. Thus, the effects of specific proteins may be autocrine or paracrine, and the site of action may be distant from adipose tissue and have been reviewed elsewhere (4,5,16–21).

3. ADIPOSE TISSUE DEVELOPMENT AND ADIPOCYTE DIFFERENTIATION

Adipocyte differentiation is a highly controlled process determined by several factors that lead to a programmed differentiation regimen.

Adipocytes begin to develop from the late embryonic stage in humans with the majority of differentiation occurring shortly after birth (7–9,22). It is well established that dormant precursors and multipotent stem cells exist in the adipose tissue throughout the life of an individual (7). All species have the ability to differentiate preadipocytes throughout their life spans in response to the body's fat storage demands (7–9).

Adipogenesis occurs in both the prenatal and postnatal states in humans, while in rodents most fat-cell development occurs postnatally. Adipogenesis results from both normal cell turnover and the requirement for additional fat mass that arises with significant calorie storage and weight gain. Adipocytes therefore form a potential tissue/organ of research interest concurrent with their disease associations.

The development of immortal preadipocyte cell lines by Green and colleagues in the 1970s was crucial in advancing the study and understanding of adipocyte development and physiology (23,24). Adipocytes develop from multipotent mesenchymal stem cells that can also give rise to muscle, bone, or cartilage. In the developing fat pad, these cells become committed to the adipocytic lineage under the influence of cues that remain undiscovered (22). It is hypothesized that these cues might be hormonal interactions or the result of cell–cell or cell–matrix interactions. The determination process results in the formation of a preadipocyte cell with fibroblastic morphology (22–27). So far, there are no known expression markers that absolutely and specifically identify a cell as a preadipocyte.

Adipocyte differentiation is characterized as the morphological transition from the undifferentiated fibroblastic PA into the mature, round lipid fat cells. In vivo studies demonstrate that PA proliferation and differentiation and their concurrent interrelations are very complex and are dependent upon a number of factors including age, hormones, species and depot (7,9,2–27).

Adipogenesis as a particular system involves two major events: preadipocyte proliferation and adipocyte differentiation (25–27). The transition between cell proliferation and cell differentiation taking place during adipocyte differentiation is a tightly regulated process where both cell cycle regulators and differentiating factors interact, creating a cascade of events leading to the commitment of the cells into the adipocyte phenotype (27–29).

The process of adipocyte development can be broadly distinguished into the following stages, as illustrated in Fig. 1: (1) cell determination; (2) exponential growth phase of adipoblasts; (3) clonal expansion, also followed by growth arrest; and (4) early, then late stages of differentiation. However, recent studies were able to dissect early differentiation events and dissociate clonal expansion from adipose conversion per se (30).

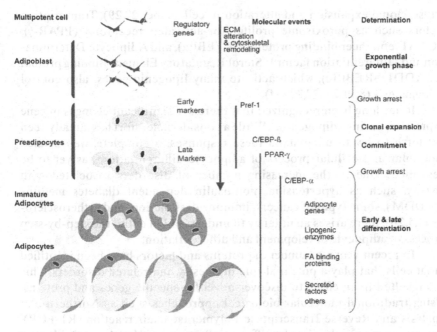

FIGURE 1 Stages of adipogenesis with sequential expression of adipogenic and lipogenic genes.

The first hallmark of adipogenesis is a dramatic alteration in cell shape as the cells convert from fibroblastic to spherical morphology. The morphological modifications are paralleled by changes in the level and type of extracellular matrix (ECM) components and the level of cytoskeletal components (27,29–32). Recent findings indicate that these events are key for regulating adipogenesis as they may promote expression of critical adipogenic transcription factors, including CCAAT/enhancer binding protein-α (C/EBP α) and/or peroxisome proliferator-activated receptor-γ (PPARγ) (25,33). Mediation of the proteolytic degradation of the stromal ECM of preadipocytes by the plasminogen cascade is required for cell-shape change, adipocyte-specific gene expression, and lipid accumulation (28). These morphological and transcriptional changes are accompanied by changes in gene expression profiles as adipocyte differentiation progresses (30–32).

4. REGULATION OF ADIPOCYTE DIFFERENTIATION

Important factors determining adipogenesis include those required for the process, and those that modulate components of the process, such as growth arrest, clonal expansion and alteration in cell shape (27,29). Transcription factors such as peroxisome proliferator-activated receptor-γ (PPAR-γ), CCAAT/enhancer binding protein-α (C/EBPα), and Adipocyte Determination and Differentiation factor 1/Sterol Regulatory Element Binding protein 1c (ADD1/SREBP1c), which activate many lipogenic genes, also control adipogenesis (8,19,25–27,30–34).

It has long been recognized that there are significant changes in gene expression during adipogenesis. While a considerable effort has already been put forth to try to understand these responses, a complete, well-defined molecular and cellular process of adipocyte differentiation has yet to be described. Due to the increasing number of disorders associated with obesity, such as hypertension, noninsulin dependent diabetes mellitus (NIDDM), some types of cancers, immune dysfunction, and artherosclerosis (4,5,20), there arises an urgency to understand in detail the step-by-step process of adipocyte development and differentiation.

In recent years, a number of proteins and factors have been identified in fat cells that play a potential role in obesity and related disorders. This has resulted in an effort to discover novel fat-specific genes and proteins. Using traditional molecular biological approaches such as Northern blot analysis and Reverse Transcriptase-Polymerase chair reaction (RT-PCR), several genes were shown to be differentially expressed in PA vs AD. Identification of additional and novel adipocyte genes requires a large-scale analysis tool such as microarray.

5. GENE EXPRESSION PROFILING: MICROARRAY ANALYSES

Traditional approaches used to investigate regulation of gene expression such as northern blot analysis and RT-PCR only allowed analysis of one or a very few genes simultaneously and required a prior knowledge of genes of interest. With the recent publication of partial or complete sequence of genomes of several organisms, all individual genes will soon be identified and assigned a function. Most genes do not function in isolation and biological systems and pathways function coordinately and interactively in networks. Thus identifying these networks is key to understanding the genome, gene function and regulation in health and disease and ultimately will impact our understanding as we develop better interventions for optimal health.

Microarray technology is an innovative and relatively comprehensive technology used to study gene expression at a large scale. Major options commonly used for large scale analysis of gene expression are reviewed in this book by St-Onge et al. (35) and others (36–48) as subsequently reviewed. Briefly, microarray approaches employed to characterize gene expression

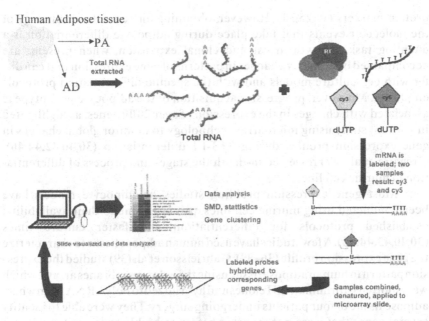

FIGURE 2 Schematic overview of a microarray experiment: from RNA labeling of two treatment groups to image and data analysis.

profiles during adipogenesis include the cDNA microarrays (Fig. 2) or Affymetrix gene chips containing probes derived from full-length or annotated genes and expressed sequence tag (EST) clusters. Figure 2 provides an overview of one of these methods.

Limited microarray studies related to adipocyte gene expression have been published to date. The model systems used for these studies have included mice, rat, hamster, and human (36–48). In this chapter we next discuss some of these studies that report gene expression regulation during adipocyte differentiation. Emphasis is then placed on the recent findings from our laboratory related to human adipocyte differentiation.

6. GENE EXPRESSION ANALYSIS DURING ADIPOGENESIS

As discussed previously, preadipocyte differentiation is characterized by altered expression levels of many genes required for the transition from a fibroblast-like preadipocyte to a lipid-storing fat cell. Several reports have attempted to schematize the stages of adipocyte differentiation into a simple hierarchy of molecular events. Genes differentially regulated during adipogenesis have been categorized into early, intermediate, and late mRNA/

protein markers (19,25,30). However, obtaining an accurate chronology of the molecular events that take place during adipocyte differentiation is a daunting task. Growth arrest and clonal expansion, when present, are accompanied by complex changes in patterns of gene expression that can differ with cell culture models and with the specific differentiation protocols employed. Moreover, progressive acquisition of the adipocyte phenotype is associated with changes in the expression of over 2000 genes, as highlighted in a recent study using microarray technology to monitor global changes in gene expression profiles during 3T3-L1 differentiation (30,40,42,44,46). Gregoire et al. (27) reviewed in-depth the stages and process of differentiation in murine cell lines.

Most gene expression profiling studies in adipocyte models have been conducted using murine cell lines, primarily due to their availability, established protocols for differentiation, and faster culture times (30,40,42,44,46). A few studies have used human adipose tissue to characterize the gene expression profile (36–39). Gabrielsson et al. (39) studied the expression pattern in human adipose tissue using the nylon membranes arrays, which were spotted with human EST clones and hybridized to PolyA RNA from whole adipose tissue of four patients undergoing surgery. They were able to identify several genes that were not reported before as highly expressed in adipose tissue, and many of these genes were distributed on chromosomes 6 and 22. Their study provided a global perspective of the genes expressed in high levels in adipose tissue and their chromosomal clustering (39).

The nylon membrane form of microarrays was also used by Guo and Liao to identify genes that were up- or downregulated in 3T3-L1 adipocytes as a consequence of differentiation (40). Cells collected at the initiation of differentiation were compared to those harvested 6 days after induction. After applying filtering criteria to reject genes with intensity levels not distinguishable from background, 1163 and 1521 genes were shown to be expressed in preadipocytes and adipocytes, respectively, for a total of 2230 unique genes (approximately 13% of genes represented on the arrays), 194 of which were specific to adipocytes and 49 to preadipocytes. The increased number of genes expressed in adipocytes versus preadipocytes was paralleled by the fact that the majority of differentially expressed genes were induced rather than repressed with differentiation. Among the 2230 genes that were expressed, 20% (447) displayed a 10-fold or greater difference between the two cell types; of these, 345 were upregulated in adipocytes. While many of the elements on the arrays represented uncharacterized ESTs, these authors reported that many of the known genes upregulated with differentiation corresponded to transcription factors and signaling molecules. Almost half (989) of all expressed genes differed by threefold or more between preadipocytes and mature cells (40).

Expression profiling during 3T3-L1 differentiation was explored more thoroughly by Jessen et al. (44) by Affymetrix arrays, using triplicate hybridizations of RNA from three time points after differentiation. Genes that passed a filter criterion of five-fold or greater change across three experiments were further classified according to functional group and time course of response. Survey of 24 hours, 4 days, and 1 week of differentiation resulted in differential expression of 24, 186, and 70 genes, respectively. The majority of genes at all time points were unknown genes, ranging from 12–22% of all differentially expressed genes. Among the known genes, the most predominant classes present at 24 hours were those involved in the acute phase inflammatory response, apoptosis, signal transduction, and hormone metabolism. After 4 days of treatment, the major categories represented were signal transduction and transcription, and genes related to the extracellular matrix. By 7 days, the expression levels of many genes involved in lipid synthesis had regressed, paralleled by greater representation of genes required for lipid transport and secreted gene products. There was little overlap in terms of the genes represented on these arrays and those on the arrays used by Guo and Liao (40) so direct comparisons between the results could not be adequately made.

Both of the studies with 3T3-L1 cells reported previously, described genes with differential expression after cells had acquired the mature adipocyte phenotype, and its experimental design was not likely to identify genes important in the early stages of cell conversion. This window in cell development was carefully profiled by Burton et al. using the more comprehensive Affymetrix system (42). Again 3T3-L1 cells were used as the model, and samples were collected in a time course of 0, 2, 8, 16, and 24 hours after treatment with the standard differentiation cocktail (insulin, dexamethasone, and isobutyl-methylxanthine) at confluence. Samples were collected from two independent cell culture experiments and hybridizations were performed in duplicate. The arrays represented 13,179 cDNA/EST clones, and 6946 (52%) of these gave a detectable signal in at least one of the time points in both experiments. This increase in expressed genes over the 13% reported by Guo and Liao (40) reflects the significantly increased sensitivity of Affymetrix arrays over conventional P_{32}-based detection. Compared to the control cells (time 0), 2.2% (285) of the genes exhibited a fivefold or greater change in expression level with differentiation. A cutoff of two-fold or greater yielded 1156 differentially expressed genes. These authors furthered their data analysis using a combination of hierarchial clustering and self-organizing maps (SOM) applied to genes meeting the five-fold or greater criterion. Hierarchial clustering identified a tendency for the genes to fall into five clusters, and this information was then used to set the nodal geometry for SOM at five. Detailed functional classification of genes within each cluster was not performed, but individual genes within each cluster were described with reference to their potential function. For example, cluster 1 represented genes differentially expressed at the earliest time point, and it included the downregulated gene Gadd153, a protein potentially related to the early growth arrest necessary for differentiation to proceed. Cluster 5 included cyclins A and B1, and their upregulation at the latest time point was related to the clonal expansin phase of differentiation known to occur within 24 hours of induction. Several genes known to be differentially regulated during this part of differentiation were examined by Western blot, and array-expression levels conformed to protein level data (42).

3T3-L1 cells represent a cell population that is already committed to the adipocyte phenotype and cannot differentiate into other cell types. Bone marrow mesenchymal stem cells are multipotent cells that can give rise to adipocytes, osteoblasts, chondrocytes, and myoblasts when cultured under the appropriate conditions (22). Suitable conditions prompt first the commitment to a specific lineage, followed by differentiation into the respective cell types. To identify the genes involved in adipocyte commitment, Nakamura et al. used microarrays to study human mesenchymal stem cells (hMSC) during both commitment and differentiation (37). hMSCs with a high-differentiating ability were induced to adipocyte development by treatment with dexamethasone, a potent glucocorticoid agonist and insulin. Temporal gene expression was studied using selected arrays made in-house for specific genes that are known to be expressed during adipogenesis. Their results provide information on the molecular mechanisms required for lineage commitment and maturation accompanying adipogenesis of hMSC and in part parallelled temporal changes in gene expression observed in murine cell lines. A total of 197 genes showed stage-specific gene expression changes during adipogenesis, including genes in lipid metabolism, cell cycle, gene transcription, and signal transduction. Forty-four percent of these genes were lipid metabolizing genes (fatty acid binding proteins, fatty acid synthase, stearoyl Coa desaturase and others) or transcription factors (C/EBP β and δ genes and PPAR γ). Cluster analysis of genes over time showed that those differentially expressed in the late phase (day 7 to day 14) corresponded to genes previously reported for preadipocyte differentiation. In contrast, genes regulated during the early phase of treatment differed from those reported for 3T3-L1 preadipocytes (40,42), likely reflecting genes involved in the process of commitment that precedes entry into the preadipocyte stage.

Microarray analysis has also been used to profile adipose tissue changes in gene expression with obesity. Affymetric arrays of 12,488 probe sets were used to query gene expression in white adipose tissue of C57BL/ 6J mice induced to become obese by feeding of a very high-fat diet for 8 weeks (43). In total, 472 genes were differentially expressed in obese versus lean animals using a cutoff of 1.2-fold or more, while a filter of three-fold produced 98 differentially expressed genes. Interestingly, the majority (69%) of the genes showed reduced expression in obese compared to lean controls. Moreover, several genes important to lipid metabolism were downregulated with obesity, as were markers of adipocyte differentiation. In addition, many genes encoding enzymes involved in detoxification processes also displayed significantly reduced expression in obese animals. On the other hand, most of the genes displaying increased expression in obese adipose tissue were genes involved in the acute phase response or inflammatory reactions. Interestingly, the latter finding has also been reported in genetically obese *ob/ob* mice (43).

Saito-Hisaminato et al. (38) performed a comprehensive analysis of the expression profiles in 25 adult and four fetal human tissues using a cDNA microarray consisting of 23,040 human genes. Their study revealed that 4080 genes were highly expressed (at least a fivefold expression ratio) in one or only a few tissues and 1163 of those were expressed exclusively (more than a tenfold higher expression ratio) in a particular tissue. Further, a hierarchical clustering analysis of gene-expression profiles in nerve tissues, lymphoid tissues, muscle tissues, or adipose tissues identified a set of genes that were commonly expressed among related tissues. Some of the adipose specific genes reported in this study were perilipin, fatty acid binding protein 4, lipoprotein lipase, adipose most abundant tissue protein and PPAR-γ (38).

7. GENE EXPRESSION PROFILING IN PREADIPOCYTES AND ADIPOCYTES FROM HUMAN ADIPOSE TISSUE

Our research focuses mainly on the differential gene expression in subcutaneous abdominal human adipose tissue taken from female patients (36,49–51). Our initial goal was to identify genes that were differentially expressed in the stromal vascular fraction (SVF) which contained the preadipocytes (PA) versus the mature adipocyte (AD) fraction. Ultimately, we would like to link specific gene expression profile in PA or AD to patient BMI or disease status. Human adipose tissue from female patients was digested to separate the fully differentiated adipose cells from the SVF (52,53). The SVF fraction was cultured to propagate cells and to obtain sufficient amounts of RNA for microarray analysis. Over 70% of these cells (referred to as PA) were able to differentiate into AD in the presence of dexamethsanone, insulin and thiazolidinediones, thus indicating their preadipocyte property. Mature floating adipocytes obtained from collagenase digests are referred to as adipocytes (AD) and were used in our microarray studies (36). Because of the considerable differences among patients in terms of BMI and potentially metabolic status, each patient's specimens (PA and AD) were analyzed separately in this study. Subjects were females with a mean patient age of 51.33 ± 7.31 years and a mean BMI of 29.03 ± 8.76. Our research focused on defining the underlying differences at the gene expression level between the committed preadipocytes and the fully developed adipocytes derived from the same patient. The initial results have been promising and we have narrowed the list of genes from 19,000 on the array to a number that can be evaluated gene-by-gene for a respective role during adipogenesis (36). Additional studies are underway to map gene clusters differentially regulated as a function of patient BMI.

We applied microarray analysis to better understand the process of adipocyte differentiation in human adipose tissue by identifying PA and AD specific genes. We used the Stanford microarray Database as a tool for our microarray analysis (54; Fig. 3). The microarray experiments with AD and PA from six patients showed a consistent pattern of gene expression for both cell types with the spot intensities, background, and hybridization efficiency being similar. To investigate consistency in gene expression profiling among patients, the genes that were either up or downregulated by two- or four-fold in at least four and five patients are summarized in Table 1.

The genes that were upregulated in adipocytes are directly or indirectly involved in fatty acid metabolism and include transcription factors, hormone

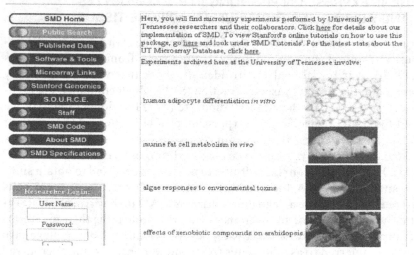

FIGURE 3   The university of Tennessee microarray database homepage.

TABLE 1 Genes Upregulated in AD or PA from Four or Five Patients by Two or Four Fold (log 2)

Fold difference	Number of patients	Number of genes upregulated	Upregulated in AD	Upregulated in PA
>4	5 or 6	2	1	1
>4	4–6	21	16	5
>2	5 or 6	51	38	13
>2	4–6	792	702	90

receptors, and metabolic genes. In the preadipocytes, the cytoskeletal associated genes were more predominantly expressed. Some of the selected genes that were upregulated in adipocytes and preadipocytes are presented in Tables 2 and 3, respectively. Using Gene Ontology Tree Machine (GOTM) developed at Oak Ridge National Laboratory ((http://genereg.ornl. gov/ gotm), we identified specific cellular pathways in which genes were preferentially expressed for both PA and AD Figs. 4 and 5). The complete tree

TABLE 2 List of Selected Genes that are Upregulated in AD.

Gene ID	Genes upregulated in adipocytes	Log$_2$ ratio of expression	Accession number
FABP4	Fatty acid binding protein 4	4.225–12.35	N78658
LPL	Lipoprotein lipase	4.601–15.425	W15543
FACL2	Fatty acid co-enzyme A ligase	1.24–13.25	AA131566
GPD1	Glycerol 3-phosphate dehydrogenase	0.68–8.02	H42536
VTN	Vitronectin	0.869–7.04	H29054
PLPN	Perilipin	1.24–17.76	T70586
AGT	Angiotensinogen	0.59–6.26	R59168
APM1	Adipose most abundant transcript 1	1.731–11.023	H28548
ADFP	Adipose differentiation relation protein	1.032–5.28	W55903
RXRA	Retinoid X receptor A	0.698–10.973	H38814
E2F5	E2F transcription factor 5	1.38–22.12	H77748
PPAR-g	Peroxisome proliferator activated receptor-gamma	1.97–5.202	H21596
SMARC	SWI/SNF	0.985–6.605	H04220
DUSP1	Dual specificity phosphatase 1	0.907–13.445	H04220
DPT	Dermatopontin	1.79–20.76	H01024

TABLE 3   List of selected genes that are upregulated in PA.

Gene ID	Genes upregulated in preadipocytes	Log$_2$ ratio of expression	Accession number
LOX	Lysyl oxidase	1.28–19.2	R07161
PPAR-δ	PPAR-delta	1.842–7.198	N30528
C/EBP-α	CCAAT enhancer binding protein-A	0.49–2.43	H25129
FN1	Fibronectin	2.653–69.94	H03906
COL3A1	Collagen type 3A1	4.02–43.029	N32802
COL4A1	Collagen type 4A1	4.32–12.25	R26967
COL5A1	Collagen type 5A1	2.504–19.25	W68613
THBS1	Thrombospondin 1	1.82–26.4	R88106
MMP2	Matrix metalloprotein 2	1.911–31.61	R48754
DCN	Decorin	2.2–22.5	N70028
SPARC	Osteonectin	1.76–19.75	R12744

of these biological processes and all related data and publication can be accessed at: http://genome.ws.utk.edu (under published data tool).

Since microarray analysis provides a general view of gene expression, it is crucial to confirm these findings for specific genes using more quantitative methods. Using RT-PCR analysis for selected genes, we confirmed that the genes found to be upregulated by microarray analysis were also similarly regulated in AD vs PA (36). It is well known that stromal vascular cells and mature adipocytes represent different cell characteristics from morphological and biochemical perspectives. Our studies provide additional insight into the genes that underlie the differences found between committed preadipocyte cells and mature, differentiated adipocytes.

Genes coding the Fatty Acid Binding Proteins (FABPs), Lipoprotein Lipase (LPL), fatty acid coA ligase (FACL2) and Glycerol 3 Phosphate Dehydrogenase (GPD1) are the main players in fatty acid metabolism. From our studies, the elevated expression of these genes is predictable in adipocytes and conforms to the predominant carbohydrate and lipid metabolic pathways in AD (7,8,27). FABP4/aP2 is an adipocyte protein involved in fatty acid uptake, transport, and metabolism in adipocytes and bind both long chain fatty acid and retinoids (55). FABP4 in particular was implicated in obesity and related disorders including insulin resistance, type II diabetes and atherosclerosis (55). Significant upregulation of such genes as aldehyde dehydrogenase, acyl coenzyme oxidase, phosphofructokinase, Adipocyte Differentiation Related Protein (ADFP, 56), Perilipin (PLPN, 57), and secreted proteins like vitronectin (VTN, 58), Adipose Most Abundant

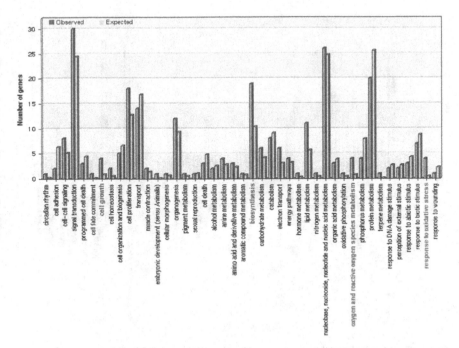

FIGURE 4 Metabolic pathways predominantly expressed in AD. The number of genes upregulated in each biological process is indicated.

Transcript 1 (APM1, 59), and angiotensinogen (AGT, 60) in adipocytes further strengthens the idea that AD possess endocrine functions and the AD themselves are physiologically active cells contributing to whole body metabolism (4,5,20,29,36). Complementing the detection of these genes and proteins involved in fatty acid metabolism is the upregulation of the transcription factors, the crucial players in adipogenesis. PPARγ and PPARδ are regulated in an adipocyte differentiation dependent manner (35). PPARδ is detectable in growing preadipocytes and is upregulated just at confluence to reach a maximal expression during the postconfluent proliferation (19,25,31–34). The γ isoform is induced at the end of clonal expansion proliferation and is maximally expressed in terminally differentiated cells (62). Our microarray studies in human adipose tissue also confirm this pattern of differential gene expression. Other important transcription factors induced during adipogenesis are the Retinoid X receptor A and E2F transcription factor 5 (26). The RXRs (A & B) form heterodimers with PPARs and regulate transcription of various genes. They bind retinoic acid, the biologically active form of vitamin A, which mediates cellular signaling in embryonic

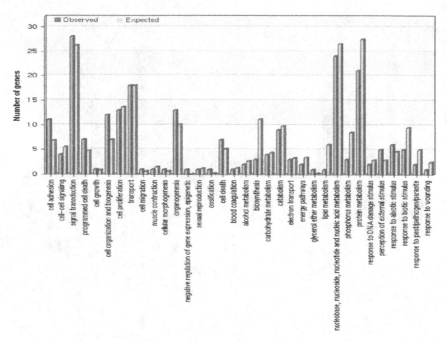

FIGURE 5 Metabolic pathways predominantly expressed in PA. The number of genes upregulated in each biological process is indicated.

morphogenesis, cell growth, and differentiation (22). The E2F family of transcription factors plays a crucial role in the control of cell cycle and regulates adipocyte differentiation (26,61). Depletion of E2F4 induces adipogenesis while E2F1 induces PPAR gamma transcription during clonal expansion and represents the link between proliferative signaling pathways, triggering clonal expansion and terminal adipocyte differentiation through regulation of PPAR gamma expression (26,61). It is interesting to note the increased expression of the gene encoding SMARCB1 in AD. Although the function of this protein is relatively unknown, members of this family (SWI/SNF family) possess helicase and ATPase activities and are thought to regulate transcription of certain genes by altering the chromatin structure around those genes (62). These results concord with the fact that adipogenesis is a very defined and complex process involving a series of changes including activation and inactivation of several genes that are coordinately regulated.

Fig. 4 illustrates the functional classification of genes significantly upregulated in AD ($p < 0.05$). The figure was generated by GOTM (75).

It was based on the fourth annotation level under biological processes. The dark/black bars represent gene numbers expected in the GO (Gene Ontology) categories using all the genes on the array as a reference. The light/white bars represent the numbers of significantly upregulated genes in AD in the GO categories. Four GO categories were found to be significantly enriched at this level based on the hypergeometric test implemented in GOTM. They are (1) cell growth, (2) biosynthesis, (3) oxygen and reactive oxygen species metabolism, and (4) response to oxidative stress. GOTM identified a total of 19 categories that were significantly enriched. The DAG (Directed Acyclic Graph) representing the 19 categories and their relative location in the GOTree can be accessed at www.genome.ws.utk.edu.

Among the genes involved in the extracellular matrix, expression of fibronectin, collagen, and MMPs in adipose tissue has been documented (63–66). These are the primary genes that were upregulated in human PA in our study. During differentiation of preadipose cells into adipose cells there is an active synthesis of collagen in the preadipose state (61), while fibronectin, the adhesive extracellular matrix (ECM) protein, is strongly expressed in preadipocytes and decrease during adipose conversion (65). Associated with fibronectin is thrombospondin (THBS), also an adhesive glycoprotein that mediates cell-to-cell and cell-to-matrix interactions (67). It facilitates the building of fibrinogen, fibronectin, laminin, and type V collagen. The increased expression of these cytoskeletal genes is therefore expected and validated. The matrix metalloproteinase (MMP) family is involved in the breakdown of extracellular matrix in normal physiological processes (63,64). However, detection of dermatopontin, an extracellular matrix protein that functions in cell–matrix interactions, matrix assembly, and mediates adhesion by cell surface integrin binding serves as a communication link between the dermal fibroblast cell surface and its extracellular matrix environment (68).

The elevated expression of LOX in PA is substantiated by the fact that it is an amine oxidase expressed and secreted by fibrogenic cells (69,70). LOX has not been previously reported in PA; however, it is known to play a critical role in the formation and repair of extracellular matrix by oxidizing lysine residues in elastin and collagen (69,70). Since in our study LOX expression is dramatically reduced in AD compared to PA, LOX is a prime candidate marker of early differentiation or committed PAs.

Osteonectin or SPARC (Secreted Protein that is Acidic and Rich in Cysteine) is a matrix-associated protein. It elicits changes in cell shape, inhibits cell-cycle progression, and influences the synthesis of extracellular matrix (71,72). Interestingly, this protein has recently been linked to obesity as SPARC knockout mice exhibit increased adiposity (71,72). Decorin is another protein that plays a functional role during growth and differentiation of adipocytes by contributing to the morphological changes occurring in the cell proteoglycan (68). Thus, in addition to LOX, we also identified SPARC, THBS and decorin as primarily expressed in PA at very high levels (more than four fold compared to AD) adding to the list of PA markers.

Figure 5 shows the functional classification of genes significantly upregulated in PA ($p < 0.05$). It was also based on the fourth annotation level under biological processes. None of the GO categories was found to be significantly enriched at this level by GOTM. However, GOTM identified totally 14 categories that were significantly enriched. The DAG representing the 14 categories and their relative location in the GOTree can be accessed from http://genome.ws.utk.edu.

Finally, it is worth noting that while several genes showed consistently high expression levels in either PA or AD across all patients, they were not statistically significant. The probable reason can be attributed to their low expression level and to variability among patients thus requiring a large number of samples to be analyzed to reach statistical power. Detailed statistical analyses of microarray data is discussed in this book in the chapters by St-Onge et al. (35) and Saxton and Moser (73). Our results support the notion that adipogenesis is a very defined and complex process involving a series of changes including activation and inactivation of several genes. This process is coordinately regulated. Many proteins secreted from adipose tissue operate in an autocrine/paracrine manner to regulate adipocyte metabolism and upon secretion into the bloodstream, act as endocrine signals at multiple distant sites to regulate energy homeostasis, e.g., resistin, angiotensin, and adiponectin (4,5,20,21,60).

In summary, our study was the first attempt to compare PA and AD derived from the same patients. We have also successfully demonstrated the installation and application of SMD for the cDNA microarray data analysis at the University of Tennessee (54). The results from the SMD analysis confirmed most of the previously reported genes in AD such as FABP, GPD1, PPAR-γ, AGT, and RXR as adipocyte markers. We were also able to identify several genes, which are selectively upregulated in adipocytes such as APM1, perilipin, VTN, LPL, SMARC, E2F5, and DPT making them selective potential adipocyte markers. Similarly, upregulated genes in the preadipocytes such as LOX, E2F4, fibronectin 1, collagen types 3A1, 5A1, and 6A1, THBS, decorin, and SPARC are the selective potential markers of committed preadipocytes. The major significance of this report is the contribution of these selective gene markers to facilitate demarcation between the preadipocyte and adipocyte state of the cells. The microarray technique is thus a very useful method to determine gene expression levels in different cell types.

8. CONCLUSION

Traditional approaches used to investigate regulation of gene expression such as northern blot analysis and reverse transcriptase polymerase chair reaction (RT-PCR) only allowed analysis of one or a very few genes simultaneously. With the recent publication of partial or complete genomes of several organisms, all individual genes will soon identified. Most genes do not function in isolation; rather biological systems and pathways function coordinately and interactively in networks. Thus identifying these networks is key to understanding the genome, gene function, and regulation in health and disease and ultimately will impact our understanding of better interventions for optimal health.

In this chapter, we presented examples of large scale analysis of genes that are expressed in a cell type-specific manner in adipose tissue. While many of the genes found to be expressed in adipocytes versus their precursors in stoma vascular fractions are genes coding for proteins and enzymes in lipid metabolism, we identified new genes with unknown function in adipose tissue metabolism or obesity. Some of these genes code for secreted proteins that may regulate extra-adipose tissues as well as whole body homeostasis, such as vitronectin and osteonectin. To further gain insight into additional proteins secreted by adipose tissue, proteomics studies will be the next logical step to identify novel proteins that may constitute an important link between adipocytes and diseases such as obesity, diabetes, and cardiovascular disease.

ACKNOWLEDGMENTS

We would like to thank Dr. Colton Smith, Bhanu Prasad Rekapalli, John Rose, and Robert Hillhouse, at the Web Services Department of the University of Tennessee who have been instrumental in installing and providing SMD services. Special thanks to Dr. Arnold Saxton for the statistical analyses of the microarray data.

This work was supported by in part by the American Heart Association, The U.S. Department of Agriculture, TN Agricultural Experiment Station, the College of Education, Health and Human Sciences and The Center of Excellence of Genomics and Bioinformatics, The University of Tennessee.

REFERENCES

1. Flegal K, Carroll M, Ogden C, Johnson C. Prevalence and trends in obesity among US adults. 1999–2000. JAMA 2002; 288:1723–1727.
2. Hill O, Wyatt H, Reed G, Peters J. Obesity and the environment: where do we go from here? Science 2003; 299(5608):853–855.

3. Chagnon Y, Rankinen T, Snyder E, Weisnagel S, Perusse L, Bouchard C. The human obesity gene map: the 2002 update. Obes Res 2003; 11(3):313–367.
4. Mohamed-Ali V, Pinkney JH, Coppack SW. Adipose tissue as an endocrine and paracrine organ. Int J Obes Relat Metab Disord 1998; 22(12):1145–1158.
5. Kim S, Moustaid-Moussa N. Secretory, endocrine and autocrine/paracrine function of the adipose. J Nut 2001; 130:3110S–3115S.
6. Nakayama T, Fujii M, Yokoyama S. Structural genomics/proteomics projects in Europe and US, and international cooperation. Tanpakushitsu Kakusan Koso 2002; 47(8):982–986.
7. Ailhaud G, Grimaldi P, Negrel R. Cellular and molecular aspects of adipose tissue development. Annu Rev Nutr 1992; 12:207–233.
8. Ntambi J, Kim Y. Adipocyte differentiation and gene expression. J Nutr 2000; 130:3122S–3126S.
9. Hausman D, DiGirolamo M, Bartness G, Hausman G, Martin R. The biology of white adipocyte proliferation. Obesity Reviews 2001; 2:239–254.
10. Abumrad N, El-Maghrabi M, Amri E, Lopez E, Grimaldi P. Cloning of a rat adipocyte membrane protein implicated in binding or transport of long chain fatty acids that is induced during preadipocyte differentiation. JBC 1993; 268:17665–17668.
11. Fried SK, Turkenkopf IJ, Goldberg IJ, Doolittle MH, Kirchgessner TG, Schotz MC, Johnson PR, Greenwood MR. Mechanisms of increased lipoprotein lipase in fat cells of obese Zucker rats. Am J Physiol 1991; 261:E653–E660.
12. Adachi H, Kurachi H, Homma H, Adachi K, Imai T, Morishige K, Matsuzawa Y, Miyake A. Epidermal growth factor promotes adipogenesis of 3T3-L1 cell in vitro. Endocrinology 1994; 135(5):1824–1830.
13. Wang S, Hwang R, Greenberg A, Yeo H. Temporal and spatial assembly of lipid droplet-associated proteins in 3T3-L1 preadipocytes. Histochem Cell Biol 2003; 120(4):285–292.
14. Holm C, Belfrage P, Osterlund T, Davis R, Schotz M, Langin D. Hormone-sensitive lipase: structure, function, evolution and overproduction in insect cells using the baculovirus expression system. Protein Eng 1994; 7(4):537–541.
15. Ailhaud G. Molecular mechanisms of adipocyte differentiation. J Endocrinol 1997; 155(2):201–202.
16. Boone C, Mourot J, Gregoire F, Remacle C. The adipose conversion process; regulation by extracellular and intracellular factors. Reprod Natu 2000; 40:325–358.
17. Zhang Y, Proenca R, Maffei M, Barone M, Leopold L, Friedman J. Positional cloning of the mouse obese gene and its human homologue. Nature 1994; 372:425–432.
18. Moustaïd-Moussa N, Claycombe K. Mechanisms of agouti-induced obesity. Obes Res 1999; 7(5):506–514.
19. Rosen E, Spielgelman B. Molecular regulation of adipogenesis. Ann Rev Cell Dev Biol 2000; 16:145–171.

20.   Trayhurn P, Beattie J. Physiological role of adipose tissue: white adipose tissue as an endocrine and secretory organ. Proc Nutr Soc 2001; 60(3):329–339.

21.   Beltowski J. Adiponectin and resistin – new hormones of white adipose tissue. Med Sci Monit 2003; 9(2):RA55–RA61.

22.   Phillips B, Vernochet C, Dani C. Differentiation of embryonic stem cells for pharmacological studies on adipose cells. Pharmacol Res 2003; 47(4):263–268.

23.   Green H, Kehinde O. Sublines of mouse 3T3 cells that accumulate lipid. Cell 1974; 1:113–116.

24.   Rubin CS, Lai E, Rosen OM. Acquisition of increased hormone sensitivity during in vitro adipocyte development. J Biol Chem 1977; 252(10):3554–3557.

25.   Rosen E, Hsu C, Wang X, Sakai S, Freeman M, Gonzalez F, Spiegelman B. C/EBPalpha induces adipogenesis through PPARgamma: a unified pathway. Genes Dev 2002; 16(1):22–26.

26.   Fajas L, Landsberg R, Huss-Garcia Y, Sardet C, Lees J, Auwerx J. E2Fs regulate adipocyte differentiation. Dev Cell 2002; 3:39–49.

27.   Gregoire F, Smas C, Sul H. Understanding adipocyte differentiation. Physiol Rev 1998; 78(3):783–808.

28.   Selvarajan S, Lund L, Takeuchi T, Craik C, Werb Z. A plasma kallikrein-dependent plasminogen cascade required for adipocyte differentiation. Nat Cell Biol 2001; 3:267–275.

29.   Gregoire F. Adipocyte differentiation: from fibroblast to endocrine cell. Exp Biol Med 2001; 226:997–1002.

30.   Liu K, Guan Y, MacNicol MC, MacNicol AM, McGehee RE Jr. Early expression of p107 is associated with 3T3-L1 adipocyte differentiation. Mol Cell Endocrinol 2002; 194:51–61.

31.   Brun R, Kim J, Hu E, Altiok S, Spiegelman B. Adipocyte differentiation: a transcriptional regulatory cascade. Curr Opin Cell Biol 1996; 8:826–832.

32.   Rangwala S, Lazar M. Transcriptional control of adipogenesis. Annu Rev Nutr 2000; 20:535–559.

33.   Grimaldi P. The roles of PPARs in adipocyte differentiation. Prog Lipid Res 2001; 40:269–281.

34.   Linhart H, Ishimura-Oka K, DeMayo F, Kibe T, Repka D, Poindexter B, Bick R, Darlington G. C/EBP alpha is required for differentiation of white, but not brown adipose tissue. Proc Natl Acad Sci 2001; 98(22):12532–12537.

35.   St-Onge MP, Page GP, DeLuca M, Zhang K, Kim K, Heymsfield SB, Allison DB. Design & analysis of microarray studies for obesity research. Moustaid-Moussa and Berdanier, Eds. In: Genomics and Proteomics in Nutrition. Marcel Dekker, 2004; 145–203.

36.   Urs S, Smith C, Campbell B, Saxton A, Taylor J, Zhang B, Snoddy J, Voy BJ, Moustaid-Moussa N. Gene expression analysis in human preadipocytes and adipocytes by microarray. J Nutr 2004; 134:762–770.

37.   Nakamura T, Shiojima S, Hirai Y, Iwama T, Tsuruzoe N, Hirasawa A, Katsuma S, Tsujimoto G. Temporal gene expression changes during adipogenesis in human mesenchymal stem cells. Biochem Biophys Res Commun 2003; 303(1):306–312.

38. Saito-Hisaminato A, Katagiri T, Kakiuchi S, Nakamura T, Tsunoda T, Nakamura Y. Genome-wide profiling of gene expression in 29 normal human tissues with a cDNA microarray. DNA Res 2002; 9(2):35–45.

39. Gabrielsson B, Carlsson B, Carlsson L. Partial genome scale analysis of gene expression in human adipose tissue using DNA array. Obesity Res 2000; 8(5):374–384.

40. Guo X, Liao K. Analysis of gene expression profile during 3T3-L1 preadipocyte differentiation. Gene 2000; 251:45–53.

41. Lopez I, Marti A, Milagro F, Zulet Md Mde L, Moreno-Aliaga M, Martinez J, De Miguel C. DNA microarray analysis of gene differentially expressed in diet-induced (Cafeteria) obese rats. Obesity Res 2003; 11(2):188–194..

42. Burton G, Guan Y, Nagarajan R, Jr. McGehee. Microarray analysis of gene expression during early adipocyte differentiation. Gene 2002; 293:21–31.

43. Nadler S, Stoehr J, Schueler K, Tanimoto G, Yandell B, Attie A. The expression of adipogenic genes is decreased in obesity and diabetes mellitus. Proc Natl Acad Sci 2000; 97(21):11371–11376.

44. Jessen B, Stevens G. Expression profiling during adipocyte differentiation of 3T3-L1 fibroblasts. Gene 2002; 299:95–100.

45. Kim S, Urs S, Massiera F, Wortmann P, Joshi R, Heo YR, Andersen B, Kobayashi H, Teboul M, Ailhaud G, Quignard-Boulange A, Fukamizu A, Jones BH, Kim JH, Moustaid-Moussa N. Effects of High-Fat Diet, Angiotensinogen (agt) Gene Inactivation, and Targeted Expression to Adipose Tissue on Lipid Metabolism and Renal Gene Expression. Horm Metab Res 2002; 34(11–12): 721–725.

46. Sottile V, Seuwen K. A high-capacity screen for adipogenic differentiation. Anal Biochem 2001; 293:124–128.

47. Albrektsen T, Richter H, Clausen J, Fleckner J. Identification of a novel integral plasma membrane protein induced during adipocyte differentiation. Biochem J 2001; 359(Pt 2):393–402.

48. Boeuf S, Klingensor M, van Hal N, Schneider T, Keijer J, Klaus S. Differential gene expression in white and brown adipocytes. Physiol Genomics 2001; 7:15–25.

49. Moustaid-Moussa N, Jones B, Taylor J. Insulin increases lipogenic enzyme activity in human adipocytes in primary culture. J Nutr 1996; 126(4):865–870.

50. Claycombe K, Jones B, Standridge M, Guo Y, Chun J, Taylor J, Moustaid-Moussa N. Insulin increases fatty acid synthase gene transcription in human adipocytes. Am J Physiol 1998; 274(5 Pt 2):R1253–1259.

51. Wang Y, Voy BJ, Urs S, Kim S, Bejnood M, Quigley N, Heo YR, Standridge M, Andersen B, Dhar M, Joshi R, Taylor JW, Chun J, Leuze M, Claycombe K, Moustaid-Moussa N. Cloning of the human fatty acid synthase gene and coordinate regulation of its gene expression and de novo lipogenesis in human adipose tissue. J Nutr. In press.

52. Rodbell M, Scow R, Chernick S. Removal and metabolism of triglycerides by perfused liver. J Biol Chem 1964; 239:385–391.

53. Fried S, Moustaid-Moussa N. Culture of human adipose tissue and adipocytes. In: Adipose tissue Protocols. Gerard Ailhaud, ed. The Humana Press, 2001: 197–212.

54. Gollub J, Ball CA, Binkley G, Demeter J, Finkelstein DB, Hebert JM, Hernandez-Boussard T, Jin H, Kaloper M, Matese JC, Schroeder M, Brown PO, Botstein D, Sherlock G. The Stanford Microarray Database: data access and quality assessment tools. Nucleic Acids Res 2003; 31:94–96.

55. Maeda K, Uysal K, Makowski L, Gorgun C, Atsumi G, Parker R, Bruning J, Hertzel A, Bernlohr D, Hotamisligil G. Role of fatty acid binding protein mal1 in obesity and insulin resistance. Diabetes 2003; 52:300–307.

56. Brasaemle DL, Barber T, Wolins NE, Serrero G, Blanchette-Mackie EJ, Londos C. Adipose differentiation-related protein is an ubiquitously expressed lipid storage droplet-associated protein. J Lipid Res 1997; 38(11): 2249–2263.

57. Brasaemle DL, Rubin B, Harten IA, Gruia-Gray J, Kimmel AR, Londos C, Perilipin C. A increases triacylglycerol storage by decreasing the rate of triacylglycerol hydrolysis. J. Biol Chem 2000; 275:38486–38493.

58. Crandall DL, Busler DE, McHendry-Rinde B, Groeling TM, Kral JG. Autocrine regulation of human preadipocyte migration by plasminogen activator inhibitor-1. J Clin Endocrinol Metab 2000; 85(7):2609–2614.

59. Maeda K, Okubo K, Shimomura I, Funahashi T, Matsuzawa Y, Matsubara K. cDNA cloning and expression of a novel adipose specific collagen-like factor, apM1 (AdiPose Most abundant Gene transcript 1). Biochem Biophys Res Commun 1996; 221(2):286–289.

60. Jones B, Standridge M, Taylor J, Moustaid-Moussa N. Angiotensinogen gene expression in adipose tissue: analysis of obese models and hormonal and nutritional control. Am J Physiol 1997; 273:236–242.

61. Gaubatz S, Lindeman GJ, Ishida S, Jakoi L, Nevins JR, Livingston DM, Rempel RE. E2F4 and E2F5 play an essential role in pocket protein-mediated G1 control. Mol Cell 2000; 6(3):729–735.

62. Ring HZ, Vamegli-Meyers V, Wang W, Crabtree GR, Francke U. Five SWI/NF Srelated, matrix associated, actin dependent regulator of chromatin (SMARC) genes are dispersed in the human genome. Genomics 1998; 51:140–143.

63. Maquoi E, Munaut C, Colige A, Collen D, Lijnen H. Modulation of adipose tissue expression of murine matrix metalloproteinases and their tissue inhibitors with obesity. Diabetes 2002; 51:1093–1101.

64. Croissandeau G, Chretien M, Mbikay M. Involvement of matrix metalloproteinases in the adipose conversion of 3T3-L1 preadipocytes. Biochem J 2002; 364:739–746.

65. Kamiya S, Kato R, Wakabayashi M, Tohyama T, Enami I, Ueki M, Yajima H, Ishii T, Nakamura H, Katayama T, Takagi J, Fukai F. Fibronectin peptides derived from two distinct regions stimulate adipocyte differentiation by preventing fibronectin matrix assembly. Biochemistry 2002; 41:3270–3277.

66. Ibrahimi A, Bertrand B, Bardon S, Amri E, Grimaldi P, Ailhaud G, Dani C. Cloning of alpha 2 chain of type VI collagen and expression during mouse development. Biochem J 1993; 289:141–147.
67. Dardik R, Lahav J. Functional changes in the conformation of thrombospondin-1 during complexation with fibronectin or heparin. Exp Cell Res 1999; 248(2):407–414.
68. Takemoto S, Murakami T, Kusachi S, Iwabu A, Hirohata S, Nakamura K, Sezaki S, Havashi J, Suezawa C, Ninomiya Y, Tsuji T. Increased expression of dermatopontin mRNA in the infarct zone of experimentally induced myocardial infarction in rats: comparison with decorin and type I collagen mRNAs. Basic Res Cardiol 2002; 97(6):461–468.
69. Kagan H, Li W. Lysyl oxidase:Properties, specificity, and biological roles inside and outside of the cell. J Cell Biochem 2003; 88:660–672.
70. Smith-Mungo L, Kagan H. Lysyl oxidase: properties, regulation and multiple functions in biology. Matrix Biol 1998; 16:387–398.
71. Tartare-Deckert S, Chavey C, Monthouel M, Gautier N, Van Obberghen E. The matricellular protein SPARC/osteonectin as a newly identified factor up-regulated in obesity. J Biol Chem 2001; 276(25):22231–22237.
72. Bradshaw A, Graves D, Motamed K, Sage E. SPARC-null mice exhibit increased adiposity without significant differences in overall body weight. Proc Natl Acad Sci 2003; 13(100):6045–6050.
73. Saxton AM, Moser EB. Statistical principles for analysis of array experiments. In Genomics and Proteomics in Nutrition. N Moustaid-Moussa, and CD Berdanier. eds. Marcel Dekker, 2004; 473–492.

11

Phytochemicals and Gene Expression

Orsolya Mezei and Neil F. Shay*

Department of Biological Sciences, University of Notre Dame,
Notre Dame, Indiana, U.S.A.

1. INTRODUCTION

The practice of self-medicating with botanical products is probably as old as humankind itself. Depicted in literature and film, we are well-familiar with the herbal remedies prepared and administered by the healers of ancient Egypt, medieval physicians, and American Indian cultures. Only in the past century has modern chemistry produced a distinction: a choice between synthetically derived drugs vs. natural products. With the great advances made by the pharmaceutical industry, our modern Western culture has distanced itself from the notion that phytochemicals may have specific medicinal actions. Most synthetically derived medicines are considered to have a specific target: as an enzyme inhibitor (e.g., cholesterol-lowering statins), receptor agonists, or receptor antagonists (the antidiabetic drug rosiglitazone is an agonist of the peroxisome proliferator activated receptor). Synthetic drugs are generally considered to have high specificity; hallmark examples may be action on specific serotonin receptor subtypes or the

*Corresponding author

selective estrogen receptor (ER) modulators (SERMs) that act more potently on ERα or ERβ.

Understanding the specific effects of phytochemicals represents distinct challenges. Most often, commercially available products represent solvent-soluble extracts prepared from a botanical. These products usually contain mixtures of a wide number of compounds. The exact number and relative abundance of these products often varies from preparation to preparation. Growing conditions and the geographical source of a botanical may ultimately affect the potency of the extract. Evaluating the effect of a mixture on a biological system—whether a cell or animal model—becomes a more difficult task compared to evaluating the effect of a single compound. Different compounds present in the extract may have distinct effects, and may even interact negatively or positively. This interaction may vary in significance depending on the relative abundance of the different compounds in the extract.

To evaluate the effect of phytochemicals on gene expression, one may use both cell and animal models. Further, if a specific compound is presumed to be the bioactive compound present in a botanical extract, that compound can be studied alone, as a purified compound, in parallel with studies evaluating a mixture containing the putative active compound. One would expect to see the same effect on the model system, if the putative compound is indeed the active factor.

If a candidate phytochemical is being evaluated as a potential regulator of gene expression, one might predict specific sites where gene regulation would be affected. For example, a phytochemical acting as a ligand for a specific nuclear receptor might enhance the trancription rate of a specific gene. Alternatively, a phytochemical might repress gene transcription if it interferes with coactivator recruitment. Phytochemicals acting as agonists or antagonists of kinases or phosphatases involved in a signal transduction pathway may ultimately affect the expression of a gene or sets of genes regulated by that specific pathway.

A wide number of research reports have detailed the interaction between various phytochemicals. Rather than attempt to provide an encyclopedic collection of those reports, we have chosen a number of examples where phytochemicals have been shown to affect gene expression in different model systems.

2. ST. JOHN'S WORT

St. John's wort is used to treat mild to moderate depression and anxiety (1). It is composed of numerous constituents, although its active component remains speculative. One component, hyperforin, inhibits synaptic uptake of various neurotransmitters such as serotonin and dopamine in vitro, although in vivo studies suggest other potential modes of action not fully dependent on hyperforin (1). St. John's wort serves a wide array of therapeutic uses, not only historically, but also currently in both the United States and Europe (2). Recent reports have suggested that St. John's wort decreases the efficacy of medications metabolized by cytochrome P450 3A4 (CYP3A4), a member of the cytochrome P450 monooxygenase family (CYP) (Table 1) (3). The CYP family of enzymes is responsible for clearing the majority of prescription drugs and ingested pollutants (4). These enzymes are located in hepatocytes and intestinal cells and are capable of metabolizing many endogenous compounds along with exogenous drugs and pollutants (5). Several CYP families exist, although CYP1, CYP2, and CYP3 metabolize most xenobiotic substrates (5). Moore et al. investigated the effect of St. John's wort on the human pregnane X receptor (PXR), a nuclear receptor responsible for regulating CYP3A4 transcription (Figure 1) (3). They found that three different commercial brands of St. John's wort extract activated PXR. Of the components tested, hyperforin induced PXR at half-maximal effective concentration (EC$_{50}$) of 23 nM. Importantly, this concentration is well below the 200 nM level observed in plasma of individuals using St. John's wort on a regular basis. Hyperforin was also found to directly bind PXR, as confirmed by competition binding assays. Finally, Moore et al. found that the St. John's wort extract and hyperforin induced the expression of CYP3A4, validating the premise that this botanical could interfere with and increase the metabolism of drugs metabolized by CYP3A4 (5). Wentworth et al. also found that St. John's wort and hyperforin activated the ligand-binding domain of the steroid X receptor (SXR), known as the human PXR (6). This occurs via the activation function site 2 (AF-2) of SXR. Steroid receptor coactivator-1 is a coactivator recruited by various nuclear receptors including SXR. St. John's wort and hyperforin mediated the SXR-SRC1 association, further confirming of activation of CYP3A transcription by this phytochemical (6). Hyperforin has recently been confirmed by crystal structure analysis to bind to the ligand-binding domain of PXR (7). St. John's wort has also been implicated to increase the expression of CYP1A2, the second most abundant CYP accounting for over 10% of human hepatic CYP content (4). However, a human in vivo study did not find evidence of increased CYP1A2 activity after two weeks of St. John's wort intake (300 mg three times per day), although CYP3A4 activity was significantly induced in the intestinal wall (8). These studies concluded that St. John's wort increases the metabolism of certain medications by increasing the expression of CYP3A4 via hyperforin binding and activating human PXR.

3. GUGGULSTERONE

Guggulsterone is the active, lipid-lowering fraction of gugulipid, a gum resin extract from the Commiphora mukul tree used in India for thousands of years to treat hyperlipidemia (9). Numerous animal studies and clinical studies have been conducted since the 1960s when the initial scientific studies were begun on the hypolipidemic effect of the gum resin and its extracts. Most of the clinical studies found that gugulipid or guggulsterone reduced serum cholesterol levels by an average of 30% (9). Upon further investigation, Urizar et al. proposed that the lipid-lowering mechanism of this gum resin occurs via the antagonistic activity of guggulsterone for the farnesoid X receptor (FXR) (10). FXR heterodimerizes with the retinoid X receptor (RXR) upon ligand binding (11) and is known as the "bile acid sensor" because it is responsible for repressing bile acid synthesis via transcription of ileal bile acid-binding protein (I-BABP). Ligands of the nuclear receptor FXR include bile acids such as chenodeocycholic acid (CDCA). Activation of FXR also increases bile acid recirculation due to elevated bile acid concentrations within the cell (11). In the recent study by Urizar et al., the Z-guggulsterone isomer had no effect on FXR alone, although the E- and Z-guggulsterone isomers were able to inhibit CDCA activation of FXR along with FXR-regulated genes (10). The isomers were also able to inhibit CDCA activation of small heterodimeric partner (SHP). Small heterodimeric partner is a nuclear receptor that heterodimerizes with other nuclear receptor complexes such as the active FXR-RXR complex, although it does not have a DNA-binding motif, as do most other nuclear receptors (11). Guggulsterone inhibited transactivation of FXR-RXR and not DNA-binding of these complexes, indicating that the isomers were exerting an

TABLE 1 List of Botanicals and Their Active Ingredients that are Able to Activate Nuclear Receptors[a]

Botanical	Active ingredient	Targeted nuclear receptor
St. John's wort	Hyperforin	PXR / SXR (hPXR)
Guggulipid	E-Guggulsterone	FXR
	Z-Guggulsterone	
Soy isoflavones	Genistein	PPARα/PPARγ
	Daidzein	
Isoflavones	Genistein	ER
	Resveratrol	

[a]ER, estrogen receptor; FXR, farnesoid X receptor; PPAR, peroxisome proliferators activated receptor; PXR, pregnane X receptor; SXR, steroid X receptor (human pregnane X receptor).

inhibitory effect via the ligand-binding domain of the FXR (10). This effect was confirmed by using a fluorescence resonance energy transfer (FRET)-based coactivator binding assay, in which guggulsterone was found to directly compete with CDCA for the ligand-binding domain, inhibiting the recruitment of a necessary coactivator, SRC-1. Finally, FXR-null mice did not respond to the cholesterol-lowering effect of guggulsterone seen in the wild-type mice, indicating that FXR is necessary for the hypolipidemic effect of guggulsterone (10). Wu et al. also found that guggulsterone had an antagonistic activity of FXR in the presence of FXR activators and was able to decrease gene expression of FXR-regulated genes (12). More recently, a study found that guggulsterone induced the expression of the FXR-regulated bile salt export pump gene (BSEP) in vitro in the presence of two different FXR ligands, CDCA and GW4064 (13). This induction was also evident in rats fed a diet containing either 2.8 or 5.6% guggulsterone, in which both BSEP and SHP mRNA were elevated compared to the control-fed rats. However, mRNA levels of other FXR-regulated genes tested—such as cholesterol 7α-hydroxylase (CYP7α1), sterol 12α-hydroxylase (CYP8b1), and I-BABP—were unaffected. In this study, guggulsterone blocked coactivator recruitment of p120 and PBP as well as SRC-1, consistent with the prior report (13). These results indicate that guggulsterone may have selective antagonistic activity on required coactivator recruitment for FXR-mediated transcription, but also agonist-enhancing activity on selective FXR-regulated genes.

4. SOY ISOFLAVONES

4.1. Genistein and PPARs

Soy isoflavones are phytochemicals often termed "phytoestrogens" due to the estrogenic properties of these botanically derived products (14). Soy iso-flavones have been credited to have antiatherosclerotic, antidiabetic, and anticarginogenic properties, although the specific physiological and cellular mechanisms affected by isoflavones are an area of controversy and debate (15–17). Recent studies found that soy isoflavones were able to activate two isoforms of the peroxisome-proliferator-activated receptors (PPARα and PPARγ), proposing a novel way in which the isoflavones may be exerting their antiatherosclerotic and antidiabetic properties (Figure 2) (18,19). The PPARs are nuclear receptors involved in cellular lipid homeostasis (20). They have a promiscuous ligand-binding domain able to bind a variety of lipophilic ligands, resulting in receptor activation. Activation of PPARα results in increased expression of genes involved in fatty acid catabolism, whereas activation of PPARγ results in increased expression of genes

TABLE 2   Effect of Various Flavanoids on Signaling Pathway
Proteins

Flavanoid	Affected signaling protein
Apigenin	Ap-1, Elk-1, c-Jun
Flavone	Ap-1
Chalcone	Ap-1, Elk-1, c-Jun, CHOP
Kaempferide	Elk-1, c-Jun
Genistein	P38MAPK, ERK-1, ERK-2
Naringenin	P13K, Akt

involved in cellular differentiation and insulin sensitization (20). Dang et al.
found that the soy isoflavone genistein was able to activate PPARγ in a
dose-dependent manner, and genistein also increased the expression of
PPARγ-regulated genes and adipogenesis in KS483 cells at a dose of 25μM
(19). Genistein interacts directly with the nuclear receptor, as verified by a
membrane-bound PPARγ binding assay. However, a lower dose of genistein
(1 μM) actually had an inhibitory effect on PPARγ-regulated genes as well
as on adipogenesis. This is most likely due to the ability of low concentrations of genistein to activate estrogen receptor-mediated activity, resulting
in a decrease of PPARγ activation (19). Mezei et al. found that both genistein
and the soy isoflavone daidzein were able to activate PPARγ-mediated
transcription (18). Furthermore, female obese Zucker rats fed a high-
isoflavone-containing soy diet had significantly improved glucose tolerance
compared to casein and low-isoflavone-containing diets, consistent with
effects of PPARγ activation. Genistein and daidzein were also able to activate PPARα-mediated transcription. Both male and female obese Zucker
rats fed a high-isoflavone-containing diet had reduced liver cholesterol, liver
triglycerides, and total liver weight, consistent with effects of PPARα
activation (18). Harmon and Harp found an opposing effect of genistein on
PPARγ (21). In this study, genistein was found to inhibit PPARγ expression
as well as adipogenesis in adipocytes, a well-characterized consequence of
PPARγ activation. However, these inconsistent effects may be due to the
elevated concentration used in this study. Other studies found that a
genistein concentration of 50 μM was enough to induce apoptosis in certain
culture models such as colon carcinoma cell lines, whereas Harmon et al.
used a genistein concentration of 100 μM (22,23). Genistein and daidzein
are not the only phytochemicals with PPAR-activating ability. Takahashi et
al. discovered that farnesol and geranylgeraniol, two common fruit and herb
isoprenols, are able to activate both PPARα and PPARγ along with several
PPAR-regulated genes (24). Therefore, these studies give new insight on

the mechanism by which soy isoflavones and other botanicals exert their favorable consequences.

4.2. Isoflavones and Estrogen Receptors

It is estimated that 80% of women over the age of 45 use some type of non-prescription therapy to manage menopause symptoms, ranging from the consumption of soy or evening primrose oil to acupuncture (25). Of these therapies, the use of isoflavones as an alternative to hormone replacement therapy may be an attractive alternative to classical "hormone replacement therapy" (HRT) for many postmenopausal women, although their potential side effects and long-term health implications are still not fully understood (26). One of the soy isoflavones, genistein, has been shown in studies to have estrogenic activity (27,28). Because some postmenopausal women with estrogen-dependent breast tumors may be consuming genistein as an alternative to HRT, Ju et al. studied the effect of genistein on estrogen-dependent breast cancer growth (29). In this study, mice with estrogen-dependent tumors had a significant, dose-dependent increase in tumor presenelin-2 (pS2) mRNA levels when provided dietary genistein. The level of induction seen with the higher genistein doses was similar to the induction produced by the subcutaneous 17β-estradiol pellet. The pS2 gene is an estrogen-responsive gene and indicative of estrogen-dependent growth. The ability of genistein to induce estrogen-dependent growth in these tumors was also observed in tumor size and proliferation; both were significantly increased with genistein ingestion or 17β-estradiol supplementation (29). Therefore, the results of this study indicate that genistein consumption may promote the growth of certain estrogen-dependent breast tumors. Another phytochemical with known estrogenic activity is resveratrol, a polyphenolic compound in grapes and wine (30). Resveratrol has also been attributed to have cancer-preventative properties in colon cancer cell lines (31,32). However, the effect of resveratrol on breast cancer growth is controversial, especially with respect to estrogen-dependent tumors (33). Levenson et al. found that resveratrol was able to induce gene expression of the estrogen-responsive gene tumor growth factor α (TGFα) in a dose-dependent manner in breast cancer cells expressing wild-type estrogen receptor (33). Higher doses of resveratrol were needed to mimic this effect in breast cancer cells expressing a mutant form of the estrogen receptor. However, resveratrol did not further stimulate TGFα expression when 17β-estradiol was present in its optimal concentration. Resveratrol inhibited the growth of the breast cancer cells regardless of the presence of the estrogen receptor or the antiestrogen ICI, indicating that growth inhibition by resveratrol is, at least in part, estrogen receptor-independent. Estrogen receptor protein levels were analyzed in both wild-type and mutant estrogen receptor-expressing cells. Both 17β-estradiol and resveratrol decreased the wild-type estrogen receptor levels. Finally, resveratrol and 17β-estradiol both increased the protein levels of p21^{cip}/WAF1, a cyclin-dependent kinase inhibitor, although this increase appears to be an estrogen-mediated effect (33). Both resveratrol and genistein have estrogenic effects and are able to regulate many estrogen receptor-mediated genes. This activity may explain some of the beneficial effects of these phytochemicals, but also warrants further investigation due to possible harmful side effects.

4.3. Genistein and Gene Expression Patterns

Recent studies utilizing microarray technology reveal that genistein affects the regulation of many genes, including those involved in reproductive development and prostate cancer (34–36). Naciff et al. found that genistein had a gene expression profile similar to an estrogen (17 α-Ethynyl estradiol) and a weak estrogenic chemical (bisphenol A) in the developing uterus and ovary of the rat (34). Genes involved in cell growth (growth hormone receptor), differentiation (progesterone receptor), stress response (glutathione S-transferase M5), and apoptosis (interleukin 4 receptor) were regulated similarly by all three compounds. RT-PCR confirmed some of these results, such as increased expression of the progesterone receptor by 17 α-Ethynyl estradiol, bisphenol A, and genistein (34). It is important to note that although the three compounds have a similar gene expression profile in the developing reproductive system, the gene expression profiles of 17 α-Ethynyl estradiol and bisphenol A were more similar to each other than to genistein. This may be due to the mainly "estrogenic" activity of these compounds, whereas genistein has other known activities, such as tyrosine kinase and topoisomerase-II inhibition along with activity as a PPAR agonist profiled earlier (34). Two recent studies also analyzed the gene expression profile of genistein using a human prostate cancer cell line (35,36). Li and Sarkar found that genistein downregulated genes involved in angiogenesis, such as vascular endothelial growth factor and its receptor, and upregulated genes inhibiting angiogenesis, such as connective tissue growth factor and connective tissue activation peptide (35,36). Furthermore, genistein also downregulated genes necessary for tumor cell invasion and metastasis (MMP-9/type IV collagenase, urokinase plasminogen activator, and urokinase plasminogen activator receptor). These results indicate that genistein may inhibit tumor metastasis and growth. Another study by Li and Sarkar found that genistein caused a difference in expression profiles of genes involved in cell cycle control, apoptosis, and cell signaling (36). Genistein downregulated cell cycle promoter genes such as cyclin

A and cyclin B and induced genes that inhibit cell cycle progression, such as cyclin G2 in human prostate cancer cultured cells. Genes involved in the inhibition of apoptosis (survivin) and genes involved in cell growth (pescadillo) were also downregulated in genistein-treated cells (36,37). Genistein also downregulated signaling genes such as NF-κB-inducing kinase and MAP kinase kinase potentially resulting in decreased cell proliferation (36). Therefore, microarray analysis revealed that genistein is able to affect many genes involved in biological processes such as cell growth, cell cycle control, differentiation, stress response, angiogenesis, tumor cell invasion, metastasis, signaling, and apoptosis.

5. SOY DIET STUDIES

Soy consumption may have many advantageous outcomes, such as an improved management of blood lipids and a decreased risk of cancer (38–40). In one rodent study, Tovar-Palacio et al. found that soy-fed gerbils had significantly reduced levels of circulating apolipoprotein B and significantly increased circulating levels of apolipoprotein A-I after a 28-day feeding study (41). However, apolipoprotein A-I gene expression was significantly reduced in gerbils fed a soy diet containing various amounts of isoflavones. This reduction was not reflected in the circulating protein content. This discrepancy may be a result of decreased circulating lipoprotein turnover or a downregulation of apolipoprotein A-I synthesis due to its elevated level in circulation in the soy-fed animals. It is important to note that the mRNA levels of apolipoprotein E an apolipoprotein also synthesized in the liver and similar in abundance to apolipoprotein A, remained unchanged (41). Two other genes also affected by soy consumption in gerbils are phosphoribosylpyrophosphate synthetase-associated protein (PAP) and a member of the cytochrome P450 2A family (CYP2A) (42). In a study by Mezei et al., gerbils fed a soy diet with increasing levels of isoflavones had a dose-dependent increase in both PAP and CYP2A gene expression. PAP is a protein that negatively regulates phosphoribosylpyrophosphate synthetase (PRPP-synthetase) activity, an enzyme involved in nucleotide synthesis. Therefore, soy may be able to decrease nucleotide synthesis and cell proliferation via PAP regulation. CYP2A belongs to a family of enzymes used to metabolize endogenous and exogenous toxins and other xenobiotics (43). Therefore, upregulation of this CYP might decrease mutagenic threat to the cell. Ronis et al. found an induction in CYP3A protein levels in dexamethasone-treated (DEX-treated), soy-fed rats relative to DEX-treated, casein-fed rats (44). This induction in protein level appears to be due in part to the increased expression of CYP3A2 mRNA. CYP3A1, CYP3A9, and CYP3A18 did not have increased mRNA levels in

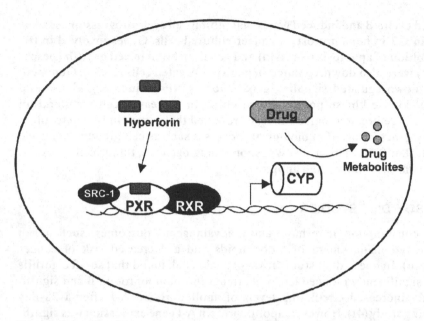

FIGURE 1 Increased drug metabolism results from hyperforin-induced PXR activation within the nucleus. Hyperforin binds directly to PXR and SRC-1 associates with the PXR-RXR heterodimer. This activation of PXR results in increased expression of several cytochrome P450s including CYP3A4. Increased CYP3A4 levels result in increased metabolism of certain prescription drugs, thereby increasing drug clearance and decreasing drug half-life and efficacy. CYP: cytochrome P450, PXR: pregnane X receptor, RXR: retinoid X receptor, SRC-1: steroid receptor coactivator-1. (Adapted from Refs. 3, 5–7.)

DEX-treated, soy-fed rats compared to the DEX-treated, casein-fed rats. Neither CYP2B1 protein levels nor mRNA levels were different between these soy-fed and casein-fed rats. However, the enzymatic activity of CYP2B1 was greater in the DEX-treated, soy-fed rats. This effect was also seen in CYP3A enzymatic activity except when CYP3A18-specific lithocholic acid was used as substrate (44). Therefore, soy had other stimulatory effects on CYP enzymes besides increased gene and subsequent protein expression. A second study by Ronis et al. focused on CYP1A induction and activity resulting from a casein, whey, or soy protein source in 3-methylcholanthrene- (3-MC) or isosafrole- (ISO) induced rats (45). 3-MC is an environmental carcinogen that induces CYP1A expression via the aryl-hydrocarbon receptor (AhR) located in the promoter of CYP1A genes, and ISO is a common phytochemical component of foodstuffs that induces CYP1A in an AhR-independent manner (45). 3-MC induced CYP1A1 gene

A

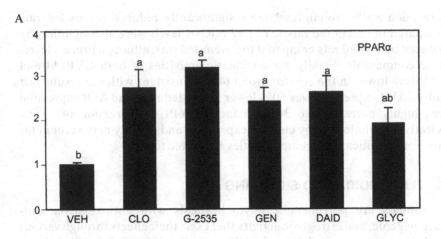

B

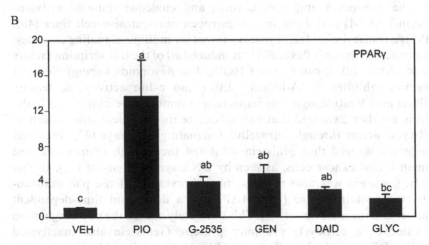

FIGURE 2 The effect of a soy extract and various soy components on PPRE-directed luciferase activity induced by the PPRE-containing segment of the acyl CoA oxidase promoter in RAW 264.7 cells using the PPARα and PPARγ expression plasmids, respectively. Cells were incubated with either a vehicle (VEH), (A) clofibrate (CLO) or (B) pioglitazone (PIO), or 2.5 mg/L of the following: G-2535 (U-SOY), Prevastein HC (C-SOY), genistein (GEN, 9.3 μmol/L), daidzein (DAID, 9.8 μmol/L), or glycitein (GLYC 8.8 μmol/L). Values are means ± SEM, n = 7–8. Values do not share a letter differ (P < 0.05). (Adapted from Ref. 18.)

expression while protein levels were significantly reduced in soy-fed rats compared to casein-fed rats. CYP1A2 mRNA levels were also significantly reduced in soy-fed rats compared to casein-fed rats, although protein levels were comparable. Finally, the enzymatic activities of both CYP1A1 and 1A2 were lower in the soy-fed group (45). Consistent with the results just noted, AhR expression was 50% lower in soy-fed rats, and AhR expression was highly correlated to 3-MC-induced CYP1A1 expression (45). The selective regulation of soy on CYP expression and activity may account for some of the anticargenogenic activities attributed to soy.

6. FLAVONOIDS AND SIGNALING

Flavonoids are polyphenolic phytochemicals with antioxidative, anti-carginogenic, and estrogenic activity that exert their effects through various biological processes including signaling cascades (46–48). Frigo et al. found that the flavonoids apigenin, flavone, and chalcone induced activator protein-1 (AP-1) activation in two estrogen-unresponsive cell lines (46). AP-1 is a transcription factor that is a target for multiple signaling cascades. Chalcone was the only flavonoid that induced all of the transcription factors tested (AP-1, Elk-1, c-Jun, and CHOP). The flavonoids kaempferide and apigenin inhibited PMA-induced Elk-1 and c-Jun activity, decreasing cellular proliferation signaling important in tumor prevention (46). Genistein is another flavonoid that can moderate its biological effects such as cell-cycle arrest through intracellular signaling pathways (47). Frey and Singletary showed that genistein inhibited the growth of immortalized human breast cancer cells, as seen by DNA synthesis arrest (47). In this study, genistein was able to cause phosphorylation of the p38 mitogen-activated protein kinase (p38 MAPK) in a dose- and time-dependent manner and increase its activity. This ultimately caused the downregulation of Cdc25C, a cell-cycle promoter protein. Genistein also inactivated ERK1/ERK/2 and had no effect on SAPK/JNK activity, indicating that this isoflavone has a selective action on MAPK signaling pathways that may be dependent on cell type (47). Another flavonoid that affects signaling cascades is naringenin, a flavonoid found in grapefruit (48). Harmon and Patel found that this flavonoid inhibited insulin-mediated glucose uptake in adipocytes (48). Naringenin arrested Akt activation, but had no effect on the insulin receptor (IRβ), insulin receptor substrate-1 and -2 (IRS-1, IRS-2), or phosphoinositide 3-kinase (PI3K) phosphorylation status. Although naringenin did not affect the phosphorylation state of PI3K, it did inhibit the activity of PI3K, resulting in the observed decrease in Akt phosphorylation (48). In conclusion, flavonoids such as chalcone, genistein, and naringenin are able to mediate biological processes such as cell-cycle arrest or altered gene transcription via intracellular signaling cascades.

7. SUMMARY

One of the most exciting advances in the field of regulation of gene expression by dietary constituents is the explosion of information becoming available regarding nuclear receptors such as PPAR, FXR, PXR, AhR, and many others. A particularly important aspect that must be considered is that many nuclear receptors are "promiscuous" receptors, having the ability to bind many different endogenous and exogenous ligands. Thus, is becomes immediately apparent that many phytochemicals have such lipophilic properties that they may make excellent ligands for one or more of these nuclear receptors. In previous sections, we have discussed some of the recent examples just beginning to show how phytochemicals interact with these promiscuous receptors. One of the exciting challenges of future research will be to identify further phytochemical ligands for these receptors and to study phytochemical/phytochemical and drug/phytochemical interactions. The ability to point out both negative interactions and potentially promising positive interactions between drugs and phytochemicals may provide great practical health benefits to the consumer.

REFERENCES

1. De Smet PA. Herbal remedies. N Engl J Med 2002; 347:2046–2056.
2. Bilia AR, Gallori S, Vincieri FF. St. john's wort and depression: efficacy, safety and tolerability-an update. Life Sci 2002; 70:3077–3096.
3. Moore LB, Goodwin B, Jones SA, Wisely GB, Serabjit-Singh CJ, Wilson TM, Collins JL, Kliewer SA. St. john's wort induces hepatic drug metabolism through activation of the pregnane X receptor. Proc Natl Acad Sci USA 2000; 97:7500–7502.
4. Karyekar CS, Eddington ND, Dowling TC. The effect of st. john's wort extract on intestinal expression of cytochrome P4501A2: studies in LS180 cells. J Postgrad Med 2002; 48:97–100.
5. Nebert DW, Russell DW. Clinical importance of the cytochromes P450. Lancet 2002; 360:1155–1162.
6. Wentworth JM, Agostini M, Love J, Schwabe JW, Chatterjee VKK. St. john's wort, a herbal antidepressant, activates the steroid x receptor. J Endocrinol 2000; 166:R11–R16.
7. Watkins RE, Maglich JM, Moore LB, Wisely GB, Noble SM, Davis-Searles PR, Lambert MH, Kliewer SA, Redinbo MR. 2.1 A crystal structure of human PXR in complex with the st. john's wort compound hyperforin. Biochemistry 2003; 42:1430–1438.

8. Wang Z, Gorski JC, Hamman MA, Huang SM, Lesko JL, Hall SD. The effects of st. john's wort (hypericum perforatum) on human cytochrome P450 activity. Clin Pharmacol Ther 2001; 70:317–326.

9. Urizar N, Moore DD. Gugulipid: a natural cholesterol-lowering agent. Ann Rev Nutr 2003; 23:303–313.

10. Urizar NL, Liverman AB, Dodds DT, Silva FV, Ordentlich P, Yan Y, Gonzalez FJ, Heyman RA, Mangelsdorf DJ, Moore DD. A natural product that lowers cholesterol as an antagonist ligand for FXR. Science 2002; 296:1703–1706.

11. Fitzgerald ML, Moore KJ, Freeman MW. Nuclear hormone receptors and cholesterol trafficking: the orphans find a new home. J Mol Med 2002; 80: 271–281.

12. Wu J, Xia C, Meier J, Li S, Hu X, Lala DS. The hypolipidemic natural product guggulsterone acts as an antagonist of the bile acid receptor. Mol Endocrinol 2002; 16:1590–1597.

13. Cui J, Huang L, Zhao A, Lew JL, Yu J, Sahoo S, Meinke PT, Royo I, Pelaez F, Wright SD. Guggulsterone is a farnesoid X receptor antagonist in coactivator association assays but acts to enhance transcription of bile salt export pump. J Biol Chem 2003; 278:10214–10220.

14. Lampe JW. Isoflavonoid and lignan phytoestrogens as dietary biomarkers. J Nutr 2003; 133(suppl 3):956S-964S.

15. Clarkson TB. Soy, soy phytoestrogens and cardiovascular disease. J Nutr 2002; 132:566S–569S.

16. Jayagopal V, Albertazzi P, Kilpatrick ES, Howarth EM, Jennings PE, Hepburn DA, Atkin SL. Beneficial effects of soy phytoestrogen intake in postmenopausal women with type 2 diabetes. Diabetes Care 2002; 25:1709–1714.

17. Lamartiniere CA, Cotroneo MS, Fritz WA, Wang J, Mentor-Marcel R, Elgavish, A. Genistein chemoprevention: timing and mechanisms of action in murine mammary and prostate. J Nutr 2002; 132:552S–558S.

18. Mezei O, Banz WJ, Steger RW, Peluso MR, Winters TA, Shay N. Soy isoflavones exert antidiabetic and hypolipidemic effects through the PPAR pathways in obese zucker rats and murine RAW 264.7 cells. J Nutr 2003; 133:1238–1243.

19. Dang ZC, Audinot V, Papapoulos SE, Boutin JA, Lowik CW. Peroxisome proliferator-activated receptor gamma (PPARgamma) as a molecular target for the soy phytoestrogen genistein. J Biol Chem 2003; 278:962–967.

20. Neve BP, Fruchart JC, Staels B. Role of the peroxisome proliferator-activated receptors (PPAR) in atherosclerosis. Biochem. Pharmacol 2000; 60: 1245–1250.

21. Harmon AW, Harp JB. Differential effects of flavonoids on 3T3-L1 adipogenesis and lipolysis. Am J Physiol Cell Physiol 2001; 280:C807–C813.

22. Wilson LC, Baek SJ, Call A, Eling TE. Nonsteroidal anti-inflammatory drug-activated gene (NAG-1) is induced by genistein through the expression of p53 in colorectal cancer cells. Int J Cancer 2003; 105:747–753.

23. Salti GI, Grewal S, Mehta RR, Das Gupta TK, Boddie Jr AW, Constantinou AI. Genistein induces apoptosis and topoisomerase II-mediated DNA breakage in colon cancer cells. Eur J Cancer 2000; 36:796–802.

24. Takahashi N, Kawada T, Goto T, Yamamoto T, Taimatsu A, Matsui N, Kimura K, Saito M, Hosokawa M, Miyashita K, Fushiki T. Dual action of isoprenols from herbal medicines on both PPARgamma and PPARalpha in 3T3-L1 adipocytes and HepG2 hepatocytes. FEBS Lett 2002; 514:315–322.

25. Kang HJ, Ansbacher R, Hammoud MM. Use of alternative and complementary medicine in menopause. Int J Gynecol Obstet 2002; 79:195–207.

26. Wuttke W, Jarry H, Westphalen S, Christoffel V, Seidlova-Wuttke D. Phytoestrogens for hormone replacement therapy?. J Steroid Biochem Mol Biol 2002; 83:133–147.

27. Rickard DJ, Monroe DG, Ruesink TJ, Khosla S, Riggs BL, Spelsberg TC. Phytoestrogen genistein acts as an estrogen agonist on human osteoblastic cells through estrogen receptors alpha and beta. J Cell Biochem 2003; 89:633–646.

28. Santell RC, Chang YC, Nair MG, Helferich WG. Dietary genistein exerts estrogenic effects upon the uterus, mammary gland and the hypothalamic/pituitary axis in rats. J Nutr 1997; 127:263–269.

29. Ju YH, Allred CD, Allred KF, Karko KL, Doerge DR, Helferich WG. Physiological concentrations of dietary genistein dose-dependently stimulate growth of estrogen-dependent human breast cancer (MCF-7) tumors implanted in athymic nude mice. J Nutr 2001; 131:2957–2962.

30. Gehm BD, McAndrews JM, Chien PY, Jameson JL. Resveratrol, a polyphenolic compound found in grapes and wine, is an agonist for the estrogen receptor. Proc. Natl. Acad. Sci. USA 1997; 94:14138–14143.

31. Wolter F, Akoglu B, Clausnitzer A, Stein J. Downregulation of the cyclin D1/Cdk4 complex occurs during resveratrol-induced cell cycle arrest in colon cancer cell lines. J Nutr 2001; 131:2197–2203.

32. Briviba K, Pan L, Rechkemmer G. Red wine polyphenols inhibit the growth of colon carcinoma cells and modulate the activation pattern of mitogen-activated protein kinases. J Nutr 2002; 132:2814–2818.

33. Levenson AS, Gehm BD, Pearce ST, Horiguchi J, Simons LA, Ward 3rd JE, Jameson JL, Jordan VC. Resveratrol acts as an estrogen receptor (ER) agonist in breast cancer cells stably transfected with ER alpha. Int J Cancer 2003; 104:587–596.

34. Naciff JM, Jump ML, Torontali SM, Carr GJ, Tiesman JP, Overmann GJ, Daston GP. Gene expression profile induced by 17alpha-ethynyl estradiol, bisphenol A, and genistein in the developing female reproductive system of the rat. Toxicol Sci 2002; 68:184–199.

35. Li Y, Sarkar FH. Down-regulation of invasion and angiogenesis-related genes identified by cDNA microarray analysis of PC3 prostate cancer cells treated with genistein. Cancer Lett 2002; 186:157–164.

36. Li Y, Sarkar FH. Gene expression profiles of genistein-treated PC3 prostate cancer cells. J Nutr 2002; 132:3623–3631.

37. Suzuki K, Koike H, Matsui H, Ono Y, Hasumi M, Nakazato H, Okugi H, Sekine Y, Oki K, Ito K, Yamamoto T, Fukabori Y, Kurokawa K, Yamanaka H. Genistein, a soy isoflavone, induces glutathione peroxidase in the human prostate cancer cell lines LNCaP and PC-3. Int J Cancer 2002; 99:846–852.

38. Kerckhoffs DAJM, Brouns F, Hornstra G, Mensink RP. Effects on the human serum lipoprotein profile of ß-glucan, soy protein and isoflavones, plant sterols and stanols, garlic and tocotrienols. J Nutr 2002; 132:2494–2505.

39. Zhou J-R, Gugger ET, Tanaka T, Guo Y, Blackburn GL, Clinton SK. Soybean phytochemicals inhibit the growth of transplantable human prostate carcinoma and tumor angiogenesis in mice. J Nutr 1999; 129:1628–1635.

40. Lamartiniere CA, Zhang JX, Cotroneo MS. Genistein studies in rats: potential for breast cancer prevention and reproductive and developmental toxicity. Am J Clin Nutr 1998; 68(suppl 6):1400S–1405S.

41. Tovar-Palacio C, Potter SM, Hafermann JC, Shay NF. Intake of soy protein and soy protein extracts influences lipid metabolism and hepatic gene expression in gerbils. J Nutr 1998; 128:839–842.

42. Mezei O, Chou CN, Kennedy KJ, Tovar-Palacio C, Shay NF. Hepatic cytochrome p450-2A and phosphoribosylpyrophosphate synthetase-associated protein mRNA are induced in gerbils after consumption of isoflavone-containing protein. J Nutr 2002; 132:2538–2544.

43. Honkakoski P, Negishi M. Regulation of cytochrome P450 (CYP) genes by nuclear receptors. Biochem J 2000; 347:321–337.

44. Ronis MJ, Rowlands JC, Hakkak R, Badger TM. Altered expression and gluco-corticoid-inducibility of hepatic CYP3A and CYP2B enzymes in male rats fed diets containing soy protein isolate. J Nutr 1999; 129:1958–1965.

45. Ronis MJ, Rowlands JC, Hakkak R, Badger TM. Inducibility of hepatic CYP1A enzymes by 3-methylcholanthrene and isosafrole differs in male rats fed diets containing casein, soy protein isolate or whey from conception to adulthood. J Nutr 2001; 131:1180–1188.

46. Frigo DE, Duong BN, Melnik LI, Schief LS, Collins-Burow BM, Pace DK, McLachlan JA, Burow ME. Flavonoid phytochemicals regulate activator protein-1 signal transduction pathways in endometrial and kidney stable cell lines. J Nutr 2002; 132:1848–1853.

47. Frey RS, Singletary KW. Genistein activates p38 mitogen-activated protein kinase, inactivates ERK1/ERK2 and decreases Cdc25C expression in immortalized human mammary epithelial cells. J Nutr 2003; 133:226–231.

48. Harmon AW, Patel YM. Naringenin inhibits phosphoinositide 3-kinase activity and glucose uptake in 3T3-L1 adipocytes. Biochem Biophys Res Commun 2003; 305:229–234.

12

Gene Expression Profiling of Immune Cells: Application for Understanding Aging of Immune Cells

Sung Nim Han

Jean Mayer USDA Human Nutrition Research Center on Aging at Tufts
University, Boston, Massachusetts, U.S.A.

Simin Nikbin Meydani

Jean Mayer USDA Human Nutrition Research Center on Aging at Tufts
University, and Department of Pathology, Sackler Graduate School
of Biomedical Sciences, Tufts University, Boston, Massachusetts, U.S.A.

1. INTRODUCTION

Aging is associated with the dysregulation of immune and inflammatory responses. This is believed to contribute to the higher morbidity and mortality from infection, neoplastic, and inflammatory diseases. Studies indicate that a multitude of defects involving different immune cells are responsible for the decline of immune function and dysregulation of inflammatory responses observed with aging (1,2). However, understanding the underlying mechanisms of these changes has progressed slowly because the activation and response of immune cells involve several pathways with complex interactions.

Recent developments in microarray analysis make it feasible to measure mRNA abundance for large sets of genes, using a small number of cells. In the past few years, there has been a surge in the use of the microarray technique in an effort to identify and characterize the genetic regulation of the immune system including development, differentiation, maturation, lineage commitment, and the activation of various immune cells. However, few studies have utilized microarray analysis in determining the effect of aging on the immune cells.

This chapter, discusses recent investigations on gene expression, the profiling of the immune cells as well as the application of microarray analysis to understand the molecular mechanisms of age-associated immune dysregulation.

2. IMMUNE RESPONSE AND IMMUNE CELLS

The immune response involves an interaction between various immune cells. Immune responses are classified as innate and adaptive, or as specific immune responses. The innate immune response is the first line of defense against many microorganisms and is provided by phagocytes (macrophages and neutrophils) and natural killer (NK) cells. The specific immune response provides long-term protection against specific antigens and is mediated by the lymphocytes. The specific immune responses can be divided into humoral and cell-mediated immune response based on the types of immune cells involved. Humoral immunity is mediated by antibodies produced by plasma cells (effector cells of B lymphocytes) and is responsible for eliminating extracellular microbes. Cell-mediated immunity is mediated by T lymphocytes. T lymphocytes activate macrophages to kill intracellular microbes and become effector cells that destroy virus-infected cells or certain tumor cells. The T cells play a regulatory role in the function of other cells. The various types of immune cells and their functions are shown in Table 1.

2.1. Large-Scale Gene Expression Profiling of Immune Cells

2.1.1. Changes in Gene Expression Profiles During Development of Immune Cells

T Cells. T cells are derived from hematopoietic tissue and mature in the thymus to various types of differentiated T cells. Goh et al. (3) compared gene expression profiles of immature $CD3^-$, 4^-, 8^- triple-negative, $CD4^+$, 8^+ double-positive, and $CD4^+$, 8^- single-positive human thymocytes to analyze the changes during T cell development by analyses of expressed

TABLE 1   Immune Cells and Their Functions

Types of immune cells		Functions
Granulocytes	Neutrophil	Phagocytosis
		Killing extracelluar pathogen
	Basophil	Induction of allergic inflammatory response
	Eosinophil	Killing of antibody coated parasites
	Mast cell	Induction of allergic reaction
Dendritic cell		Antigen presentation
Macrophage		Phagocytosis
		Antigen presentation
Natural killer cell		Lysis of virus-infected cells and tumor cells
T lymphocytes	Cytotoxic lymphocyte	Lysis of virus-infected cells and tumor cells
		Macrophage activation
	T helper 1	Macrophage activation
	T helper 2	Stimulate B cell growth and differentiation
B lymphocytes		Antibody production
		Antigen presentation

sequence tags. Polymerase chain reaction (PCR)-based cDNA libraries were constructed, a total of 1477 randomly selected clones were analyzed by automated single-pass sequencing and the sequences were matched to known genes. Expression of genes involved in cell division/DNA synthesis and gene expression/protein synthesis was highly elevated in the double-positive stage, whereas genes related to metabolism were expressed highly in triple-negative and $CD4^+$ single-positive stages. This study had limited scope. Only a small number of sequences (392) were compared and cells representing different developmental stages were obtained from a single subject.

Most of the gene expression profiling experiments on the development of T cells have so far focused on the lineage commitment or polarization of T helper (Th) 1 and Th2 cells (4–8). Differentiation of $CD4^+$ T cells to Th1 or Th2 cells has implications for protection against different microbial pathogens and the development of different types of chronic diseases. Th1 cells mainly produce IFN-γ and protect against intracellular pathogens while Th2 cells produce IL-4, 5, and 13 and provide protection against extracellular pathogens. Th1 cells are involved in chronic inflammatory diseases such as autoimmune diseases and Th2 cells are associated with allergic diseases. Rogge et al. (4), Hamalainen et al. (8), Bonecchi et al. (6), and Nagai et al. (7) used Th1 and Th2 lines derived from human cord blood leukocytes. Cord blood leukocytes were initially stimulated with phytohemagglutinin in the presence of IL-12 (with or without anti-IL-4) for Th1 development or IL-4 (with or without anti-IL-12) for Th2 development, and then expanded further. Chtanova et al. (5) used purified T cells from mice and cultured with anti-CD3, anti-CD28, IL-6, IL-2, and IL-12 (Th1 culture) or IL-4 (Th2 culture) for five days and then restimulated with anti-CD3 for 24 hrs. These studies differ in the method used in determining the gene expression profile of the immune cells. Rogge et al. (4) used HuGeneFL array from Affymetrix, Hamalainen et al. (8) used Roche PA-1 oligonucleotide array, and Nagai et al. (7) used serial analysis of gene expression (SAGE). Rogge et al. (4) identified 215 genes, which were differentially expressed, at a confidence level of 95% and whose change in expression levels was at least twofold, between Th1 and Th2 cells collected at an early stage of polarization (three days). Transcription factors, which previously were not known to be associated with polarization of T helper cells such as RORα2, IRF-7A, and c-fos were identified. In addition, Th1 cells were suggested to be more susceptible to activation-induced cell death (AICD). The higher expression of Th1 genes is related to apoptosis and proteolysis, including tumor necrosis factor-related apoptosis-inducing ligand (TRAIL), proapoptotic Bcl-2 family member BAK-2, caspase-8, perforin (PRF1), and granzyme B (Fig. 1). On the other hand, Hamalainen et al. (8) found increased expression of granzyme B as well as caspase-1 and clusterin in Th1 cells and higher expression of caspase-6 in Th2 cells. A distinct pattern of chemokine receptor (CCR and CXCR) expression was observed in Th1 and Th2 cells. Chemokines are key components of the lymphocyte recruitment process. Therefore, differential expression of their receptors leads to a differential response to chemokines and the migration of Th1 and Th2 cells. Th1 cells have been shown to preferentially express CCR1, CCR2, CCR5, and CXCR3, whereas Th2 cells were shown to preferentially express CCR3 and CCR4.

Chtanova et al. (5) used Mu11K array from Affymetrix and reported expression patterns that were different from those by Rogge et al. (4) and Hamalainen et al. (8). For example, they reported higher expression of CCR1 and CCR5 in Th2 cells and identified more type II-biased genes than Rogge et al. (4) and Hamalainen et al. (8). Rogge (9) attributed this discrepancy in findings to the difference in culture condition to obtain polarized cells, especially the use of IL-6. These differences may also reflect the disparity in gene expression between human and mouse.

cytokines, growth factors and receptors

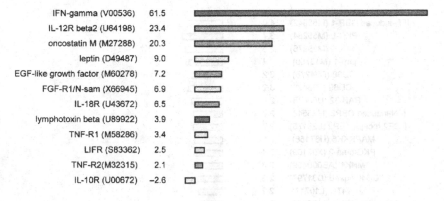

IFN-gamma (V00536)	61.5
IL-12R beta2 (U64198)	23.4
oncostatin M (M27288)	20.3
leptin (D49487)	9.0
EGF-like growth factor (M60278)	7.2
FGF-R1/N-sam (X66945)	6.9
IL-18R (U43672)	6.5
lymphotoxin beta (U89922)	3.9
TNF-R1 (M58286)	3.4
LIFR (S83362)	2.5
TNF-R2(M32315)	2.1
IL-10R (U00672)	−2.6

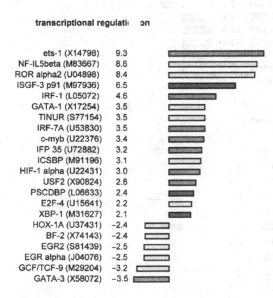

transcriptional regulation

ets-1 (X14798)	9.3
NF-IL5beta (M83667)	8.6
ROR alpha2 (U04898)	8.4
ISGF-3 p91 (M97936)	6.5
IRF-1 (L05072)	4.6
GATA-1 (X17254)	3.5
TINUR (S77154)	3.5
IRF-7A (U53830)	3.5
c-myb (U22376)	3.4
IFP 35 (U72882)	3.2
ICSBP (M91196)	3.1
HIF-1 alpha (U22431)	3.0
USF2 (X90824)	2.6
PSCDBP (L06633)	2.4
E2F-4 (U15641)	2.2
XBP-1 (M31627)	2.1
HOX-1A (U37431)	−2.4
BF-2 (X74143)	−2.4
EGR2 (S81439)	−2.5
EGR alpha (J04076)	−2.5
GCF/TCF-9 (M29204)	−3.2
GATA-3 (X58072)	−3.5

FIGURE 1 Gene expression patterns in human Th1 and Th2 cells. Bars represent "fold change" of the mRNA level of a particular gene when comparing Th1 with Th2 cells (mean of five experiments). Positive values indicate that the transcript is more abundant in Th1 than in Th2 cells and negative values indicate the opposite. Colors Indicate the 'absolute' expression level of a gene (arbitrary fluorescence units). Red = high level of expression (>1000); orange = medium level of expression (200–1000); yellow = low transcript abundance (<200). The column next to the bar diagram indicates the fold change. (Adapted from Ref. 4).

enzyme and other signaling molecules

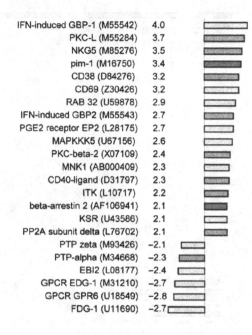

metabolic pathways

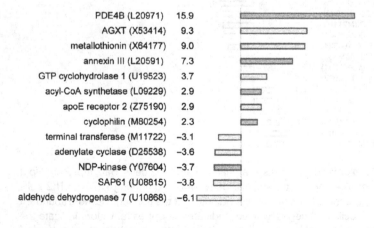

FIGURE 1   (Continued).

enzyme and other signaling molecules

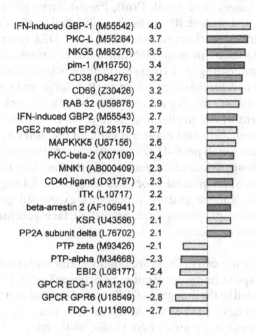

IFN-induced GBP-1 (M55542)	4.0
PKC-L (M55284)	3.7
NKG5 (M85276)	3.5
pim-1 (M16750)	3.4
CD38 (D84276)	3.2
CD69 (Z30426)	3.2
RAB 32 (U59878)	2.9
IFN-induced GBP2 (M55543)	2.7
PGE2 receptor EP2 (L28175)	2.7
MAPKKK5 (U67156)	2.6
PKC-beta-2 (X07109)	2.4
MNK1 (AB000409)	2.3
CD40-ligand (D31797)	2.3
ITK (L10717)	2.2
beta-arrestin 2 (AF106941)	2.1
KSR (U43586)	2.1
PP2A subunit delta (L76702)	2.1
PTP zeta (M93426)	-2.1
PTP-alpha (M34668)	-2.3
EBI2 (L08177)	-2.4
GPCR EDG-1 (M31210)	-2.7
GPCR GPR6 (U18549)	-2.8
FDG-1 (U11690)	-2.7

metabolic pathways

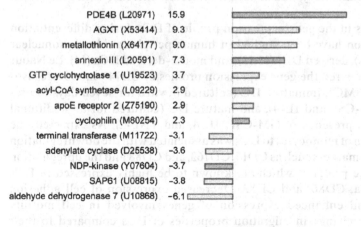

PDE4B (L20971)	15.9
AGXT (X53414)	9.3
metallothionin (X64177)	9.0
annexin III (L20591)	7.3
GTP cyclohydrolase 1 (U19523)	3.7
acyl-CoA synthetase (L09229)	2.9
apoE receptor 2 (Z75190)	2.9
cyclophilin (M80254)	2.3
terminal transferase (M11722)	-3.1
adenylate cyclase (D25538)	-3.6
NDP-kinase (Y07604)	-3.7
SAP61 (U08815)	-3.8
aldehyde dehydrogenase 7 (U10868)	-6.1

Expression level   > 1000   ▢ 200-1000   ▢ <200

FIGURE 1   (Continued).

B Cells. B cells develop in bone marrow from hematopoietic stem cells. Their developmental stages (Pre-ProB, ProB, PreB-I, large preB-II, small preB-II, immature B, and mature B) can be distinguished by their differences in expression of surface markers such as CD19 and CD25, rearrangement machinery, and status of immunoglobulin gene rearrangement. Hoffmann et al. (10) investigated the gene expression profiles of five consecutive stages of mouse B cell development including preB-I, large and small preB-II, immature B, and mature B, using Affymetrix Mu11k gene chips. The gene expression patterns of the preB-I and large preB-II cells on the one hand, and the resting immature and mature B cells on the other hand, were the closest to each other. Small preB-II cells displayed a pattern that is transitional between two groups. Most of the genes expressed in early B cell precursors were involved in such general processes as protein folding or cell-cycle regulation, whereas more mature precursors expressed genes involved in specific molecular programs such as cell-surface receptors, secreted factors, and adhesion molecules.

Dendritic Cells. Dendritic cells (DCs) are potent antigen-presenting cells that can initiate the adaptive immune response by priming T cells. Functions of DCs are closely related to their degree of differentiation and maturation. Immature DCs primarily take up antigens by phagocytosis, macropinocytosis, and receptor-mediated endocytosis, while mature DCs become highly efficient T cell activators as their expression of surface level of MHC, costimulatory, and adhesion molecules increase with maturation (11).

Changes in the gene expression profile of DCs during differentiation and maturation have been studied in human peripheral blood monuclear cells (PBMCs), derived DCs (12,13), and mouse-derived DCs (14). Le Naour et al. (12) compared the gene expression profiles of CD14$^+$ monocytes isolated from PBMCs, immature DC (cultured in vitro for seven days in presence of GM-CSF and IL-4), and mature DC (cultured for an additional seven days in presence of GM-CSF, IL-4, and TNF-α). As expected, the differentiation of monocytes to DCs was associated with the downregulation of monocytic markers such as CD14, CD163, and CD88 and the upregulation of cell-surface proteins, which are known to be highly expressed in DCs such as CD1a, CD86, and CD83. Decreased expression of cell adhesion molecules and enhanced expression of genes involved in cell motility indicated the change in migration properties of DCs compared to their precursor, monocytes. Differentiation of DCs was accompanied by upregulation of anti-inflammatory proteins such as cyclophilin C and TSG-6 and downregulation of proinflammatory cytokines and their receptors, including TNF-α, IL-6 receptor, TNF-α receptor, and IL-8. Genes involved in lipid metabolism such as apolipoprotein E, apolipoprotein C-I, ABCG1, and lipo-protein lipase were upregulated in DCs. On the other hand, there were only few differences in overall gene expression between immature and mature DCs. Granucci et al. (15) used mouse DC line and D1 (splenic, myeloid, and growth factor-dependent DC line in the immature state) to determine the effects of different stimuli, lipopolyssacharide (LPS) or TNF-α, on gene expression during maturation. The authors concluded that there is an impor-tant difference between LPS- and TNF-α-activated D1 cells in the control of cell-cycle progression. Terminal differentiation usually results in the growth arrest, and only LPS-treated cells showed a pattern of gene expression com-patible with a definitive growth arrest; suppression of cyclin A and clyclin B1, B2, upregulation of cyclin D2, and upregulation of antiproliferative genes such as B cell translocation gene 1 (BTG1) and growth arrest-specific (GAS) gene. Upon exposure to Escherichia coli (Gram-negative bacteria), many genes involved in cytoskeleton rearrangements, antigen processing, control of migration and apoptosis, and regulation of inflammatory response were modulated in DC1 cells (14). Of particular interest was that DCs pro-duced IL-2 in a tightly regulated timeframe, first at 4–8 h after bacterial uptake and second at 14–18 h after activation, which was compatible with the timing of appearance of MHC molecules at the cell surface. This finding provides one of the mechanisms by which DCs activate T lymphocytes in addition to their regulatory role in innate and adaptive immunity.

2.1.2. Changes in Gene Expression During Activation of Immune Cells

Macrophages. Macrophages are involved in the innate immune response by eliminating extracellular and intracellular pathogens through phagocytosis, and contribute to the specific immune response by presenting antigens and producing chemokines and cytokines. Developed from mono-cytes, mature macrophages are widely distributed throughout the body and exhibit considerable heterogeneity in expression of markers and receptors (16).

Macrophages can be activated by microbial stimuli, cytokines and che-mokines, and through cellular interaction. Upon stimulation, macrophages can secrete many products, including enzymes involved in antimicrobial resistance, prostaglandins and leukotrienes, cytokines, and reactive oxygen and nitrogen intermediates (16). Lipopolyssacharide stimulation of macro-phages induces many genes, including those of cytokines IL-1, IL-6, TNF-α, and GM-CSF; chemokines such as IL-8 and MCP-1; transcription factors such as p50, c-Rel, IRF-1, Egr-1, and IκBα; and inducible nitric oxide synthase (iNOS) and cyclooxygenase 2 (17).

Locati et al. (18) examined the gene expression profiles induced by the CC chemokine ligand 5/RANTES or LPS in human monocytes using Affymetrix HuGeneFL array. Of 5600 transcripts examined, 42 were consistently induced by CCL5 and none were suppressed. Cytokine and receptors such as IL-1β, CCL2/monocyte chemotactic protein-1, and CCL5 receptor, and molecules involved in extracellular matrix recognition and digestion such as CD44 splice transcripts, urokinase-type plasminogen activator receptor, and matrix metalloprotease (MMP)-9 and -19 were upregulated. The chemokine-induced gene profile was distinct from that activated by LPS and showed more restricted activation compared with those induced by LPS. Rosenberger et al. (19) studied the expression of genes during <u>Salmonella typhimurium</u>, a Gram-negative bacteria with an outer membrane rich in LPS, infection, or LPS stimulation of murine macrophage cell line RAW 264.7 using Atlas mouse cDNA expression array from Clontech. Overall patterns of macrophage gene expression observed during <u>S. typhimurium</u> infection and LPS activation overlapped considerably. Many of the genes were associated with proinflammatory or direct antimicrobial properties. Highly elevated expression levels were observed for (1) the chemokines (MIP-1α, MIP-1β, and MIP-2α), which recruit other effector cells to infection sites (2) NF-κB inhibitory factors (IκB-α and IκB-β), which downregulate the transcriptional program initiated by the translocation of NF-κB, (3) signaling molecules involved in cell death (caspase 1, TNF receptor 1, Fas, TRAIL), and (4) genes involved in macrophage differentiation [leukemia inhibitory factor (LIF), Egr-1]. Stimulation by LPS seemed to enhance the macrophages' ability to interact with other cells by upregulating expression of receptors such as CD40 and ICAM-1.

T Cells. The activation of T cells involves multiple signal transduction pathways that eventually result in the transcription of a large variety of genes. The early events that follow TCR engagement include phosphorylation of receptor-associated tyrosine kinases and downstream signaling molecules, activation of phospholipase C and subsequent hydrolysis of inositol phospholipids, and increases in intracellular calcium. These signaling events culminate in the activation of well-characterized transcription factors such as activator protein (AP)-1, nuclear factor (NF)-κB, and NF-AT. Although some targets of these transcription factors have been identified, a comprehensive view of transcriptional events occurring during T cell activation requires the ability to simultaneously monitor the levels of all tanscripts (20). In addition, it has become increasingly clear that receptor-mediated signal transduction events leading to the proliferation and acquisition of differentiated functions by the T cells are not the result of a single wave of gene-activation events. Rather, these are likely to result from a regulated cascade of sequential gene-activation events that may be conditionally regulated (21).

Teague et al. (22) examined the patterns of genes expressed in resting T cells and T cells 8 and 48 hrs after activation using Affymetrix gene arrays. T cells were activated in vivo by the injection of superantigen to C57BL mice. The authors reported that resting T cells expressed large diversity of genes and the patterns of gene expression showed dramatic changes within 8 hr, but returned to a pattern more like that of resting T cells within 48 hr of exposure. Many genes contributing to cell division, such as DNA polymerase, primase, cyclins, and enzymes involved in synthesis of DNA precursors, were expressed at higher levels in activated compared with resting T cells. Diehn et al. (23) examined the gene expression responses in primary human purified T cells to stimulation by anti-CD3 alone or anti-CD3 and anti-CD28 (costimulatory molecule) at seven time points (0, 1, 2, 6, 12, 24, and 48 hrs). A microarray with 4359 cDNA elements representing 2926 genes was used in this study. The following general features were discerned from this study: (1) genes encoding cytokines and chemokines, cytokine receptors, cell adhesion molecules, as well as cytotoxic effector molecules, such as granzyme B, granulysin, and fas ligand, were induced in a temporally choreographed pattern, (2) a significant fraction of the genes that were induced at intermediate and late time points had direct roles in promoting proliferation and progression through the cell cycle, (3) the gene expression pattern reflected the increased metabolic demand and macromolecular biosynthesis of T cells, (4) many of the genes, products of which are involved in transducing signals from the T cell receptor (TCR) such as phospholipase C, LAT, LCK, and genes encoding subunits of TCR were repressed with T cell activation, and, (5) expression of NFAT target genes was enhanced by CD28 costimulation.

2.2. Age-Associated Changes in Immune Function and Immune Cells

2.2.1. Macrophage and Aging

Macrophages play a key role in inflammatory responses by releasing a variety of inflammatory mediators including prostaglandins and proinflammatory cytokines (24). Prostaglandins are generated from arachidonic acid, which is released from the membrane phospholipid by Phospholipase A_2, by the action of the enzyme cyclooxygenase. We have shown that macrophages from old mice have a significantly higher production of n PGE_2 compared to young mice. The higher PGE_2 production was due to increased Cox-2 activity, which was in turn due to higher protein and mRNA expression of Cox-2 (25) in old compared to young macrophages. PGE_2 has a direct inhibitory effect on the early stages of T cell activation (26) and can modulate Th1/Th2 cytokine secretion (27).

Pro-inflammatory cytokines produced by macrophages have been shown to contribute to pathogenesis of several age-associated diseases such as atherosclerosis and arthritis, however, there is inconsistency in the literature regarding changes in their production/regulation with aging (28).

2.2.2. T Cell and Aging

Aging is associated with reduced T cell function, as demonstrated by decreased T cell proliferation and IL-2 production. One of the hallmarks of age-related changes is a shift toward greater proportions of antigen (Ag)-experienced memory T cells with fewer T cells of naive phenotype. Naive T cells have different response kinetics to Ag challenge than memory T cells, with memory T cells responding faster and to a lower Ag dose than naive T cells (29). Recent evidence indicates that naive T cells show an age-related functional decline in the earliest stages of activation induced by peptide/ MHC complexes (1). We showed that T cells from old mice go through lower activation-induced cell division, have fewer IL-2^+ cells, and produce less IL-2 per cell. These age-associated changes in T cells were only observed within naive T cell subpopulations (30). Other researchers reported that IL-2 receptor expression is decreased in cells from elderly individuals (31), and functional disruption of the CD28 gene transcriptional initiator is observed in senescent T cells (2). The adverse effects of age are observed in various steps involved in the T cell activation pathway, including tyrosine kinases such as ZAP-70, calcium/calmodulin-dependent protein kinases, and adaptor proteins (32).

Although several advances have been made in our understanding of changes in T cell response with aging, it has been difficult to construct a comprehensive picture of molecular changes, which lead to a functional dysregulation of T cells because the activation of T cells involves simultaneous up- or downregulation of many different signaling pathways.

The global view of the gene expression patterns of the immune cells and comparison of those between immune cells originated from young and old host will help identify the pathways or changes most profoundly affected with aging. With a growing knowledge of the functions of individual genes, microarrays can provide comprehensive and dynamic molecular pictures of the immune cells to better understand the changes occurring with aging.

2.3. Changes in Gene Expression Profile with Aging

2.3.1. Immune Cells

Thus far, only a limited number of studies have investigated the age associated changes in the gene expression profiles of immune cells (33–35). Mo et al. (33) stimulated purified CD4 T cells from young and old mice with anti-CD3 and anti-CD28 for 72 hrs and determined their gene expression profiles using Affymetrix microarray chips. They found significantly higher expression of CCR2, CCR5, and CXCR5 and lower expression of CCR7 in unstimulated CD4 T cells from old animals compared with those from the young. Stimulation resulted in decreased expression of CCR2, CCR5, and CXCR5 in both young and old T cells. As mentioned previously, a distinct pattern of chemokine receptor expression was associated with polarization of Th1 and Th2 cells. A shift toward the Th2 profile has been reported in aging T cells (36). However, in this study, the authors found that both Th1- and Th2-associated chemokine receptors were increased in aged T cells. Table 2 shows a gene expression profile of a selected list of chemokines and their receptors and proteins associated with apoptosis and the proteolytic system, differentially expressed in Th1 and Th2 cells. The table also shows their expression pattern with aging. Results reported in this study were from two separate experiments using RNAs pooled from 5–15 animals and from two different chips, µ11k and U74A, used for each experiment. Rao et al. (34) examined the response of lymphocytes purified from two young and two old individuals to heat shock (1 hr at 42°C) using in-house microarray chips representing 4032 genes. Authors concluded that some genes associated with signal transduction and mitochondrial respiration were upregulated and some genes associated with heat shock response and cell survival were downregulated in lymphocytes from old individuals after heat shock treatment. We (35) investigated the effect of aging on gene expression profiles of purified T cells stimulated with anti-CD3 and anti-CD28 for 2 hrs. In unstimulated T cells, 43 genes were expressed significantly higher in old compared to young, mainly immunoglobulin genes, whereas 22 genes, including T cell receptor-related genes, were expressed significantly higher in young compared to old. Response to stimulation was significantly affected by age in expression of 40 genes. Significantly lower expression of T cell receptor alpha chain and factors involved in the signal transduction pathways were observed with aging.

2.3.2. Other Cells and Tissues

Global changes in the gene expression profile with aging have been reported for brain, muscle, colon, and liver tissues (37–42).

TABLE 2   Gene Expression Profile of Selected List of Chemokines and Their Receptors and Proteins Associated with Apoptosis and Proteolytic System, Which are Differentially Expressed in Th1 and Th2 Cells and Their Expression Pattern with Aging

	Hamalainen et al. (8)	Bonecchi et al. (6)	Rogge et al. (4)	Chtanova et al. (5)	With aging (32)
Chemokines					
CCR1	Th1		↑ with IL-12	Th2	
CCR2	Th1				↑
CCR3		Th2			
CCR4	Th2	Th2			
CCR5	Th1	Th1	↑ with IL-12	Th2	↑
CCR7					↓
Chemokine receptors					
CXCR1					
CXCR2					
CXCR3		Th1	Th1		
CXCR4				Th2	
CXCR5					↑
Apoptosis and proteolytic systems					
Perforin			Th1		
Caspase 1	Th1				
Caspase 6	Th2				
Caspase 8			Th1		
Granzyme H			Th1		
Granzyme B	Th1		Th1		
Granzyme C				Th1	
Adhesion and migration					
Integrin alpha4				Th1	
Integrin beta7			Th2	Th2	

Lee et al. (38) observed an increase in genes involved in inflammatory and stress responses and a decrease in genes involved in protein turnover and growth and trophic factors in the neocortex and cerebellum part of the brain with aging. Subsequently, Blalock et al. (39) showed that synaptic structural plasticity, extracellular matrix formation/turnover, activity-regulated signaling, and protein chaperone functions were downregulated and myelin turnover, cholesterol synthesis/transport, lipid metabolism, iron utilization, tyrosine/tryptophan/monoamine metabolism, and cytoskeletal reorganization were upregulated in the hippocampus of aged rat brain. Most of these changes in gene expression pattern began by midlife (14 months), but cognition was not clearly impaired until late in life (24 months).

Lee et al. (43) analyzed the age-related changes in the gene expression profile of the gastrocnemius muscle using Affymetrix oligonucletide array. Aged muscle showed: (1) increased stress response evidenced by induction of heat shock response, DNA damage-inducible genes, and oxidative stress-inducible genes, (2) decreased energy metabolism evidenced by reduced glycolysis and mitochondrial dysfunction, and, (3) increased neuronal injury evidenced by reinnervation and neurite extension and sprouting. Welle et al. (41) compared the age-associated changes in gene expression in muscle between mice and men by comparing gene expression profiles of muscles from young (mean age 23 yr) and old (mean age 71 yr) men obtained by the SAGE method and by using Affymetrix HG-U95A microarray to the results of gene expression profiles in mouse muscle by Lee et al. (43). Seventeen genes showed a similar age-related change between men and mice and 32 genes showed a difference in the effect of age on the level of expression between men and mice. The authors did not find any evidence of an increased stress response in older human muscle. They stated that the goal of the study was to evaluate the overall degree of consistency between the results reported for mice and those observed in humans, rather than to determine age-related global changes across species. Nevertheless, they cautioned against generalizing the reported effects of age on gene expression in muscle of inbred mice to that of humans. In the aged colon from rats, increased expression of genes involved in energy-generating pathways and lipid oxidation, and genes that show an aberrant regulation in colon cancer (CD44, ras, and mapsin) were observed (40).

3. CONCLUSION

Use of the microarray technique to investigate global changes in immune cells in response to environmental stimuli has improved our understanding of immune cell regulation. Several previously unknown genes have been found to be involved in immune cell regulation. Whereas several studies have utilized this technique to determine the age-associated global changes in other tissues, limited investigation has focused on the effect of aging on gene expression of immune cells.

Several considerations need to be taken into account before meaningful conclusions can be drawn from the gene expression profile analysis using the microarray technique. First, the nature of the microarray analysis requires multiple comparisons between limited numbers of samples, and therefore raises a challenge in discerning true changes from false-positive or false-negative results. Changes in expression levels observed from microarray analysis need to be validated by other methods and need to be associated with functional changes. Second, the cost of the chips and other reagents to conduct microarray experiments pressures the investigators to compromise by analyzing a small number of samples. Some of the previously published papers used as few as two samples per group. This is particularly important, considering the day-to-day, chip-to-chip, and animal/subject variability inherent in these experiments. Third, it is important to carefully examine the chips used, the statistical methods applied, and the study design when results from different studies are compared. Although more and more researchers are using commercial chips such as the Affymetrix microarray, many chips made in house with different gene profile and probe efficiency have been used in the past. This could affect the outcome of the study and should be taken into consideration when results from different investigations are compared.

REFERENCES

1. Garcia GG, Miller RA. Single-cell analyses reveal two defects in peptide-specific activation of naive T cells from aged mice. J Immunol 2001; 166:3151–3157.
2. Vallejo AN, Weyand CM, Goronzy JJ. Functional disruption of the CD28 gene transcriptional initiator in senescent T cells. J Biol Chem 2001; 276:2565–2570.
3. Goh SH, Park JH, Lee YJ, Lee HG, Yoo HS, Lee IC, Kim YS, Lee CC. Gene expression profile and identification of differentially expressed transcripts during human intrathymic T-cell development by cDNA sequencing analysis. Genomics 2000; 70:1–18.
4. Rogge L, Bianchi E, Biffi M, Bono E, Chang SY, Alexander H, Santini C, Ferrari G, Sinigaglia L, Seiler M, Neeb M, Mous J, Sinigaglia JF, Certa U. Transcript imaging of the development of human T helper cells using oligonucleotide arrays. Nat Genet 2000; 25:96–101.
5. Chtanova T, Kemp RA, Sutherland AP, Ronchese F, Mackay CR. Gene microarrays reveal extensive differential gene expression in both CD4(+) and CD8(+) type 1 and type 2 T cells. J Immunol 2001; 167:3057–3063.
6. Bonecchi R, Bianchi G, Bordignon PP, D'Ambrosio D, Lang R, Borsatti A, Sozzani S, Allavena P, Gray PA, Mantovani A, Sinigaglia F. Differential expression of chemokine receptors and chemotactic responsiveness of type 1 T helper cells (Th1s) and Th2s. J Exp Med 1998; 187:129–134.
7. Nagai S, Hashimoto S, Yamashita T, Toyoda N, Satoh T, Suzuki T, Matsushima K. Comprehensive gene expression profile of human activated T(h)1- and T(h)2-polarized cells. Int Immunol 2001; 13:367–376.
8. Hamalainen H, Zhou H, Chou W, Hashizume H, Heller R, Lahesmaa R. Distinct gene expression profiles of human type 1 and type 2 T helper cells. Genome Biol 2001; 2:RESEARCH0022.1–0022.11.

9. Rogge L. A genomic view of helper T cell subsets. Ann N Y Acad Sci 2002; 975:57–67.
10. Hoffmann R, Seidl T, Neeb M, Rolink A, Melchers F. Changes in gene expression profiles in developing B cells of murine bone marrow. Genome Res 2002; 12:98–111.
11. Richards J, Le Naour F, Hanash S, Beretta L. Integrated genomic and proteomic analysis of signaling pathways in dendritic cell differentiation and maturation. Ann N Y Acad Sci 2002; 975:91–100.
12. Le Naour F, Hohenkirk L, Grolleau A, Misek DE, Lescure P, Geiger JD, Hanash S, Beretta L. Profiling changes in gene expression during differentiation and maturation of monocyte-derived dendritic cells using both oligonucleotide microarrays and proteomics. J Biol Chem 2001; 276:17920–17931.
13. Lapteva N, Ando Y, Nieda M, Hohjoh H, Okai M, Kikuchi A, Dymshits G, Ishikawa Y, Juji T, Tokunaga K. Profiling of genes expressed in human monocytes and monocyte-derived dendritic cells using cDNA expression array. Br J Haematol 2001; 114:191–197.
14. Granucci F, Vizzardelli C, Pavelka N, Feau S, Persico M, Virzi E, Rescigno M, Moro G, Ricciardi-Castagnoli P. Inducible IL-2 production by dendritic cells revealed by global gene expression analysis. Nat Immunol 2001; 2:882–888.
15. Granucci F, Vizzardelli C, Virzi E, Rescigno M, Ricciardi-Castagnoli P. Transcriptional reprogramming of dendritic cells by differentiation stimuli. Eur J Immunol 2001; 31:2539–2546.
16. Gordon S. Macrophages and the immune response. In: Paul WE, ed. Fundamental Immunology. 4th ed. Philadelphia: Lippincott-Raven Publishers, 1999:533–545.
17. Guha M, Mackman N. LPS induction of gene expression in human monocytes. Cell Signal 2001; 13:85–94.
18. Locati M, Deuschle U, Massardi ML, Martinez FO, Sironi M, Sozzani S, Bartfai T, Mantovani A. Analysis of the gene expression profile activated by the CC chemokine ligand 5/RANTES and by lipopolysaccharide in human monocytes. J Immunol 2002; 168:3557–3562.
19. Rosenberger CM, Scott MG, Gold MR, Hancock RE, Finlay BB. Salmonella typhimurium infection and lipopolysaccharide stimulation induce similar changes in macrophage gene expression. J Immunol 2000; 164:5894–5904.
20. Ellisen LW, Palmer RE, Maki RG, Truong VB, Tamayo P, Oliner JD, Haber DA. Cascades of transcriptional induction during human lymphocyte activation. Eur J Cell Biol 2001; 80:321–328.
21. Weiss A. T-lymphocyte activation. In: Paul WE, ed. Fundamental Immunology. 4th ed. Philadelphia: Lippincott-Raven, 1999:411–447.
22. Teague TK, Hildeman D, Kedl RM, Mitchell T, Rees W, Schaefer BC, Bender J, Kappler J, Marrack P. Activation changes the spectrum but not the diversity of genes expressed by T cells. Proc Natl Acad Sci U S A 1999; 96:12691–12696.
23. Diehn M, Alizadeh AA, Rando OJ, Liu CL, Stankunas K, Botstein D, Crabtree GR, Brown PO. Genomic expression programs and the integration of the

CD28 costimulatory signal in T cell activation. Proc Natl Acad Sci U S A 2002; 99:11796–11801.

24. Nathan CF. Secretory products of macrophages. J Clin Invest 1987; 79: 319–326.

25. Hayek MG, Mura C, Wu D, Beharka AA, Han SN, Paulson KE, Hwang D, Meydani SN. Enhanced expression of inducible cyclooxygenase with age in murine macrophages. J Immunol 1997; 159:2445–2451.

26. Vercammen C, Ceuppens JL. Prostaglandine E_2 inhibits T-cell proliferation after crosslinking of the CD3-Ti complex by directly affecting T cells at an early step of the activation process. Cell Immunol 1987; 104:24–36.

27. Gately MK, Renzetti LM, Magram J, Stern AS, Adorini L, Gubler U, Presky DH. The interleukin-12/interleukin-12-receptor system: role in normal and pathologic immune responses. Ann Rev Immunol 1998; 16:495–521.

28. Han SN, Meydani SN. Antioxidants, cytokines, and influenza infection in aged mice and elderly humans. J Infect Dis 2000; 182:S74–S80.

29. Rogers PR, Dubey C, Swain SL. Qualitative changes accompany memory T cell generation: faster, more effective responses at lower doses of antigen. J Immunol 2000; 164:2338–2346.

30. Adolfsson O, Huber BT, Meydani SN. Vitamin E-enhanced IL-2 production in old mice: naive but not memory T cells show increased cell division cycling and IL-2-producing capacity. J Immunol 2001; 167:3809–3817.

31. Nagel JE, Chopra RK, Chrest FJ, McCoy MT, Schneider EL, Holbrook NJ, Adler WH. Decreased proliferation, interleukin 2 synthesis, and interleukin 2 receptor expression are accompanied by decreased mRNA expression in phyto-hemagglutinin-stimulated cells from elderly donors. J Clin Invest 1988; 81:1096–1102.

32. Chakravarti B. T-cell signaling–effect of age. Exp Gerontol 2001; 37:33–39.

33. Mo R, Chen J, Han Y, Bueno-Cannizares C, Misek DE, Lescure PA, Hanash S, Yung YL. T cell chemokine receptor expression in aging. J Immunol 2003; 170:895–904.

34. Rao DV, Boyle GM, Parsons PG, Watson K, Jones GL. Influence of ageing, heat shock treatment and in vivo total antioxidant status on gene-expression profile and protein synthesis in human peripheral lymphocytes. Mech Ageing Dev 2003; 124:55–69.

35. Han SN, Adolfsson O, Lee CK, Dallal GE, Prolla TA, Ordovas JM, Meydani SN. Effect of age and vitamin E on gene expression profiles of T cells. Faseb J 2003; 17:A280.

36. Globerson A, Effros RB. Ageing of lymphocytes and lymphocytes in the aged. Immunol Today 2000; 21:515–521.

37. Weindruch R, Kayo T, Lee CK, Prolla TA. Gene expression profiling of aging using DNA microarrays. Mech Ageing Dev 2002; 123:177–193.

38. Lee CK, Weindruch R, Prolla TA. Gene expression profile of the ageing brain. Nat Genet 2000; 25:294–297.

39. Blalock EM, Chen KC, Sharrow K, Herman JP, Porter NM, Foster TC, Landfield PW. Gene microarrays in hippocampal aging: statistical profiling

identifies novel processes correlated with cognitive impairment. J Neurosci 2003; 23:3807–3819.

40. Lee HM, Greeley Jr GH, Englander EW. Age-associated changes in gene expression patterns in the duodenum and colon of rats. Mech Ageing Dev 2001; 122:355–371.

41. Welle S, Brooks A, Thornton CA. Senescence-related changes in gene expression in muscle: similarities and differences between mice and men. Physiol Genomics 2001; 5:67–73.

42. Cao SX, Dhahbi JM, Mote PL, Spindler SR. Genomic profiling of short- and long-term caloric restriction effects in the liver of aging mice. Proc Natl Acad Sci 2001; 98:10630–10635.

43. Lee CK, Klopp RG, Weindruch R, Prolla TA. Gene expression profile of aging and its retardation by caloric restriction. Science 1999; 285:1390–1393.

13

Improving the Nutritional Value of Cereal Grains Using a Genomic Approach

Liang Shi and Tong Zhu
Torrey Mesa Research Institute, Syngenta, San Diego, California, U.S.A.

1. INTRODUCTION

Cereal grains are a major food and feed source. The top three stable food grains—rice, corn, and wheat—represent over 50% of the human food source. In a low-fat, high-energy diet, these grains can provide over 30% of the total energy intake. They also contribute protein, vitamins, minerals, and dietary fiber in the human diet. In addition, cereal grain products are used in animal feeds, indirectly providing protein and energy resources to human nutrition. The health benefits of consuming grain products have been demonstrated to extend beyond nutrients the grains provide. It is well-recognized that a high-bulk diet, high in nonfermentable insoluble fiber, is beneficial to the human in terms of the prevention and treatment of numerous disorders.

Optimization of the composition and yield of the grains could significantly improve their nutritional value (1,2). For example, increasing phytase would allow more complete degradation of phytate during fermentation, thus increasing the bioavailability of the minerals and vitamins contained by the grains. Another example, more directly related to human nutrition, is the transgenic expression of vitamin A pathways in endosperm of golden rice (3). The success of this approach provides vitamin A in an otherwise deficient diet. Populations consuming processed rice grain lose significant amounts of vitamin A found in the aleurone layer of the grain because this layer is removed during processing. The genomic approach relocates the vitamin from the aleurone layer to the endosperm. However, increasing biousable vitamins and minerals is not the only goal. In fact, the major targets for grain improvement include increasing the total protein content (4) and improving protein and carbohydrate quality by increasing the amount of lysine or increasing the contents of branched starch (5–8). Changing the biochemical property of fatty acid, starch or protein could lead to specialty products with industrial or biopharmaceutical applications (9–12).

The nutritional value of the grains is determined by the nutrient amount and composition in the grains that accumulate during grain filling and by the bioavailability of micronutrients (7,13,14). Thus, improvement of the nutritional value of the grain depends on our understanding of the biological systems that control the nutrient partitioning in the grain-filling process.

Despite their importance, designing and optimizing nutrient partitioning during grain-filling remains a challenge, because of the complexity of the grain-filling process. Grain filling involves reproductive development and nutrient biosynthesis through multiple metabolic pathways. Although the biochemical pathways and the many genes participating in this complex process have been studied, the regulation and coordination of different participants is poorly understood. Because grain-filling is a complex trait controlled by a genetic network rather than a single gene, single-gene manipulation through breeding and biotechnology has had limited success. For instance, to deliver multiple transgenes and to reassemble a metabolic pathway requires coordinated in vivo expression of each transgene (3). However, due to the position effect of the transgene insertion in the genome, such coordination in gene expression is limited. An alternative approach is to identify and engineer the key regulators that control these multiple pathways, thus optimizing the quantity and quality of the grain nutrients. This requires a detailed understanding of the genetic network that modulates the gene expression. The available genome drafts, the EST (Expressed Sequence Tag) sequences in many crop species, and the technology for genome-wide gene expression analysis allow for the study and understanding of global transcription regulation of grain development and nutrient partitioning. This chapter describes our efforts to use a genomic approach to improve the nutritional value of rice.

Our initial effort has been characterization of the transcriptome of grains during their development, with a focus on the coupling nutrient-partitioning process. Specifically, we wanted to answer four questions. How many nutrient-partitioning genes are involved in the grain-filling and what are they? How are these grain-filling genes expressed and regulated? How are genes belonging to different pathways coordinately expressed? Finally, can we identify and engineer the key genes that regulate grain-filling genes in different pathways?

We used rice as our model system for studying grain-filling and nutrient partitioning. Rice is a valuable human food source. Rice is an ideal grain to use in answering our genomic questions. It has the smallest genome size (420 Mb) among the cereal crops, yet shows a high level of synteny to maize, wheat, barley, and other cereal crops (15). Rice also has typical grain structural and developmental features, thus many observations obtained from this species can be readily applied to other cereals. In addition, the genome drafts are available (16,17) and there is a continuing effort in sequencing and assembling the complete genome (18–20).

Microarrays have been proven to be a powerful tool for genome-wide gene expression analysis. The measured transcript level of all genes related to a certain process or condition reflects the molecular events underlying the cell physiological status, and can be integrated computationally to identify the responsive genes and dissect their relationship network. In order to identify the unknown key regulators of the grain transcriptome, it is necessary to survey the entire transcriptome to obtain the global transcriptome image.

Two high-density and high-capacity oligonucleotide-probe GeneChip microarrays have been designed to date. The first rice GeneChip array was designed based on an earlier assembly of the genome draft and consists of approximately 400,000 unique oligonucleotide probes for 21,000 rice genes (21). Since then, a second generation of GeneChip microarray has been designed, based on the recently published rice genome draft (16,22). This single GeneChip microarray contains 500,000 unique probes for 51,000 rice genes and covers almost the entire transcriptome (22). The high coverage of the transcriptome is enabled by the modification of the original GeneChip standard design. In both rice GeneChip arrays, each probe set contains 11–16 perfect-match oligonucleotide probes for one gene target. Mismatch probes were omitted from this modified design based on several simulation and statistical studies. A custom algorithm was also developed to accommodate such a change. The modified design and resultant high capacity means that the GeneChip data can be validated by a real-time quantitative polymerase chain reaction (PCR) assay with correlation coefficient of 0.92 (22).

2. GRAIN FILLING IS A COMPLEX BIOLOGICAL PROCESS

By examining the transcript level of the 51,000 genes in 33 samples collected throughout rice development, genes that participate in grain filling and development were identified (Fig 1). Similar to the results obtained in a previous survey, the majority are nutrient-partitioning genes. These include genes that encode products involved in carbohydrate metabolism, fatty acid metabolism, amino acid metabolism, and protein synthesis, as well as various transporters, storage proteins, and membrane proteins. The involvement of various transporter genes in grain development has been previously reported. For example, it was observed that sucrose transporter genes are expressed in developing cereal grains (23). In addition, the genes encoding DNA-binding proteins and other transcription regulators, genes encoding kinases, and phosphatases that are involved in signal transductions are also significantly presented. However, as expected, the majority of the genes responsive to grain filling are encoding products with unknown function. Of the 1553 genes encoding known and predicted transcription factors, 135 are upregulated. It is necessary to point out that a significant number of genes are downregulated. These genes may also play important roles in grain intermediary metabolism. Initially, only genes that are upregulated during the grain-filling process, including those genes involving nutrient partitioning and their regulators, were studied. These genes are defined as grain-filling genes. The diversified functions of these grain-filling genes suggest that the grain filling is a complex biological process.

The grain-filling genes were identified by the following approach. First, all of the nutrient-partitioning genes were identified based on the available sequence annotation, regardless of their involvement in grain filling. These genes include those involved in the biosynthesis of starch, storage proteins, and fatty acids and their transporters. Second, upregulated genes were identified among the nutrient-partitioning genes based on cluster analysis (24) of the expression pattern throughout the rice development. Third, the grain nutrient-partitioning genes were used as baits to identify (through pattern matching) the genes with similar upregulated expression pattern among all of the genes surveyed.

In order to increase the confidence of this selection, the microarray data were reanalyzed by an alternative method—singular vector decomposition (SVD) (25), a pattern recognition method (26). The SVD analysis categorized the expression profiles of all nutrient-partitioning genes into 35 distinct expression patterns or signals. Among these patterns, the most frequently occurring patterns were the grain-filling signals. These are upregulated in expression during either the early or late developmental stage of grain development. When the genes associated with these patterns identified by SVD are compared to the grain-filling genes previously identified by clustering analysis, 80% overlapped. The rest of the previously identified grain-filling genes had upregulation patterns that were less significant. SVD analysis also identified an additional 20 genes with grain-filling patterns. Thus, the alternative pattern recognition method computationally validated the selection of the grain-filling genes.

To identify the genes that are involved in the basic grain-filling mechanism in other cereal crops, the expression patterns of the identified rice grain-filling genes were examined in other cereals using the rice GeneChip microarray. The feasibility of using rice GeneChip for genome expression analysis of other cereals was demonstrated previously (27). The comparative genomic DNA hybridization of microarray was used to identify the probes with conserved sequences that could hybridize to targets from different cereal species. For example, if genomic DNA of rice and corn (maize) both hybridize to one probe, this probe is a usable probe. Only the usable probes were included in the analysis. The results showed that most of the rice oligonucleotide probes can cross-hybridize to DNA or RNA targets from other cereal species (27). Based on this observation, 10 maize samples were analyzed using the rice GeneChip microarray. Among the 269 grain-filling genes identified in rice, 151 showed a similar upregulation expression pattern in maize, suggesting that they may have similar functions in grain filling (28).

Although the grain-filling genes were primarily selected based on the temporal pattern of gene expression, that is, upregulation or downregulation during the grain development, the spatial expression pattern of these genes further supports their grain-filling related functions. The grain-filling genes are mainly expressed in the endosperm, a premier nutrient-storage tissue in rice grain, as predicted based on the previous knowledge. However, genes in different pathways may have different spatial patterns: The starch biosynthesis-related genes are preferentially expressed in the endosperm, whereas the storage protein biosynthesis genes are preferentially expressed in the aleurone, a special layer of the endosperm. Genes involved in fatty acid biosynthesis have a more balanced expression between the embryo and endosperm (21).

To experimentally validate the expression pattern of selected grain-filling genes, the promoters of these genes were isolated and linked to a b-glucuronidase (GUS) reporter gene for in vivo expression analysis. It was found that the in vivo spatial expression pattern of these genes correlates well with the GeneChip data (29).

3. GRAIN FILLING REQUIRES COORDINATION OF TRANSCRIPTION AMONG DIFFERENT PATHWAYS

Although the grain-filling genes have diverse functions and belong to different metabolic pathways, the temporal expression of these genes shows a common pattern, that is, the upregulation during the grain development. This common expression pattern could be the result of the selection for the grain genes, or an indication of a coregulated synchronous expression pattern.

To prove that these genes are coregulated, it is necessary to demonstrate that there is a mechanism for the coregulation, such as common regulatory motif in the promoters of these genes. A total of 16 known regulatory motifs were examined in the grain-filling genes for their representations. These motifs included Cold (CCGAC), G-Box (CACGTG), AACA motif (AACAAAA), GCN4 [TGA(G/C)TCA], C-Box (TGACGTCA), GCC-Box (TAARAGCCGCC), Amylase (TATCCAY), GARE (RTAACRRANTCYGG), GT1 box (GGTTAA), In amylase (CGACG), BS1 (AGCGGG), DPBF-1 (ACACNNG), ASF1 like (TGACGT), E-box (CANNTG), Prolamin-box (TGYAAAG), and I-box (core) (GATAAG). Among the 117 promoters examined, 75 genes contained the AACA motif. The bootstrap analysis was used to evaluate its statistical significance. The results demonstrated that the AACA element is overrepresented among the promoters for the grain gene cluster at 95.4% confidence interval (P-value 0.046).

The existing overrepresented motif in the grain-filling genes strongly suggested that there is a regulatory mechanism involved in coordinating the transcription of the grain-filling genes in several metabolic pathways. The AACA element is known to be presented in genes encoding the seed storage proteins GluB-1 (30,31). However, the 75 genes are involved in diverse functions, including carbohydrate and fatty acid metabolism, nutrient transportation, transcription, and translation. If this can be confirmed, it will support the fact that these genes are coregulated, and the AACA element may play a critical role in the regulation of these grain-filling genes. Furthermore, it will provide direct evidence to support the hypothesis that a high degree of coordination of gene expression among many important pathways in major biological processes such as cereal grain filling could be achieved by transcription regulation through interactions between a transcription factor and a cis-regulatory element.

4. MULTIPLE PATHWAYS MAY BE REGULATED BY COLLECTIVE TRANSCRIPTION FACTOR GENES DURING GRAIN FILLING

To identify transcription factor genes involved in the grain-filling process, the expression of genes belonging to this category was investigated. A total of 57 genes that showed the grain-filling pattern were identified. Among them, nine genes encode MYB transcription factors. Because the consensus-binding sequence of type II MYB is similar to the AACA element (30,31) and their typical grain-filling expression pattern, it was hypothesized that these MYB transcription factor genes may be involved in regulating gene expression during rice grain filling (21).

4.1. Modeling Transcription Control of the Grain-Filling Process

Although mutations in a single transcription factor gene could result in extensive changes in gene expression (32), it has been considered that specific gene expression during grain development requires the involvement and coordinated action of multiple transcription factors (regulatory modules). For instance, endosperm-specific expression of the rice glutelin gene needs the presence of Prolamin-box, ACGT, and ACAA motifs in the promoter region (31). In order to dissect the genetic network controlling the grain filling and identify the key regulators controlling multiple pathways, a straightforward approach is to investigate the composition and orders of various motifs in the regulatory region (33). Alternatively, the correlation of temporal expression patterns of all genes involved in grain filing could be interrogated in a time-delayed fashion (34). Genes with delayed positive or negative correlations indirectly suggested their regulatory relationships. Based on these relationships, a gene interaction network could be constructed.

These gene interaction relationships could be further characterized by examining the temporal expression patterns among genes with potential regulatory relationships predicted based on coregulation analysis and promoter sequence analysis. For example, genes encoding a DNA-binding protein, such as the transcription factors, could be identified based on studies of sequence-binding domains, and annotations based on a previous molecular biology study. Their relationships to their downstream target genes could be further revealed by their temporal and spatial expression correlation to the transcription factor genes, and the correlation between the promoter-binding sequences and binding transcription factors (Fig. 2).

Phylogenetic footprints of the regulatory regions in the grain-filling genes of different crops may also provide valuable reference. Recent studies discovered that sequences at regulatory gene sites are relatively conserved among human and mouse (35). Based on these observations, a method was developed to detect the regulatory sequences directing liver-specific transcription (36). It would be interesting to test this hypothesis and gain insights of transcription control of grain filling. It is important to recognize the potential regulatory mechanism at a higher order, such as at the chromatin level (22).

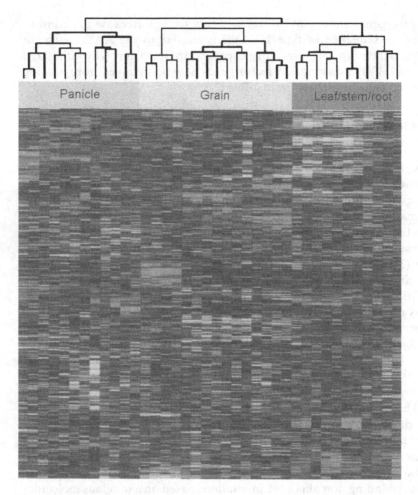

FIGURE 1 Hierarchical clustering analysis of gene expression data obtained from 33 rice samples collected through various developmental stages. A total of 51,000 genes, representing over 90% of the rice genome, were analyzed. Genes involved in grain development and grain filling formed tight clusters.

4.2. Use Functional Genomics Resource to Validate In Silico Predictions

The transcriptional profiling and computational modeling suggested a comprehensive view of the molecular events underlying the grain-filling process and provided an opportunity to dissect the gene regulatory network that controls these events. However, the large-scale analysis and in silico

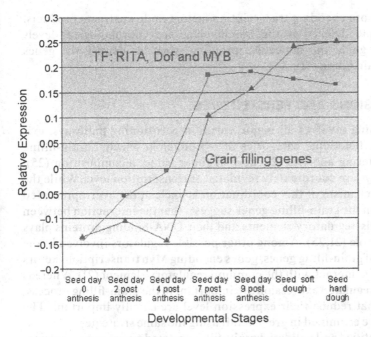

FIGURE 2 Genes encoding different classes of transcription factors that are potentially involved in grain filling were identified based on their correlated expression patterns to the expression patterns of grain-filling genes during grain development.

modeling results have to be experimentally validated. Recent developed functional genomics tools and resources, such as genome-wide DNA-protein interaction analysis and reverse genetics resources are increasingly used for this purpose. The identified gene network and key regulators could be validated experimentally by various approaches. By altering the expression of genes encoding putative regulators and examining the phenotypic changes, the associated function could be validated. The alteration of gene expression could be achieved by gene knockout via either the small interfering RNA (siRNA) or double-strand RNA (dsRNA) approach (37,38) or by overexpression through the introduction of an extra copy of a full-length gene driven by a strong promoter. Once their involvements in grain filling are established, the interaction among genes could be confirmed by in vitro or in vivo DNA-protein interaction experiments, as demonstrated by Ren et al. in yeast (39).

Recently developed reverse genetics resources, such as rice insertion or deletion of mutant collections, could be used to investigate gene functions in a systematical way (40). This, of course, will benefit the study of gene functions of those encoding transcription factors or downstream products. The phenotypic defects at the biochemical and morphological levels could further guide the molecular profiling work to identify the genes involved in grain filling (41).

5. CONCLUSIONS AND PERSPECTIVES

Rice grain filling involves all major nutrition-partitioning pathways, evidenced by the functional categories of reproducibly selected grain-filling genes by clustering analysis (21) and singlular value decomposition (25). These pathways are coordinately regulated at transcription level. While the regulatory mechanism of this coordination is unclear, the overrepresented AACA motif in the grain-filling genes suggests that the interaction between the common cis-regulatory elements and their DNA-binding proteins plays an important role (21,25). Among other possible regulators involved in the coregulation of grain-filing genes, genes encoding Myb transcription factors may potentially interact with the common regulatory element (21). Whereas genes with upregulated expression are important to the grain-filling process, those genes that reduce their expression level are equally important. The genes should be examined in great detail using the same strategies.

The identified and validated grain-filling-related target genes and possible regulatory mechanism could be utilized for improving grain quality and yield in cereal crops. It will be especially useful when combined with the information from proteome analysis (42) and quantitative trait locus (QTL) analysis (43). By combining the information, it is now possible not only to identify the genes responsible for simple traits, but also to identify genes responsible for complex quantitative traits. The orthologues of these genes could be identified and cloned by screening mutant lines in other crops. Alternatively, the orthologues could be identified by heterologous hybridization of microarray. Because the hybridization signals reflected both sequence homology and transcript abundance, the entire experiment should be conducted using samples from the same species to emphasize the detected gene expression pattern. The key regulator for multiple pathways could be optimized to improve the complex traits by adjusting their expression level in elite lines using the biotechnology approach. It has been shown that altering a single gene that encodes granule-bound starch synthase can dramatically affect the culinary quality (8), resulting in softer, less sticky cooked rice. It is also possible to screen the breeding lines to select the ones with favorable variations in their regulatory genes or regulatory domains. For example, polymorphisms in the coding regions of key regulators among different breeding lines that potentially modify protein structure and binding characteristics could be identified based on their sequence alignment and expression profiles. Polymorphisms in the promoters of key regulators/regulates among different breeding lines that may alter gene expression could also be identified and used as genetic markers to assist the selection (44).

REFERENCES

1. Mazur B, Krebbers E, Tingey S. Gene discovery and product development for grain quality traits. Science 1999; 285:372–375.
2. Mazur BJ. Developing transgenic grains with improved oils, proteins and carbohydrates. Novartis Found Symp 2001; 236:233–241.
3. Ye X, Al-Babili S, Kloti A, Zhang J, Lucca P, Beyer P, Potrykus I. Engineering the provitamin A (beta-carotene) biosynthetic pathway into (carotenoid-free) rice endosperm. Science 2000; 7:303–305.
4. Zhang Y, Darlington H, Jones HD, Halford NG, Napier JA, Davey MR, Lazzeri PA, Shewry PR. Expression of the gamma-zein protein of maize in seeds of transgenic barley: effects on grain composition and properties. Theor Appl Genet 2003; 106:1139–1146.
5. Shewry PR, Tatham AS, Halford NG, Barker JH, Hannappel U, Gallois P, Thomas M, Kreis M. Opportunities for manipulating the seed protein composition of wheat and barley in order to improve quality. Transgenic Res 1994; 3:3–12.
6. Shewry PR, Halford NG. Cereal seed storage proteins: structures, properties and role in grain utilization. J Exp Bot 2002; 53:947–958.
7. Tabe L, Hagan N, Higgins TJ. Plasticity of seed protein composition in response to nitrogen and sulfur availability. Curr Opin Plant Biol 2002; 5:212–217.
8. Umemoto T, Nakamura Y, Ishikura N. Activity of starch synthase and the amylose content in rice endosperm. Phytochemistry 1995; 40:1613–1616.
9. Larkins BA, Lending CR, Wallace JC. Modification of maize-seed-protein quality. Am J Clin Nutr 1993; 58:264S–69S.
10. Lehrer SB, Reese G. Recombinant proteins in newly developed foods: identification of allergenic activity. Int Arch Allergy Immunol. 1997; 113:122–124.
11. Murphy DJ. Manipulation of plant oil composition for the production of valuable chemicals. Progress, problems, and prospects. Adv Exp Med Biol 1999; 464:21–35.
12. Thelen JJ, Ohlrogge JB. Metabolic engineering of fatty acid biosynthesis in plants. Metab Eng 2002; 4:12–21.
13. Hirel B, Bertin P, Quillere I, Bourdoncle W, Attagnant C, Dellay C, Gouy A, Cadiou S, Retailliau C, Falque M, Gallais A. Towards a better understanding of the genetic and physiological basis for nitrogen use efficiency in maize. Plant Physiol 2001; 125:1258–1270.
14. Lopez CG, Banowetz G, Peterson, CJ, Kronstad WE. Differential accumulation of a 24-kd dehydrin protein in wheat seedlings correlates with drought stress tolerance at grain filling. Hereditas 2001; 135:175–181.
15. Gale MD, Devos KM. Comparative genetics in the grasses. Proc Natl Acad Sci USA 1998; 95:1971–1974.

16. Goff SA, Ricke D, Lan TH, Presting G, Wang R, Dunn M, Glazebrook J, Sessions A, Oeller P, Varma et al. H. A Draft Sequence of the Rice Genome (Oryza sativa L. ssp. Japonica). Science 2002; 296:92–100.

17. Yu J, Hu S, Wang J, Wong GK, Li S, Liu B, Deng Y, Dai L, Zhou Y, Zhang X, et al. A draft sequence of the rice genome (Oryza sativa L. ssp. indica). Science 2002; 296:79–92.

18. Sasaki T, Matsumoto T, Yamamoto K, Sakata K, Baba T, Katayose Y, Wu J, Niimura Y, Cheng Z, Nagamural Y et al. The genome sequence and structure of rice chromosome 1. Nature 2002; 420:312–316.

19. Feng Q, Zhang Y, Hao P, Wang S, Fu G, Huang Y, Li Y, Zhu J, Liu Y, Hu X et al. Sequence and analysis of rice chromosome 4. Nature 2002; 420:316–320.

20. Rice Chromosome 10 Sequencing Consortium. In-depth view of structure, activity, and evolution of rice chromosome 10. Science 300:1566–1569, 2003.

21. Zhu T, Budworth P, Chen W, Provart N, Chang HS, Guimil S, Estes B, Zou G, Wang X. Transcriptional control of nutrient partitioning during rice gain filling. Plant Biotechnol J 2003; 1:59–70.

22. Zhu T. Global analysis of gene expression using GeneChip microarrays. Curr Opin Plant Biol 2003; 6:418–425.

23. Aoki N, Whitfeld P, Hoeren F, Scofield G, Newell K, Patrick J, Offler C, Clarke B, Rahman S, Furbank RT. Three sucrose transporter genes are expressed in the developing grain of hexaploid wheat. Plant Mol Biol 2002; 50:453–462.

24. Eisen MB, Spellman PT, Brown PO, Botstein D. Cluster analysis and display of genome-wide expression patterns. Proc Natl Acad Sci USA 1998; 95:14863–14868.

25. Anderson A, Hudson M, Chen W, Zhu T. Identification of nutrient partitioning genes participating in rice grain filling by singular value decomposition (SVD) of genome expression data. BMC Genomics 2003; 4:26.

26. Alter O, Brown PO, Botstein D. Singular value decomposition for genome-wide expression data processing and modeling. Proc Natl Acad Sci. USA 2000; 18:10101–10106.

27. Zhu T, Chang HS, Schmeits J, Gil P, Shi L, Budworth P, Zou G, Chen X, Wang X. Gene expression microarrays: Improvements and applications towards agricultural gene discovery. J Assoc Lab Automation 2001; 6:95–98.

28. Gil P, Budworth P, Zhu T. Unpublished results.

29. Kononova K, White C, Brown D, Guimel S, Harper B, Nelson A, Zhu T, Chang H-S, Budworth P, Lawton K. Rice GeneChip microarrays a useful tool for novel monocot promoter discovery. Proceedings of the 7th International congress of plant molecular biology. S05-1. Barcelona June 23–28, 2003.

30. Washida H, Wu CY, Suzuki A, Yamanouchi U, Akihama T, Harada K, Takaiwa F. Identification of cis-regulatory elements required for endosperm expression of the rice storage protein glutelin gene GluB-1. Plant Mol Biol 1999; 40:1–12.

31. Wu C, Washida H, Onodera Y, Harada K, Takaiwa F. Quantitative nature of the Prolamin-box, ACGT and AACA motifs in a rice glutelin gene promoter: minimal cis-element requirements for endosperm-specific gene expression. Plant J 2000; 23:415–421.

32. Hunter BG, Beatty MK, Singletary GW, Hamaker BR, Dilkes BP, Larkins BA, Jung R. Maize opaque endosperm mutations create extensive changes in patterns of gene expression. Plant Cell 2002; 14:2591–2612.

33. Berman BP, Nibu Y, Pfeiffer BD, Tomancak P, Celniker SE, Levine M, Robin GM, Eisen MB. Exploiting transcription factor binding site clustering to identify cis-regulatory modules involved in pattern formation in the Drosophila genome. Proc Natl Acad Sci. USA 2002; 99:757–762.

34. Luan Y, Li H. Clustering of time-course gene expression data using a mixed-effects model with B-splines. Bioinformatics 2003; 19:474–482.

35. Wasserman WW, Palumbo M, Thompson W, Fickett JW, Lawrence CE. Human-mouse genome comparisons to locate regulatory sites. Nat Genet 2000; 26:225–228.

36. Krivan W, Wasserman WW. A predictive model for regulatory sequences directing liver-specific transcription. Genome Res 2001; 11:1559–1666.

37. Hamilton AJ, Baulcombe DC. A species of small antisense RNA in posttranscriptional gene silencing in plants. Science 1999; 286:950–952.

38. Baulcombe D. RNA silencing. Curr Biol 2002; 12:82–84.

39. Ren B, Robert F, Wyrick JJ, Aparicio O, Jennings EG, Simon I, Zeitlinger J, Schreiber J, Hannett N, Kanin E, et al. Genome-wide location and function of DNA binding proteins. Science 2000; 290:2306–2309.

40. Cyranoski D. Rice genome: A recipe for revolution. Nature 2003; 422:796–798.

41. Maitz M, Santandrea G, Zhang Z, Lal S, Hannah LC, Salamini F, Thompson RD. rgfl, a mutation reducing grain filling in maize through effects on basal endosperm and pedicel development. Plant J 2000; 23:29–42.

42. Finnie C, Melchior S, Roepstorff P, Svensson B. Proteome analysis of grain filling and seed maturation in barley. Plant Physiol 2002; 129:1308–1319.

43. Consoli L, Lefevre A, Zivy M, D de Vienne, Damerval C. QTL analysis of proteome and transcriptome variations for dissecting the genetic architecture of complex traits in maize. Plant Mol Biol 2002; 48:575–581.

44. Rafalski A. Applications of single nucleotide polymorphisms in crop genetics. Curr Opin Plant Biol 2002; 5:94–100.

14

Proteomics and Genomics to Detail Responses to Extracellular Stimuli

James A. Carroll

Department of Molecular Genetics and Biochemistry, University of
Pittsburgh School of Medicine, Pittsburgh, Pennsylvania, U.S.A.

Clayton E. Mathews

Diabetes Institute, Department of Pediatrics, University of Pittsburgh School
of Medicine, Pittsburgh, Pennsylvania, U.S.A.

1. INTRODUCTION

The concept of comparing protein profiles of treated versus control samples
is not novel, yet the concept of comparing the global protein profile from an
organism or tissue in one biological state versus another is clearly innovative. This model has given rise to a new era in biological science, the proteomic era. The term "proteome" was realized at a conference in 1994 when a
graduate student Marc Wilkins was searching for a word that would convey
the protein equivalent of the genome (1). In truth, the proteome is a much
more complex and dynamic entity than the genome. The genome of an
organism is relatively static in the short term, whereas the proteome is a
constantly changing mosaic of peptides and proteins that can be degraded,
modified, cleaved, and increased or decreased in synthesis. This symphony
of alterations occurs in response to numerous signals, both internal as well

as external. For the purpose of studying the proteome, these changes in protein amount or modification are typically visualized by two-dimensional polyacrylamide gel electrophoresis (2D-PAGE), a technique developed simultaneously by the two independent labs of Klose (2) and O'Farrell (3) in the 1970s.

Two-dimensional-polyacrylamide gel electrophoresis relies on two fundamental properties of polypeptides: They can be separated by (1) charge and/or (2) molecular mass. Proteins are charged biomolecules, and when an electric field is applied, they will migrate through an established pH gradient until the net charge of the polypeptide is zero. The position in the pH gradient where a protein has a net zero charge, resulting in a loss of migration, is the isoelectric point (pI). Two general types of 2D-PAGE have been developed, which differ in the way proteins are separated in the first dimension (based on pI). One establishes a pH gradient by use of free-moving ampholytes (2,3) or by immobilized pH gradients (IPG) (4) prior to the application of a sample for separation. This system is called isoelectric focusing (IEF), and the use of IPG in proteomics has grown in popularity in recent years due to its ease of use and reproducibility. The disadvantages of IEF include a deficit in the ability to resolve extremely acidic or basic proteins and longer focusing times (around 50,000 Volt hours). The second 2-D application is termed nonequilibrium pH gradient gel eletrophoresis (NEPHGE) (5). In NEPHGE, a pH gradient is not formed prior to sample loading, allowing for the resolution of both acidic and very basic proteins on the same gel. In theory, the acidic proteins will focus to completion, but the basic proteins are moving toward their pI, never actually reaching it. Advantages to this system include a more complete picture of the proteome on a single gel and shorter focusing times (around 3000–4000 Volt hours). Once focusing is complete in the first dimension, proteins are subjected to standard SDS-PAGE in the second dimension to separate proteins by molecular mass.

Isoelectric focusing and NEPHGE are hampered their inability to resolve the entire proteome. There are several components at work that define this innate inability. Low abundant proteins are difficult to identify by both methods. By current silver staining methods, more than 1000 copies of a protein per eukaryotic cell are needed for it to be visible (6). Furthermore, protein solubility is a tremendous factor that affects focusing by either method. Typically, a nonionic detergent such as CHAPS or NP-40 is used to assist in the solubilization of proteins for focusing, but many proteins have a tendency to precipitate once they reach their pI. This propensity to fall out of solution can be problematic with proteins of high abundance and can actually inhibit the migration of other proteins in the electric field. Furthermore, many membrane proteins are extremely difficult to solubilize in the detergents that are commonly used with 2D-PAGE. Recent advancements in protein labeling (7–9) and the development of better nonionic detergents have allowed for the separation and identification of highly hydrophobic proteins. Moreover, some researchers are moving away from 2-D gels all together and are separating trypsin-digested peptides of a proteome by liquid chromatography (LC) for identification via mass spectrometry (MS) using a technique termed LC/MS (10). Recent advancements in MS technology, coupled with rapid advances in whole genome sequencing and annotation, have created novel proteomic approaches offering reasonable alternatives to answer difficult questions.

A review of the literature will reveal the many variations of MS that are available. As is characteristic of many technologies in science, each technique has benefits and shortcomings. The decision to use a given technique will call into question many variables, such as which exact instrument should be employed for MS. Two popular methods of ionization of peptides for MS analysis are electrospray ionization (ESI) (11) and matrix-assisted laser desorption/ionization-time of flight (MALDI-TOF) (12). These technological breakthroughs were developed in the 1980s and were instrumental in overcoming the problems associated with producing ions of proteins and peptides, molecules that are large and nonvolatile, for transfer into the gas phase for the MS analyzer. Electrospray ionization and MALDI are considered "soft-ionization" methods because they generate minimal fragmentation of the analyte. MALDI-TOF as a method to generate a mass fingerprint of a particular protein of interest has gained popularity for several reasons, including its simplicity, sensitivity, high mass range, increased accuracy and resolution, and tendency to form singly charged ions.

The steps in the process of comparative proteomics are as such: (1) A sample is exposed to two different conditions, and protein samples are prepared for analysis, (2) proteins are then separated by 2D-PAGE and visualized, (3) differences are determined by eye or by image analysis, (4) protein spots are then excised and prepared for in situ digestion, (5) proteins are enzymatically cleaved into smaller peptides by timed exposure to proteases (usually trypsin), (6) the peptides are extracted from the gel and, as is the case for MALDI-TOF, are mixed with a matrix that will aid in the transfer of energy to the peptides to enhance ionization, (7) the peptide-matrix mix is then spotted on a target plate for MS analysis and ionized by use of a laser, (8) a mass spectra is generated that can be used to search protein databases for potential matches using analysis software (i.e., SEQUEST or Protein-Prospector) that compares the query mass spectra to all known proteins in the database digested in silico, and (9) a report is compiled listing, in order of significance, the probable identity of the unknown. It is essentially up to the investigator to perform quality control at this stage to verify the findings.

2. PROTEOMICS & DISCOVERY

Using proteomics to identify changes in protein synthesis patterns under one condition versus another is ideally employed in prokaryotic systems. The number of sequenced bacterial genomes that are now available in the NCBI database is astounding. Prokaryotes, although having a simpler genetic makeup compared to eukaryotes, have evolved various complex mechanisms to sense and adapt to their ever-changing environments (Fig. 1). Microorganisms do not possess the five senses that mammals are accustomed

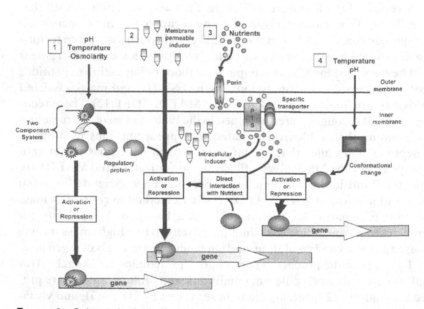

FIGURE 1 Schematic depicting some of the pathways that allow microorganisms to respond to alterations in the environment. Pathway 1 is a two-component system with a sensor kinase situated in the inner membrane that undergoes autophosphorylation. When the appropriate signal is received, the sensor kinase phosphorylates a response regulator, which in turn elicits the effector function. Pathway 2 demonstrates an inducer molecule that is membrane-permeable (i.e., the quorum-sensing molecule N-acetyl-L-homoserine lactone), which transduces a signal to a regulatory protein that then controls gene expression. Pathway 3 shows nutrients (i.e., metals, sugars, amino acids) that are actively transported by a phosphotransferase system (PTS) or by a high-affinity, specific transporter. Nutrients are brought into the cell and can be converted to an inducer or act directly upon a regulatory protein to alter gene expression. Finally, pathway 4 depicts how environmental cues such as temperature and pH can cause a conformational change in a regulator when the internal milieu is altered, causing a response in the bacteria cell. All of these adaptive responses lead to changes in gene expression, often leading to drastic changes in levels of specific proteins within the microorganism.

to; rather, they often rely on their ability to sense changes in nutrient levels, temperature, pH, and/or osmolarity to indicate their surroundings. This, in turn, triggers alterations in genes' expression that often lead to changes in protein synthesis. Bacterial pathogens rely on this adaptive response to regulate the expression of virulence determinants that assist in the infection process.

How do environmental stimuli yield alterations in gene expression and protein synthesis? There are four general mechanisms that microorganisms use to sense the environment and regulate gene expression. First, a change in the environment can cause a conformational alteration in a protein that binds or interacts with DNA or RNA polymerase. Second, an allosteric mechanism affecting the oligomeric state of a regulatory protein can be influenced by environmental stimuli. Third, many proteins undergo covalent modification (i.e., phosphorylation, proteolysis, methylation, or oxidation/reduction) in response to the environment. Finally, many proteins experience noncovalent modification such as activation, inhibition, or titration. These changes in response to environmental cues can be monitored by several methods including proteomics.

In this chapter, we share some observations stemming from our experiences in both prokaryotic and eukaryotic systems. We have successfully initiated proteomic approaches to determine pH-regulated proteins in Borrelia burgdorferi (the cause of Lyme disease); to identify substrates for proteases, which may play a role in disease progression, that are produced by B. burgdorferi and Bartonella quintana; to identify accessory regulatory factors that bind specific promoters in B. burgdorferi; to investigate serum protein changes in Scrapie disease (a transmissible spongiform encephalopathy caused by a prion protein) in mice; and to determine changes that occur in pancreatic islets due to oxidative stress and exposure to various cytokines. The use of proteomics as a research tool can often help to quicken the discovery process and increase productivity.

In 1997, the B. burgdorferi genome was completed, allowing for the use of MALDI-TOF to identify proteins of interest (13). In that same year, we began to investigate what environmental cues, other than temperature, may influence differential expression of virulence determinants in B. burgdorferi as the spirochete is transmitted from the tick vector to a mammalian host. The tick and the mammal propose extremely different environments for B. burgdorferi in temperature, pH, and nutrients. Upon shifting the pH of the bacterial medium and separation of membrane proteins by 2D-NEPHGE, we observed well over 30 alterations in the protein profile in spirochetes cultured at pH 7.0 compared to pH 8.0 (14) (Fig. 2). It was then that we first embraced the soft-ionization technique, MALDI-TOF MS, to assist in identifying these pH-regulated proteins (15).

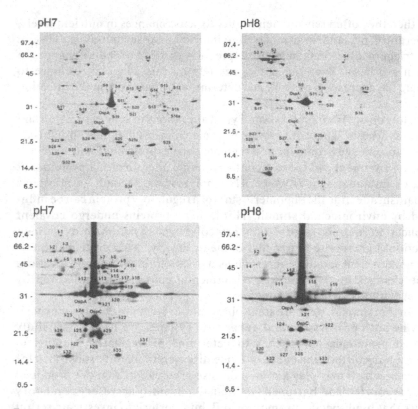

FIGURE 2   B. burgdorferi membrane proteins separated by 2D-NEPHGE and stained
with silver (top two panels) or probed with hyperimmune serum (bottom two panels).
Changes in the protein profiles from samples prepared from spirochetes grown in pH
7.0 (compared to pH 8.0) are indicated and numbered. Major outer surface protein
(Osp) A and OspC are marked. Relative molecular masses in kiloDaltons are indicated
to the left of each panel. The acidic end is to the left of the gels.

Using this technique, we were able to identify more than 20 membrane
proteins that were influenced by changes in the external pH. Determination
of their identity has enabled better comprehension of which proteins and
protein families may be important in invasion, infection, and ultimately in
pathogenesis. Rather than subject the entire cellular proteome to separation,
we took to the task of fractionating the spirochetes into the subcellular components of membrane-associated proteins and soluble proteins. Cells were
separated by lyses and ultracentrifugation ($\sim$300,000 $\times$ g). This pellets the
membranes (and associated proteins) and leaves the soluble proteins in suspension. Fractionation can assist in simplifying the overall protein pattern

and allow the visualization of some lower-abundance proteins. The membrane-associated proteins can be further partitioned (1) by adding potassium chloride to the membrane sample to remove loosely associated membrane proteins from anchored/integral membrane proteins, or (2) by the differential solubility of proteins in various detergents (i.e., triton-X 114 phase partitioning). Each fraction can then be subjected to separation by 2D-IEF or 2D-NEPHGE.

Proteomics has been implemented to determine not only the differences among bacteria exposed to different environments, but to determine the substrates for specific proteases within a cell as well. A recent article details the use of a molecular trap combined with MS to identify putative substrates for the ClpXP protease complex in E. coli (16). We are currently involved in collaborative studies of B. quintana (cause of trench fever) and B. burgdorferi where the gene encoding a C terminal protease (CtpA) has been disrupted. Two-diemensional NEPHGE analysis of wild-type compared to mutant has revealed changes in the mobility of multiple proteins. The ability to identify these putative substrates of the CtpA protease by MALDI-TOF has assisted in defining a putative cleavage motif for the CtpA protease. Moreover, it has expanded our understanding of the specificity and necessity of this enzyme. Our proteomic studies suggest that CtpA is important in processing the C terminus of many membrane-bound proteins that are highly positively or negatively charged. The role that CtpA plays in infectivity has not been determined, but it is interesting that CtpA is associated with a pathogenicity island in Bartonella spp (17).

Hopes for the future of proteomics include the ability to compare disease and nondisease states of a particular tissue, insight into the disease process, and the determination of specific markers for disease for diagnostic purposes. This is an ambitious endeavor when considering the number of potential proteins encoded by the genome of mammals, but success could have a monumental impact on the way diseases are diagnosed and treated. With the human and mouse genome projects nearing completion, several institutes and pharmaceutical companies have launched large-scale proteomic projects in an attempt to identify molecular markers for diagnostic purposes that are associated with disease states. As stated earlier, many have moved away from the labors of 2D-PAGE and have invested in analyzing complex mixtures of proteins by MS to increase throughput. Still, these groups face difficult challenges due to the sheer number of possible protein species present in the cell or tissue. There are over 100 types of posttranslational modifications recognized to date that include phosphorylation, glycosylation, and amino- and carboxy-terminus processing. It is estimated the human genome could potentially express well over one million different protein species and isoforms (1). Sifting through such an incredible array of material looking for significant differences will undoubtedly be a difficult endeavor. The overrepresented proteins in the proteome are typically associated with housekeeping functions (i.e., chaperones, matrix proteins, and cytoskeletal proteins), and it seems plausible that the most interesting and more disease-relevant proteins will likely be of low abundance (i.e., phosphotases, kinases, and regulators), which will make identification all the more complicated.

2.1. Proteomics to Determine Cell Signaling Differences in Eukaryotic Systems

Validation of proteomic approaches in prokaryotic systems has provided and will continue to provide a basis for detailing protein changes in response to environmental stimuli. In the case of prokaryotes, environmental cues will generally arise in an organism-independent manner, whereas mammalian cells abide in a temperature-controlled environment and the chemical environment is dependent on products made or released by the cell (Autocrine), neighboring cells (Paracrine), or distant cell/tissue (Endocrine). Eukaryotic systems therefore present a great many more challenges due to diversity, specifically in vivo systems. Proteomic analysis on in vivo systems where whole tissues are sampled will introduce differences possibly due only to populations of cell types in the tissue sampled. Primary cell culture is also not immune from this variable, as cell clusters such as pancreatic islets can contain 6–10 different cells types. A further modifier when using samples from animal models or human patients is the genetic diversity present in the populations. Therefore, it is critical that experimental design take into account the many types of differences inherent to the system or model. Work in vitro can aid to suppress differences; moreover, use of a single cell line can aid in eliminating background differences. In cases where an in vivo system is necessary, the use of inbred animal models that are essentially genetically identical can eliminate genetic variation, allowing the investigator to focus on treatment-induced differences.

In the study of insulin-dependent diabetes mellitus (IDDM) or type 1 diabetes mellitus (T1D), the most widely used animal model has been the NOD mouse. The pathology of T1D in NOD mice resembles that in the human patient and is a spontaneous autoimmune T cell-mediated disease (18). The distinguishing pathologic lesion in T1D is a destructive, immune cell infiltrate (Insulitis) initiated after weaning, preceded by entry of macrophages and dendritic cells. In the islets of young NOD mice, macrophages and dendritic cells produce the proinflammatory cytokines tumor necrosis factor alpha (TNFα) and interleukin 1 beta (IL-1β). IL-1β and TNFα are detrimental alone; in combination they have synergistic effects on human and rodent islet structure, function, and viability (19). Further, the combination of these two cytokines primes the β cell for T cell killing.

Several years ago, we identified a mouse strain in which the islet cells were refractory to the effects of proinflammatory cytokines. This strain was specifically developed to resist the beta cell-specific free radical-generating toxin, alloxan (AL). Alloxan, a glucose analogue, gains entrance into the β cell via the glucose transporter 2 (GLUT2) and spontaneously decomposes (20), generating superoxide radicals and H_2O_2 in the presence of an iron catalyst (21,22). Reactive oxygen species (ROS), produced by AL, mediate β cell necrosis and a permanent insulin-dependent diabetes mellitus syndrome. ALR is closely related to the T1D-susceptible NOD, and maintains remarkable resistance to chemical and cell-mediated β cell destruction. This resistance entails a systemic ability to dissipate free radicals that extends to the level of the pancreatic islet. Alloxan resistant islets express high levels of antioxidant enzymes, including glutathione peroxidase (GPX) and glutathione reductase. These two enzymes are not normally expressed in islets. Free-radical production by ALR islets compared to controls (NOD and C3H) was determined by incubating islets for 5 d in IL-1β (30 U/ml), IFNγ (1000 U/ml), and TNFα (1000 U/ml). Whereas control islets showed significant decreases in viability and insulin content, and almost a complete depletion of reduced glutathione (GSH), ALR islets showed no loss in viability or insulin content, although GSH was reduced, suggesting that ALR islets are not unreactive to the effects of cytokines. Control islets produced H_2O_2 and NO, and had two- to fourfold increases in nitrotyrosine (NT). In contrast, these radicals were not elevated in ALR islets, although basal levels of nitrite and H_2O_2 were similar. As basal levels of NT were approximately 100-fold less in ALR, ALR must efficiently denitrosylate, a process catalyzed by GSH and GPX, two antioxidants that are significantly elevated in ALR islets.

The ALR genome confers systemically elevated free radical defenses, dominantly protecting their pancreatic islets from free radical-generating toxins, cytotoxic cytokines, as well as diabetogenic T cells (23–25). To establish the genetic basis for the unusually strong resistance of ALR islet cells to immune-mediated destruction, T1DM-free (NOD×ALR) F1 females were backcrossed to NOD males (BC1) (26). BC1 progeny females were monitored for T1DM and genetic linkage analysis was performed on all progeny. Backcross mapping identified three genetic loci conferring T1DM resistance to the ALR strain. The loci were identified on Chr. 17, 8, and 3. To detect epistatic interactions, simultaneous genome scans for all pairs of markers were implemented (27,28). The genome scan searched through all pairs of loci by fitting a two-way ANOVA model with an interaction item. The pairwise genome scan calculated a significant interaction between the Chr. 8 and Chr. 17

loci. Further, to identify possible contributions from the mtDNA, we performed reciprocal outcrosses between ALR and NOD, and the F1 progeny were backcrossed to NOD. A fourfold lower frequency of spontaneous T1D development occurred when ALR contributed the mitochondrial genome (29). Because of the apparent interaction between nuclear and mitochondrial (mt) genomes, a mouse strain survey was performed and the mitochondrial genomes of 14 strains were sequenced. An ALR-specific novel variant in the *mt-Nd2* gene, producing a nonconservative leucine to methionine substitution at amino acid residue 276 in the NADH dehydrogenase 2 subunit (*mt-Nd2*), was discovered.

The genetic analysis in these crosses was complicated by the fact that ALR and NOD share genetic identity at approximately 70% of the 1000 simple sequence repeat markers (SSRM) we have typed. While ALR and NOD do share important diabetes susceptibility genes on Chr. 17, most of the chromosome is rife with informative markers. The high level of polymorphisms allowed the mapping of a very narrow interval (2.7 Mb) on Chr. 17, spanning from *D17Mit61* to the gene for MHC Class 1, *H2-K*. Unfortunately, this level of genetic disparity was not present on Chr. 3 and 8. In fact, the level of polymorphisms was 16 and 10%, respectively, creating long distances between markers on both chromosomes. Therefore, in the statistical analysis, the 95% confidence intervals for the linkages on these two chromosomes are quite large, 30 cM (between 20–50 cM on Chr. 3) and 19 cM (spanning from 36–55 cM on Chr. 8) (26), which has created difficulties in shortening these two intervals.

In order to identify the ALR-derived protective genes, reciprocal congenic mouse lines for all four loci are being generated. These eight lines should allow us to confirm the genetic linkage analysis and also better understand the genetic protective phenotypes provided by each locus. Historically, the positional cloning of susceptibility genes in the NOD mouse has been difficult due to a phenotype (diabetes onset) that can be controlled by many variables (30). The unique phenotypes of the ALR have allowed us to utilize more stringent tests for diabetes protection. We have also employed surrogate tests to aid in the determination of genetic responsibilities for T1D protection for each locus. The determination of the protective responsibility should support narrowing the candidate genes in individual loci.

Combined analysis of surrogate phenotypes comparing the congenic lines to the parental strains has allowed for us to determine preliminary protective responsibilities for the three nuclear loci. Analysis of the NOD.ALRc17 congenic mice has shown that the β cells from these mice are resistant to killing by T cells by in vivo and in vitro assays (31). We have assessed that resistance of the pancreatic islets of ALR to cytokine-induced death and dysfunction is controlled by the locus on Chr. 8 (32). As stated previously, an epistatic interaction was found between the Chr. 17 and Chr. 8 linkages, and it is interesting to point out that proinflammatory cytokines prime the β cells for T cell killing. Therefore, this synergism for may be a critical link in protection from immune-mediated β cell loss in T1D. Further, we have also determined that the region on Chr. 3 linked to diabetes protection controls, in part, the upregulated free radical defenses, specifically the ability to dissipate superoxide through an increase in the activity of the enzyme superoxide dismutase 1 (SOD1). The locus has been given the symbol *Susp* (Suppressor of superoxide production) (33).

To determine whether overt differences in the islet protein profile could be detected comparing ALR and NOD strains, we have used the Amersham Ettan DIGE System for analysis of proteins run on 2-D gels. In classical analysis of proteome in different states, the comparison of a 2-D protein pattern from one gel is compared to the pattern of a second (or more) gel. A common problem with this sort of analysis is the variations that can exist between 2-D gels. To combat this predicament, the 2-D gel system was modified, giving rise to the technique commonly referred to as difference gel electrophoresis (DIGE) (34). Difference gel electrophoresis technology allows for the separation of two distinct protein samples on one acrylamide gel, effectively eliminating the need to compare gels run at different times. Protein samples are derivatized by covalently labeling the samples with one of a mass and charge-matched set of fluorphores (typically Cy2, Cy3, or Cy5). The chemistry is such that the dyes undergo a nucleophilic substitution reaction with the lysine epsilon amine group of proteins. In our experiments, ALR islet protein was labeled with Cy3 and NOD with Cy5. Labeled islet protein samples were mixed and then subjected to isoelectric focusing in Immobiline DryStrips (pI 3–10, 13 cm) for 80,000 Volt hours. The second dimension was run and the gels were then scanned using a Typhoon 9410 fluorescence laser scanner and the fluorescence intensities and image from the different channels were captured for analysis. A merging of the Cy3 and Cy5 channels allowed for comparison of each protein spot. A merge of the images showed multiple differences comparing the NOD and ALR samples. Differences in fluroscent intensities of each spot were analyzed using DeCyder Differential Analysis Software (Amersham). Areas showing differential protein composition were excised and processed for identification by MS.

Several proteins were identified. Two of these proteins are important signaling mediators of the IL-1 pathway. Beta cell death induced by cytokines is dependent upon activation of NF-☐B downstream of the IL-1 Receptor (IL-1R) for the generation of iNOS. Through DIGE, we determined a lack of IKKβ and in ALR islets, an increase in the NK-kB inhibitor IkB (I kappa B). The role of IKK [inhibitor of kappa B kinase beta (*Ikbkb*)] is critical for the activation of NF-kB via the phosphorylation and degredation of the NF-kB inhibitory subunit IkB. Phosphorylation of IkB causes it to dissociate from the p50 and p65 subunits of NF-kB. The active heterodimer of NF-kB is then free to translocate to the nucleus and begin transcriptional activity. We have confirmed the DIGE analysis via western blot that shows the constitutive levels of IKKβ in ALR islet are very low compared to the level in islets of control strains (Fig. 3A). Correlating with this is the increased level of IkB in ALR islets compared to NOD islets (Fig. 3B). The suggestion from these results in untreated islets would be that the lack of IKKβ in ALR islets would inhibit IL-1R signaling through the inability to phosphorylate IkB, thereby activating NF-kB. This difference in ALR islets compared to controls would result in a break in the signaling pathway or at least a decrease in the velocity of signaling at the point where IkB is phosphorylated and NF-kB is activated. Our analysis after 48 hours of IL-1β treatment shows just this result (32). Exposure of control islets to

(A)

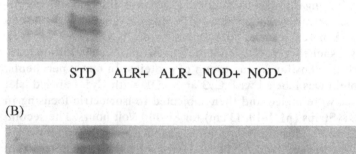

STD   ALR+ ALR- NOD+ NOD-

(B)

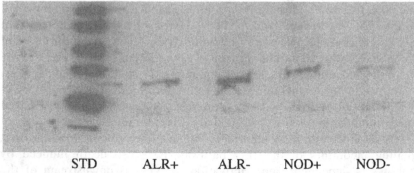

STD   ALR+   ALR-   NOD+   NOD-

FIGURE 3 Immunoblots of IKKβ and IkB in 48 hours. IL-1β treated (+) and untreated (−) ALR and NOD islets. (A) IKKβ is not present in untreated ALR islets. Levels of IKKβ were diminished with IL-1β treatment of NOD islets. (B) The constitutive level of IkB is elevated in ALR islets.

10 U/ml of IL-1β causes a decrease in the level of and IKKβ as well as elevation in IkB. In contrast, treatment of ALR islets with the same dose of IL-1 does not alter the levels of IKKs and further, IkB was decreased. Further, the incubation of control islets with IL-1 led to the induction of iNOS and the production of nitric oxide (NO), yet ALR islets produced neither iNOS nor NO.

The critical link between the IL-1R and iNOS is NF-kB. These results clearly lead us to hypothesize that NF-kB was not activated in ALR islets after incubation with IL-1β. Historically, to determine whether a transcription factor is activated, scientists have used Electrophoretic Mobility Shifts Assays (EMSA) to determine nuclear translocation of transcription factors that are constitutively present in an inactivated state in the cytoplasm. In accordance with National Institute of Health guidelines for reducing the use of animals in research, we have adopted methods that allow for a decrease in the number of animals used. The replacement of EMSA with immunofluorescence for the detection of activated transcription factors or cellular signaling components allows for a tenfold reduction in animals per experiment. This is due to the small protein mass isolated from pancreatic islets and the number of mice needed for EMSA versus the new methodology. A time course for proinflammatory, cytokine-induced translocation was performed using isolated pancreatic islets. Pancreatic islets isolated by standard methodology (35) were subjected to media containing cytokines of control media at 37°C in an atmosphere containing 5% CO_2 for a maximum of 24 hours. After the incubation period, islets were washed free of media and fixed with 2% paraformaldehyde for 15 min and embedded in a mixture of Affi-Gel Blue Gel (BioRad) and HistoGel (Richard Allen Scientific). After refrigeration for 15 min, samples were fixed again in 2% paraformaldehyde for 15 min, incubated in 30% sucrose overnight, frozen in liquid nitrogen and stored at −80. Sections for each time point were cut and the nuclei were stained with cytox green and a Phycoerythrin (PE)-labeled monoclonal antibody, specific for the p65 subunit of NF-kB. As shown in panels A and C of Fig. 4, in untreated islets, NF-kB is sequestered in the cytoplasm. Treatment of islets isolated from NOD mice with cytokines clearly induced a translocation of NF-kB into the nucleus (Fig. 4B), as indicated by the yellow nuclei in the overlay. In accord with our hypothesis, the cytokines did not induce translocation of p65 into the nucleus of ALR islets, clearly demonstrating that the NF-kB pathway is blunted in ALR islets. In fact, at all time points assayed, in the ALR islet, no translocation was extant. The fact that NF-kB does not translocate to the nucleus in cytokine-treated ALR islets proves that resistance of ALR islets to proinflammatory cytokines is mediated through the lack of IKKβ.

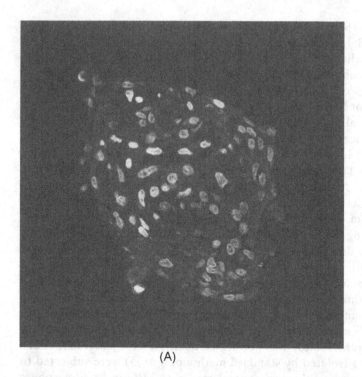

(A)

FIGURE 4 Immunofluorescent detection of NF-kB nuclear translocation NOD and ALR islets. Isolated pancreatic islets were treated with 1000 U TNFα, 1000 U IFNγ, and 30 U IL-1β (NOD 4A and ALR 4C) or left untreated (NOD 4B and ALR 4D) at 37°C in 5% CO_2 for 16 hours. At the end of the incubation period, islets were fixed and embedded in HistoGel (Richard Allen Scientific) and frozen in liquid nitrogen. Frozen gel plugs containing islets were cut, mounted on slides, and stained with Sytox green and a PE-conjugated monoclonal antibody specific for the p65 subunit of NF-kB. Sytox green labels the nucleus green and the PE-p65 NF-kB antibody labels NF-kB in red. A yellow nucleus denotes the translocation of NF-kB. In panels A (untreated NOD) and C (untreated ALR), the p65 NF-kB is sequestered in the cytoplasm. Treatment of NOD islets results in the translocation of p65 NF-kB into the nucleus and is demonstrated by the yellow nucleus on the merged image (B). In contrast, in ALR, islets do not exhibit nuclear translocation upon cytokine treatment (D).

The determination that the NF-kB signaling pathway intermediate IKKβ is present at very low levels in ALR islets compared to controls has aided our search for the candidate genes that provide protection of ALR islets to diabetogenic effectors. IKKβ maps to Chr. 8 and inside of the 95%

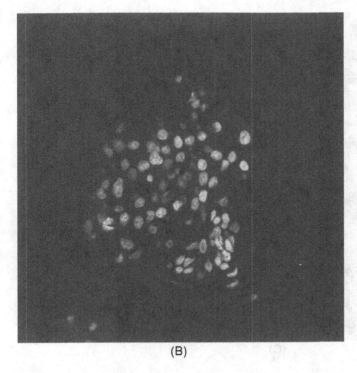

(B)

FIGURE 4   (*continued*)

confidence interval generated by the genetic analysis of the first backcross. Real Rime PCR analysis for the IKKβ transcript has revealed that the ALR islets express significantly less IKKβ compared to cDNA prepared from BALB/c, C3H/HeJ C57BL/6, or NOD islets. Whereas IKKβ showed different expression in ALR, the level of IkB transcript in ALR islets was equal to control islets. We are currently examining the promoter of the IKKβ gene to determine if mutations exist in ALR that might be the cause of the reduction in expression.

Coupling proteomic technology to our genetic analysis has allowed for the rapid identification of IKKβ as a candidate gene. The application of DIGE, while expensive, can provide important results if the reagents exist to confirm the findings. Although these methods can be a powerful tool, the use of DIGE may not be appropriate in every model system. As is the case with the ALR-derived locus on Chr. 3, we have applied a different proteomics tool to aid in the identification of the mechanism of protection.

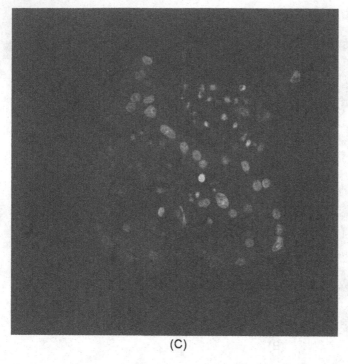

(C)

FIGURE 4   (continued)

The clear elevation in SOD1 activity in the ALR mouse compared to controls lead us to initially examine the levels of SOD1 via western blot and via gene expression via Quantitative-Real Time-PCR (Q-RT-PCR) analysis (24,25). When compared to other antioxidant enzymes, such as Thioredoxin, Glutathione Reductase, or Glutathione Peroxidase, which are all expressed at elevated levels in ALR, SOD1 expression was identical compared to controls (32). Western blot analysis also revealed no difference in the level of SOD1 protein. Therefore, the increase in SOD1 activity in ALR tissues is not due to mechanisms involving elevated transcription or translation. A possible reason for the striking SOD1 activity difference between ALR and ALS could be a mutation or variation in the *Sod1* gene, leading to a difference in the primary amino acid sequence. Yet, sequencing of *Sod1* cDNA showed that there were no differences in the primary sequence. A result came from the examination of liver homogenate from ALR and controls via cellulose acetate gels. On these gels, it was clear that the Sod1 of ALR ran slower than expected and seemed to be a novel allele. Thus, this difference in mobility of

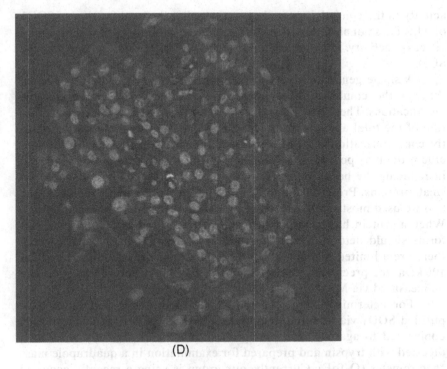

(D)

FIGURE 4 (continued)

Sodl from ALR was hypothesized to be the result of posttranslational modification (PTM), either enzymatic or nonenzymatic, between ALR and control strains tested, and further, the dominant isoform in ALR confers a higher enzymatic activity (36,37).

Currently, very little work has been done to characterize the SOD1 isoforms extant in mice. This knowledge is based on work in avian species and from human samples. To preliminarily determine whether the modification(s) to the SOD1 of ALR resulted in a change in pI, we performed nondenaturing isoelectric focusing on liver homogenates from ALR versus controls. Analysis of mobility on immobilized pH gradient gel strips (pH 4–7) showed that there was no difference in the pI of the ALR variant. We also examined the pI of SOD1 from samples that had been reduced by incubation with 1mM dithiothreitol (DTT) for 2 hr at 37°. These samples underwent isoelectric focusing for 80,000 Volt hours and were developed via activity assay. Under reducing conditions, SOD1 from ALR maintained activity, albeit at a reduced level, compared to a complete abolition of activity in the controls. These data suggest that the modification of SOD1 of ALR does not alter the charge of the molecule, but does have a profound effect on activity, even under conditions that ablate activity in wild-type SOD.

A single gene can give rise to a large number of protein products, through the combination of alternative splicing and varying possible modifications. Therefore, a specific species may exist as only a small fraction of the total amount of the gene product. A critical consideration in the characterization of PTM is the need for as large an amount of the specific protein as possible (38). Further, purification of the protein will aid in reducing the background and will allow a direct analysis of PTM on small proteins. Protein isolation can be performed by many methods; the two we used most often are column purification or immunoprecipitation. When a protein has been purified, the size and predicted number of isoforms should determine the methodology followed. In instances where there are a limited number of gene products and the protein is less than 100 kDa, the precise molecular weight of the intact purified fraction can be measured via MS.

For determination of the modifications on ALR SOD1, we have purified SOD1 via immunoprecipitation with an SOD1-specific antibody conjugated to agarose. The precipitated SOD1 and antibody were then digested with trypsin and prepared for examination in a quadrupole mass spectrometer (Q-ToF). Currently, our group is using a recently acquired ThermoFinnigan ProteomeX, MDLC nanoESI Ion Trap with a quadrupole/octapole ion guidance system at the front end to focus the ions into the ion trap. A Q-ToF consists of four parallel rods oriented like the four poles on a compass. The north-south pair has a radio frequency (rf) and direct current (dc) voltage of one polarity and the east-west pair has rf and dc voltages of opposite polarity. Hexapoles and octapoles are similar, but have six and eight rods, respectively. They are used as collision and/or storage cells. In tandem mass spectrometers such as triple quadrupoles or Q-Tofs, the quadrupole functions as either a mass spectrometer or in an ion transmission mode. To acquire MS spectra, the quadrupole is operated in the so-called rf mode, in which it acts to transmit the ions from the ion source to the third quadrupole or the TOF where the spectrum is recorded. In daughter ion/product ion MS/MS mode, the quadrupole is used in a static mode to select a particular ion for CAD (collision-activated dissociation). The resulting MS spectrum can be used to determine a partial amino acid sequence of a peptide. In parent ion/precursor ion MS/MS, and neutral loss MS/MS, the quadrupole is scanned in a particular relationship to the third quadrupole or the TOF. These two types of scans may both be used to detect the type and position of a modification.

3. SUMMARY

In this chapter, we have shown that proteomic technologies can be used to detail the effects of extracellular stimuli. During the life cycle of the B. burgdorferi environmental cues including pH changes can have a profound effect on gene expression and the resulting protein profile. Clearly, pH changes that mimic movement from the tick midgut (pH $\sim$ 8.2) to the tick salivary gland (pH $\sim$ 9.5) and the salivary gland to host (pH $\sim$ 6.2) involve protein changes that are critical for adaptation, infection, and persistence. The Borrelia genome encodes approximately 1400 genes, and the possible number of protein species has been estimated to be <4000. Whereas this is considerably less when compared to both the number of genes and protein species in mammals, nonetheless it provides an excellent model system to understand whole-proteome changes. Borrelia provides an excellent system for testing hypotheses and developing new technologies such as isotope-coded affinity tag (ICAT) (39), tandem-affinity purification (TAP) (40), and multidimensional protein identification technology (MudPIT) (41). Further, combining these techniques with cell fractionation, improved detergents, and more sensitive MS technology will likely shift the equilibrium for identifying differences in favor of the investigator.

Inherent technical roadblocks of mammalian systems via the complexities of the genome size and presence of numerous protein species as well as genetic variation present in populations force the development of an experimental design to minimize the possible interpretations of results obtained. We have tailored our proteomic experiments based on the phenotypes of our mouse models (23,25,42,43), the genetic analyses that have identified and positioned loci protective from T1DM (25,26,33), and expertise in specific proteomic techniques (14,15,44). Clearly, use of DIGE has accelerated our results and has successfully identified a valid candidate gene for T1DM protection on Chr. 8. Further, the pursuit of the PMT in the SOD1 of ALR that has increased both the stability and activity of the protein species would not be possible without technologies such as MALDI-ToF, SELDI-ToF, or Q-Tof with CID.

ACKNOWLEDGMENTS

We gratefully acknowledge support of the Juvenile Diabetes Foundation International, National Institutes of Health grants [AI055178 (JAC) and AI056374 (CEM)], and Department of Defense grant ERMS 00035010. We thank Dr. Dale Patterson, Mr. Richard Pintal, Mr. Jeff Baust, and Mrs. Audra Ziegenfuss, as well as Dr. Billy Day and the staff of the Genomics and Proteomics Core Laboratory of The University of Pittsburgh School of Medicine for excellent technical assistance.

REFERENCES

1. Huber LA. Is proteomics heading in the wrong direction? Nat Rev Mol Cell Biol 2003; 4:74–80.
2. Klose J. Protein mapping by combined isoelectric focusing and electrophoresis of mouse tissues. A novel approach to testing for induced point mutations in mammals. Humangenetik 1975; 26:231–243.
3. O'Farrell PH. High resolution two-dimensional electrophoresis of proteins. J Biol Chem 1975; 250:4007–4021.
4. Bjellqvist B, Ek K, Righetti PG, Gianazza E, Gorg A, Westermeier R, Postel W. Isoelectric focusing in immobilized pH gradients: principle, methodology and some applications. J Biochem Biophys Methods 1982; 6:317–339.
5. O'Farrell PZ, Goodman HM, O'Farrell PH. High resolution two-dimensional electrophoresis of basic as well as acidic proteins. Cell 1977; 12:1133–1141.
6. Gygi SP, Corthals GL, Zhang Y, Rochon Y, Aebersold R. Evaluation of two-dimensional gel electrophoresis-based proteome analysis technology. Proc Natl Acad Sci U S A 2000; 97:9390–9395.
7. Gygi SP, Rist B, Gerber SA, Turecek F, Gelb MH, Aebersold R. Quantitative analysis of complex protein mixtures using isotope-coded affinity tags. Nat Biotechnol 1999; 17:994–999.
8. Smolka MB, Zhou H, Purkayastha S, Aebersold R. Optimization of the isotope-coded affinity tag-labeling procedure for quantitative proteome analysis. Anal Biochem 2001; 297:25–31.
9. Gavin AC, Bosche M, Krause R, Grandi P, Marzioch M, Bauer A, Schultz J, Rick JM, Michon AM, Cruciat CM, Remor M, Hofert C, Schelder M, Brajenovic M, Ruffner H, Merino A, Klein K, Hudak M, Dickson D, Rudi T, Gnau V, Bauch A, Bastuck S, Huhse B, Leutwein C, Heurtier HA, Copley RR, Edelmann A, Querfurth E, Rybin V, Drewes G, Raida M, Bouwmeester T, Bork P, Seraphin B, Kuster B, Neubauer G, Superti-Furga G. Functional organization of the yeast proteome by systematic analysis of protein complexes. Nature 2002; 415:141–147.
10. Hunt DF, Henderson RA, Shabanowitz J, Sakaguchi K, Michel H, Sevilir N, Cox AL, Appella E, Engelhard VH. Characterization of peptides bound to the class I MHC molecule HLA-A2.1 by mass spectrometry. Science 1992; 255:1261–1263.
11. Fenn JB, Mann M, Meng CK, Wong SF, Whitehouse CM. Electrospray ionization for mass spectrometry of large biomolecules. Science 1989; 246:64–71.
12. Karas M, Hillenkamp F. Laser desorption ionization of proteins with molecular masses exceeding 10,000 daltons. Anal Chem 1988; 60:2299–2301.
13. Fraser CM, Casjens S, Huang WM, Sutton GG, Clayton R, Lathigra R, White O, Ketchum KA, Dodson R, Hickey EK, Gwinn M, Dougherty B, Tomb JF, Fleischmann RD, Richardson D, Peterson J, Kerlavage AR, Quackenbush J, Salzberg S, Hanson M, van Vugt R, Palmer N, Adams MD, Gocayne J, Weidman J, Utterback T, Watthey L, McDonald l, Artiach P, Bowman C, Garland S, Fujii C, Cotton MD, Horst K, Roberts K, Hatch B, Smith HO,

Venter JC. Genomic sequence of a Lyme disease spirochaete, Borrelia burgdorferi. Nature 1997; 390:580–586.

14. Carroll JA, Garon CF, Schwan TG. Effects of environmental pH on membrane proteins in Borrelia burgdorferi. Infect Immun 1999; 67:3181–3187.

15. Carroll JA, Cordova RM, Garon CF. Identification of 11 pH-regulated genes in Borrelia burgdorferi localizing to linear plasmids. Infect Immun 2000; 68:6677–6684.

16. Flynn JM, Neher SB, Kim YI, Sauer RT, Baker TA. Proteomic discovery of cellular substrates of the ClpXP protease reveals five classes of ClpX-recognition signals. Mol Cell 2003; 11:671–683.

17. Mitchell SJ, Minnick MF. A carboxy-terminal processing protease gene is located immediately upstream of the invasion-associated locus from Bartonella bacilliformis. Microbiology 1997; 143:1221–1233.

18. Leiter EH, Atkinson MA. NOD Mice and Related Strains: Research Applications in Diabetes, AIDS, Cancer, and Other Diseases. In Medical Intelligence Unit Austin. R.G. Landes Company, 1998;208.

19. Rabinovitch A. An update on cytokines in the pathogenesis of insulin-dependent diabetes mellitus. Diab Metab Rev 1998; 14:129–151.

20. Lenzen A, Panten U. Alloxan: History and mechanism of action. Diabetologia 1988; 31:337–342.

21. Heikkila R, Cohen G. Inhibition of biogenic amine uptake by hydrogen peroxide: A mechanism for toxic effects of 6-hydroxydopamine. Science 1971; 172:1257.

22. Murata M, Imada M, Inoue S, Kawanishi S. Metal-mediated DNA damage induced by diabetogenic alloxan in the presence of NADH. Free Radical Biology and Medicine 1998; 25:586–595.

23. Mathews CE, Graser R, Savinov A, Serreze DV, Leiter EH. Unusual resistance of ALR/Lt beta cells to autoimmune destruction: role for beta cell expressed resistance determinants. Proc Natl Acad Sci 2001; 98:235–240.

24. Mathews CE, Leiter EH. Constitutive differences in anti-oxidant defense status distinguish Alloxan Resistant (ALR/Lt) and Alloxan Susceptible (ALS/Lt) mice. Free Radical Biology and Medicine 1999; 27:449–455.

25. Mathews CE, Leiter EH. Resistance of ALR/Lt Islets to free radical mediated diabetogenic stress is inherited as a dominant trait. Diabetes 1999; 48:2189–2196.

26. Mathews CE, Graser RT, Bagley RJ, Caldwell JW, Li R, Churchill GA, Serreze DV, Leiter EH. Genetic analysis of resistance to Type-1 Diabetes in ALR/Lt mice, a NOD-related strain with defenses against autoimmune-mediated diabetogenic stress. Immunogenetics 2003; 55:491–496.

27. Sugiyama F, Churchill G, Higgins D, Johns C, Makaritsis K, Gavras H, Paigen B. Concordance of murine quantitative trait loci for salt-induced hypertension with rat and human loci. Genomics 2001; 71:70–77.

28. Sen S, Churchill G. A statistical framework for quantitative trait mapping. Genetics 2001; 159:371–387.

29. Mathews CE, Leiter EH, Spirina O, Bykhovskaya Y, Fischel-Ghodsian N. The ALR/Lt Mouse: Contributions of the mitochondrial genome in resistance against both chemically-induced and autoimmune diabetes. Submitted PNAS USA.

30. Atkinson MA, Leiter EH. The NOD mouse model of type 1 diabetes: as good as it gets? Nat Med 1999; 5:601–604.

31. Mathews CE, Bagley RJ, Holl T, Serreze DV, Leiter EH. Congenic analysis of H2gx-linked diabetes protection. Diabetes Metab Res Rev 2002; 18:S22.

32. Mathews CE, Suarez-Pinzon W, Baust JJ, Rabinovitch A, Leiter EH. Mechanisms underlying resistance of ALR islets to cytokine destruction. Diabetes Metab Res Rev 2002; 18:S23.

33. Mathews CE, Dunn BD, Hannigan MO, Huang CK, Leiter EH. Genetic control of neutrophil superoxide production in diabetes-resistant ALR/Lt mice. Free Radic Biol Med 2002; 32:744–751.

34. Unlu M, Morgan ME, Minden JS. Difference gel electrophoresis: a single gel method for detecting changes in protein extracts. Electrophoresis 1997; 18:2071–2077.

35. Ablamunits V, Elias D, Cohen IR. The pathogenicity of islet-infiltrating lymphocytes in the non-obese diabetic (NOD) mouse. Clin Exp Immunol 1999; 115:260–267.

36. Schinina ME, Carlini P, Polticelli F, Zappacosta F, Bossa F, Calabrese L. Amino acid sequence of chicken Cu, Zn-containing superoxide dismutase and identification of glutathionyl adducts at exposed cysteine residues. Eur J Biochem 1996; 237:433–439.

37. Yano S. Multiple isoelectric variants of copper, zinc-superoxide dismutase from rat liver. Arch Biochem Biophys 1990; 279:60–69.

38. Mann M, Jensen O. Proteomic analysis of post-translational modifications. Nat Biotechnol 2003; 21:255–261.

39. Gygi S, Rist B, Gerber S, Turecek F, Gelb M, Aebersold R. Quantitative analysis of complex protein mixtures using isotope-coded affinity tags. Nat Biotechnol 1999; 17:994–999.

40. Gavin AC, Bosche M, Krause R, Grandi P, Marzioch M, Bauer A, Schultz J, Rick JM, Michon AM, Cruciat CM, Remor M, Hofert C, Schelder M, Brajenovic M, Ruffner H, Merino A, Klein K, Hudak M, Dickson D, Rudi T, Gnau V, Bauch A, Bastuck S, Huhse B, Leutwein C, Heurtier HA, Copley RR, Edelmann A, Querfurth E, Rybin V, Drewes G, Raida M, Bouwmeester T, Bork P, Seraphin B, Kuster B, Neubauer G, Superti-Furga G. Functional organization of the yeast proteome by systematic analysis of protein complexes. Nature 2002; 415:141–147.

41. Washburn M, Wolters D, Yates 3rd JR. Large-scale analysis of the yeast proteome by multidimensional protein identification technology. Nat Biotechnol 2001; 19:242–247.

42. Mathews CE, Leiter EH. Constitutive differences in anti-oxidant defense status distinguishes Alloxan Resistant (ALR/Lt) and Alloxan Susceptible (ALS/Lt) mice. Free Rad Biol Med 1999; 27:449–455.

43. Leiter EH, Serreze DV, Prochazka M. The genetics and epidemiology of diabetes in NOD mice. Immunol Today 1990; 11:147–149.
44. Carroll JA, El-Hage N, Miller JC, Babb K, Stevenson B. Borrelia burgdorferi RevA antigen is a surface-exposed outer membrane protein whose expression is regulated in response to environmental temperature and pH. Infect Immun 2001; 69:5286–5293.

15

2-D Gel Proteomics Studies on Leukocytes

Ivan C. Gerling and Michael J. Pabst
University of Tennessee Health Science Center, Memphis, Tennessee, U.S.A.

1. INTRODUCTION

Sequencing of the human genome is a critically important advance in medicine. However, the genome of a person is largely static. The effects of infection, trauma, stress, or poor diet do not appear in the genome. The proteome is the set of all proteins actually present in a cell or tissue at a given time. The proteome includes all the posttranslational modifications present in mature proteins, which are often required for the proteins to perform their functions. These modifications also record the effects on proteins of the cell or tissue interacting with its environment. Analysis of the proteome—or proteomics—should provide evidence of pathology and suggest target proteins for drug or nutritional therapy (1).

Circulating leukocytes (or peripheral blood mononuclear cells) are not the cell type most often used for nutritional research, yet they have advantages for studying proteomes. Circulating leukocytes are easy to obtain in large numbers from both normal controls and people with diseases. Furthermore, these cells are nucleated cells, with normal, active metabolic pathways and receptors (such as glucose, lipid, cytokine, and insulin receptors) that make them useful for nutritional research.

Although many promising gel-free approaches to proteome characterization and comparative analysis are being developed, 2-D gel electrophoresis (2) remains the most used method. In this method, proteins are first separated according to their isoelectric points (pI), and then in the second dimension according to their molecular weights. The resulting two-dimensional array of protein spots are then visualized by staining. In comparative studies, the intensities of matched spots in gels from different samples (e.g., control vs. treated) are used to evaluate any possible differential expression of individual proteins. The identity of the protein in a spot of interest can then be determined by mass spectrometry (3).

2-D gel technology has the following advantages compared to nongel approaches: (1) highly complex mixtures like whole-cell homogenates can be studied with a minimum of sample preparation (and associated artifacts and protein loss), (2) mixtures with total protein content from several milligrams to less than a microgram can be studied, (3) 2-D gels produce sample fractionation with high resolution compared to most other protein separation procedures, and thereby allow detection of small changes associated with posttranslational modifications, (4) the individual steps in a 2-D gel proteome analysis can be separated in time and space, allowing flexibility in performing the work, (5) gels are efficient fraction collectors and sample-storage devices that can keep thousands of separated proteins stored indefinitely on a single page (gel) in a notebook, (6) a simple visual inspection can immediately reveal problems in quality or reproducibility of the sample fractionation process, although in many cases, proteome characterization can proceed in spite of slight distortions in separation, and (7) the methodology and equipment is relatively simple and inexpensive, allowing individual investigators to plan, conduct, and finance studies on their regular individual research grants (as long as 2-D gel maps or spot identification services by mass spectrometry are available).

The normal in vivo physiological state of leukocytes is one of floating in liquid suspension. This means that they can be kept in tissue culture under nearly normal physiological conditions. In addition, isolation of leukocytes does not involve harsh manipulations that may damage function or alter the molecular state of cells.

The ability to culture the leukocyte sample is a great advantage for comparative proteomics studies. The most sensitive visualization methods for proteins in 2-D gels are based on radioactive labeling of the proteins. By culturing cells in media containing isotope-labeled amino acids (usually ^{35}S) with high specific activity, it is possible to visualize protein spots with a sensitivity several orders of magnitude higher than the most sensitive protein stains. Alternatively, the use of radioactive phosphate can be used to specifically label phosphorylated proteins, again with an unsurpassed level of sensitivity. Isotope labeling can be combined with pulse-chase approaches to study turnover rates of proteins, and to determine if the change in overall concentration of a protein is due to increased synthesis or decreased degradation.

By combining radioisotope labeling (4) with cutting edge mass spectrometry (5), it is in theory possible to study proteins expressed at as few as 10 copies per cell using cells from <1 ml of blood (see Table 1).

When using blood leukocytes, the main obstacle to studying proteins expressed at as few as 10 copies per cell is not obtaining sufficient sample nor is it the sensitivity of protein visualization and mass spectrometry identification methods. The problem is our inability to simultaneously study proteins whose expression levels differ by a factor of $>10^5$ (6). In 2-D gels of cell homogenates, this problem rears up in the neighborhood of a high-abundance protein like actin. With general protein staining methods, actin covers a fairly large area of the gel around pI 5.5 and mass 55,000 Daltons, yet it is clear that a number of other proteins must also be present in that particular area of the gel. Indeed, if the gel is stained using a phosphoprotein stain (which does not stain actin because that protein is not phosphorylated), a number of spots appear that are not visible using a general protein stain (see Fig. 1, upper-right corner). Thus, the high-abundance proteins may mask and impede the study of less abundant proteins.

Therefore, the main challenge of studying the estimated 50,000 different protein types in a tissue (including posttranslationally modified versions of the same protein) is the separation of high-abundance proteins from low-abundance proteins to allow quantitative study of the latter. In 2-D

TABLE 1 Number of Human Mononuclear Cells Needed to Detect Medium-to Low-Abundance Proteins.*

Sensitivity	1,000 Copies/cell	100 Copies/cell	10 Copies/cell
10 femtomole (10×10^{-15}) (using protein stains and current standard MS)	6×10^6 cells Mononuclear cells from 1 ml of blood	60×10^6 cells Mononuclear cells from 10 ml of blood	600×10^6 cells Mononuclear cells from 100 ml of blood
10 attomole (10×10^{-18}) (using isotope labeling and cutting edge MS)	6×10^3 cells Mononuclear cells from 1 µl of blood	60×10^3 cells Mononuclear cells from 10 µl of blood	600×10^3 cells Mononuclear cells from 100 µl of blood

*Numbers are based on a theoretical calculation of how many molecules are in a mole and presume 100% recovery and use in the detection signal.

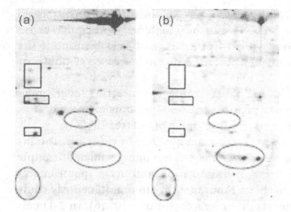

FIGURE 1   Section of a 2-D gel of mouse leukocyte proteins. The large spot in the upper-right corner is actin. The gel was sequentially stained by a phosphoprotein fluorescent stain (Pro-Q-Diamond, Molecular Probes) and SYPRO ruby fluorescent (general protein) stain. (a) The result of the SYPRO ruby image scan. (b) The result of the phosphoprotein image scan. The rectangles indicate some of the proteins that stain substantially stronger with general (SYPRO ruby) protein stain. The ovals indicate some of the proteins that stain substantially weaker with general protein stain. These results illustrate the power and advantages of using multiple stains on the same gel.

gel proteome studies, the two main strategies for dealing with this problem are: (1) sample preseparation and (2) the use of high-resolution gels (so-called zoom gels that expand a narrow range of pI). For studies of proteins expressed at lower levels, sample protein preseparation allows the loading of proteins from a large enough sample to study less abundant proteins without overloading the gel with abundant proteins. Although the 600 million mononuclear cells from 100 ml of blood would, in theory, allow studies of protein that existed at the 10 copies/cell level (see Table 1), we routinely apply proteins from only 10–20 million cells, and we would not recommend loading regular gels with proteins from much more than that number of cells to avoid problems with gel overloading. Yet, if we could enrich a sample by a factor of 60 and remove interfering high-abundance proteins, current methodology/technology would allow us to study proteins expressed at the 10 copies/cell level.

2. EXPERIMENTAL DESIGN

An essential consideration in planning the overall proteomic experiment, and in deciding which samples should be chosen for analysis, is statistics.

Inherent biological variability and the vicissitudes of complex experimentation require that an investigator recognize the importance of good statistics. To have confidence in the results of a proteomics study, there should be at least $N = 3$ to 5 independent experiments, not just multiple gels from the same experiment. Statistical significance is, or course, the minimum requirement. One must consider carefully whether a statistically significant difference is biologically meaningful. On the other hand, a result cannot be biologically meaningful if the difference is not statistically significant. (Like roulette, science is stacked against the player.)

Because there are many proteins under study, one must use analysis of variance and post hoc tests to determine the significance of differences in the abundance of individual proteins. It is invalid to use multiple t-tests, for example, because with multiple t-tests, one in 20 differences will appear significant by chance alone, if P is set at 0.05. We also recommend that the investigator extract the numerical results from the gel analysis program output and run independent statistics. The gel analysis programs are probably accurate, but the programs move too fast for the human eye and brain to follow. It is important for the investigator to look carefully at the data and make judgments about the validity of the analysis and conclusions. As with other discovery approaches to gaining molecular insights, major conclusions cannot be drawn from a single experiment. It is much better to have several experimental indicators supporting a conclusion, preferably from independently obtained samples and different experimental methods.

We have suggestions for the overall experimental design. Do not attempt to study too large a biological change by proteomics, or there will be too many protein changes to rationalize. Consider the following:

1. Comparing healthy liver to necrotic liver would be pointless.
2. Comparing healthy liver to cirrhotic liver would be difficult, because it would be hard to decide which changes represented a cause of cirrhosis, and which changes are central effects, and which are tangential.
3. On the other hand, comparing liver from an animal fed alcohol for two weeks with a control liver may show something interesting about the early changes that ultimately damage the liver.

Be aware that proteins with molecular weight (MW) <10,000 Da will not separate on standard 2-D gels, but will run with the dye front. So, if you want to study peptide hormones or chemokines, for example, you will have to modify the gels or use another proteomics technique.

Do not attempt too many different types of sample in a given 2-D gel experiment. Four or fewer is good; lengthy time courses or scores of drugs are impractical. A good experiment with statistical validity needs multiple replicates of each sample. Current high-capacity, 2-D gel instruments are designed to run a maximum of 12 gels at a time. It has been our experience that each run has its own characteristics, and that it is much easier to compare gel images from within the same run than from different runs. There is a practical limit to the number of gels that can be compared in a computer analysis. Twelve appears to be the practical limit, despite what is claimed by some software makers. Twenty-four may be possible with the latest versions of software and a fanatic computer operator. There is a practical limit to the number of gels that a human can examine carefully, and ensure that the spots are matched, and that the bubbles and other artifacts are corrected. Headaches and personality disorders can result from too many hours of spot analysis. There is also a practical limit to the number of samples and gels that a human can comprehend and think about critically. So, it is important to think through the experiment carefully, and to keep it simple, preferably to a maximum of 12 samples.

3. SAMPLE PREPARATION

When considering which samples to submit to proteomic analysis, it is important to make sure that the underlying biological experiment is good. 2-D gel proteomics is labor-intensive and expensive, and should not be wasted on dubious experimental samples. A typical 2-D gel experiment takes a month and costs at least $1000. There should be assays to prove that the cells showed "activation" or "differentiation" in culture, or that the drug worked. Or there should be good histology to show that the dissection was clean, and that the samples are truly comparable.

Leukocytes are sensitive to becoming activated by mitogens and bacterial products, and this activation will induce substantial changes in the proteome. To avoid problems with inadvertent activation, one must either take steps to avoid contamination, or take steps to ensure a normalization of such contamination to equal levels in all samples. Impure water is the most likely source of contamination, and we recommend the use of bottled water from the pharmacy. Indeed, the changes in the monocyte proteome after exposure to bacterial lipopolysaccaride (LPS) have been a focus of a research program of author Michael Pabst (7).

If cells are placed in tissue culture, LPS contamination is not the only concern regarding culture medium preparation. The choice of culture medium and supplements may drastically influence the function of cells and therefore their expressed proteome. Traditionally, a complete medium such as RPMI 1640 has been popular for long-term cultures of leukocytes. Yet simpler media such as DMEM or EMEM has also been successfully used for studies of cultured leukocytes. Indeed, our studies of monocyte activation by LPS were conducted in a modified Earle's balanced salt solution without any serum supplementation (7). Although the use of as much as 10% fetal bovine serum supplementation of culture medium has been used to enrich culture medium, we recommend the least possible use of serum. The serum proteins may stick to the cells and become a substantial contaminant in the cell proteome studied. If the biological effect of interest does not show up in serum-free media, we suggest use of a low % serum and/or serum-free supplements such as insulin/transferrin/selenium.

Even when working with single-cell suspensions, the release and solubilization of proteins from the sample is not a trivial question. When cells are dissolved directly in first-dimension electrophoresis buffer, a precipitate of material can be seen after centrifugation. Such undissolved material is not observed if the cells are extracted using a tri-reagent such as TRIzol® (Invitrogen). The tri-reagent extracts DNA, RNA, and proteins into separate fractions that can be used for further study. Indeed, in many of our studies, we are interested in characterizing both the transcriptome (the mRNA) and the proteome from the same sample in a process we call "molecular phenotyping." The use of a tri-reagent to extract proteins from cells overcomes the problem of undissolved material, and has allowed us to recover 30–40% more protein from the cell sample than if the cells are dissolved directly into electrophoresis buffer.

Although as many as 10,000 proteins can be separated in specialized, large-format gels (8), most 2-D gel studies on unfractionated samples only study the 1000–2000 most abundant proteins. The technical challenges of large-format, high-resolution gels are often too high to warrant the effort. Similar consideration should be given to the issue of fractionating the sample before 2-D gel studies (9). Although this preseparation will allow the study of less abundant proteins, it may also produce technical challenges and isolation artifacts whose solution may require considerable effort. An individual judgment must be made about how important it is to dig deep into the proteome of a given sample in a given project.

Fractionating the sample can be done at either the cellular or the protein level. Standard methods for isolating membranes, organelles, nuclear fractions, etc. can help focus proteome characterization to a specific subcellular component of interest, and gain higher resolution of proteome characterization of that component (10). Membrane proteins are difficult to analyze, mostly due to issues of solubility, so modified protocols are highly recommended for that particular subcellular fraction (11). If proteomes of several subcellular compartments are characterized, it is possible to gain information about changes in intracellular location of individual proteins that cannot be gained from whole-cell proteome studies.

Protein fractionation can also be done using traditional biochemical methods. Usually, there is little gain in sensitivity by separating according to the same or similar biochemical parameters as those by which proteins are separated in 2-D gel electrophoresis. A number of different affinity separation methods can be used to either fractionate the proteome of a sample or to specifically remove certain abundant proteins like albumin or actin. Subproteomes can also be created based on protein solubility (12,13). A number of commercial kits and matrixes are available, and new products are introduced constantly.

Gels can hold several milligrams of protein, but gels are sensitive to excess salts, which can be a problem with cultured cells. If salt levels vary among samples, the gels will not run alike. The solution to such problems can be to use tri-reagent or acetone precipitation to concentrate proteins and remove salts. If samples are put directly into the first dimension buffer, be sure to centrifuge the samples before starting, to minimize problems associated with DNA and connective tissue.

During sample preparation and throughout the proteomic experiment, wear vinyl gloves and a hairnet to avoid keratin contamination. A mask to prevent sample contamination by breath-borne materials is also useful. This is critical in preparing samples, especially up to the point of running the first dimension. After that point, contaminants will add to background, but will not form spots. However, a general background contamination with keratin may ensure that all protein spots are identified as keratin by mass spectrometry. We confess to having samples in which mass spectrometry showed about a dozen forms of keratin, in addition to the protein of interest.

4. GEL ELECTROPHORESIS

The use of immobilized pH gradient strips has improved both inter- and intralaboratory reproducibility of 2-D gel electrophoresis (14). These precast first-dimension strips have been a critical improvement for 2-D gel proteomics; they are available from Bio-Rad and from Amersham Biosciences. For the first-dimension electrophoresis, it is possible to either: (1) use a strip and buffer that cover a large pI range such as pH 3–10 (linear or nonlinear gradient), or (2) spread the proteins out more in the first dimension by using a more narrow range (zoom gels) that covers only one or a few units of pI (15). The advantage of gaining better protein separation comes at the cost of having to run and analyze more gels, or studying a smaller range of the pI scale. Furthermore, if the choice is to run multiple, narrow-range gels rather than one broad pI-range gel, this will usually consume more sample. For first-dimension separation, a dry pI gradient strip is rehydrated in the dissolved sample and then focusing is performed. New integrated systems simplify this procedure by performing rehydration and focusing in one step (16).

In the second dimension, which is usually SDS-PAGE gel electrophoresis, proteins are separated according to molecular weight. The choices here are between gradient or homogeneous gels and choosing the percentage of acrylamide to use. Precast, plastic-backed gels are commercially available, but can be expensive. The plastic backing can be particularly helpful for handling low-percentage gels that otherwise have a tendency to break during fixing and staining. Manipulation of the percentage of acrylamide in the gels can help increase separation within a specific molecular weight range at the expense of separation in another range. With standard gels using 12–13% acrylamide/bisacrylamide, the separation in the lower MW range may be unnecessarily high and separation in the higher MW range may be less than desired. Use of low-percentage gels or gradient gels can help with this problem, but at the cost of being more technically challenging to produce and handle. An alternative is to use a homogeneous 12 or 13% gel, but to let it run for one or more hours past the time that the dye line exits the bottom of the gel. Although the lowest molecular weight (<15,000 Da) proteins are lost by this approach, those proteins usually do not yield good and reproducible results in 2-D gel proteome studies anyway. The use of several different second-dimension gels for each sample is another possibility, provided that enough sample is available.

5. STAINING PROCEDURES

After proteins have been separated by 2-D gel electrophoresis, they must be visualized in a quantitative or semiquantitative way that allows judgment of relative expression levels between gels (17). As mentioned earlier, the ability to culture leukocytes allows for use of radioisotopes to metabolically label and visualize individual proteins. ^{35}S-Methionine for general proteins and ^{33}P-phosphate for phosphoproteins are most often used to label proteins (18). They are available at high specific activities, and their radiation is of low enough energy to produce small, well-defined spots in the image process. The main disadvantage of these labeling procedures is the need to work with fairly high levels of radioisotopes. Another potential disadvantage is that the isotopes in themselves may induce radiation artifacts (19). Furthermore, in most labs, the ability to visualize proteins by isotope labeling far outperforms the ability to identify faint spots by mass spectrometry. So although a larger number of protein spots may be visualized, few of the additional spots can be identified. Therefore, most labs prefer to use regular protein staining methods for visualization.

The traditional protein-staining procedures for gels have been Coomassie blue stain and silver stain. Both provide a visible image that can be captured using a regular flat-bed computer scanner. Coomassie blue is not as sensitive as silver staining, but silver staining is more prone to overstaining artifacts that may interfere with the relative quantification. The initial problems of compatibility of silver staining with the mass spectrometry methods for protein identification have been solved by modifying the protocols for silver staining (20,21). Colloidal Coomassie blue preparations—such as GelCode Blue (Pierce, Rockford, IL)—give nice uniform staining with low background (22), so that gels loaded with high amounts of protein give a good spot pattern. If abundant sample is available, such gels can be loaded with several mg of proteins, making it easier to identify spots by mass spectrometry.

The new fluorescent protein stains such as SYPRO ruby (Molecular Probes, Eugene, OR) have several advantages relative to traditional staining procedures (23,24). Although their sensitivity is not significantly different from that of silver stains, the dynamic range of these stains is much larger (>1000). As a result, losing the ability to conduct relative quantification is a much less frequent problem with the fluorescent stains. In our hands, the background staining of fluorescent stains is both less intense and more uniform than that of silver stain. This allows for a more precise quantification of the more weakly stained spots in a gel.

In addition to general protein stains, there are several stains that specifically stain subsets of the proteome. Many of the older glycoprotein stains or phosphoprotein stains have the disadvantage of being not very sensitive. Newer fluorescence-based glycoprotein and phosphoprotein stains appear to have higher sensitivity (25,26). Indeed, in experiments using both phosphoprotein fluorescent stain (Pro-Q-Diamond, Molecular Probes) and SYPRO ruby fluorescent (general protein) stain, we found that certain protein spots appeared to stain more intensely with the phosphoprotein stain than with the SYPRO (general protein) stain (Fig. 1).

In some cases, it may be advantageous to conduct several staining procedures on the same gel. In a recent multiplexed proteomics study, we conducted four sequential stains on one gel to gain the maximum information (unpublished data): First, phosphoprotein stain; then glycoprotein stain; followed by SYPRO ruby stain; and at the end, silver stain. This allowed a maximum collection of information from a single gel. Based on these stains, certain spots in the SYPRO-stained images could be assigned preliminary status as phosphoproteins and/or glycoproteins. Using both SYPRO ruby and silver general protein stains allows study of approximately 20–30% more protein spots, because each stain does uniquely stain some spots that are not visible with the other. It is important to always use the silver stain as the last stain in a staining sequence, because silver, for some unknown reason, appears to block the fluorescence-based stains.

In an alternative approach to comparative proteomics in 2-D gels, two samples are stained before electrophoresis with two different fluorescent dyes (Cy dyes, Amersham Biosciences), and are then mixed and applied together to a 2-D gel (27,28). The images from this differential in-gel electrophoresis (DIGE) are then captured at the two corresponding wavelengths. Because the proteins have been subjected to electrophoretic separation in the same gel, the images are a perfect match. An internal standard of a third color can be added to help normalize spot intensities between several of these multicolor gels. For simple comparison of two samples, this method has the advantage that image comparison is not complicated by different local distortions in different gels. If multiple sample sets are analyzed (e.g. five controls vs. five treated), several gels have to be run, and the issues of alignment of images from gels with their own individual distortions again become a complication. The use of an internal third-color standard sample added to all gels may help with such problems. Another problem is that either a low percentage of each protein is bound to the dye, limiting sensitivity; or a high percentage of each protein is bound to dye, causing chemical modification of the bulk of the protein, disturbing movement in the gel and complicating identification by mass spectrometry.

6. IMAGE ANALYSIS

At its best, 2-D gel electrophoresis can be used to judge the upregulation or downregulation of thousands of proteins in one set of samples vs. another set of samples. For this to work, the staining method must be at least semi-quantitative for each protein spot evaluated, such that increased staining intensity is seen always and only when there is an increase in the protein content of that spot. The previously discussed issue of dynamic range of the protein dye is important for that quantification to work. Furthermore, it is important that the images from all gels can be aligned such that it is known which spots are corresponding and which are different between all spots on all gels. With local gel distortions and differences in overall staining intensities of gels, it may become difficult for the computer to assign which spots are identical between several gels. An example of this problem is shown in Fig. 2, in which gels containing proteins from the islets of Langerhans from three different strains of mice are compared. In the upper-left corner of the gel, local distortions make it difficult to judge to what extent protein spots have moved in relative position, whether one spot has disappeared and another reappeared nearby. Another example is shown in Fig. 3, where staining intensities are similar, but a distortion in the second dimension of

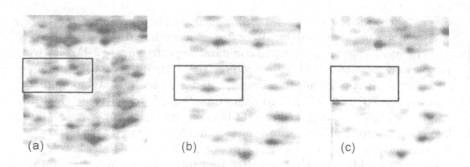

(a)   (b)   (c)

FIGURE 2 Section of silver-stained 2-D gels of mouse islet of Langerhans proteins. Samples obtained from (a) four-week-old NOD, (b) NOD-scid, and (c) NON female mice (Jackson Laboratory, Bar Harbor Maine). Note the stronger overall stain of the gel with the NOD sample (a) relative to the other gels. The box encloses an area where the differences in staining intensity, combined with a local distortion of gel separation, make it difficult for the computer to determine which spots correspond to each other in the three gels. Some manual inspection and spot assignment is required in that area.

these images of the leukocyte proteome from two different strains of mice at two different ages, makes spot assignment difficult for some spots. Ultimately, the identity of two protein spots from different gels may have to be proven by mass spectrometry, but even that may not be enough. A slight change in posttranslational modification of a protein may change its position in a gel, and in reality, create what should be considered a new spot, yet the new spot and the old spot would be identified as identical with most mass spectrometry methods. It is in those situations that the DIGE approach has its greatest strength, because the spots in images from these gels will only separate if there is a real difference in their biochemical properties, not if there is a distortion in the gel.

Although the human eye is adept at analyzing which spots correspond to each other on different gels, this task is remarkably difficult for a computer. Only recently have image analysis programs progressed to the point where they can align images without extensive input (landmarking) and correction by human evaluators. The problem is that many images are imperfect due to small and highly localized imperfections in the protein separation. The image analysis program will often either take these distortions at face value and consider them true differences (because the exact x, y coordinates differ), or be programmed to accept a certain misalignment, which then, in other parts of the gel, will lead to assignment of a match between two spots that are truly different. Ultimately, the most important differences should always be confirmed by a critical visual inspection of the gel images, and not just accepted from a bar graph produced by a computer.

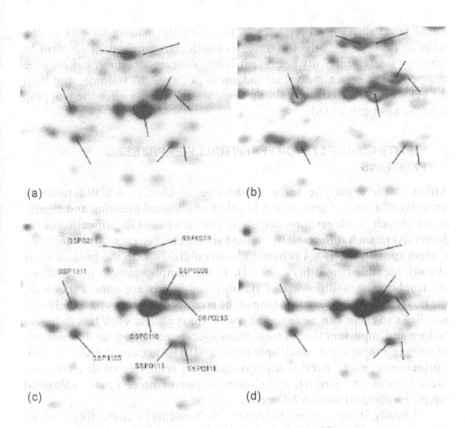

FIGURE 3 Section of silver-stained 2-D gels of mouse leukocyte proteins. Samples were obtained from two-week-old (a, b) and four-week-old (C, D) female NOD (a, c), and NON (b, d) mice (Jackson Laboratory, Bar Harbor Maine). A distortion in the second dimension (b) has changed the relative distances between proteins in the top (ssp0211), middle (ssp0118), and bottom (ssp0111) of the image. This makes it more difficult for the computer to assign which spots in (b) correspond to which spot on the three other gels. But a visual inspection of the spot "constellations" can solve the puzzle.

The computer is superior to the human eye when it comes to the ability to evaluate the relative intensity of spots in gels, in particular if there are significant differences in overall staining intensities between gels (see Fig. 2). Most image analysis programs allow the normalization of spot intensity to the total intensity of all stained spots in the whole gel. This serves the purpose of compensating for overstaining or understaining of one gel relative to other gels, or for small differences in protein loading. Without such normalization, a large number of differential expression differences would be artifacts of differences in total protein loading on gels or the staining procedures. Silver staining is particularly difficult to control to the same overall intensity because it is not a saturation stain, but must be stopped based on an individual judgment of "enough" staining, and then the gel will often continue gaining some stain intensity after the stopping solution has been added.

7. IDENTIFICATION OF DIFFERENTIALLY EXPRESSED PROTEINS

Although there may be some satisfaction in observing a difference in intensity of a stained spot on a 2-D gel, the biological meaning and significance of such an observation cannot be gained without identification of the protein(s) in such differentially stained spots. The fastest way to identify a protein spot on a 2-D gel is to use a map published with the protein spots already identified by others (29–31). If the sample, sample preparation, electrophoresis conditions, and staining conditions are almost identical, the images are usually similar. One of the largest collections of 2-D gel maps and links to them can be found at www.expasy.ch. Yet even this database lacks many important tissues from many species, and often the descriptions of sample preparation, electrophoresis, and staining are sketchy at best. Furthermore, many published maps contain a relatively small number of identified spots, making it less likely that all spots of interest can be identified simply by comparison to a 2-D gel map.

Usually, it is necessary to identify the proteins in interesting spots by mass spectrometry (32). This is done by cutting the spot of interest out of the gel, and cleaving the protein or proteins into fragments by digestion with trypsin or another protease. The masses of the peptide fragments are then characterized using mass spectrometry. Two different approaches are usually used to identify the proteins that produced the peptide fragments. In peptide mass fingerprinting, the masses of all the peptides are determined with precision as high as possible. Information from gel position and mass spectrometry regarding pI and molecular weight (both intact protein and tryptic peptides) can be used to determine the identity of the protein(s) in the spot (33). Two of the most widely used protein databases for this identification are the NCBInr (nonredundant) database maintained by the National Center for Biotechnology Information, and the SWISSPROT database maintained by the University of Geneva and the European Bioinformatics Institute. Several search engines and other proteomics tools are available on the Internet (34), e.g., the PeptIdent on the ExPASy site (www.expasy.ch) or MS-fit and MS-Tag at the Protein Prospector site (prospector.ucsf.edu). These sites also provide links to other databases, and allow scientists to access and process information in an efficient and integrated manner (35,36). Several criteria are used to evaluate the database search output and judge the confidence of the protein identification by peptide mapping (keck.med.yale.edu/prochem/procmald.htm) (37). These criteria are: (1) the number of matching peptides and their deviation from the calculated mass, (2) the percentage of the protein's sequence covered by the matching peptides, (3) the difference in number of matched peptides between each candidate protein, and (4) the agreement of the experimental (from position on the gel) and theoretical pI and MW of the protein. When more than one credible candidate protein is retrieved by the search, further investigation is conducted (e.g., obtaining sequence information on "key" peptides or digestion with a different enzyme). Peptides are selected for sequencing based on whether their amino acid sequence can help rule in or rule out the presence of one or more of the candidate proteins.

The second approach to identify a protein in a spot via its peptide fragments is to use tandem mass spectrometry (MS/MS) to obtain amino acid sequence information on the peptides (38). In this approach, the peptide bonds between amino acids in the peptide are broken, and the masses of the resulting fragments determined. With information on the exact masses of the fragments corresponding to breaking each of the peptide bonds in the peptide, an analysis of the masses from the lowest to the highest mass from either end of the peptide will allow prediction of the amino acid sequence of the peptide. Each step from lowest to highest peptide mass will correspond to a mass increase of exactly the mass of the corresponding amino acid. If masses are determined with high enough accuracy, the only uncertainty left is that of leucine and isoleucine, because these two amino acids have identical mass. In real-world experiments, it is rare to obtain masses of all the fragments of a peptide. But incomplete information may still be enough to get partial amino acid sequences of peptides. A relatively short amino acid sequence is often enough to get an unequivocal identification of a protein. The combination of amino acid sequence and peptide fragment sizes provides a very powerful tool for identification of the protein or proteins in a spot on a gel. An example of a monocyte protein identified by a combination of peptide mass fingerprinting and peptide sequencing by tandem mass spectrometry is shown in Fig. 4 .

Regardless of the method used to identify a protein, it should be remembered that the exact isoform and the degree of posttranslational modification of a given protein in a given spot often remain unclear. In most cases, the investigator does not know whether or where the protein is phosphorylated, glycosylated, etc. Sometimes it is not even possible to determine which of several alternative splicing variants of a protein is present in a specific spot (39). Although such information can be gained by

```
MYTAIPQSGS PFPGSVQDPG LHVWRVEKLK PVPVAQENQG VFFSGDSLYV LHNGPEEVSH LHLWIGQQSS RDEQGACAVL
AVHLNTLLGE RPVQHREVQG NESDLFMSYF PRGLKYQEGG VESAFHKTST GAPAAIKKLY QVKGKKNIRA TERALNWDSF
NTGDCFILDI GQNIFAWCGG KSNILERNKA RDLALAIRDS ERQGKAQVEI VTDGEEPAEM IQVLGPKPAL KEGNPEEDLT
ADKANAQAAA LYKVSDATGQ MNLTKVADSS PFALELLISD DCFVLDNGLC GKIYIWKGRK ANEKERQAAL QVAEGFISRM
QYAPNTQVEI LPQGRESPIF KQFFKDWK
```

MALDI-TOF MS	LCQ MS/MS	Sequence Coverage	
EVQGNESDLFMSYFPR	ANAQAAALYK		
MQYAPNTQVEILPQGR	EGNPEEDLTADK	MALDI-TOF Fingerprint	20%
ERQAALQVAEGFISR	ESPIFK		
QAALQVAEGFISR	SNILER	LCQ Sequence	18%
DSER	TSTGAPAAIK		
		Both	5%
YQEGGVESAFHK =	YQEGGVESAFHK		
DLALAIR =	DLALAIR	Total Coverage	31%

FIGURE 4 Amino acid sequence of Macrophage Capping Protein. Tryptic peptides detected by peptide mass fingerprinting, using the MALDI-TOF mass spectrometer, are shown in *italics*. The identification of Macrophage Capping Protein was made by comparing the masses of the tryptic peptides with the theoretical masses of tryptic peptides from all known human proteins, using Pept-Ident software and the Swiss-Prot database. Peptides sequenced by the Liquid Chromatography Electrospray Quadrupole Ion Trap mass spectrometer (LCQ) are shown in **bold type**. In this instrument, tryptic peptides were separated by liquid chromatography and sprayed into the mass spectrometer, where the masses of the various peptides were measured, and the ions of each peptide were collected and fragmented by collision with helium gas. The masses of the ion fragments were measured for each peptide, and compared with theoretical masses of the fragments predicted from the sequences of tryptic peptides from all known human proteins, using Sequest software. Peptides identified in both instruments are shown in ***bold italics***. The percentages of the total amino acid covered by the peptides detected by each method are shown. On 2-D gels, the spot containing Macrophage Capping Protein was increased 4.4-fold in monocytes exposed to bacterial lipopolysaccharide, making this protein spot interesting for understanding inflammation and resistance to infection (7).

mass spectrometric methods, those investigations may require substantial further effort.

8. SUMMARY

Blood leukocytes are among the easiest cellular biopsies to obtain in humans, and leukocytes can be harvested in large amounts. They are nucleated cells with many different types of relevant receptors and a normal metabolism. They are in their natural environment in liquid cell culture suspension, and can therefore provide physiologically relevant data in pulse-chase and radiolabeling experiments. Although these cells are often used by immunologists, investigators in other fields may find them useful due to the advantages mentioned earlier in the chapter.

Although many interesting new gel-free approaches to proteome studies are being developed, the approach most often used to characterize and compare proteomes is 2-D gel electrophoresis. 2-D gels provide a relatively simple and cost-effective way to characterize and compare expression of over 1000 proteins, as well as detect some of their posttranslational modifications. Although 2-D gel proteome studies may take some learning, the instrumentation and available products for sample preparation, electrophoresis, gel staining, and image analysis have, even in the last few years, made this technique a lot easier to use for people without previous 2-D gel experience. Although the final spot identification does require some specialized knowledge and expensive mass spectrometry equipment, it is not beyond the capability of nonspecialist investigators to conduct protein identification, if they are unable to find a 2-D gel proteome map or a core service to do the work.

Proteomics technology is still developing; currently we can easily study only the most abundant proteins. The techniques of proteomics will continue to evolve rapidly, however, because this science is so important. As mentioned in the introduction, DNA is stable and tells us little about our interaction with the environment. mRNA analysis correlates only partially with protein expression, and mRNA cannot reveal protein modifications. Therefore, the answers to the critical questions in medicine and nutrition are hidden in the proteins. For the foreseeable future, 2-D gel-based proteomics will be an important technique to discover these secrets.

REFERENCES

1. Patterson SD, Abersold RH. Proteomics: the first decade and beyond. Nat Genet 2003; 33 suppl:311–323.
2. Gorg A, Postel W, Domscheit A, Gunther S. Two-dimensional electrophoresis with immobilized pH gradients of leaf proteins from barley (Hordeum vulgare): method, reproducibility and genetic aspects. Electrophoresis 1988; 11:681–692.
3. Patterson SD, Abersold RH. Mass spectrometric approaches for the identification of gel-separated proteins. Electrophoresis 1995; 16:1791–1814.
4. Vuong GL, Weiss SM, Kammer W, Priemer M, Vingron M, Nordheim A, Cahill MA. Improved sensitivity proteomics by postharvest alkylation and radioactive labeling of proteins. Electrophoresis 2000; 21:2594–2605.
5. Andren P, Emmett M, Caprioli R. Micro-Electrospray: zeptomole/attomole per microliter sensitivity for peptides. Am Soc Mass Spectrom 1994; 5:867–869.
6. Corthals GL, Wasinger VC, Hochstrasser DF, Sanchez JC. The dynamic range of protein expression: a challenge for proteomic research. Electrophoresis 2000; 21:1104–1115.
7. Gadgil HS, Pabst KM, Giorgianni F, Umstot ES, Desiderio DM, Beranova-Giorgianni S, Gerling IC, Pabst M. Proteome of monocytes primed with

lipopolysaccharide: analysis of the abundant proteins. Proteomics 3, 2003 (in press).

8. Jungblut P, Thiede B, Zimny AU, Muller E, Scheler C, Wittmann LB, Otto A. Resolution power of two-dimensional electrophoresis and identification of proteins from gels. Electrophoresis 1996; 17:839–847.

9. Badock V, Steinhusen U, Bommert K, Otto A. Prefractionation of protein samples for proteome analysis using reversed-phase high-performance liquid chromatography. Electrophoresis 2001; 22:2856–2864.

10. Huber LA, Pfaller K, Vietor I. Organelle proteomics: implications for subcellular fractionation in proteomics. Circ Res 2003; 92:962–968.

11. Santoni V, Molloy M, Rabilloud T. Membrane proteins and proteomics: un amour impossible? Electrophoresis 2000; 21:1054–1070.

12. Molloy MP, Herbert BR, Walsh BJ, Tyler MI, Traini M, Sanchez JC, Hochstrasser DF, Williams KL, Gooley AA. Extraction of membrane proteins by differential solubilization for separation using two-dimensional gel electrophoresis. Electrophoresis 1998; 19:837–844.

13. Cordwell SJ, Nouwens AS, Verrills NM, Basseal DJ, Walsh BJ. Subproteomics based upon protein cellular location and relative solubilities in conjunction with composite two-dimensional electrophoresis gels. Electrophoresis 2000; 21:1094–1103.

14. Corbett JM, Dunn MJ, Posch A, Gorg A. Positional reproducibility of protein spots in two-dimensional polyacrylamide gel electrophoresis using immobilised pH gradient isoelectric focusing in the first dimension: an interlaboratory comparison. Electrophoresis 1994; 15:1205–1211.

15. Gorg A, Obermaier C, Boguth G, Harder A, Scheibe B, Wildgruber R, Weiss W. The current state of two-dimensional electrophoresis with immobilized pH gradients. Electrophoresis 2000; 21:1037–1053.

16. Gorg A, Obermaier C, Boguth G, Weiss W. Recent developments in two-dimensional gel electrophoresis with immobilized pH gradients: wide pH gradients up to pH 12, longer separation distances and simplified procedures. Electrophoresis 1999; 20:712–717.

17. Patton WF. Detection technologies in proteome analysis. J Chromatogr B Analyt Technol Biomed Life Sci 2002; 771:3–31.

18. Link AJ. Autoradiography of 2-D gels. Methods Mol Biol 1999; 112:285–290.

19. Hu VW, Heikka DS, Dieffenbach PB, Ha L. Metabolic radiolabeling: experimental tool or Trojan horse? (35)S-Methionine induces DNA fragmentation and p53-dependent ROS production. FASEB J 2001; 15:1562–1568.

20. Shevchenko A, Wilm M, Vorm O, Mann M. Mass spectrometric sequencing of proteins silver-stained in polyacrylamide gels. Anal Chem 1996; 68: 850–858.

21. Yan JX, Wait R, Berkelman T, Harry RA, Westbrook JA, Wheeler CH, Dunn MJ. A modified silver staining protocol for visualization of proteins compatible with matrix-assisted laser desorption/ionization and electrospray ionization-mass spectrometry. Electrophoresis 2000; 21:3666–3672.

22. Neuhoff V, Stamm R, Pardowitz I, Arold N, Ehrhardt W, Taube D. Essential problems in quantification of proteins following colloidal staining with coomassie brilliant blue dyes in polyacrylamide gels, and their solution. Electrophoresis 1990; 11:101–117.

23. Steinberg TH, Haugland RP, Singer VL. Applications of SYPRO orange and SYPRO red protein gel stains. Anal Biochem 1996; 239:238–245.

24. Lopez MF, Berggren K, Chernokalskaya E, Lazarev A, Robinson M, Patton WF. A comparison of silver stain and SYPRO ruby protein gel stain with respect to protein detection in two-dimensional gels and identification by peptide mass profiling. Electroporesis 2000; 21:3673–3683.

25. Steinberg TH, Pretty On Top K, Berggren KN, Kemper C, Jones L, Diwu Z, Haugland RP, Patton WF. Rapid and simple single nanogram detection of glycoproteins in polyacrylamide gels and on electroblots. Proteomics 2001; 1:841–855.

26. Martin K, Steinberg TH, Goodman T, Schulenberg B, Kilgore JA, Gee KR, Beechem JM, Patton WF. Strategies and solid-phase formats for the analysis of protein and peptide phosphorylation employing a novel fluorescent phosphorylation sensor dye. Comb Chem High Throughput Screen. 2003; 6:331–339.

27. Unlu M, Morgan ME, Minden JS. Difference gel electrophoresis: a single gel method for detecting changes in protein extracts. Electrophoresis 1997; 18:2071–2077.

28. Tonge R, Shaw J, Middleton B, Rowlinson R, Rayner S, Young J, Pognan F, Hawkins E, Currie I, Davison M. Validation and development of fluorescence two-dimensional differential gel electrophoresis proteomics technology. Proteomics 2001; 1:377–396.

29. Appel RD, Sanchez JC, Bairoch A, Golaz O, Ravier F, Pasquali C, Hughes GJ, Hochstrasser DF. Federated two-dimensional electrophoresis database: a simple means of publishing two-dimensional electrophoresis data. Electrophoresis 1996; 17:540–546.

30. Celis JE, Ostergaard M, Jensen NA, Gromova I, Rasmussen HH, Gromov P. Human and mouse proteomic databases: novel resources in the protein universe. FEBS Lett. 1998; 430:64–72.

31. Sanchez JC, Chiappe D, Converset V, Hoogland C, Binz PA, Paesano S, Appel RD, Wang S, Sennitt M, Nolan A, Cawthorne MA, Hochstrasser DF. The mouse SWISS-2D PAGE database: a tool for proteomics study of diabetes and obesity. Proteomics 2001; 1:136–163.

32. Gevaert K, Vandekerckhove J. Protein identification methods in proteomics. Electrophoresis 2000; 21:1145–1154.

33. Henzel WJ, Billeci TM, Stults JT, Wong SC, Grimley C, Watanabe C. Identifying proteins from two-dimensional gels by molecular mass searching of peptide fragments in protein sequence databases. Proc Natl Acad Sci USA 1993; 90:5011–5015.

34. Pennington SR, Wilkins MR, Hochstrasser DF, Dunn MJ. Proteome analysis: from protein characterization to biological function. Trends Cell Biol 1997; 7:168–173.

35. Bairoch A. Proteome databases. In: Wilins MR, Williams KL, Appel RD, Hochstrasser DF, eds. Proteome Research: New frontiers in Functional Genomics. Berlin: Springer, 1997:93–148.

36. Hoogland C, Sanchez JC, Walther D, Baujard V, Baujard O, Tonella L, Hochstrasser DF, Appel RD. Two-dimensional electrophoresis resources available from ExPASy. Electrophoresis 1999; 20:3568–3571.

37. Jensen ON, Wilm M, Shevchenko A, Mann M. Sample preparation methods for mass spectrometric peptide mapping directly from 2-DE gels. Methods Mol Biol 1999; 112:513–530.

38. Lamond AI, Mann M. Cell biology and the genome project – a concerted strategy for characterizing multiprotein complexes by using mass spectrometry. Trends Cell Biol 1997; 7:139–142.

39. Rappsilber J, Mann M. What does it mean to identify a protein in proteomics? Trends Biochem Sci 2002; 27:74–78.

16

Proteomics: Tools for Nutrition Research in the Postgenomic Era

Arthur Grider

Department of Foods and Nutrition, University of Georgia,
Athens, Georgia, U.S.A.

1 INTRODUCTION

The term proteome originated in 1995 as a descriptor for the full protein complement of the genome of a cell or tissue (1). Proteomic technologies are increasingly recognized as necessary for understanding the role nutrients and drugs play in nutrient and cellular metabolism. With the publishing of the human genome in 2001, the answer to the question, "Where do we go next?" needs to be addressed (2–4). Many believe the answer lies in the proteome. Research utilizing proteomic techniques will augment the data obtained from studying gene expression in response to experimental cellular conditions, and will significantly increase our understanding of the role nutrients play in cellular physiology and biochemistry (5). The use of proteomic techniques allows investigators to analyze global protein expression, and define the functions and interrelationships of proteins within cells or tissues (6).

The transcriptome represents those genes that are expressed in response to the cellular conditions at a given time, and becomes the link between the genotype and the phenotype, as expressed by the proteome (7).

The connection between genotype and phenotype is not linear. Complex epigenetic interactions involving the environment, genes, and gene products result in the cell or tissue phenotype (2). Researchers are fully aware that a single gene product is capable of producing a protein that, due to posttranslational modification, could potentially result in various forms with different isoelectric points and/or masses. The regulation of the expression and post-translational modification of proteins within various cellular compartments will differ. In addition, variations in start and stop sites, or frameshifting, result in providing a single mRNA with the means to produce multiple proteins. Frameshifting has been proposed as a mechanism for HIV-1 synthesis of viral selenoproteins such as glutathione peroxidase, thioredoxin reductase, and env-fs (8–10). Further, there are differences in the turnover of mRNA and protein, resulting in ratios of mRNA: protein significantly different from unity (11). Analysis of mRNA and protein from human liver found that, though β actin exhibited mRNA abundance less than that of γ actin, the protein abundance of β actin was greater than γ actin (mRNA: β = 0.189% vs. γ = 0.215%; protein:β = 1.41% vs. γ = 0.65%). Carbamyl phosphate synthase mRNA comprised only 0.139% of the message, yet the protein abundance level was 2.83% (11). Thus, the disparity between mRNA and protein levels indicates the need for use of proteomic techniques to obtain a more complete picture of cellular and molecular metabolic processes.

2   USE OF PROTEOMIC TECHNIQUES IN NUTRITION RESEARCH

A recent review article proposes that the development of nutrition science in the future will require the use of proteomic techniques (5). Adaptation of these techniques will yield exciting results as novel protein interactions are identified and characterized. The goals for any proteomic technique are to visualize global protein expression of cells or tissues, and to analyze the dynamic changes protein expression undergoes following treatment with various nutrients or pharmacological agents, or to investigate changes in global protein expression occurring in cells and tissues carrying genetic mutations. Two-dimensional gel electrophoresis, combined with mass spectrometry, is currently the primary tool for proteomic studies. However, other approaches are being used, including multidimensional protein identification utilizing liquid chromatography (6,12), isotope-coded affinity tag (ICAT), electrophoretic prefractionation (13–15), and surface-enhanced laser desorption/ionization protein chip arrays combining the retention of proteins on solid-phase chromatographic surfaces with time-of-flight mass spectrometry (16).

Several investigators interested in understanding the mechanism(s) involved in β-cell destruction associated with diabetes have utilized the proteomic techniques, two-dimensional gel electrophoresis and mass spectrometry. Interleukin 1β is thought to play a significant role in the loss of β-cells from pancreatic islets. The global protein expression pattern of rat islets was altered in rats administered interleukin 1β, or chemicals that induce nitric oxide production (17–20). The proteins belonged to several biochemical pathways, including those involved in energy transduction, glycolysis, protein expression and posttranslational modification, signal transduction, and apoptosis (19). Others have used these proteomic techniques to identify disease markers for type I diabetes (21), compositional changes in protein in rat cerebral microvessels (22), and the regulation of gene and protein expression in β-cells by glucose (23).

There are also several reports in the literature utilizing proteome analysis in obesity research. 2DE and mass spectrometry have been used to create proteome maps of adipose tissue (24,25). Others have used proteome analysis to study fatty acid metabolism in obese mice. The nuclear transcription factors—peroxisome proliferator-activated receptors (PPARs)—are activated by fatty acids and their metabolites. PPARα is expressed in the liver and kidney, and its activation induces the proliferation of peroxisomes, and increases mitochondrial and peroxisomal fatty acid oxidation. Proteome analysis of livers from obese diabetic mice (ob/ob) indicated that 16 of 1500 spots detected following 2DE were upregulated with treatment with the peroxisome proliferator WY14,643. Fourteen of these proteins were components of peroxisomal fatty acid metabolism (26). Treatment of these mice with the PPARγ agonist, rosiglitazone, increased peroxisomal enzymes in obese but not lean mice. In comparison, WY14,643 increased peroxisomal enzymes in both lean and obese mice (27,28).

3 USE OF PROTEOMIC TECHNIQUES IN ZINC RESEARCH

My laboratory has applied two-dimensional gel electrophoresis to identify the proteins that are influenced by genes affecting, or affected by, zinc nutriture and metabolism. The general methodology involves the separation of proteins according to their isoelectric point (pI) using either carrier ampholyte or immobilized pH gradient isoelectric focusing (IEF). The IEF gel is placed across the top of a slab gel for the separation of the proteins according to their mass. Following the running of the second dimension, the gel is stained and analyzed for differential protein expression. The differentially expressed proteins are then excised and prepared for mass spectral analysis. The sequences within protein sequence databases are queried for peptide matches and the identities of the proteins of interest are determined. Initial proteomic research from this lab has shown that a human mutation affecting cellular zinc metabolism alters protein expression in human fibroblasts (29). Another study with young rats indicated that the expression of the hippocampal purinergic receptor subunit P2X6 was enhanced with dietary zinc deficiency (30). In these studies, proteins not previously linked with zinc status were identified as differentially expressed. The application of proteomic techniques in nutrition research will be essential to understanding the complex interactions between nutrient status, proteins, and cellular metabolism.

4 DEVELOPMENT OF A PROTEOME FOR ACRODERMATITIS ENTEROPATHICA (AE)

The acrodermatitis enteropathica mutation affects intestinal zinc absorption, and is treated with supplemental zinc (31,32). Skin lesions, hair loss, growth retardation, and diarrhea characterize the disease (33–36). Abnormal lipid metabolism has also been observed (37,38). Partial penetrance of the mutation has also been suggested (39,40). Recent advances in genomic technology have identified hZIP4 as a potential candidate for the biochemical defect in AE (41,42), the protein sequence having been derived from the mRNA sequence in GenBank. Data from pooled fibroblast cell lysates indicating the presence of several proteins affected by the AE mutation are presented susequently. This is an expanded analysis using fibroblasts from three normal and four AE individuals.

4.1 Methods

Human fibroblast cells were purchased from the Coriell Institute for Medical Research Genetic Mutant Cell Repository (Camden, NJ) and from the Montreal Children's Hospital Cell Repository (Montreal, Canada). Normal cells were GM5659D, GM5756A, and GM8680 from the Coriell Institute for Medical Research. The AE cells were GM2814 from the Coriell Institute for Medical Research and WG0575 and WG0576 from the Montreal Children's Hospital Cell Repository. The cells were grown in minimal essential medium (Eagle's) containing 20% fetal calf serum, penicillin, streptomycin, and double the concentrations of essential and nonessential amino acids and vitamins. The fibroblasts were subcultured in 75 cm^2 cell culture flasks at a density of 1.3×10^4 cells/cm^2 to confluence.

The method for the preparation of the total cell lysate has been published previously (29). The composition of the buffers used for preparing the cell lysate is listed in Table 1. The confluent cell layer of each flask was rinsed three times with 10 ml ice-cold rinsing buffer, and then scraped into 0.24 ml boiling sample buffer 1. The cell lysate was transferred into 1.5 ml microfuge tubes, heated for 5 min at 100°C, and chilled on ice for 5 min. Sample

TABLE 1   Buffers for Preparation of Total Cell Lysate

• Rinsing buffer	10 mM Tris HCl, pH 7.4
	150 mM NaCl
• Sample buffer 1	50 mM Tris HCl, pH 8.0
	200 mM DTT
	10.4 mM SDS
• Sample buffer 2	500 mM Tris HCl, pH 8.0
	50 mM $MgCl_2$
	1 g/L DNAase 1
	250 mg/L RNAase A
• Acetone precipitation	13.8 M final concentration
• Sample buffer mix	22.4 mM Tris HCl
	17.6 mM Tris
	7.92 M urea
	2.1 mM SDS
	17.6 mM ampholytes (pH 3–10)
	120 mM DTT
	51 mM Triton X-100

buffer 2 (0.024 ml) was added to digest the nucleic acids and incubated with the cell lysate for 8 min. Acetone was used to precipitate the cellular proteins. Following centrifugation, the supernatant was discarded, the pellet was dried at room temperature for 5 min, and then resuspended in 0.24 ml sample buffer mix. The protein concentration from each cell lysate preparation was determined using the Bradford method (43). The protein concentrations obtained from individual cell lysate preparations were 10 to 12 µg/µL. Pooled samples were prepared by combining equal amounts of protein together per genotype.

Isoelectric focusing was performed using precast carrier ampholyte tube gels (180 mm×1.2 mm; Genomic Solutions, Ann Arbor, MI). Their composition is listed in Table 2 along with the buffers used for this first dimension. Sample overlay buffer was first applied to the tubes, and then 100 µg protein in no more than 50 µL was applied under the overlay buffer. The gels were run at 100 µA per gel for 17.5 h.

Precast Tris/Tricine slab gels were used for the second dimension (Genomic Solutions, Ann Arbor, MI). The buffers used are listed in Table 3. The IEF gels were extruded into equilibration buffer, incubated for 2 min at room temperature, and loaded onto large-format 10% acrylamide precast gels (22 cm×22 cm×1 mm). The gels were run at 25 W/gel at 4°C until the dye front reached within 1 cm of the gel bottom. The separated proteins were

TABLE 2   Carrier Ampholyte Isoelectric Focusing Buffers

• IEF gel buffer	580 mM acrylamide
	9.5 M urea
	32 mM Triton X-100
	5 mM CHAPS
	0.58 g/L ampholytes (pH 3–10)
• Sample overlay buffer	500 mM urea
	3.2 mM Triton X-100
	1 g/L ampholytes (pH 3–10)
	50 mM DTT
• Cathode buffer	100 mM NaOH
• Anode buffer	10 mM phosphoric acid

visualized by silver staining using a modified Rabilloud method (44). Molecular weight (MW) and pI standards were also run to estimate the mass and pI of differentially expressed proteins.

The gels were removed from their glass plates and fixed with buffer containing 6.9 M ethanol and 1.7 M acetic acid for 1 h, then overnight with buffer containing 50 mM glutaraldehyde, 5.2 M ethanol, 8.3 mM potassium tetrathionate, and 500 mM sodium acetate. Following rinsing with 18 MΩ water (4×15 min), the gels were incubated with a solution containing 5.9 mM silver nitrate and 8.3 mM formaldehyde for 30 min. The gel image was developed for 30 min in a solution containing 200 mM potassium carbonate, 5 mM formaldehyde, and 0.05 mM sodium thiosulfate. Development was stopped by incubating the gels for 10 min in a solution containing 410 mM Tris and 350 mM acetic acid. The gels were stored in 270 mM glycerol until scanning to digitize the image. The gels were visually inspected and differentially

TABLE 3   Second-Dimension Slab Gel Buffers

• IEF gel equilibration buffer	112 mM Tris acetate
	5% SDS
	50 mM DTT
	0.01% bromophenol blue
• Cathode buffer	200 mM Tris
	200 mM Tricine
	0.4% SDS
• Anode buffer	25 mM Tris acetate, pH 8.3

expressed proteins were noted. The relative density of the differentially expressed protein spots was measured using the 2-D Advanced densitometry program (Advanced American Biotechnology, Fullerton, CA).

The gel analysis protocol for identifying consistent differences between the different cell lysate preparations was to first run the pooled samples and identify differentially expressed proteins. Following this gel analysis, the individual cell lysate samples were run and each gel was compared to the pooled gel to confirm the expression levels within and between the two cell lysate genotypes. The Student's t-test was used to determine significant differences in expression levels ($p < 0.05$; Statmost, Dataxiom, Los Angeles, CA).

4.2 Results and Discussion

Figure 1 shows the two-dimensional gel images from pooled normal and AE fibroblast cell lines. Thirty-six protein spots were identified as potentially different between normal and AE fibroblast's cell lysates. Of these, six were consistently different between genotypes following individual cell lysate's analysis (Fig. 2; Table 4). The relative densities of these spots are shown in Fig. 3. The differences in expression of these spots was 50% or greater. The spots were excised and submitted for peptide mapping using mass spectrometry; however, the peptide fragments obtained were not sufficient for a peptide mass fingerprint. The silver staining technique that was used was not mass spectrometry-compatible. Future gels will be stained with Sypro Ruby, which is as sensitive as to silver staining and is compatible with mass spectrometry protocols (45,46). The carrier ampholyte IEF system used here exhibited a total protein capacity of 100 µg. Presented in the next section are proteomic data obtained following the use of immobilized pH gradient strips, which have a loading capacity of mg quantities of protein (47,48).

The amino acid sequences of several human zinc transport proteins are present in the National Center for Biotechnology Information database (Table 5). The hZIP4 protein has been identified in a Jordanian family to be defective in those suffering from AE (41,42). The protein that is closest in MW and pI to the theoretical values for hZIP4 is protein number 2. However, peptide mass fingerprinting and/or tandem mass spectrometry sequencing will be necessary to definitively identify any of the differentially expressed proteins and determine whether any of the proteins are hZIP4 or other zinc transporter proteins.

4.3 Caloric Restriction and Dietary Zinc Deficiency Affects Hippocampal Protein Expression

The hippocampus is involved in spatial learning and memory, and contains the highest zinc concentration of all the brain regions (49–51). Zinc is found

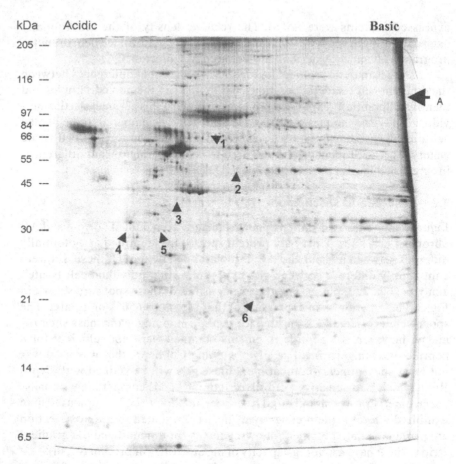

FIGURE 1 Pooled normal fibroblast cell lysate 2DE. Carrier ampholyte IEF was used for the first dimension and a 10% acrylamide denaturing gel was used for the second dimension. Over 800 protein spots were visible following staining of the gel with silver. The presence of cathodic drift is indicated by the black arrowhead labeled A. Following visual inspection of the gels, six protein spots were identified as exhibiting differential expression (arrowheads numerically labeled). The densities of these spots were determined using Universal Software 2-D Advanced (Advanced American Biotechnology, Fullerton CA).

in the mossy fiber system of the hippocampus, and is found with the synaptic vesicles of hippocampal neurons (52–54). One role for zinc in hippocampal function is as a modulator for the NMDA and GABA receptors (55–58). Other roles for zinc include those involving gene expression and protein

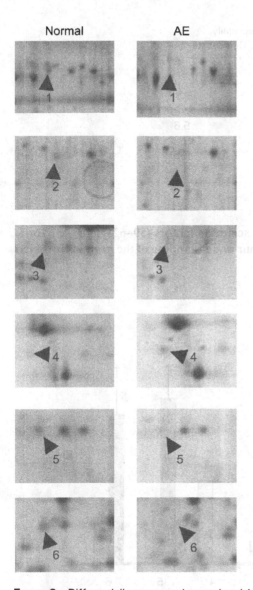

FIGURE 2 Differentially expressed normal and AE cell lysate proteins. Individual normal and AE cell lysate samples were analyzed by two-dimensional gel electrophoresis. Six protein spots were observed to be consistently up- of downregulated between the two genotypes.

TABLE 4 MW and pl of Proteins Differentially
Expressed in Normal and AE Fibroblast Cell Lysate

Protein #	MW (kDa)	pl
1	82	6.1
2	60	6.3
3	43	5.6
4	37	5.3
5	38	5.5
6	22	6.4

synthesis through zinc-finger transcription factors (59–62). The use of proteomic technologies will expand our understanding of the proteins involved in hippocampal zinc metabolism.

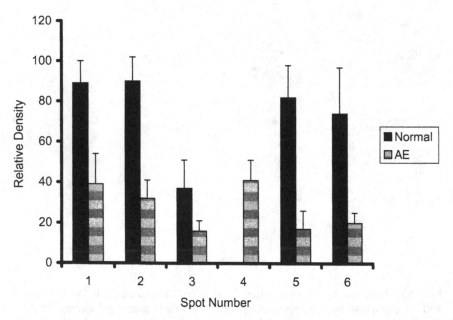

FIGURE 3 Relative densities of differentially expressed proteins from normal and AE cell lysates. The densities of the six spots in Fig. 2 were measured by the 2-D Advanced program (Advanced American Biotechnology, Fullerton, CA). The data are expressed as the mean ± SE. For each spot, the values were significantly different between the genotypes (P ≤ 0.05).

TABLE 5 Human Zinc Transporter Proteins in the National Center for Biotechnology Information Database[a]

Transporter	Number of amino acids	MW	pl
hZIP2	309	33	6.0
hZIP4	626	66	5.7
hZTL1	523	57	6.8
hZNT4	429	47	6.1

[a]MW and pl are theoretical.

In the previous section, carrier ampholyte isoelectric focusing was used in the first dimension. In this section, immobilized pH gradient (IPG) strips were used for the first dimension isoelectric focusing. There are several benefits to using IPG strips, including cathodic drift (compare Fig. 1 to Fig. 4), limitations on the protein loading capacity, and the ease of handling the gels when loading the IPG strip onto the slab gel for the second dimension separation. The reduced resolution of proteins at the basic end of carrier ampholyte isoelectric focusing gels is called cathodic drift (47). The loading capacity of carrier ampholyte isoelectric focusing gels is from 100 μg for analytical gels to 300 μg for preparative gels. Immobilized pH gradient strips can handle protein loads a magnitiude higher (47,48). Finally, the plastic backing on the IPG strips facilitates their easy transfer to the slab gels, without stretching and physical distortion of the gel, including breakage.

4.4 Methods

Male Sprague-Dawley rats (Harlan, Indianapolis, IN) 37–41 days old were divided into three dietary groups (CT, PF and ZD). The ZD group was fed an AIN-93G egg-white-based diet with less than 1 ppm Zn (Dyets, Bethlehem, PA), whereas the PF and CT groups were given AIN-93G with 20 ppm Zn (Dyets, Bethlehem, PA) for 24 days. Upon sacrifice, the hippocampal tissue was quickly dissected out and frozen in liquid nitrogen. Hippocampal tissue was pulverized with protease inhibitors and extracted in sample buffer I (Table 6). The sample mixes were vortexed and incubated at 100°C for 5 min, and sample buffer II (Table 6) was added and incubated on ice for 10 min. Urea was added (1:1 of total weight of the mixture). The sample mix was stirred at room temperature for 30 min and centrifuged at 80,000 g for 30 min. Acetone was added into the supernatant to precipitate the protein. The protein pellet was resuspended in sample loading buffer (Table 6). Protein concentration of the sample was measured by the Bio-Rad protein assay (Bio-Rad, Richmond, CA).

Acidic   Linear IPG pH 3-10   Basic

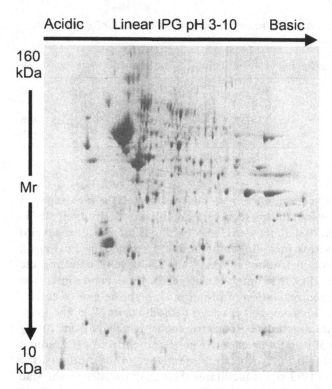

FIGURE 4 Immobilized pH 2DE gradient of rat hippocampus. This gel is from the hippocampus of weanling male rats fed a normal diet ad libiutm containing 20 ppm zinc. Approximately 500 protein spots were visible following staining of the gel with Coomassie blue. Unlike Fig. 1, there is no cathodic drift evident at the basic region of the gel. Gels from control, pair-fed, and zinc-depleted diet groups were analyzed for differential protein expression using Phoretix 2-D software (Nonlinear Dynamics Ltd., Newcastle, UK).

First-dimension isoelectric focusing was performed using immobilized pH gradient gels (IPG strip 17 cm, pH 3–10; Bio-Rad, Richmond, CA). Two mg protein per dietary group were loaded onto the IPG strips. Isoelectric focusing was performed at 250 V for 15 min, followed by 10,000 V for 6 hr at 20°C using the Protean IEF cell (Bio-Rad, Richmond, CA). Immediately prior to loading the IPG strips onto the second-dimension slab gels, the strips were equilibrated in 5 ml equilibration buffer (Table 7) for 20 min. The strip was then dipped into the cathode buffer (Table 7), placed on top of the slab gel, and sealed with a layer of 1% agarose. Second-dimension electrophoresis was performed (12.5% polyacrylamide with dimensions of 22×22×0.1 cm) at 20 W/gel. The gel images were analyzed with Phoretix

TABLE 6 Immobilized pH Gradient Isoelectric Focusing Buffers

• Sample buffer I	0.3% SDS
	200 mM DTT
	28 mM Tris HCl
	22 mM Tris base
	Protease inhibitors
• Sample buffer II	24 mM Tris base
	476 mM Tris HCl
	50 mM MgCl2
	1 mg/ml DNAse I
	0.25 mg/ml RNAse A
• Sample loading buffer	7 M urea
	2 M thiourea
	4% CHAPS
	1% DTT
	2% Pharmalyte 3–10

software (Nonlinear Dynamics Ltd, Newcastle, UK). Spots with staining levels of change greater than 50% were excised manually and processed for mass spectrometry.

The excised gel plugs were incubated with 150 μL of 50 mM NH_4HCO_3 containing 12.2 M acetonitrile. The gel plugs were then dried and rehydrated with 5 μL of 50 mM NH_4HCO_3 containing 0.5 μg trypsin. When the gel piece

TABLE 7 Slab Gel Buffers for Immobilized pH Gradient 2DG

• Equilibration buffer	6 M urea
	30% glycerol
	2% SDS
	0.05 M Tris HCl
	2 mM tributylphosphine
	Bromophenol blue
• Cathode buffer	50 mM Tris base
	384 mM glycine
	6.9 mM SDS
• Anode buffer	25 mM Tris base
	192 mM glycine
	3.5 mM SDS
• Agarose gel	1% agarose in cathode buffer

had swelled to its original size, the plug was covered with 50 mM NH_4HCO_3 and incubated overnight at 30°C. After the addition of 1.5 µL of 880 mM trifluoroacetic acid, the peptides were extracted with 50 mM NH_4HCO_3 containing 14.6 M acetonitrile. The supernatants were dried to approximately 10 µL and analyzed by mass spectrometry.

4.5 Results and Discussion

Figure 4 shows a two-dimensional gel stained with Coomassie blue. Approximately 500 spots were detected and 20 were observed to exhibit differential staining intensities between the control and dietary treatment groups. Currently, four of the 20 spots were further analyzed by mass spectrometry (Figure 5; Table 8).

Spot 1 was identified as µ-crystallin (accession number NP446407), which contains 313 amino acids. Although crystallins make up the lens proteins, they are also found in other tissues. The mouse tissues in which

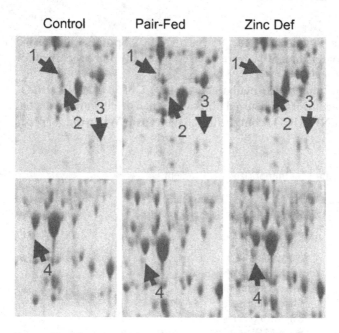

FIGURE 5 Hippocampal protein expression from control, pair-fed, and zinc-deficient male weanling rats. Four protein spots, indicated by the numbered arrows, have identified by mass spectrometry. Each of the spots in this figure exhibited a change in density of 50% or greater. This identities of the spots, along with their normalized volumes, are listed in Table 8.

TABLE 8 Differentially Expressed Proteins of the Hippocampus[a]

Spot #	Control	Pair-fed	Zinc def	Identity
1	1.5	2.1	0.7	μ-cystallin
2	1.2	2.4	1.2	Guanine nucleotide binding protein G_q α-subunit
3	0.2	0.6	0.5	Proteosome subunit
4	0.7	0.7	1.3	Homeobox protein

[a]Volumes normalized.

μ-crystallin mRNA transcripts are found include the hair follicle (highest levels), eye, brain, kidney, heart, lung, and liver (63). In the human, μ-crystallin is also a NADP-regulated, thyroid-binding protein in the kidney (64). The expression of this protein was increased in the calorie-restricted, pair-fed rats and reduced in the zinc-deficient animals. The specific role this protein may play in hippocampal function is not known; however, a link between zinc status and thyroid hormone metabolism has been observed (65,66).

Guanine nucleotide-binding protein G_q α-subunit (accession number P82471) was identified as spot 2. It is a 353-amino acid polypeptide involved in signal transduction. Antidepressant drugs modulate the expression of these G proteins in the rat brain (67). Its expression was increased in the pair-fed, calorie-restricted rat hippocampus and may be part of a neuroadaptive mechanism for this condition. Consumption of a zinc-depleted diet had no apparent effect on the expression of this protein, suggesting that its expression was in response to the stress associated with food restriction.

Spot 3 was identified as a proteasome subunit α-type 6 (accession number NP036098; 246 amino acids). The proteasome, 700 kDa MW, functions as a nonlysosomal proteinase in eukaryotic cells and consists of 28 protein subunits stacked as heptameric rings of α and β subunits (68). Seven copies of the α subunit form two outer rings, which direct assembly, while seven copies of the β subunit form the two inner rings, the catalytic core for the proteasome (69). Increased expression of this protein was observed in both the pair-fed and zinc-deficient rats. However, caloric restriction (pair-feeding) is the control for the reduced food consumption observed in rats consuming a zinc-depleted diet. Therefore, the primary factor for the expression of this proteasome subunit is likely to be due to caloric restriction.

Spot 4 was identified as a homeobox protein HOXC-10 or HOX-3.6 (accession number P31257). The protein has 342 amino acids. These DNA-binding proteins serve to control pattern formation in Drosophila, and their sequences are conserved across species including plants (70). The homeobox proteins reside in the nucleus, with the homeodomain exhibiting similarity with DNA-binding proteins and transcription factors (71,72). This particular protein is not considered to be a zinc-finger transcription factor; however, its increased expression in the zinc-deficient animals may be associated with the transcriptional regulation of other proteins involved in hippocampal development or plasticity.

In this study, the combined use of two-dimensional gel electrophoresis and mass spectrometry resulted in the identification of proteins whose expression was affected by caloric restriction and/or consumption of a zinc-depleted diet. Upon the identification of all 20 of the proteins whose expression was altered by the dietary treatment, a clearer picture of their relationship to each other and to the hippocampal response to diet will arise.

REFERENCES

1. Wasinger VC, Cordwell SJ, Cerpapoljak A, Yan JX, Gooley AA, Wilkins MR, Duncan MW, Harris R, Williams KL, Humphery-Smith I. Progress with gene-product mapping of the mollicutes-mycoplasma-genitalium. Electrophoresis 1995; 16:1090–1094.
2. Anderson NG, Anderson NL. Twenty years of two-dimensional electro-phoresis: past, present and future. Electrophoresis 1996; 17:443–453.
3. Humphery-Smith I, Cordwell SJ, Blackstock WP. Proteome research: Complementary and limitations with respect to the RNA and DNA worlds. Electrophoresis 1997; 18:1217–1242.
4. Fields S. Proteomics – Proteomics in genomeland. Science 2001; 291: 1221–1224.
5. Daniel H. Genomics and proteomics: importance for the future of nutrition research. Br J Nutr 2002; 87(Suppl 2):S305–S311.
6. Lin D, Tabb D, Yates J. Large-scale protein identification using mass spectrome-try. Biochim Biophys Acta 2003; 1646:1–10.
7. Velculescu VE, Zhang L, Zhou W, Vogelstein J, Basrai MA, Bassett Jr DE, Hieter P, Vogelstein B, Kinzler KW. Characterization of the yeast transcriptome. Cell 1997; 88:243–251.
8. Zhao L, Cox AG, Ruzicka JA, Bhat AA, Zhang W, Taylor EW. Molecular model-ing and in vitro activity of an HIV-1-encoded glutathione peroxidase. Proc Natl Acad Sci 2000; 97:6356–6361.
9. Taylor EW, Cox AG, Zhao L, Ruzicka JA, Bhat AA, Zhang W, Nadimpalli RG, Dean RG. Nutrition, HIV, and drug abuse: the molecular basis of a unique role for selenium. J Acquir Immune Defic Syndr 2000; 25(Suppl 1):S53–S61.
10. Taylor EW, Bhat A, Nadimpalli RG, Zhang W, Kececioglu J. HIV-1 encodes a sequence overlapping env gp41 with highly significant similarity to selenium-dependent glutathione peroxidases. J Acquir Immune Defic Syndr Hum Retrovirol 1997; 15:393–394.

11. Anderson L, Seilhamer J. A comparison of selected mRNA and protein abundances in human liver. Electrophoresis 1997; 18:533–537.

12. Link AJ. Multidimensional peptide separations in proteomics. Trends Biotechnol 2002; 20:S8–S13.

13. Righetti PG, Castagna A, Antonucci F, Piubelli C, Cecconi D, Campostrini N, Zanusso G, Monaco S. The proteome: anno domini 2002. Clin Chem Lab Med 2003; 41:425–438.

14. Patton WF, Schulenberg B, Steinberg TH. Two-dimensional gel electrophoresis; better than a poke in the ICAT?. Curr Opin Biotech 2002; 13:321–328.

15. Flory MR, Griffin TJ, Martin D, Aebersold R. Advances in quantitative proteomics using stable isotope tags. Trends Biotechnol 2002; 20:S23–S29.

16. von Eggeling F, Junker K, Fiedler W, Wollscheid V, Durst M, Claussen U, Ernst G. Mass spectrometry meets chip technology: A new proteomic tool in cancer research?. Electrophoresis 2001; 22:2898–2902.

17. Andersen HU, Larsen PM, Fey SJ, Karlsen AE, Mandruppoulsen T, Nerup J. 2-Dimensional gel-electrophoresis of rat islet proteins-interleukin 1-beta-induced changes in protein expression are reduced by L-arginine depletion and nicotinamide. Diabetes 1995; 44:400–407.

18. John NE, Andersen HU, Fey SJ, Larsen PM, Roepstorff P, Larsen MR, Pociot F, Karlsen AE, Nerup J, Green IC, Mandrup-Poulsen T. Cytokine- or chemically derived nitric oxide alters the expression of proteins detected by two-dimensional gel electrophoresis in neonatal rat islets of Langerhans. Diabetes 2000; 49:1819–1829.

19. Larsen PM, Fey SJ, Larsen MR, Nawrocki A, Andersen HU, Kahler H, Heilmann C, Voss MC, Roepstorff P, Pociot F, Karlsen AE, Nerup J. Proteome analysis of interleukin-1 beta-induced changes in protein expression in rat islets of Langerhans. Diabetes 2001; 50:1056–1063.

20. Sparre T, Christensen UB, Larsen PM, Fey SJ, Wrzesinski K, Roepstorff P, Mandrup-Poulsen T, Pociot F, Karlsen AE, Nerup J. IL-1 beta induced protein changes in diabetes prone BB rat islets of Langerhans identified by proteome analysis. Diabetologia 2002; 45:1550–1561.

21. Karlsen AE, Sparre T, Nielsen K, Nerup J, Pociot F. Proteome analysis - A novel approach to understand the pathogenesis of Type 1 diabetes mellitus. Dis Markers 2001; 17:205–216.

22. Mooradian AD, Pinnas JL, Lung CC, Yahya MD, Meredith K. Diabetes-related changes in the protein-composition of rat cerebral microvessels. Neurochem Res 1994; 19:123–128.

23. Schuit F, Flamez D, De Vos A, Pipeleers D. Glucose-regulated gene expression maintaining the glucose-responsive state of beta-cells. Diabetes 2002; 51:S326–S332.

24. Lanne B, Potthast F, Hoglund A, von Lowenhielm HB, Nystrom AC, Nilsson F, Dahllof B. Thiourea enhances mapping of the proteome from murine white adipose tissue. Proteomics 2001; 1:819–828.

25. Sanchez JC, Chiappe D, Converset V, Hoogland C, Binz PA, Paesano S, Appel RD, Wang S, Sennitt M, Nolan A, Cawthorne MA, Hochstrasser DF. The mouse

SWISS-2D PAGE database: a tool for proteomics study of diabetes and obesity. Proteomics 2001; 1:136–163.

26. Edvardsson U, Alexandersson M, von Lowenhielm HB, Nystrom AC, Ljung B, Nilsson F, Dahllof B. A proteome analysis of livers from obese (ob/ob) mice treated with the peroxisome proliferator WY14,643. Electrophoresis 1999; 20:935–942.

27. Edvardsson U, Bergstrom M, Alexandersson M, Bamberg K, Ljung B, Dahllof B. Rosiglitazone (BRL49653), a PPARgamma-selective agonist, causes peroxisome proliferator-like liver effects in obese mice. J Lipid Res 1999; 40:1177–1184.

28. Edvardsson U, Von Lowenhielm HB, Panfilov O, Nystrom AC, Nilsson F, Dahllof B. Hepatic protein expression of lean mice and obese diabetic mice treated with peroxisome proliferator-activated receptor activators. Proteomics 2003; 3:468–478.

29. Grider A, Mouat MF. The acrodermatitis enteropathica mutation affects protein expression in human fibroblasts: analysis by two-dimensional gel electrophoresis. J Nutr 1998; 128:1311–1314.

30. Chu Y, Mouat MF, Coffield JA, Orlando R, Grider A. Expression of P2X(6), a purinergic receptor subunit, is affected by dietary zinc deficiency in rat hippocampus. Biol Trace Elem Res 2003; 91:77–87.

31. Atherton DJ, Muller DPR, Aggett PJ, Harries JT. A defect in zinc uptake by jejunal biopsies in acrodermatitis enteropathica. Clin Sci 1979; 56:505–507.

32. Weismann K, Hoe S, Knudsen L, Sorensen SS. [65]Zinc absorption in patients suffering from acrodermatitis enteropathica and in normal adults assessed by whole-body counting technique. Br J Dermatol 1979; 101:573–579.

33. Margileth AM. Acrodermatitis enteropathica: Case report and review of literature. Am J Dis Child 1963; 105:285–291.

34. Rodin AE, Goldman AS. Autopsy findings in acrodermatitis enteropathica. Am J Clin Pathol 1969; 51:315–322.

35. Moynahan EJ. Acrodermatits enteropathica: A lethal inherited human zinc-deficiency disorder. Lancet 1974; 8:399–400.

36. Schneider JR, Fischer H, Feingold M. Picture of the month: Acrodermatitis enteropathica. Am J Dis Child 1991; 145:211–212.

37. Walldius G, Michaelsson G, Hardell L-I, Aberg H. The effects of diet and zinc treatment on the fatty acid composition of serum lipids and adipose tissue and on serum lipoproteins in two adolescent patients with acrodermatitis entero-pathica. Am J Clin Nutr 1983; 38:512–522.

38. Cunnane SC, Krieger I. Long chain fatty acids in serum phospholipids in acrodermatitis enteropathica before and after zinc treatment: A case report. J Am Coll Nutr 1988; 7:249–250.

39. Garretts M, Molokhia M. Acrodermatitis enteropathica without hypozincemia. J Pediatr 1977; 91:492–494.

40. Mack D, Koletzko B, Cunnane S, Cutz E, Griffiths A. Acrodermatitis entero-pathica with normal serum zinc levels: diagnostic value of small bowel biopsy and essential fatty acid determination. Gut 1989; 30:1426–1429.

41. Wang K, Pugh EW, Griffen S, Doheny KF, Mostafa WZ, al-Aboosi MM, el-Shanti H, Gitschier J. Homozygosity mapping places the acrodermatitis enteropathica gene on chromosomal region 8q24.3. Am J Hum Genet 2001; 68:1055–1060.

42. Wang K, Zhou B, Kuo YM, Zemansky J, Gitschier J. A novel member of a zinc transporter family is defective in acrodermatitis enteropathica. Am J Hum Genet 2002; 71:66–73.

43. Bradford MM. A rapid and sensitive method for the quantitation of microgram quanities of protein utilizing the principle of protein-dye binding. Anal Biochem 1976; 72:248–254.

44. Rabilloud T. A Comparison between low background silver diammine and silver-nitrate protein stains. Electrophoresis 1992; 13:429–439.

45. Lopez MF, Berggren K, Chernokalskaya E, Lazarev A, Robinson M, Patton WF. A comparison of silver stain and SYPRO Ruby protein gel stain with respect to protein detection in two-dimensional gels and identification by peptide mass profiling. Electrophoresis 2000; 21:3673–3683.

46. Berggren K, Chernokalskaya E, Steinberg TH, Kemper C, Lopez MF, Diwu Z, Haugland RP, Patton WF. Background-free, high sensitivity staining ofproteins in one- and two-dimensional sodium dodecyl sulfate-polyacrylamide gels using a luminescent ruthenium complex. Electrophoresis 2000; 21:2509–2521.

47. Bjellqvist B, Ek K, Righetti PG, Gianazza E, Gorg A, Westermeier R, Postel W. Isoelectric focusing in immobilized pH gradients: principle, methodology and some applications. J Biochem Biophys Methods 1982; 6:317–339.

48. Hanash SM, Strahler JR, Neel JV, Hailat N, Melhem R, Keim D, Zhu XX, Wagner D, Gage DA, Watson JT. Highly resolving two-dimensional gels for protein sequencing. Proc Natl Acad Sci 1991; 88:5709–5713.

49. Frederickson CJ, Suh SW, Silva D, Thompson RB. Importance of zinc in the central nervous system: the zinc-containing neuron. J Nutr 2000; 130: 1471S–1483S.

50. Morris RG, Garrud P, Rawlins JN, O'Keefe J. Place navigation impaired in rats with hippocampal lesions. Nature 1982; 297:681–683.

51. Steele RJ, Morris RG. Delay-dependent impairment of a matching-to-place task with chronic and intrahippocampal infusion of the NMDA-antagonist D-AP5. Hippocampus 1999; 9:118–136.

52. Frederickson CJ. Neurobiology of zinc and zinc-containing neurons. Int Rev Neurobiol 1989; 31:145–238.

53. Hesse GW. Chronic zinc deficiency alters neuronal function of hippocampal mossy fibers. Science 1979; 205:1005–1007.

54. Slomianka L. Neurons of origin of zinc-containing pathways and the distribution of zinc-containing boutons in the hippocampal region of the rat. Neurosci 1992; 48:325–352.

55. Smart TG, Moss SJ, Xie X, Huganir RL. GABAA receptors are differentially sensitive to zinc: dependence on subunit composition. Br J Pharmacol 1991; 103:1837–1839.

56. Xie X, Smart TG. A physiological role for endogenous zinc in rat hippocampal synaptic neurotransmission. Nature 1991; 349:521–524.

57. Westbrook GL, Mayer ML. Micromolar concentrations of Zn++ antagonize NMDA and GABA responses of hippocampal neurons. Nature 1987; 328:640–643.
58. Huang EP. Metal ions and synaptic transmission: Think zinc. Proc Natl Acad Sci 1997; 94:13386–13387.
59. Honkaniemi J, States BA, Weinstein PR, Espinoza J, Sharp FR. Expression of zinc finger immediate early genes in rat brain after permanent middle cerebral artery occlusion. J Cereb Blood Flow Metab 1997; 17:636–646.
60. Nahm WK, Noebels JL. Nonobligate role of early or sustained expression of immediate-early gene proteins c-fos, c-jun, and Zif/268 in hippocampal mossy fiber sprouting. J Neurosci 1998; 18:9245–9255.
61. Honkaniemi J, Sharp FR. Prolonged expression of zinc finger immediate-early gene mRNAs and decreased protein synthesis following kainic acid induced seizures. Eur J Neurosci 1999; 11:10–17.
62. Fenster SD, Chung WJ, Zhai R, Cases-Langhoff C, Voss B, Garner AM, Kaempf U, Kindler S, Gundelfinger ED, Garner CC. Piccolo, a presynaptic zinc finger protein structurally related to bassoon. Neuron 2000; 25:203–214.
63. Aoki N, Ito K, Ito M. mu-Crystallin, thyroid hormone-binding protein, is expressed abundantly in the murine inner root sheath cells. J Invest Dermatol 2000; 115:402–405.
64. Segovia L, Horwitz J, Gasser R, Wistow G. Two roles for mu-crystallin: a lens structural protein in diurnal marsupials and a possible enzyme in mammalian retinas. Mol Vis 1997; 3:9.
65. Nishiyama S, Futagoishi-Suginohara Y, Matsukura M, Nakamura T, Higashi A, Shinohara M, Matsuda I. Zinc supplementation alters thyroid hormone metabolism in disabled patients with zinc deficiency. J Am Coll Nutr 1994; 13:62–67.
66. Freake HC, Govoni KE, Guda K, Huang C, Zinn SA. Actions and interactions of thyroid hormone and zinc status in growing rats. J Nutr 2001; 131:1135–1141.
67. Lesch KP, Manji HK. Signal-transducing G proteins and antidepressant drugs: evidence for modulation of alpha subunit gene expression in rat brain. Biol Psychiatry 1992; 32:549–579.
68. Elenich LA, Nandi D, Kent AE, McCluskey TS, Cruz M, Iyer MN, Woodward EC, Conn CW, Ochoa AL, Ginsburg DB, Monaco JJ. The complete primary structure of mouse 20S proteasomes. Immunogenetics 1999; 49:835–842.
69. Zwickl P, Kleinz J, Baumeister W. Critical elements in proteasome assembly. Nat Struct Biol 1994; 1:765–770.
70. Singh G, Kaur S, Stock JL, Jenkins NA, Gilbert DJ, Copeland NG, Potter SS. Identification of 10 murine homeobox genes. Proc Natl Acad Sci 1991; 88:10706–10710.
71. Ko HS, Fast P, McBride W, Staudt LM. A human protein specific for the immunoglobulin octamer DNA motif contains a functional homeobox domain. Cell 1988; 55:135–144.
72. Nelson C, Albert VR, Elsholtz HP, Lu LI, Rosenfeld MG. Activation of cell-specific expression of rat growth hormone and prolactin genes by a common transcription factor. Science 1988; 239:1400–1405.

17

Proteomic Analysis of the Adipocyte Secretome

Nieves Ibarrola and Akhilesh Pandey

McKusick-Nathans Institute of Genetic Medicine and the Department of
Biological Chemistry, Johns Hopkins University, Baltimore,
Maryland, U.S.A.

Chaerkady Raghothama

Institute of Bioinformatics, International Technology Park Ltd.,
Bangalore, India

1. INTRODUCTION

Adipose tissue constitutes the main site for the storage of energy in the form
of fat. Growing concerns about the increase of obesity and associated metabolic disorders has intensified the need to understand the molecular and
cellular mechanisms regulating adipose tissue biology. Adipose tissue is
now clearly accepted as a major secretory and endocrine organ. Proteins
secreted by adipocytes are intimately involved in a host of paracrine and
autocrine functions, including the regulation of preadipocyte proliferation
and differentiation. The study of adipocyte secretome is therefore of prime
interest for the development of preventive and therapeutic strategies that
seek to target diseases such as diabetes and obesity.

The purpose of this chapter is provide an overview of the adipocyte as a secretory organ and to describe the genomic and proteomic approaches that have been used for the study of the adipocyte secretome. Finally, we will discuss some of the novel advances in proteomics that could be employed to dissect the adipocyte secretome in greater detail.

2. ADIPOSE TISSUE AS AN ORGAN

Two types of adipose tissues exist in mammals: white adipose tissue (WAT) and brown adipose tissue (BAT). Although BAT in hibernating mammals plays an important role in thermogenesis and is maintained throughout the life span, in humans its importance seems to be limited to the early periods of life where it plays a role in compensatory thermogenesis (1). During human fetal and newborn life, BAT constitutes between 2–5% of the body weight and is localized to very defined areas; however, its abundance decreases in most sites in the adult. In contrast, WAT represents as much as 20% of the body weight in men and 25% of the body weight in women, thus being the largest storage of energy in the body. Until now, this tissue was considered to be a mere fat depot, but the last decade has proven this view to be far from reality. The discovery of a large variety of secreted proteins that exert pleiotropic endocrine and metabolic functions has raised the adipose organ to a central position in the regulation of energy balance and body homeostasis. Other secondary roles attributed to WAT are thermal regulation and protection against trauma.

A large deposit for energy storage has clear advantages for survival and reproduction during long periods of starvation; however, an excessive energy intake relative to energy expenditure leads to obesity and associated disorders such as diabetes, hypertension, cancer, atherosclerosis, and gall bladder disease (2,3). The increasing incidence and prevalence of obesity in western societies has made it necessary to understand adipocyte biology, as it may help in the design and development of appropriate therapies and prevention strategies (3–5).

Food provides energy mainly in the form of carbohydrates and fat. Glucose can be readily used by all organs and is the only source of energy for the brain. Except for the brain, most organs can store only a limited amount of glucose in the form of glycogen for their own consumption. The excess of glucose in blood is converted into fatty acids and subsequently into triglycerides mostly by the liver. Triglycerides from the diet or produced by the liver are transported in the blood in the form of chylomicrons and very low density lipoproteins. In the adipose tissue, insulin-dependent lipoprotein lipase catalyzes the hydrolysis of triglycerides to glycerol and free fatty acids (FFAs), which enter the adipocyte and are converted into triacylglycerides, the main form of fat storage. Under fasting conditions, adipose tissue supplies the energy necessary to maintain the energy balance. The fat stored in the adipocyte is converted by triacylglycerol lipase into glycerol and FFAs. The liver converts FFAs into glucose for release into blood.

3. BIOLOGY OF ADIPOGENESIS

The principal constituent of WAT is the adipocyte. In addition, adipose tissue contains stromal and vascular cells including fibroblasts, leukocytes, macrophages, and preadipocytes. The fat is largely stored in the form of triacylglycerides in a single large droplet surrounded by a ring of cytoplasm. The cell nucleus and the remaining organelles are displaced against the plasma membrane. Fat constitutes approximately 60–85% of the weight of white adipose tissue, with 90–99% as triacylglycerides and the rest as diacylglycerides, FFAs, cholesterol, phospholipids, and small quantities of cholesterol esters and monoglycerides.

Obesity can result from adipocyte hypertrophy as well as hyperplasia. It is now well recognized that adipogenesis can occur throughout the lifetime of humans. The capacity to increase the adipocyte number is retained in adulthood, as has been shown by several rodent obesity models and suggested by several studies in humans where preadipocytes were isolated from adult tissues (6–8).

Clonal cell lines and primary cultures have been extremely helpful in the elucidation of the molecular and developmental pathways of adipogenesis. The details of the commitment process of embryonic stem cell precursors to the adipose lineage is obscure due to the difficulty in studying embryonic stem cells and the lack of specific markers for adipocyte precursors. Stem cells derived from tissues of mesenchymal origins such as bone marrow and muscle have been shown to retain the potential to generate adipocytes (9–11). Other studies have shown that stromal cells derived from adipose tissue can also give rise to a variety of tissues of mesenchymal origin (12,13). Studies using cell lines or stem cells of mesenchymal origin have shown different lineage plasticity that depends on the differentiation state. Cultures of the fibloblastlike CH310T1/2 and Swiss 3T3 cell lines or the osteoclastlike BMS2 cell line have different cell lineage differentiation capabilities, including the potential to differentiate into adipocytes (14,15).

The most understood adipogenic process is the process of adipocyte differentiation. The study of this process has been enormously facilitated by the use of cell lines that resemble the fibloblastic stromal preadipocytes that have been used to reproduce the events occurring during differentiation in vivo (16–19). Primary cell cultures have also been used to confirm the results obtained with cell lines and to study the regional differences of the adipocyte differentiation process. Studies using 3T3-L1 and 3T3-F442A cell lines have provided much of the knowledge about the adipocyte differentiation process. 3T3-L1 and 3T3-F442A cell lines were isolated from Swiss 3T3 mouse embryo lines and are thought to be already committed preadipocytes (20,21). 3T3-L1 can differentiate into clusters of adipocytes if maintained long enough in cell culture medium containing fetal calf serum. This process can be accelerated by the addition to a postconfluent cell culture of a differentiation-inducing cocktail (DMI) containing dexamethasone, methylisobutylxanthine (a phosphodiesterase inhibitor that increases the concentration of cAMP), and insulin (22). The fact that subcutaneously implanted 3T3-F442A cells give rise to adipose tissue indistinguishable from the endogenous WAT pads (21,23,24) proves the value of these in vitro models and represents a new approach to the study adipose tissue biology.

The differentiation process of preadipocytes can be divided into three well-defined steps: exit from cell cycle, clonal expansion, and differentiation into adipocytes. Exit from cell cycle is achieved in vitro by growing postconfluent preadipocyte cultures for two days, followed by treatment with the DMI cocktail described earlier. Clonal expansion occurs during the ensuing two days in which the cells undergo two rounds of division and stop dividing. Although this process is required in vitro, the requirement of clonal expansion in vivo is not clear (25,26). During the next several days, the cells gradually acquire the phenotypic features that are typical of adipocytes, as shown in Fig. 1. Shown here are adipocytes stained by oil red at different time points of their differentiation process.

Day 1   Day 3   Day 5

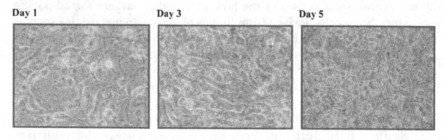

FIGURE 1 Differentiation of 3T3-L1 preadipocytes into adipocytes. 3T3-L1 cells were cell cycle-arrested by growth in medium with serum until confluence and further cultured for two days. Clonal expansion was induced by treatment with a differentiation-inducing cocktail: dexamethasone, methylisobutylxanthine, and insulin (DMI). After two days of clonal expansion, the cells were grown for an additional two days in medium containing serum and insulin. Cells were grown in medium with serum for an additional day. Parallel cell cultures at days 1, 3, and 5 were stained with Oil-red to visualize lipid accumulation.

4. THE ADIPOCYTE SECRETOME

Our perception of adipose tissue has changed drastically from its being a mere storage site for fat to a highly active endocrine and paracrine organ. Although lipoprotein lipase has been known to be secreted by adipocytes for a long time, and several studies have suggested endocrine roles of the adipose tissue, it was not until the discovery of leptin in 1994 that adipose tissue caught the attention of the scientific community (27–29). Table 1 shows a listing of the various molecules secreted from adipocytes that have autocrine and/or paracrine effects. The central nervous system and several other organ systems are directly modulated by molecules secreted by the adipocytes as shown in Fig. 2.

One of the modes of regulating the energy balance is the action of these molecules on the central nervous system—a major center for integrating different metabolic signals from different organs. We will briefly discuss the effects of leptin on the regulation of food intake. At the hypothalamus, several neuronal centers are responsible for the integration of signals and the activation of effector pathways that regulate food intake and energy homeostasis. Neurons of the arcuate nucleus are the primary targets of leptin and insulin. These two potent signals form part of a negative feedback loop that results in the modulation of feeding behavior and energy expenditure. Leptin was identified as a product of the obese gene that is truncated in the obese homozygous (*ob/ob*) mutant mice. Leptin is secreted almost exclusively by adipocytes and senses the body energy reserves. Activation of leptin receptors induces the expression of two anorexigenic peptides, POMC (pro-opiomelanocortin), which is the precursor of the anorexigenic peptide, α-melanocyte-stimulating hormone (α-MSH), and CART (cocaine- and amphetamine-related transcript), at the same time reducing the expression of two orexigenic neuropeptides, neuropeptide Y (NPY) and agouty-related protein (AgRP) (30–33). Thus, an increase of leptin levels leads to a reduction of appetite and an increase in energy expenditure (34–37).

5. GENOMIC METHODS FOR ANALYZING THE ADIPOCYTE SECRETOME

The use of high-throughput genomic methods such as microarrays and signal sequence trapping have contributed greatly to the identification of the factors secreted by the adipocytes and also to the discovery of new genes and their expression patterns in adipose tissue. DNA microarrays are powerful tools to study mRNA expression levels of thousands of genes simultaneously and changes in the expression profile under different biological

TABLE 1   Molecules Secreted by Adipose Tissue[a]

Category	Molecule
1. Extracellular matrix and related proteins	Procollagen alpha-2(I)
	Procollagen alpha C-proteinase enhancer protein
	Type I, III, IV, VI, XV collagen alpha-1
	Type IV collagen alpha-2
	Fibronectin
	Cysteine-rich glycoprotein (SPARC/osteonectin)
	Laminin
	Entactin/Nidogen
	Fibulin-2
	Lysyl oxidase
	Dystroglycan
	Matrix metalloproteinases 2–9, 13, and 14
	Tissue inhibitor of matrix metalloproteases (TIMP-2,3)
2. Acute phase response	α1-Acid glycoprotein
	Haptoglobin
	Serum amyloid A3
3. Complement pathway	Adipocyte complement-related protein (ACRP30, GBP-28, apM1, AdipoQ, Adiponectin)
	Acylation-stimulating protein (ASP)
	Complement factor B
	Complement factor C3
	Complement H
	Complement factor D (Adipsin)
4. Cardiovascular	Angiotensinogen
	Angiopoietin-2
	Plasminogen activator inhibitor-1 (PAI-1)
	PPARgamma angiopoietin-related (PGAR/FIAF)
5. Lipid metabolism enzymes, binding and transfer proteins.	Lipoprotein lipase (LPL),
	Cholesterol ester transfer protein (CETP)
	Phospholipid transfer protein (PLPT)
	Apoprotein E (Apo E)
	FK506-binding protein (FKBP23)

(continued)

TABLE 1 (Continued)

Category	Molecule
6. Obesity	Leptin
	Agouty signaling protein (ASIP)
7. Cytokines	Tumor necrosis factor (TNF)
	Interleukins: IL-6, IL-16
	Macrophage migration inhibitory factor (MIF)
8. Growth factors	Transforming growth factor β (TGFβ)
	Insulin-like growth factor (IGF-1)
	Vascular endothelial growth factor (VEGF)
	Macrofage colony-stimulating factor (MCSF)
9. Other secreted proteins	Resistin
	Cystatin C (cysteine protease inhibitor)
	Neutrophil gelatinase-associated lipocalin precursor (NGAL)
	PEDF/SDF-3, SDF-1
	Calumenin
	Gelsolin
	Colligin-1/Hsp 47
	Hippocampal cholinergic-neurostimulating peptide precursor protein (HCNP)
	Sulfated glycoprotein (Sgp1)
	Epithelin
	Disulfide isomerase-related protein (Erp72)
	Interferon receptor-soluble isoform (IFNAR2)
10. Miscellaneous molecules	
Ecosanoids	Prostaglandin E2 (PGE2)
	Prostaglandin I2 (PI2)
Fatty acids	Free fatty acids
Corticosteroids	Cortisol
Sex steroids	Testosterone
	Oestradiol

[a]A representative compilation of proteins secreted by adipocytes has been classified according to molecular, biochemical, and functional relationships. (Adapted from Refs. 50, 61, 77, and 78)

conditions. This technology has been applied to the study of the adipocyte transcriptome, its changes during the differentiation process, and the regulation by different growth factors and pathological conditions (38–47). Classification of mRNAs expressed by the adipose tissue has shown that the proportion of mRNA-encoding secreted proteins is much higher in adipose

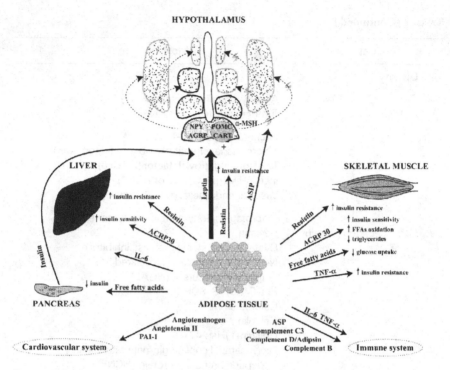

FIGURE 2 Metabolic roles of adipose tissue. An overview of proteins secreted by the adipocytes that are involved in the regulation of food intake and energy homeostasis. The arrows indicate the main target organs or systems. The negative-feedback loop exerted by leptin and insulin over the neurons at the arcuate nucleus of the hypothalamus is represented. (Adapted from Ref. 36, 79, and 80)

tissue than in a wide range of other human tissues, again revealing this organ as the biggest endocrine organ in the body. The signal sequence trapping method takes advantage of the presence of signal peptides, allowing the specific identification of secreted and transmembrane proteins (48,49). Tsuruga and colleagues used this method to identify 50 secreted proteins from a 3T3-L1 adipocyte cDNA library demonstrating the utility of this approach (50).

6. PROTEOMICS METHODS FOR ANALYZING THE ADIPOCYTE SECRETOME

Proteomic approaches are necessary to complement the study of the adipocyte secretome. With respect to the identification of secreted proteins, genomic studies do not account for protein heterogeneity derived from posttranslational modifications and protein cleavage, both important in the determination of the protein activity. In addition, expression profiling studies based on DNA microarrays are biased by the fact that the levels of mRNA and protein are not always correlated (51–53). Therefore, proteomic approaches are needed to complement genomic methods to characterize the adipocyte secretome fully.

Several different major proteomic approaches have been used for large-scale studies of the adipocyte proteome and secretome. These include subtractive antibody screening and various mass spectrometry-based approaches. Subtractive antibody screening relies on the generation of antibodies that specifically recognize cell surface or secreted proteins. A polyclonal antiserum is raised against target cells and subsequently subtracted by incubation with different tissues or cells. This has been used to clone cDNAs encoding secreted and plasma membrane proteins induced during adipocyte differentiation (54). Mass spectrometry-based approaches can be broadly divided into two categories based on whether two-dimensional (2-D) gel electrophoresis is used for separation of the protein mixture.

In the method involving 2-D gel electrophoresis, the spots are visualized by staining, followed by excision and digestion with a specific endoprotease and identification of the proteins by mass spectrometry. The development of databases of 2-D-gel electrophoresis images and software tools for cross comparison of samples has supplied additional values and favored its use in differential proteomics (55,56). Analysis of the proteome of WAT and BAT using 2-D-gel electrophoresis (57,58) is available in the SWISS-2D PAGE database (59). A major drawback of this technique is the difficulty in detecting certain classes of proteins such as very large or very small proteins, or very acidic or basic proteins. Other limitations, when different protein samples are compared, are the low reproducibility between gels and the fact that several proteins might comigrate, giving rise to unreal differences between two samples. These problems have been improved by the introduction of labeling techniques that allow for the mixing of protein samples and will be discussed subsequently.

The aforementioned limitations have spurred the development of new separation and identification techniques. Sophisticated HPLC systems can be used to concentrate and separate the digested peptides and coupled online with mass spectrometers. The resolution and sensitivity of mass spectrometric analysis have also been improved by the use of tandem mass spectrometry that allows identification of thousands of peptides in a single experiment (60). This combination of reverse-phase liquid chromatography with tandem mass spectrometry (LC-MS/MS) is the basis of high-throughput studies that aim to analyze complex protein mixtures without the use of gel electrophoresis.

Taking advantage of this system, we have described and carried out a proteomic approach to identify differentially secreted proteins between 3T3-L1 preadipocytes and adipocytes (61). In this study, the conditioned medium from preadipocytes and adipocytes at various differentiation stages was collected, and the proteins were separated by one-dimensional (1-D)-gel electrophoresis. After silver staining of the proteins, the differentially expressed protein bands were digested with trypsin and identified by LC-MS/MS. In the future, differential identification of proteins using such approaches with preadipocytes/adipocytes, BAT/WAT, or adipose tissue from different regions will provide great progress toward the understanding of the role of the adipocyte secretome.

7. NOVEL PROTEOMIC APPROACHES FOR STUDYING THE ADIPOCYTE SECRETOME

Recent efforts to obtain quantitative profiles using mass spectrometry have resulted in several protein-labeling techniques that could be applied to the study of proteins secreted under different biological situations (62). The general advantage of using a labeling technique is that the reproducibility of the experiments is improved because the protein samples are mixed after labeling and the remaining experimental procedure is carried out on the mixed sample.

7.1. In Vitro Labeling with Dyes

One promising labeling strategy in 2-D-gel electrophoresis differential proteomics is difference gel electrophoresis (DIGE) technology (63). By using cyanine-2, -3, or -5 fluorescent dyes to covalently label the protein samples, up to three differently labeled protein extracts can be mixed and separated in a single 2-D gel electrophoresis run. The gel can be scanned at different wavelengths, avoiding problems with spot matching between gels. This method has improved the reproducibility of 2-D electrophoresis, however, it .presents some serious limitations when it is used for the quantification of protein abundance changes. It has been shown that changes in low-abundance proteins can be overestimated, whereas changes in highly abundant proteins can be missed (64,65).

7.2. In Vitro Labeling with Stable Isotopes

The incorporation of stable isotopes after cell lysis or during cell growth, coupled with high resolution mass spectrometry, provides a robust foundation for large-scale quantitative approaches. The basis for the quantification of the relative changes in protein abundance is the use of stable isotopes to label the samples with a "light" or a "heavy" isotope. The mass difference between the peptides containing each of the isotopes allows their distinction in the mass spectrum in which they appear as pairs of isotopic distributions separated by an expected distance corresponding to the mass difference (Fig. 3). The relative abundance between proteins derived from two different samples is obtained by estimating the ratio between the integrated areas of the isotopic distributions of the peptide pairs.

Several in vitro labeling methods have been developed, the most frequently used ones being ICAT and ^{18}O-labeling. The chemical method of isotope labeling using isotope-coded affinity tags (ICAT) was first pioneered by Aebersold and coworkers (66). The labeling is performed by the covalent binding of a linker group that contains a heavy or a light isotope to the reduced cysteines of peptides. After combining both light- and heavy-labeled samples, the addition of a biotin moiety enables the selection of the labeled peptides by avidin affinity chromatography, reducing the complexity of the sample and thereby simplifying its analysis. The peptides are subsequently separated by liquid chromatography and analyzed by mass spectrometry. Two major ICAT technological improvements have emerged since its first description: the use of cleavable linkers, which avoids the problems derived from the presence of bulky tags, and the use of linkers labeled with isotopes of carbon instead of deuterium. This results in the coelution of the peptides in the reverse-phase chromatography, which facilitates their comparison (67,68). The major constraint of this method is the partial coverage of proteins, as only the peptides containing cysteines can be labeled.

Labeling with heavy water ($H_2{}^{18}O$) during endoprotease digestion can also be applied to the qualitative and quantitative comparison of two protein extracts (69–71). In this procedure, one of the systems is completely shifted to [^{18}O] water during endoprotease digestion, while the other is digested in [^{16}O] water (normal water). Labeling of the peptides is catalyzed by the endoprotease, usually trypsin, during digestion, resulting in the complete incorporation of two ^{12}O atoms into the C-terminal carboxyl group of the cleaved peptide (72). The samples are mixed before mass spectrometry analysis. The two ^{18}O atoms incorporated in one peptide population lead to a 4-Da difference with respect to the peptides cleaved in normal water, which can be detected easily using LC-MS/MS. Limitations of ^{18}O-labeling arise when quantifying the abundance of highly charged peptides (bearing more than three charges) because the separation between the isotopic distributions is too small to be differentiated unambiguously.

A

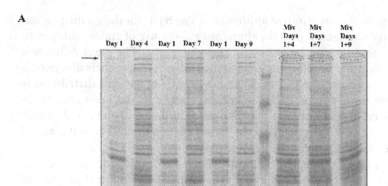

Profile of proteins secreted by adipocytes

B

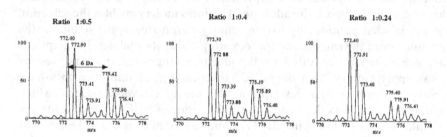

Fibronectin secretion is downregulated

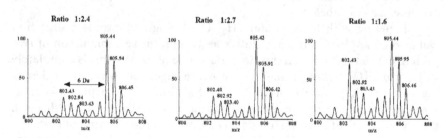

Collagen alpha 3 secretion is upregulated

7.3. In Vivo Labeling with Stable Amino Acid Isotopes

Labeling of growing cells by stable isotope-containing amino acids has been successfully employed for quantitative proteomics and, although limited to culture systems, offers several advantages over in vitro systems (73–75): it is highly reproducible as the samples can be mixed earlier in the experimental procedure (following cell harvesting), and labeling only requires the growth of cells in special media; and it is simple to perform and versatile, as several amino acid isotopes or different isotopic versions of the same amino acid can be used.

Because in vitro models for the study of adipose tissue biology and the preadipocyte differentiation process are well established, this labeling method combined with LC-MS/MS represents a promising approach for the study of the changes in the adipocyte secretome. As an example, we present an experiment where we have used stable isotope labeling by $[^{13}C_6]$lysine to study the secretome changes during the preadipocyte differentiation process into adipocytes.

In this experiment, 3T3-L1 preadipocytes were grown in media containing $[^{12}C_6]$lysine (light) or $[^{13}C_6]$lysine (heavy) as the only source of lysine. Conditioned media from $[^{12}C_6]$lysine-labeled adipocytes on day 1 and $[^{13}C_6]$lysine-labeled adipocytes on days 4, 7, and 9 after the addition of DMI were collected. Equal volumes were used independently or mixed 1:1 (v/v) (day 1 vs. the rest of the days) for protein separation by 1-D gel electrophoresis (Fig. 3A). Following visualization by Coomassie blue staining, several protein bands were excised, trypsin-digested, and identified by LC-MS/MS. Because trypsin cleaves after arginine or lysine, the lysine-containing peptides from the mixture of two experimental samples could be visualized as pairs separated by a mass difference of 6 Da. Figure 3B shows the mass spectrometric analysis of a band where we observed an increase in

---

FIGURE 3 Quantitative analysis of the adipocyte secretome during differentiation. To compare the amount of proteins secreted during adipocyte differentiation, conditioned medium from adipocytes grown in $[^{12}C_6]$ lysine (light) was collected on day 1 and an equal volume of conditioned medium from $[^{13}C_6]$ lysine (heavy)-labeled adipocytes was collected on days 4, 7, and 9 of the adipocyte differentiation process. These samples were loaded independently or mixed 1:1 (v/v) for protein separation by 1-D gel electrophoresis, as indicated in panel A. Following Coomassie blue staining, the protein bands were excised, digested by trypsin, and analyzed by LC-MS/MS. Panel B presents the mass spectrometric analysis of the circled bands in panel A. Sequencing of several peptide pairs by mass spectrometry indicated that two proteins were present in this band: fibronectin and collagen alpha-3. A comparison of intensities of the light vs. the heavy peptides on different days of the differentiation process is shown.

the intensity of one of the bands as the differentiation progressed. Analysis by mass spectrometry revealed that two proteins, fibronectin and collagen alpha-3, were present in this band. Quantification of the relative amounts of each of the proteins in the different experimental mixtures indicated that while the amount of secreted fibronectin was downregulated during the differentiation process, the amount of secreted collagen alpha-3 was upregulated as indicated by the respective decrease or increase in the ratios between day-1 adipocytes and day-4, -7, and -9 adipocytes. By combination of $[^{13}C_6]$lysine and $[^{13}C_6]$arginine isotopes, all the peptides generated by trypsin digestion will be labeled and can be used for quantification purposes. Moreover, by using several amino acid isotopes, more than two systems could be compared simultaneously, which could be especially useful in time-course experiments.

8. CONCLUSION

A number of labeling, fractionation and quantitative methods can be used in gel-free systems or in conjunction with 1-D or 2-D gel electrophoresis, as exemplified in the study by Smolka and colleagues where ICAT was coupled to 2-D gel electrophoresis (76). The importance of secreted proteins as modulators of body homeostasis makes their identification and study especially relevant in the biomedical field. Quantitative approaches to examining the expression levels of proteins and posttranslational modifications may be the key to obtaining a more complete understanding of adipocyte biology.

REFERENCES

1. Gunn TR, Gluckman PD. Perinatal thermogenesis. Early Hum Dev 1995; 42:169–183.
2. Must A, Spadano J, Coakley EH, Field AE, Colditz G, Dietz WH. The disease burden associated with overweight and obesity. Jama 1999; 282:1523–1529.
3. Kopelman PG. Obesity as a medical problem. Nature 2000; 404:635–643.
4. James PT, Leach R, Kalamara E, Shayeghi M. The worldwide obesity epidemic. Obes Res 2001; 9(Suppl 4):228S–233S.
5. Lean ME. Obesity: Burdens of illness and strategies for prevention or management. Drugs Today (Barc) 2000; 36:773–784.
6. Halvorsen YD, Bond A, Sen A, Franklin DM, Lea-Currie YR, Sujkowski D, Ellis PN, Wilkison WO, Gimble JM. Thiazolidinediones and glucocorticoids synergistically induce differentiation of human adipose tissue stromal cells: biochemical, cellular, and molecular analysis. Metabolism 2001; 50:407–413.
7. Sen A, Lea-Currie YR, Sujkowska D, Franklin DM, Wilkison WO, Halvorsen YD, Gimble JM. Adipogenic potential of human adipose derived stromal cells from multiple donors is heterogeneous. J Cell Biochem 2001; 81:312–319.

8. Zuk PA, Zhu M, Mizuno H, Huang J, Futrell JW, Katz AJ, Benhaim P, Lorenz HP, Hedrick MH. Multilineage cells from human adipose tissue: implications for cell-based therapies. Tissue Eng 2001; 7:211–228.

9. Pittenger MF, Mackay AM, Beck SC, Jaiswal RK, Douglas R, Mosca JD, Moorman MA, Simonetti DW, Craig S, Marshak DR. Multilineage potential of adult human mesenchymal stem cells. Science 1999; 284:143–147.

10. Minguell JJ, Erices A, Conget P. Mesenchymal stem cells. Exp Biol Med (Maywood) 2001; 226:507–520.

11. Wada MR, Inagawa-Ogashiwa M, Shimizu S, Yasumoto S, Hashimoto N. Generation of different fates from multipotent muscle stem cells. Development 2002; 129:2987–2995.

12. Zuk PA, Zhu M, Ashjian P, De Ugarte DA, Huang JI, Mizuno H, Alfonso ZC, Fraser JK, Benhaim P, Hedrick MH. Human adipose tissue is a source of multipotent stem cells. Mol Biol Cell 2002; 13:4279–4295.

13. De Ugarte DA, Morizono K, Elbarbary A, Alfonso Z, Zuk PA, Zhu M, Dragoo JL, Ashjian P, Thomas B, Benhaim P, Chen I, Fraser J, Hedrick MH. Comparison of multi-lineage cells from human adipose tissue and bone marrow. Cells Tissues Organs 2003; 174:101–109.

14. Taylor SM, Jones PA. Multiple new phenotypes induced in 10T1/2 and 3T3 cells treated with 5-azacytidine. Cell 1979; 17:771–779.

15. Gimble JM, Robinson CE, Wu X, Kelly KA. The function of adipocytes in the bone marrow stroma: an update. Bone 1996; 19:421–428.

16. Hwang CS, Loftus TM, Mandrup S, Lane MD. Adipocyte differentiation and leptin expression. Annu Rev Cell Dev Biol 1997; 13:231–259.

17. Mandrup S, Lane MD. Regulating adipogenesis. J Biol Chem 1997; 272: 5367–5370.

18. Gregoire FM, Smas CM, Sul HS. Understanding adipocyte differentiation. Physiol Rev 1998; 78:783–809.

19. Wu Z, Puigserver P, Spiegelman BM. Transcriptional activation of adipogenesis. Curr Opin Cell Biol 1999; 11:689–694.

20. Green H, Meuth M. An established pre-adipose cell line and its differentiation in culture. Cell 1974; 3:127–133.

21. Green H, Kehinde O. Spontaneous heritable changes leading to increased adipose conversion in 3T3 cells. Cell 1976; 7:105–113.

22. Student AK, Hsu RY, Lane MD. Induction of fatty acid synthetase synthesis in differentiating 3T3-L1 preadipocytes. J Biol Chem 1980; 255: 4745–4750.

23. Mandrup S, Loftus TM, MacDougald OA, Kuhajda FP, Lane MD. Obese gene expression at in vivo levels by fat pads derived from s.c. implanted 3T3-F442A preadipocytes. Proc Natl Acad Sci U S A 1997; 94:4300–4305.

24. Rosen ED, Walkey CJ, Puigserver P, Spiegelman BM. Transcriptional regulation of adipogenesis. Genes Dev 2000; 14:1293–1307.

25. Entenmann G, Hauner H. Relationship between replication and differentiation in cultured human adipocyte precursor cells. Am J Physiol 1996; 270: C1011–1016.

26. Tang QQ, Otto TC, Lane MD. Mitotic clonal expansion: a synchronous process required for adipogenesis. Proc Natl Acad Sci U S A 2003; 100:44–49.

27. Stewart JE, Whelan CF, Schotz MC. Release of lipoprotein lipase from fat cells. Biochem Biophys Res Commun 1969; 34:376–381.

28. Siiteri PK. Adipose tissue as a source of hormones. Am J Clin Nutr 1987; 45:277–282.

29. Zhang Y, Proenca R, Maffei M, Barone M, Leopold L, Friedman JM. Positional cloning of the mouse obese gene and its human homologue. Nature 1994; 372:425–432.

30. Schwartz MW, Baskin DG, Bukowski TR, Kuijper JL, Foster D, Lasser G, Prunkard DE, Porte, Jr. D, Woods SC, Seeley RJ, Weigle DS. Specificity of leptin action on elevated blood glucose levels and hypothalamic neuropeptide Y gene expression in ob/ob mice. Diabetes 1996; 45:531–535.

31. Schwartz MW, Seeley RJ, Campfield LA, Burn P, Baskin DG. Identification of targets of leptin action in rat hypothalamus. J Clin Invest 1996; 98: 1101–1106.

32. Elias CF, Lee C, Kelly J, Aschkenasi C, Ahima RS, Couceyro PR, Kuhar MJ, Saper CB, Elmquist JK. Leptin activates hypothalamic CART neurons projecting to the spinal cord. Neuron 1998; 21:1375–1385.

33. Elias CF, Aschkenasi C, Lee C, Kelly J, Ahima RS, Bjorbaek C, Flier JS, Saper CB, Elmquist JK. Leptin differentially regulates NPY and POMC neurons projecting to the lateral hypothalamic area. Neuron 1999; 23:775–786.

34. Schwartz MW, Woods SC, Porte, Jr. D, Seeley RJ, Baskin DG. Central nervous system control of food intake. Nature 2000; 404:661–671.

35. Elmquist JK. Hypothalamic pathways underlying the endocrine, autonomic, and behavioral effects of leptin. Physiol Behav 2001; 74:703–708.

36. Spiegelman BM, Flier JS. Obesity and the regulation of energy balance. Cell 2001; 104:531–543.

37. Hillebrand JJ, de Wied D, Adan RA. Neuropeptides, food intake and body weight regulation: a hypothalamic focus. Peptides 2002; 23:2283–2306.

38. Maeda K, Okubo K, Shimomura I, Mizuno K, Matsuzawa Y, Matsubara K. Analysis of an expression profile of genes in the human adipose tissue. Gene 1997; 190:227–235.

39. Guo X, Liao K. Analysis of gene expression profile during 3T3-L1 preadipocyte differentiation. Gene 2000; 251:45–53.

40. Wabitsch M, Brenner RE, Melzner I, Braun M, Moller P, Heinze E, Debatin KM, Hauner H. Characterization of a human preadipocyte cell strain with high capacity for adipose differentiation. Int J Obes Relat Metab Disord 2001; 25:8–15.

41. Ross SE, Erickson RL, Gerin I, DeRose PM, Bajnok L, Longo KA, Misek DE, Kuick R, Hanash SM, Atkins KB, Andresen SM, Nebb HI, Madsen L, Kristiansen K, MacDougald OA. Microarray analyses during adipogenesis: understanding the effects of Wnt signaling on adipogenesis and the roles of liver X receptor alpha in adipocyte metabolism. Mol Cell Biol 2002; 22:5989–5999.

42. Nakamura T, Shiojima S, Hirai Y, Iwama T, Tsuruzoe N, Hirasawa A, Katsuma S, Tsujimoto G. Temporal gene expression changes during adipogenesis in human mesenchymal stem cells. Biochem Biophys Res Commun 2003; 303:306–312.

43. Yang YS, Song HD, Li RY, Zhou LB, Zhu ZD, Hu RM, Han ZG, Chen JL. The gene expression profiling of human visceral adipose tissue and its secretory functions. Biochem Biophys Res Commun 2003; 300:839–846.

44. Ruan H, Hacohen N, Golub TR, Van Parijs L, Lodish HF. Tumor necrosis factor-alpha suppresses adipocyte-specific genes and activates expression of preadipocyte genes in 3T3-L1 adipocytes: nuclear factor-kappaB activation by TNF-alpha is obligatory. Diabetes 2002; 51:1319–1336.

45. Ruan H, Miles PD, Ladd CM, Ross K, Golub TR, Olefsky JM, Lodish HF. Profiling gene transcription in vivo reveals adipose tissue as an immediate target of tumor necrosis factor-alpha: implications for insulin resistance. Diabetes 2002; 51:3176–3188.

46. Mulligan C, Rochford J, Denyer G, Stephens R, Yeo G, Freeman T, Siddle K, O'Rahilly S. Microarray analysis of insulin and insulin-like growth factor-1 (IGF-1) receptor signaling reveals the selective up-regulation of the mitogen heparin-binding EGF-like growth factor by IGF-1. J Biol Chem 2002; 277: 42480–42487.

47. Nadler ST, Attie AD. Please pass the chips: genomic insights into obesity and diabetes. J Nutr 2001; 131:2078–2081.

48. Tashiro K, Nakano T, Honjo T. Signal sequence trap. Expression cloning method for secreted proteins and type 1 membrane proteins. Methods Mol Biol 1997; 69:203–219.

49. Kojima T, Kitamura T. A signal sequence trap based on a constitutively active cytokine receptor. Nat Biotechnol 1999; 17:487–490.

50. Tsuruga H, Kumagai H, Kojima T, Kitamura T. Identification of novel membrane and secreted proteins upregulated during adipocyte differentiation. Biochem Biophys Res Commun 2000; 272:293–297.

51. Gygi SP, Rochon Y, Franza BR, Aebersold R. Correlation between protein and mRNA abundance in yeast. Mol Cell Biol 1999; 19:1720–1730.

52. Chen G, Gharib TG, Huang CC, Taylor JM, Misek DE, Kardia SL, Giordano TJ, Iannettoni MD, Orringer MB, Hanash SM, Beer DG. Discordant protein and mRNA expression in lung adenocarcinomas. Mol Cell Proteomics 2002; 1:304–313.

53. Griffin TJ, Gygi SP, Ideker T, Rist B, Eng J, Hood L, Aebersold R. Complementary profiling of gene expression at the transcriptome and proteome levels in Saccharomyces cerevisiae. Mol Cell Proteomics 2002; 1:323–333.

54. Scherer PE, Bickel PE, Kotler M, Lodish HF. Cloning of cell-specific secreted and surface proteins by subtractive antibody screening. Nat Biotechnol 1998; 16:581–586.

55. Appel RD, Sanchez JC, Bairoch A, Golaz O, Miu M, Vargas JR, Hochstrasser DF. SWISS-2DPAGE: a database of two-dimensional gel electrophoresis images. Electrophoresis 1993; 14:1232–1238.

56. Celis JE, Ostergaard M, Jensen NA, Gromova I, Rasmussen HH, Gromov P. Human and mouse proteomic databases: novel resources in the protein universe. FEBS Lett 1998; 430:64–72.

57. Sanchez JC, Chiappe D, Converset V, Hoogland C, Binz PA, Paesano S, Appel RD, Wang S, Sennitt M, Nolan A, Cawthorne MA, Hochstrasser DF. The mouse SWISS-2D PAGE database: a tool for proteomics study of diabetes and obesity. Proteomics 2001; 1:136–163.

58. Lanne B, Potthast F, Hoglund A, Brockenhuus von Lowenhielm H, Nystrom AC, Nilsson F, Dahllof B. Thiourea enhances mapping of the proteome from murine white adipose tissue. Proteomics 2001; 1:819–828.

59. Hoogland C, Sanchez JC, Tonella L, Binz PA, Bairoch A, Hochstrasser DF, Appel RD. The 1999 SWISS-2DPAGE database update. Nucleic Acids Res 2000; 28:286–288.

60. Peng J, Gygi SP. Proteomics: the move to mixtures. J Mass Spectrom 2001; 36:1083–1091.

61. Kratchmarova I, Kalume DE, Blagoev B, Scherer PE, Podtelejnikov AV, Molina H, Bickel PE, Andersen JS, Fernandez MM, Bunkenborg J, Roepstorff P, Kristiansen K, Lodish HF, Mann M, Pandey A. A proteomic approach for identification of secreted proteins during the differentiation of 3T3-L1 preadipocytes to adipocytes. Mol Cell Proteomics 2002; 1:213–222.

62. Tao WA, Aebersold R. Advances in quantitative proteomics via stable isotope tagging and mass spectrometry. Curr Opin Biotechnol 2003; 14:110–118.

63. Unlu M, Morgan ME, Minden JS. Difference gel electrophoresis: a single gel method for detecting changes in protein extracts. Electrophoresis 1997; 18:2071–2077.

64. Patton WF. Detection technologies in proteome analysis. J Chromatogr B Analyt Technol Biomed Life Sci 2002; 771:3–31.

65. Lilley KS, Razzaq A, Dupree P. Two-dimensional gel electrophoresis: recent advances in sample preparation, detection and quantitation. Curr Opin Chem Biol 2002; 6:46–50.

66. Gygi SP, Rist B, Gerber SA, Turecek F, Gelb MH, Aebersold R. Quantitative analysis of complex protein mixtures using isotope-coded affinity tags. Nat Biotechnol 1999; 17:994–999.

67. Zhou H, Ranish JA, Watts JD, Aebersold R. Quantitative proteome analysis by solid-phase isotope tagging and mass spectrometry. Nat Biotechnol 2002; 20:512–515.

68. Hansen KC, Schmitt-Ulms G, Chalkley RJ, Hirsch J, Baldwin MA, Burlingame AL. Mass Spectrometric Analysis of Protein Mixtures at Low Levels Using Cleavable 13C-Isotope-coded Affinity Tag and Multidimensional Chromatography. Mol Cell Proteomics 2003; 2:299–314.

69. Mirgorodskaya OA, Kozmin YP, Titov MI, Korner R, Sonksen CP, Roepstorff P. Quantitation of peptides and proteins by matrix-assisted laser desorption/ionization mass spectrometry using (18)O-labeled internal standards. Rapid Commun Mass Spectrom 2000; 14:1226–1232.

70. Yao X, Freas A, Ramirez J, Demirev PA, Fenselau C. Proteolytic 18O labeling for comparative proteomics: model studies with two serotypes of adenovirus. Anal Chem 2001; 73:2836–2842.
71. Wang YK, Ma Z, Quinn DF, Fu EW. Inverse 18O labeling mass spectrometry for the rapid identification of marker/target proteins. Anal Chem 2001; 73: 3742–3750.
72. Yao X, Afonso C, Fenselau C. Dissection of proteolytic 18O labeling: endoprotease-catalyzed 16O-to-18O exchange of truncated peptide substrates. J Proteome Res 2003; 2:147–152.
73. Ong SE, Blagoev B, Kratchmarova I, Kristensen DB, Steen H, Pandey A, Mann M. Stable isotope labeling by amino acids in cell culture, SILAC, as a simple and accurate approach to expression proteomics. Mol Cell Proteomics 2002; 1:376–386.
74. Steen H, Pandey A. Proteomics goes quantitative: measuring protein abundance. Trends Biotechnol 2002; 20:361–364.
75. Goshe MB, Smith RD. Stable isotope-coded proteomic mass spectrometry. Curr Opin Biotechnol 2003; 14:101–109.
76. Smolka M, Zhou H, Aebersold R. Quantitative protein profiling using two-dimensional gel electrophoresis, isotope-coded affinity tag labeling, and mass spectrometry. Mol Cell Proteomics 2002; 1:19–29.
77. Kim S, Moustaid-Moussa N. Secretory, endocrine and autocrine/paracrine function of the adipocyte. J Nutr 2000; 130:3110S–3115S.
78. Gregoire FM. Adipocyte differentiation: from fibroblast to endocrine cell. Exp Biol Med (Maywood) 2001; 226:997–1002.
79. Mohamed-Ali V, Pinkney JH, Coppack SW. Adipose tissue as an endocrine and paracrine organ. Int J Obes Relat Metab Disord 1998; 22:1145–1158.
80. Fruhbeck G, Gomez-Ambrosi J, Muruzabel FJ, Burrell MA. The adipocyte: a model for integration of endocrine and metabolic signaling in energy metabolism regulation. Am J Physiol Endocrinol Metab 2001; 280:E827–E847.

18

Mass Spectrometry Strategies for Proteomic Studies

Maya Belghazi

Institut National de la Recherche Agronomique (INRA), Laboratory of Mass Spectrometry for Proteomics, Nouzilly, France

Hélène Rogniaux

Institut National de la Recherche Agronomique (INRA), Unité de Recherche sur les Protéines Végétales et Leurs Interactions, Laboratory of Mass Spectrometry, Nantes, France

1. INTRODUCTION

Recently, mass spectrometry has become a routine technique for the automatic identification of thousands of proteins per day, in the subpicomolar range, and thus a key technique in proteomic applications. First introduced by Wilkins, the word "proteomic" designates the systematic identification of proteins found in organs, tissues, cells, and biological fluids that are the expression products of a genome (1).

Most proteomic studies are based on two main analytical techniques: two-dimensional gel electrophoresis (2DE) and mass spectrometry (MS). The common way of processing consists of three steps (Fig. 1). First, the proteins from a crude sample (of varied origins such as tissues, cells, or

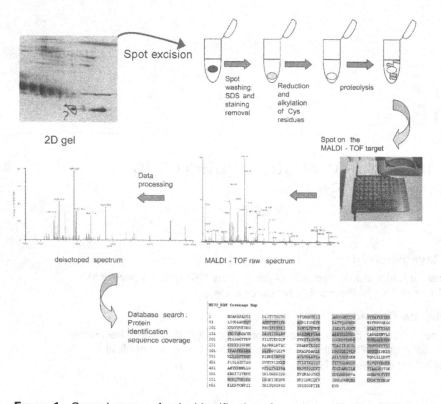

FIGURE 1 General strategy for the identification of proteins purified by 2DE. Protein spots are excised from the gel, either manually or by dedicated robots. Spots are then washed to remove detergents, and proteins are reduced and alkylated. Note that this step is often conserved even if the alkylation has been done on the samples prior to electrophoresis. Then samples are submitted to proteolytic cleavage and the supernatant is directly analyzed by mass spectrometry. Several protocols using an additional step called peptide extraction, sometimes followed by a desalting step, can be found in the literature (see, for example, the protocol used at EMBL at http://www.mann.embl-heidelberg.de/GroupPages/PageLink/activities/protocols/ingeldigest.html). In our proteomic lab, we have chosen to by-pass these steps due to extensive sample loss during desalting. For a detailed protocol, see Fig. 2. After these sample preparation steps, mass analysis can be performed, followed by database searching.

biological fluids) are separated by 2DE; second, individual spots are excised from the gel and submitted to a protease treatment, usually trypsin hydrolysis—trypsin cleaves specifically the C-terminal side of lysine and arginine residues—to generate protein specific peptides (Fig. 2); and third, these peptide mixtures are measured by MS, and the resulting peptide mass lists

IN GEL PROTEOLYSIS FOR PROTEOMICS

Products

Trypsin sequencing grade (Roche Boehringer Manheim)
Iodoacetamide (Sigma)
DTT (Sigma)
Distilled water (Merck)

1 – spots excision

➢ Cut the spots and place them in either a polypropylene 96 well plate or eppendorff tubes.

2– Spot washing

➢ Wash the gel pieces in distilled water two times (10 mi) under agitation.
➢ Remove water and add 50 µl of acetonitrile
➢ Remove ACN and add 50 µl of NH_4HCO_3 100mM (2g in 250 mL of distilled water)
➢ Agitate 5 min and add 50 µl of ACN
➢ Remove supernatant and replace by 50 µl of pure ACN. Wait until the gel pieces become opaque.

3 – Reduction-Alkylation

➢ hydrate gel pieces in 10 mM Dithiothreitol (1,5 mg/mL) in 100mM NH_4HCO_3
➢ Incubate 45 min at 56°C
➢ Remove DTT anda dd 50 µl ACN
➢ Replace by 55mM Iodoacetamide (10,2 mg/mL) in 100 mM NH_4HCO_3 and incubate at room temperature in the dark
➢ Remove iodoacetamide and add 50 µl of NH_4HCO_3 100 mM , agitate 5 min
➢ Add 50 µl of ACN and agitate 5 min
➢ Replace by 50 µl of pure ACN and wait until the gel pieces become opaque.

4 –tryptic cleavage

➢ Hydrate the gels pieces in 40 µl in a solution containing 12.5 ng/µL of trypsin in 25mM NH_4HCO_3. Place 45 min in ice.
➢ Remove trypsin and replace by 25mM NH_4HCO_3 without trypsin and leave overnight at 37°C
➢ Add TFA to a final concentration of 0,1 % and deposit on MALDI plate.

FIGURE 2 Protocol for the proteolysis of 2DE gel spot prior to mass spectrometry analysis. (Adapted from Ref. 82.)

are compared to database entries, using appropriate search engines. These experiments can be completed by tandem mass spectrometry (MS/MS) in order to obtain sequence information on the peptides. This is performed either by de novo sequencing (the deduction of a peptide sequence from a raw fragmentation spectrum) or by straight comparison of the fragmentation pattern with the theoretical fragmentation spectra of all proteins of a database. Database searches constitute a key step for protein identification: large-scale nucleotide sequencing of Expressed Sequence Tags (EST) or of genomic DNA has provided much information on the amino acid sequences of proteins now contained in databases, and complete genome sequencing has been achieved for a wide variety of organisms (http://www.ncbi.nlm.nih.gov/entrez/query.fcgi?db=Genome).

The purpose of this chapter is to introduce recent mass spectrometry applications in the field of proteomics. The basics of proteomics experiments will be described, with particular emphasis on MS contributions both to large-scale proteomic projects and to fine protein characterization such as posttranslational modifications. The reader will be given an overview of what has become possible in this emerging field.

2. MASS SPECTROMETRY

2.1. Common Strategies for Protein Identification by MS

To identify a protein by MS, two strategies can be employed: peptide mass fingerprinting (PMF) and sequencing by MS/MS.

The finger printing strategy PMF, also called peptide mass mapping, is the most commonly employed strategy for large-scale protein identification by MS. The principle is based on the matching of an experimentally measured peptide mass list with masses obtained by in silico digesting the proteins of a database using the same protease specificity as that used for the sample. Eventually, partial cleavage by the protease and chemical modifications of the protein can be taken into account. Crucial points for the confidence of protein identification using the PMF approach are the mass accuracy (50 ppm is a typical error window if internal calibration has been done on trypsin autolysis peaks) and the number of peptide masses that match with theoretical fragments of the protein. Several free algorithms allow database searching by PMF, for example, Profound (http://prowl.rockefeller.edu/cgi-bin/ProFound), Mascot (http://www.matrixscience.com), PeptIdent (http://us.expasy.org/tools/peptident.html), and MS-Fit (http://prospector.ucsf.edu/ucsfhtml4.0/msfit.htm).

Most MS/MS experiments are based on the selection of a given ("parent") peptide and its sequence-dependent fragmentation in a collision cell. The rules for the fragmentation of peptides are well established (2–4): most fragments are produced by the breakage of the amide bond (...–CO–NH–...) along the peptide backbone. Therefore, MS/MS spectra usually contain easily identifiable series of fragment ions from which a sequence tag can be deduced; the mass difference between two consecutive

fragment peaks corresponds to the mass of an amino acid (Fig. 3) and, therefore, one can deduce a short stretch—commonly 3 to 4 residues—of amino
acid sequence. The sequence tag is part of this short sequence of amino acids,
completed by the start and the end masses of the fragments. This tag provides
additional information to the peptide mass list obtained by PMF and can
be used for database searching in order to reach more specific protein identification (5). Actually, many algorithms do not reconstitute a sequence tag
but rather proceed by comparison of the whole fragmentation pattern
observed on the MS/MS spectrum with predictable ions generated by virtual

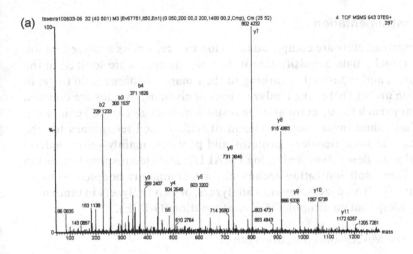

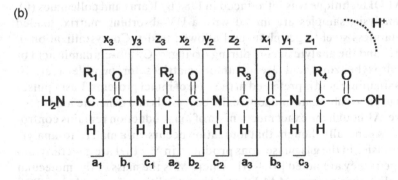

FIGURE 3   (a) MS/MS spectrum obtained on the Global Q-TOF (Waters) for the
peptide IDAALAQVDTLR of flagellin from *Salmonella typhimurium* (NCBI entry
gi|96744) separated by 2DE after tryptic cleavage. Corresponding fragments (y- or
b-) are indicated on the peaks. (b) Nomenclature of the peptide fragmentation. (From
Ref. 4.)

fragmentation of all possible peptides from a database. This is a powerful tool when the protein record is in the database; however, if the protein is not in the database, one has to reconstitute sequence tags for several peptides of the protein and use them for an MS-BLAST (Basic Local Alignment Search Tool) search (http://dove.embl-heidelberg.de/Blast2/msblast.html) or for a BLAST search (search for short nearly exact matches) if the sequence stretch is long enough (http://www.ncbi.nlm.nih.gov/BLAST/). The third possibility is the de novo sequencing of the peptide that exists within the interpretation of the complete fragmentation spectrum without any prior knowledge of the peptide sequence.

2.2. Instrumentation

Mass spectrometers are composed of an ion source, a mass analyzer, an ion detector, and a data acquisition unit. Sample molecules are ionized in the ion source, and separated according to their mass-to-charge ratio (m/z, in Thomson units (Th)) in the analyzer. Ions of given m/z values are counted when they reach the detector and the result is an ion current (IC) composed of all individual mass spectra. The most widely used techniques for the ionization of biomolecules—proteins and peptides, mainly—are: matrix-assisted laser desorption ionization (MALDI) and electrospray ionization (ESI). These soft ionization techniques are commonly coupled to time-of-flight (TOF) analyzers, ion trap analyzers, or two analyzers in tandem as in the widespread quadrupole-TOF configuration (Q-TOF).

2.2.1. The MALDI-Based Instruments

The MALDI technique was introduced in 1988 by Karas and colleagues (6). Before analysis, samples are mixed with a UV-absorbing matrix, loaded onto a stainless steel target plate, and allowed to dry. Cocrystallization of the matrix and the analyte occurs during drying. α-Cyano-cinnamic acid or 2,5-dihydroxybenzoic acid are common matrices for peptide analysis, whereas sinapinic acid is preferred in the case of intact proteins. Laser pulses are then used to sublimate and ionize the cocrystals and generate ions in the gas phase. Although the exact mechanism of ion production remains controversial, it is generally thought that ionization occurs via a matrix-to-analyte proton transfer in the gas phase. Ions produced in MALDI are generally singly charged (they are noted $[M + H]^+$ where M is the mass of the molecular species). For that reason, MALDI has been traditionally coupled to TOF analyzers, enabling the measurement of high—theoretically unlimited—m/z ratios. The fact that singly charged ions are preferentially produced by MALDI implies that most molecular species will give only one mass peak in the spectrum, allowing the interpretation of even complex mixture spectra. Such analysis can be easily automated, and modern MALDI sources accommodate 96- or 384-well plates, making the technique especially attractive for large-scale proteomic projects.

Until recently, most instruments with a MALDI source had a single analyzer and were not very adaptable to MS/MS measurement. Fragmentation of ions could be performed by post-source decay (PSD) and the fragments analyzed in a TOF tube equipped with an electrostatic reflector; however, performance of PSD experiments is complex and difficult to automate. Therefore, proteomic studies using MALDI instruments were largely performed with the PMF approach. Yet, one difficulty in the PMF approach is when one starts with a mixture of proteins. This might happen in proteomic samples even when they are extracted from 2DE gels. In fact, after protease cleavage, the peptide mixture is even more complex than the starting sample; although MALDI easily deals with mixtures, this is not true for database search algorithms. Usually, the score calculated by these algorithms in order to sort protein hits is inappropriate if several proteins match nicely with database entries. To overcome this difficulty, newer instrument configurations with MALDI ion sources allowing fragmentation of selected peptides (MS/MS) have been introduced: MALDI-Q-TOF and MALDI-TOF-TOF instruments.

One available MALDI-MS/MS hybrid configuration is based on the use of a quadrupole in tandem with a TOF analyzer (MALDI-Q-TOF). Today, double-source Q-TOF spectrometers enable the use of both MALDI and ESI ionizations on the same instrument. In the double-source configuration proposed by Waters (Q-TOF Global), ESI and MALDI sources are rapidly interchangeable without venting the instrument. It has been shown that the fragmentation patterns differ according to the ionization method used. A study by Wattenberg and collaborators demonstrated that with MALDI-Q-TOF, both C-terminal (y-fragments) and N-terminal (b- and a-fragments) as well as neutral loss of water and ammonia were found in the spectra, whereas mainly C-terminal fragments are generated with the ESI-Q-TOF technique (7). Although attractive, current MALDI-Q-TOF instruments can hardly be used for high-throughput proteomic studies because target plates usually contain no more than a dozen spots. Nevertheless, this instrument can be helpful to complete the information provided by MALDI-TOF alone, without the need to desalt the sample, as required in the case of ESI-Q-TOF analysis.

The MALDI-TOF-TOF technique utilizes two TOF tubes in combination (in tandem). Parent ion masses are measured in the first TOF analyzer, selected by an ion gate (which allows the transmission of an ion population at a given time), and transmitted to a collision cell where fragmentation occurs. This is called collision-induced dissociation (CID) which is a high-energy fragmentation process. The fragment masses are then measured in the second analyzer. The MALDI-TOF-TOF procedure is especially adapted to high-throughput identification of proteins because it combines facile automation with the accurate mass determination of fragments and the sensitivity of TOF analyzers (8). Another advantage of this technique is the high specificity of the mass spectra obtained, allowing more confidence in protein identification, and its possible application to the analysis of complex protein mixtures without preliminary separation. Although relatively more complex than MALDI-Q-TOF spectra, the data can be used for de novo sequencing. This has been done by Yergey and coworkers for sequence determination of sea urchin egg membrane proteins. In this study, the method allowed the exact determination of peptide sequences differing by only a single mass unit (9).

2.2.2. Electrospray Ionization-Based Instruments

The application of the ESI technique for the analysis of biomolecules by MS was introduced by Fenn in 1989 (10). The sample is prepared in an acidified aqueous-organic solvent (usually, a 1:1 (v/v) mixture of water and acetonitrile containing 0.5% formic acid) and is continuously sprayed into the ESI source via a metal capillary. An electric field (ca. 3000 V/cm) is applied at the exit of the capillary, which has the effect of polarizing the liquid and directing positive droplets toward the entrance of the mass spectrometer. Due to a charge repulsion effect, the droplets undergo multiple explosions upon evaporation of the solvent. This phenomenon generates "daughter" droplets of smaller volume. Although the exact mechanism of ESI remains unknown, it is admitted that daughter droplets give rise—in the end—to completely dried (positive) ions. The positive charges that are carried by the ions in ESI are preexisting in solution; in the case of proteins and peptides, for example, positive charges occur through protonation of basic residues (arginine, lysine, histidine, N-terminus) under the acidic conditions of the sprayed solution. A consequence is that, contrary to MALDI, it is rather common to observe multiply charged ions in ESI. In the case of peptides generated by trypsin proteolysis, doubly charged ions are generally observed, because these peptides contain a C-terminal lysine or arginine residue, and become protonated at their N-terminal amino group. This situation is particularly favorable for the fragmentation of such peptides because spectra contain singly charged fragments and are thus quite easily interpreted.

Crucial advantages of ESI are that the whole ionization/desorption process occurs at atmospheric pressure, and that the analytes are infused into the ion source from a solution. Therefore, this method is compatible with separative techniques such as liquid chromatography or capillary electrophoresis. However, one drawback of ESI is that it is not tolerant to salts, making desalting of the samples absolutely necessary before analysis.

The fact that ESI produces multiply charged ions enables the coupling of this technique to mass analyzers with a moderate m/z range (e.g., 0–2000 m/z); hence, traditionally, ESI was coupled to triple quadrupole analyzers, which provided affordable MS instruments with already good performances (mass accuracy was better than 0.01%). Today, more and more laboratories involved in proteomics projects are equipped with ESI-Q-TOF instruments (e.g., a quadrupole mounted in tandem with a TOF tube), because these configurations enable high-sensitivity measurement (a few femtomole/μL), especially in the MS/MS mode. Furthermore, the excellent mass accuracy (around 50 ppm in routine measurements) and the resolution of the TOF analyzer that enables unambiguous determination of the charge states increase the confidence of peptide sequence determination. Adapted algorithms are now available that enable routine interpretation of MS/MS experiments, and MS/MS spectra acquisition can be easily automated by using the "survey MS scan" (e.g., the selection for fragmentation of every possible precursor ion on the basis of its intensity, charge state, mass range, etc., as defined by the user).

Another type of analyzer classically coupled to ESI sources is the ion trap. In such analyzers, the ions are captured in a small volume of helium (the trap), using high-frequency electric fields. By varying the value of the frequencies, ions are scanned out of the trap and detected to give a mass spectrum. The advantages of this type of analyzer are its low cost and its sensitivity, since ions can be accumulated in the trap to increase the signal-to-noise (S/N) ratio. Moreover, multiple stages of MS/MS measurements can be performed since the trap plays the role of an analyzer and a collision cell at the same time. After an MS scan, a set of ions of a given m/z can be stored in the trap to be fragmented; fragmentation then occurs by application of a voltage on the cap electrodes of the ion trap, with a frequency corresponding to the resonance frequency of the trapped parent ions. Fragment ions can be stored again in the trap, and the fragmentation process repeated. These instruments allow the realization of multiple-stage MS experiments whereas only MS/MS experiments can be done on Q-TOF or TOF-TOF instruments that have only two analyzers separated by one collision cell. Ion traps are widely used for fine determination of the chemical structure of small molecules (e.g., oligosaccharides (11,12)). In proteomics, many studies have been carried out by LC-MS/MS using ion trap analyzers (13). However, the MS/MS spectra generated during these LC-MS/MS runs are almost impossible to interpret in a de novo manner (i.e., by manually reconstituting stretches of amino acid sequence). In fact, fragmentation patterns are often

more complex than in Q-TOF mass spectrometers and are further complicated by the abundant loss of water in the traps. Above all, the resolution is
usually insufficient to separate a series of isotopes in a peptide and, therefore, does not allow the use of software deconvolution tools with good confidence (e.g., transformation of a m/z spectrum into a "real mass" spectrum).
Thus, interpretation is done by straight comparison of the experimentally
observed fragmentation pattern with in silico-simulated fragmentation
spectra for all peptides of a database, which is limited to proteins that are
present in the database. In these algorithms, a set of candidate peptides is
selected from the in silico-digested database if their masses match the experimentally measured mass of the parent ion; only these selected peptides will
be considered for the second step of the search: the comparison of the fragmentation patterns. Thus, if the protein is not in the database, there are few
chances that it will be identified with these algorithms unless a very homologous protein is recorded in the database that gives similar digestion peptides.

Recently, hybrid instruments combining triple quadrupole technology
and linear ion trap analyzers have been developed (Q-TRAP from Applied
Biosystems). These instruments allow the enhancement of multiply charged
ions (an interesting feature for the analysis of tryptic peptides) and the use
of three different scan modes. In addition to the traditional "survey scan"
mode, precursor ion scan detection can be used: this mode allows the selective detection of a modified peptide by the detection of a given fragment.
For example, phosphorylated peptides are characterized by a fragment of
$m/z = -79$ in the negative mode (loss of the phosphate PO_3^-), and tryptic
peptide spectra contain a fragment at $m/z = +147$ in the positive mode. The
third original scanning mode is the neutral loss mode which permits, for
example, the detection of phosphorylated peptides (loss of 98 Da for H_3PO_4
loss of 49 Da for doubly-charged peptides in the positive mode) without the
need to switch instrument polarity (14).

Another type of mass spectrometer using ESI ionization and also
based on the trapping of ions (but in a high magnetic field under vacuum) is
the Fourier Transform Ion Cyclotron Resonance Mass Spectrometer
(FTICR-MS). Among all the MS techniques, FTICR-MS is the most
resolutive, sensitive, and accurate. For example, with only 400 attomoles
of a protein digest, a glutamine and a lysine (representing a mass difference
of only 0.036 Da) could be discriminated by accurate mass measurement of
precursor ion and fragment ions (15). Moreover, the technique is used to
determine accurate masses for high-molecular-weight proteins, allowing
the characterization of posttranslational modifications (16). However, its
high cost, complexity of use, and poor fragmentation efficiency explain why
it is still not widely used in the field of proteomics. Nevertheless, an example
of the application of this technique to high-throughput proteomics is the

"accurate mass tag" approach, described by Smith and coworkers (17). This study used the accurate mass determination (e.g., with a precision < 1 ppm) by FTICR-MS to unambiguously validate the identity of peptides sequenced by conventional MS/MS techniques. This method provides a very high level of confidence regarding peptide identification and will probably have a great impact on the field.

2.3. Quantitative Mass Spectrometry

The determination of protein quantitative variation as a function of various parameters such as cell-cycle states, environmental conditions, genotypes, disease states, or administration of drugs represents a major interest in our field, called "differential expression proteomics." Today, with more and more studies focusing on protein differential expression profiles, such as, for example, between control and treated cells, it has become a challenge for mass spectrometry to not only identify proteins, but also to be able to quantify the differences in protein expression levels.

Yet, it must be noted that mass spectrometry has always been a poor quantitative method: Neither MALDI-MS nor ESI-MS alone can be used to obtain quantitative information due to preferential ionization of some peptides compared to others of different chemical nature. This observation, known as spectral suppression, although poorly understood, prevents the detection of some peptide species and systematically results in incomplete sequence coverage (18).

One possible way to overcome the problem is to use synthetic peptides as internal standards to realize an absolute quantification of proteins, as was done by Gerber and colleagues (19). In this example, peptides were synthesized with incorporated stable isotopes (^{13}C and ^{15}N) in order to mimic their native counterpart. The study permitted the identity of low-abundance yeast proteins as well as quantitative measurement of the phosphorylation state of a serine residue from human separase protein. Obviously, absolute quantification is rather complex by mass spectrometry, since it utilizes chemically synthesized standard materials mimicking the molecules of the analyzed samples.

Relative quantification by MS was achieved using the isotope-coded affinity tag (ICAT), a method pioneered by Gygi and coworkers in 1999 (20). In this precursor work, Gigy and his colleagues applied this technology to compare protein expression levels in the yeast *Saccharomyces cerevisiae* using either ethanol or galactose as the source of carbon. The methodology enabled the concurrent identification and comparative quantitative analysis of proteins from different biological samples such as cells, tissues, and biological fluids. Since the introduction of this methodology, proteomic studies have gained a new dimension: they not only produced extensive protein lists but could also address fundamental questions related to changes in protein expression.

The ICAT method is based on a specific chemical reagent composed of 1) a cysteine-reactive group (iodoacetic acid, commonly used as a sulfhydryl group-alkylating agent to prevent the formation of disulfide bonds between reduced cysteine residues); 2) a linker containing eight heavy (^{2}D atoms) or eight light (^{1}H atoms) isotopes; and 3) an affinity tag (biotin, for example). The principle of the method consists of labelling the control sample and the experimental sample (for instance, two extracts from cells in two different biological states) with the light and heavy ICAT reagents, respectively. The two samples are mixed together before proteolytic cleavage. The mixture is then separated on a suitable affinity column (avidin, if the biotin group has been used) to selectively purify the labelled peptides. The last step is the analysis of the mixture by reverse-phase nanocapillary liquid chromatography coupled with electrospray tandem mass spectrometry (nanoLC ESI-MS/MS). Each peptide is represented by a pair of peaks separated by 8 Da (the mass difference between the light- and heavy-tagged forms) which can be attributed to the control and experimental samples. Their relative abundances, and hence the relative abundance of the protein from which they originate, can be deduced from the peak intensities, since the light and the heavy forms of the peptide behave—e.g., ionize—in the same manner in the mass spectrometer, due to similar chemical composition (see Fig. 4).

Although the ICAT method is rather robust, some complications may appear during automated data treatment in the case of deuterated ICAT: The addition of deuterium atoms to a given peptide has some influence on its retention time, compared to the nondeuterated species, and the deuterated peptide has been shown to elute at lower organic solvent percentage, compared to its native homologue. The shift in the retention times may lead to false data interpretation with the software tools available for treating the ICAT-based experiments (21,22). Today, a new generation of ICAT reagents immobilized on beads and using ^{13}C isotopes overcomes this problem. The linker is composed of nine ^{12}C atoms or ^{13}C atoms for the light and heavy forms, respectively. One other advantage is that the use of these reagents eliminates the possible confusion between peptides that have two ICAT-labelled cysteines (mass shift of +16.10 Da) and peptides that have one oxidized methionine (mass shift of +15.99 Da). These new ICAT reagents also incorporate an acid-cleavable group between the isotope tag and the biotin affinity group in order to avoid the fragmentation of the tag during MS, which greatly simplifies mass spectral interpretation (23,24).

During an ICAT experiment, the labelled peptides are also fragmented for protein identification by MS/MS. Recent software developments enable

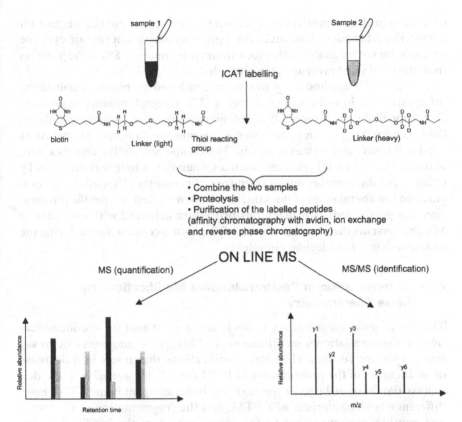

FIGURE 4 Quantification by MS: the ICAT method. This method allows the comparative quantification of the protein content of two samples obtained under different conditions (for example, two samples of bacterial cells grown in culture media with different compositions). Before mixing, each sample is labelled with two ICAT reagents differing in mass by 8 Da due to different isotope composition. Mass spectrometry analysis of the mixture allows the comparison of the intensities of a given peptide labelled with the lighter isotope and the same peptide labelled with the heavier isotope, logically, the abundances of the protein from which the peptide is derived can be deduced. (Adapted from Ref. 19.)

the selective identification of differentially expressed proteins, based on the selective fragmentation of the pairs of peptides showing a mass difference of 8 Da (in the case of deuterated and nondeuterated ICAT reagent) and differing in their abundances (25,26).

One additional benefit of the ICAT technique is that cysteine-containing peptides are selectively purified by affinity chromatography (thanks to the biotin part), thus decreasing the complexity of the mixture. Of course, the subsequent drawback is that proteins that do not contain cysteine residues are not amenable to this technique; this concerns 8% of the proteins from the total yeast proteome, for example.

The ICAT methodology has been employed to obtain quantitative information on low-abundance proteins (27), integral membrane proteins (28), subcellular structures (29), and macromolecular complexes (30). In the latter study, a complex composed of 68 subunits (a RNA polymerase II pre-initiation complex) was analyzed. The components of the complex were identified due to their higher abundances when the sample was prepared by using a specific complex purification (use of promoter DNA affinity), compared to the abundances of the same components when no specific purification was performed. This ingenious procedure achieved with quantitative MS circumvents the problem of copurifying nonspecific proteins during the isolation of macromolecular complexes.

2.4. Determination of Posttranslational Modifications by Mass Spectrometry

The use of mass spectrometry is obviously a great tool for the identification of posttranslational modifications (PTMs). It is commonplace to say that PTMs are covalent chemical modifications that result in a decrease or an increase of the protein mass (a PTM mass list is available in the deltamass database at http://www.abrf.org/index.cfm/dm.home). This mass difference is characteristic of a PTM, and the fragmentation of the modified peptide may even lead to the identification of the modified amino acid. Reality is more complicated, since careful interpretation of hundreds of mass spectral data is necessary to actually locate and identify these PTMs.

By the way, some large-scale PTM identification studies have been attempted, an example of which is the investigation of the "phosphoproteome" (e.g., all phosphorylated proteins in cells or tissues). Phosphorylation (representing a net monoisotopic mass adduct of $+79.96$ Da) is a ubiquitous PTM (about one-third of all mammalian proteins are supposed to be phosphorylated at any given time) that occurs mostly on serine or threonine residues; reversible phosphorylation is a key factor in the regulation of many biological mechanisms such as subcellular localization, enzymatic activity, protein degradation, protein complex formation, and cell apoptosis. Thus, the analysis of the phosphoproteome has raised the interest of the scientific community including mass spectrometry users.

One difficulty in phosphoproteome analysis is that phosphorylation occurs at a very low level in the cell. Thus, finding phosphopeptides in a mixture with their non-phosphorylated counterparts is like finding a needle in a haystack. One old but robust technique in order to specifically enrich samples in phosphoproteins is immobilized metal affinity chromatography (IMAC); it is based on the specific retention of negatively charged phosphate groups (PO_3^-) on the stationary phase containing immobilized positively charged metal ions (Fe^{3+}, Cu^{2+}, or Ga^{3+}). Nonspecific binding of acidic residues on the IMAC resin can be lowered by performing a methyl esterification of the side-chain carboxylates of glutamic and aspartic acids; this greatly improves the recovery of phosphorylated peptides after IMAC by diminishing the competition for the binding (31,32). Other derivatization methods can be found in the literature but they require larger sample amounts (33,34).

As illustrated for phosphoproteins, one way of undertaking the complexity of a sample is to carry out a selective enrichment of the proteins of interest. Other enrichment methods exist for other PTMs, for example, to enhance the detection of glycopeptides (35,36). The latter study describes the purification of O-linked N-acetylglucosamine peptides. More recently, a large-scale method has been developed to purify N-glycosylated proteins. The strategy termed IGOT for isotope-coded glycosylation-site-specific tagging consists of five steps: 1) the affinity capture of glycoproteins from a complex sample using a lectin column; 2) the tryptic digestion of the glycoprotein; 3) the purification of the glycopeptides using the same lectin column; 4) the treatment of the glycopeptides with an enzyme that cleaves specifically the N-glycosylation (peptide N-glycosidase F (PNGase F); the cleavage is catalyzed in $H_2^{18}O$, which leads to the incorporation of an ^{18}O atom at the N-glycosylation site; and 5) the analysis of the ^{18}O-labelled peptide by LC-LC-MS/MS. This method allowed the identification of 400 unique glycosylation sites from 250 glycoproteins purified from an extract of *Caenorhabditis elegans* (37).

Considering that it is impossible to develop selective purification methods for every single type of covalent modification—more than 200 protein PTMs have been listed so far (38)—and that reduction of sample complexity leads to biased data since all the proteins that do not contain the reactive group under study are not identified, the need of a shotgun method for PTM characterization seems necessary. MacCoss and coworkers have presented an alternative approach based on the use of a multienzymatic system, including both specific (trypsin) and nonspecific (subtilisin and elastase) enzymes for proteolytic cleavage and multidimensional chromatography coupled to tandem MS (39); they show that the redundancy obtained in the protein sequences considerably reduces the ambiguity in PTM determination. The authors were able to detect 60 unambiguous PTM sites (acetylations, oxidations, phosphorylations, and methylations) on 11 human crystalline proteins. This strategy was slightly modified by Wu and colleagues with the single nonspecific enzyme, proteinase K. Although the complete digestion of proteins using this enzyme usually leads to dipeptides, partial cleavages occurred under acidic conditions and resulted in peptides of 6 to 20 residues long. Using this method, low sequence coverage was obtained (<20% sequence coverage for the majority of the identified proteins). However, the study permitted the identification of 79 modifications (phosphorylations, acetylations, and mono-, di-, or tri-methylations) on 51 proteins from rat brain homogenate (40).

Software tools such as the "findmod" software available on the internet (http://www.expasy.ch/sprot/findmod/) might be helpful to characterize and locate PTMs from peptide mass fingerprinting data; this software compares mass differences between experimental and theoretical peptides, and if these mass differences correspond to one of the 35 referenced PTMs, some rules are applied to determine the most probable localization in the sequence (41). A similar tool is also available at the following address: http://www.expasy.ch/tools/glycomod/. This tool aims at determining the possible oligosaccharide structures occurring in glycopeptides from experimental mass data (42).

2.5. Probing Biomolecular Interactions by MS

2.5.1. The Surface Plasmon Resonance-MS method

Surface Plasmon Resonance (SPR) used in combination with MS has demonstrated its efficiency for determining biomolecular interactions in the field of proteomics. This technique is also called biomolecular interaction analysis (BIA). The methodology involves the fixation of the receptor of interest (either an antibody or any other protein for which a ligand has to be determined) on a "sensor chip" (or "biosensor"). The chip is made of glass coated with a layer of gold to which carboxymethyl dextran is attached. The carboxy termini of dextran chains can be activated by treatment with N-hydroxysuccinimide (NHS) and 1-ethyl 3-(3 dimethylaminopropyl) carbodiimide (EDC), enabling the reaction of amino, sulfhydryl, aldehyde, or carboxyl groups with the activated surface. Therefore, any protein can be coated on the chip. Upon injection of the sample, the ligand bounds to the chip with its attached protein, and their interaction can be measured by the variation of the refractive index close to the surface of the sensor chip (Fig. 5). The output is a sensorgram that represents the real-time monitoring of the association-dissociation process between the receptor and the target molecule. A very new development of the technique allows the regeneration of the chip using suitable buffer and the elution of the bound target molecule

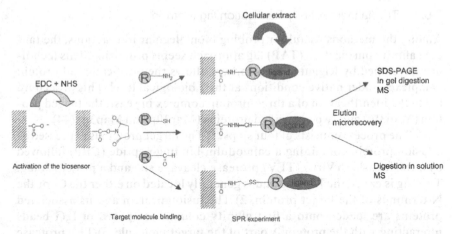

FIGURE 5 Strategy of a SPR experiment for the off-line coupling with MS. The SPR-MS method consists of the binding of a target molecule (a receptor, antibody, or any protein of interest) on a chip. After chip preparation, the crude sample can be loaded on the chip and, after washing, the specifically bound protein of interest can be eluted with adapted buffer. In the eluate solution, proteolysis can then be performed in preparation for mass spectrometry analysis. The method of SPR allows the user to measure the value of the interaction between the protein used as the "fishhook" and the unknown molecule. Although very promising, this technique only permits the recovery of a few femtomoles of biological material, a quantity which is difficult to manage, thus making mass spectrometry analysis uncertain.

in a reduced volume (4 to 7 μL). This process, called "microrecovery," has opened new insights for the identification of a molecule of interest by MS. Moreover, the microrecovery process can be automated, allowing successive recoveries of the target molecule and the pooling of several fractions for the enrichment of the target molecule. To be identified, the protein recovered in a small volume can be digested by a protease prior to mass analysis. A paper by Lopez and collaborators describes this strategy for the fishing of SHP2 tyrosine phosphatase present in cytosolic extracts using an immunoreceptor tyrosine-based inhibitory motif sequence of the sst2 somatostatin receptor. A low amount of the target protein (17 femtomoles) was recovered using 10 mM triethylamine (TEA) containing 0.5 M urea. This buffer was chosen because TEA can easily be removed by evaporation prior to the addition of ammonium hydrogenocarbonate, a buffer suitable for trypsin digestion. The authors clearly demonstrate the capabilities of the BIA-MS strategy for the specific and unambiguous identification of target molecules present at the femtomolar range even in complex mixtures (43).

2.5.2. The Tandem Affinity Purification Tag Method

Among the methods useful for probing biomolecular interactions, the tandem affinity purification (TAP) tag approach seems promising. This technique developed by Rigaut and coworkers allows the purification of protein complexes under native conditions at their biological level. This procedure led to the identification of a three-protein complex in yeast: the tagged protein (Map31p) and two copurified proteins (Map10p and Map3p) (44).

The process consists of four steps: 1) The target protein is expressed as a fusion protein containing a calmodulin-binding peptide (CBP) followed by a Tobacco-Etch Virus (TEV) protease cleavage site and a protein A tag. This tag is called the TAP tag and is generally located on either the C- or the N-terminus of the target protein, 2) The fusion protein and its associated proteins are loaded onto a first affinity column composed of IgG beads interacting with the protein A part of the target molecule, 3) TEV protease is added, which results in the cleavage of the target sequence and in the elution of the protein complex, and 4) A second affinity step is performed on calmodulin beads in the presence of Ca^{2+} to wash the TEV protease, and the complex is then eluted in ethyleneglycol-bis-(beta-amino ethyl ether) (EGTA). The eluted fraction can be analyzed by sodium dodecyl sulfate polyacrylamide gel electrophoresis (SDS PAGE) followed by the traditional in-gel digestion and MS analysis (Fig. 6).

Another recent study by Gavin and collaborators showed that the TAP tag methodology was applicable to large-scale identification of protein complexes from baker's yeast. This work allowed the identification of 232 distinct multiprotein complexes showing the applicability of this kind of approach for the study of the "interactome" (45).

Another interesting report by Taoka and coworkers illustrates the use of the TAP tag procedure for the identification in rat cerebral extracts of the V1 protein partners, namely, the α and β subunits of β-actinin (46). These two proteins where identified after in-gel tryptic digestion and tandem MS. Interestingly, SPR experiments were also performed in order to obtain kinetic data on complex formation. This technique was used after the identification of the complex molecules. Other studies on higher-eukaryote cells demonstrated the efficiency of this purification procedure coupled to the identification of the proteins by MS (47,48).

2.5.3. Selective Enhance Laser Desorption Ionization-MS

Selective enhance laser desorption ionization (SELDI) has been introduced very recently. This technique allows a more selective detection than MALDI-MS due to the use of ProteinChips^{TM} developed by the Ciphergen Company. The surface of the chip is coated with a variety of solid phases

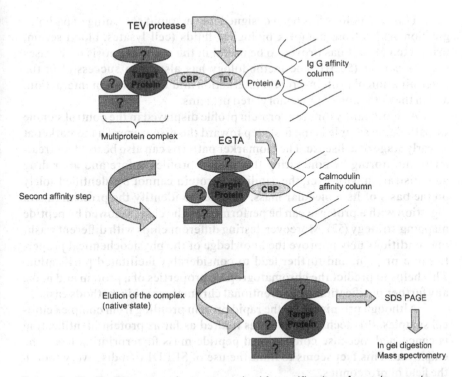

FIGURE 6 The Tap Tag approach, a new method for purification of protein complexes. This method allows the purification of protein complexes by two-step affinity chromatography. The protein is expressed with a Calmodulin Binding Peptide (CBP) part and a protein A tag separated by a TEV protease cleavage site. After a first affinity purification step based on the protein A part of the target protein (on IgG beads), the complex is eluted by the action of the TEV protease. This purification is followed by a second affinity step based on the binding of the free CBP part of the target protein. Classic 1DE and mass spectrometry can then be performed to respectively isolate and identify the individual proteins of the complex studied. (Adapted from Ref. 44).

classically used as chromatographic sorbents (hydrophobic C18, hydrophilic SiO_2, cationic, anionic, or metal affinity phases) or with an activated surface on which any customer may select the molecules to be bound (antibodies, DNA, receptors etc). The basic steps of the technique involve: 1) the selective binding of target proteins on the chip, 2) the washing of the chip to preserve specific interactions with the chip surface and elute unbound proteins and interfering substances, 3) the addition of a matrix (similar to those used in MALDI-MS) that cocrystallizes with the analytes bound onto the chip, 4) the ionization of the crystals by a laser, and 5) the detection of the analytes in a linear TOF analyzer.

This technology has been designed for the rapid screening of pathologic biomarkers from complex biological fluids (cell lysates, blood serum, urine, etc.) (49). It has proven to be useful in the early diagnosis of diseases such as cancer (50,51). The methodology has also been successful in the determination of protein-protein interaction and DNA-protein interaction, and in the detection of phosphorylated proteins.

A significantly different protein profile displayed in the control sample vs. pathologic sample is the first step toward the discovery of a biomarker at an early stage of a disease. The biomarker patterns can also be used for treatment monitoring by comparing the protein profiles before and after drug administration. However, the candidate protein cannot be identified solely on the basis of its molecular mass. In order to identify the protein, in situ digestion with a protease can be performed on the chip, followed by peptide mapping strategy (52). Moreover, testing different chips with different washing conditions may improve the knowledge of the physicochemical properties of a protein, and further lead to considerably facilitated purification. This helps to predict the chromatographic properties of a protein and make any further purification by conventional chromatographic methods easier.

Although promising for the rapid protein profiling from complex clinical samples, this technique remains limited as far as protein identification is concerned, because conventional peptide mass fingerprinting has to be employed. This fact seems to limit the use of SELDI to a discovery tool in the field of proteomics.

3. SAMPLE SEPARATION

In the classical approach, proteomic studies combine a separative method allowing the isolation of all protein components from crude samples, and a mass spectrometric method allowing their identification and, possibly, the fine characterization of their posttranslational modifications. There are basically two alternatives for sample separation prior to mass spectrometry analysis: mono- or two-dimensional gel electrophoresis (1DE or 2DE) or liquid chromatography (monodimensional or multidimensional).

3.1. Electrophoresis

Sample separation is of utmost importance for proteome analysis. Whatever mass spectrometric method is used downstream, the way samples are prepared and separated is crucial for the overall success of the experiment. Moreover, it is often the time-limiting factor for large-scale studies. Today, automation of mass spectra acquisitions as well as data treatment and databank searching is widespread. However, among the sample separation methods, the most widely used, namely 2DE, is not amenable to automation. Because this technique is also time-consuming and cumbersome, it is the limiting factor in proteomics.

The 2DE technique was introduced by O'Farrell in 1975 (53) and is widely used. It is obvious that the resolving power of 2DE (ca. 1500–3000 proteins can be separated on one gel) is not, so far, reachable in routine by any other technique. However, the technique has inherent limitations, especially regarding the separation of membrane proteins, proteins of extreme molecular weight, proteins of extreme isoelectric points (< 3,5 and > 10), or low-abundance proteins. The detractors of 2DE say that it is a good method for 50-kDa proteins within a pI range of 5–7. On the other hand, Rabilloud recently claimed that 2DE is "old, old fashioned, but it still climbs up the mountains." We feel that the technique will remain the separation technique of choice in proteomics studies for a long time (54). We will not describe this technique in detail here but focus on the staining methods, which can greatly influence the amenability of samples to MS.

Among the standard procedures for protein staining, we can enumerate the Coomassie blue staining, the photochemical staining with silver, the imidazole-zinc negative staining, and the staining with fluorescent dyes named Sypro. Although Coomassie blue staining methods are very popular and compatible with mass spectrometric analysis, the detection limit of 30–100 ng (for conventional Coomassie Blue) or 8–10 ng (for colloidal Coomassie Blue) prevents their use for low-abundance proteins. Silver staining methods are more sensitive (detection limit reaches 1 ng of protein), but the use of aldehydes as fixatives has shown poor recovery of peptides after in-gel proteolysis. Another problem of the silver staining method is that lysine and arginine residues are irreversibly modified which makes trypsinolysis difficult or even impossible. Mass spectrometry-compatible silver staining methods omitting the use of aldehydes for the sensibilization and silver impregnation steps have been extensively reported (55,56). Many alternatives to this staining method have been published during the last years, among which are some protocols using a destaining step prior to the standard gel digestion procedure (57,58).

Zinc-imidazole staining is less sensitive than silver staining, the detection limit being around 5 to 10 ng (59). This type of detection is a reverse-staining method where free Zn ions precipitate with imidazole but protein-bound Zn ions do not. The result is an opaque background gel where protein spots appear transparent. The staining method is fully compatible with trypsin digestion followed by MS analysis (60–62).

The fluorescent dye SYPRO ruby shows several attractive features for proteomic applications: first, in term of sensitivity, it is very close to silver staining (less than 10 ng of proteins), but a better sequence coverage is obtained for low-abundance proteins stained with SYPRO ruby (63); and second, fluorescent staining provides a relatively routine method to quantify protein expression levels on a gel, especially since Amersham Biosciences has put on the market special kits for labelling protein extracts with the so-called 2D difference gel electrophoresis (2D-DIGE) method (64). The principle is to label two different samples with two different cyanine fluorescent dyes (Cy3 and Cy5) exhibiting distinct fluorescent properties. This enables the separated detection of two different samples that have been loaded on a single 2D gel. One advantage is that it suppresses the differences in protein levels that come from gel-to-gel variation of the amount of protein detected. Another considerable advantage is that the linearity range of fluorescent stains is much wider than that of Coomassie blue stains or silver stains, which enables the use of fluorescent dyes to quantify changes in protein levels between two samples after image analysis. Last but not least, this staining method is fully compatible with MS. This method was applied to study the *Escherichia coli* proteome after benzoic acid treatment. A total of 179 differentially expressed proteins were unambiguously identified from 500 μg of protein recovered from control and treated bacteria. The majority of the relative change ratios (expression levels due to the acidic treatment) were rather small (< 1.4-fold), but some larger ratios were also found (15.4-fold for the largest). Although careful interpretation of the observed changes is needed to establish their biological relevance, it is clear that fluorescent dyes open new possibilities in the field of protein expression studies to reveal small quantitative changes. The use of an internal standard—the incorporation of a known quantity of proteins in the samples—has proven to improve the accuracy of the detection of small variations in protein levels of different samples (65).

Other recent developments in the field of 2DE that led to improved resolution of the technique include: larger format gel for the second dimension (24 cm), smaller pH range strips for the first dimension, and pH strips adapted to the electro-focusing of basic proteins (66).

3.2. High-Performance Liquid Chromatography Coupled to Tandem Mass Spectrometry

First described by Hunt and coworkers in 1992, high performance liquid chromatography coupled to tandem MS (LC-MS/MS) is widely used in the field of proteomics. This precursor work aimed at the characterization of peptides bound to a class I major histocompatibility complex molecule, namely, HLA-A2. The authors identified nine amino acid-long peptides present in subpicomolar amounts by MS sequencing (67).

Unfortunately, the LC-MS/MS technique cannot be directly applied to crude samples (for example, total cell lysates) because the complexity of such samples is too high compared to the separative power of HPLC. It thus has to be used after a primary fractionation step, the most widely used being 1DE or 2DE. An example of the former way of processing is represented by the work of Pflieger and coworkers on the yeast mitochondrial proteome (67). Since this organelle is composed of large numbers of both membrane-associated proteins and highly alkaline proteins, the 1DE approach was successfully chosen. Using this method, 35% of the mitochondrial proteome was identified. The authors concluded that their strategy was unbiased because of the equal representation of hydrophilic vs. hydrophobic identified proteins and the identification of extreme pI proteins. This work also points out that the 1D-LC-MS/MS approach is particularly adapted to low-molecular-weight protein analysis (68).

Today LC-MS/MS experiments are also commonly employed with 2DE-separated samples. Even though 2DE spots are supposed to contain a single protein, in several cases the LC-MS/MS analysis reveal colocalization of different proteins on the gels. By combining two separative techniques (2DE and LC), additional information can be gained for protein identification and characterization. Other major reasons for using LC-MS/MS are the easy automation of sample introduction, the on-line desalting of the samples, the better sensitivity of MS detection for separated peptides, and the up-to-now relatively easier MS/MS experiments with ESI instruments rather than with MALDI.

Another non-gel-based technique for analyzing the proteome has been proposed recently by Gevaert and coworkers (69). In order to reduce the complexity of the sample prior to LC-MS/MS analysis, the authors selectively purified N-terminal peptides from the cytosolic and membrane skeleton fraction of human thrombocytes. This imaginative off-line technique is based on: 1) the acetylation of free amine groups, 2) the tryptic digestion of the sample which leads to the generation of acetylated N-terminal peptides and free amino groups from internal peptides, 3) the labelling of internal peptides generated by trypsinolysis by a very hydrophobic chemical reagent (2,4,6-trinitrobenzenesulfonic acid or TNBS), 4) the LC fractionation of the sample to separate the N-terminal peptides from the TNBS-tagged internal peptides shifted to later elution times, and 5) the LC-MS/MS analysis of the pre-fractionated N-terminal peptides.

Another way of considerably reducing sample complexity is to analyze selectively the C-terminal peptides generated by trypsinolysis. This can be done by digesting the sample in a mixture of $H_2^{16}O/H_2^{18}O$ (1:1). The internal peptides and the N-terminal peptide of each protein present a disrupted isotopic pattern, due to ^{18}O incorporation in their C-terminal end during trypsin hydrolysis, but the C-terminal peptide peak is undisrupted. Although this does not permit the prefractionation of the C-terminal peptides, their characteristic isotopic pattern obtained by MS allows their selective selection for fragmentation (70).

3.3. Multidimensional Chromatography

As mentioned earlier, the off-line coupling of a primary separation method is necessary before an LC-MS/MS analysis can be performed, due to the complexity of the samples in common proteomic studies. Both 1DE and 2DE are classically used as upstream separation techniques. However, a complete on-line approach would enable the total automation of the proteome analysis process. Furthermore, by reducing the number of separate preparative steps on the samples, one can save a significant amount of material. The multidimensional protein identification technology (MudPIT) has shown promising results in that sense, by analyzing the proteome without the previous electrophoretic step. This is of great interest especially for proteins that are not amenable to 2DE due to their size or physicochemical properties. The term MudPIT refers to multidimensional chromatography: peptides are first separated and fractionated according to their electrostatic charges using strong cation exchange high-performance liquid chromatography (SCX HPLC). Increasing salt concentrations are applied to the column, either in a stepwise manner or as a continuous gradient of the mobile phase, which results in the elution of the peptides by increasing pI. After this first dimension, each discrete peptide fraction is further separated according to hydrophobicity using reverse-phase high-performance liquid microchromatography (RP µHPLC) with a gradient of increasing organic solvent (second dimension). These eluting peptides can be directly analyzed by ESI MS/MS. This method was first presented by Link and coworkers, who identified macromolecular protein complexes (13). Another interesting work by Washburn et al. described the first large-scale proteomic study using the MudPIT method on the whole yeast proteome. In this study, 1484 proteins were identified, among which were 131 integral or membrane-associated proteins, 29 highly basic proteins (with pI > 11), 12 acidic proteins (pI < 4.3), and 24 high-molecular-weight proteins (>190 kDa) (71). These results and others (72) clearly demonstrate the relevance of the MudPIT approach for high-throughput mapping of large proteomes.

3.4. Proteomics of Membrane Proteins

As we pointed out before, 2DE is not applicable to the separation of membrane proteins (73), mainly because of their hydrophobicity: Membrane proteins are usually hard to solubilize, and detergents have to be added in the isoelectric focusing buffer. Another characteristic of membrane proteins, again affecting their solubility, is that they easily precipitate at their isoelectric point. This latter observation has pushed toward the suppression of the first dimension and the use of classical 1DE for membrane proteins. The 1DE step is then followed by in-gel proteolysis and MS analysis, usually by LC-MS/MS because one single 1DE gel band generally contains several proteins.

As alternatives, other strategies employing direct LC-MS/MS or LC-LC-MS/MS are found in the literature. They require protein solubilization using different methods based on the selective extraction of membrane proteins with either organic solvents (74,75) or non-ionic or zwitterionic detergents (76).

Another difficulty encountered in proteomic analysis of membrane proteins concerns the trypsin cleavage step: These proteins often lack sufficient possible cleavage sites (lysine and arginine) preventing the production of reasonably sized peptides (in the range of 900 to 3000 Da) for their analysis by MS. Moreover, the hydrophobicity of the peptides is also a limit for their ionization in MS. Chemical cleavage by cyanogen bromide (CNBr) presents an alternative method to overcome this problem and enhance the sequence coverage in protein identification. For example, Van Monfort and coworkers used CNBr treatment after trypsinolysis on the same gel fragments, allowing the generation of peptides of mass 2500 Da or lower for the membrane protein lactose transporter (LacS) of *Streptococcus thermophilus* (77). This study showed that neither trypsin cleavage nor CNBr treatment alone could lead to suitable peptide sizes for MS analysis.

The development of multidimensional chromatography-based strategy has made possible the identification of membrane proteins by global proteomic approaches. In the study by Wu and colleagues (78), the use of proteinase K, a protease with poor amino acid specificity, permitted the identification of membrane proteins (representing 28% of a total of 1610 identified proteins). An interesting aspect of this study is the topological information obtained on membrane proteins by the digestion of extracellular proteins and of the soluble part of membrane proteins using neutral pH proteinase K treatment, and the re-isolation of the membranes, followed by an acidic pH digestion step that unseals the membranes, giving the protease access to the intracellular proteins and to the luminal part of membrane proteins (78).

4. CONCLUDING REMARKS

Today, proteomics is a tremendously promising field that is rapidly producing large amounts of data. One pillar of proteomic studies is MS techniques, which have the capability of being fully automated, thus enabling the analysis of thousands of proteins per day. Because of the high cost of mass spectrometers, the trend is for the creation of huge MS facilities that offer their knowledge as a service for a community of laboratories (or technology platforms). For these laboratories, the bottleneck remains the interpretation of enormous sets of data. Although software—mainly provided with spectrometers (for example, BioAnalyst™ for Applied Biosystems, Protein Lynx Global Server for Waters, Turbo-SEQUEST for Finnigan)—greatly facilitates the interpretation of mass data, a careful manual verification is often needed to validate the results. The interpretation must take into account the sample preparation method used. Indeed, the simple filtering of peptide mass fingerprinting results based on the scores calculated for the database search (based on a probability) can lead to incorrect peptide assignment, often due to poor spectral quality for some of the samples. This is less true for MS/MS data that usually contain fewer false positive results when the protein is in the databank. When the protein or the EST sequence is not available, de novo sequencing is needed. The feature of de novo sequencing is not a simple achievement and automation of sequence determination is still an unresolved problem, although some software packages do propose this functionality. Note that a very recent algorithm that generates de novo sequences from a tandem MS spectrum (from data file format) has been developed by Lu and colleagues (79). The computer program is available at http://hto-c.usc.edu:8000/msms/menu/denovo.htm. Suitable data processing and careful interpretation of MS/MS spectra can lead to unambiguous sequence determination (but not in all cases). Indeed, the deciphering of complete peptide sequences is possible only if the fragmentation has occurred at each peptide bond, an ideal case which is rarely accomplished. Common situations include the determination of the best combination for two consecutive amino acids (usually for the N-terminal determination since the corresponding y-fragments are usually missing), or dealing with other missing fragments, such as, for example, the fragment corresponding to a cleavage at the C-terminal side of a proline residue. Moreover, internal fragment ions or neutral loss fragments can appear on the spectrum and be misinterpreted as being b- or y-ions. The digestion of the protein in a $H_2^{16}O/H_2^{18}O$ (1:1) mixture was proposed a few years ago with the aim of helping the interpretation of MS/MS spectra: Carboxy termini of tryptic peptides are labelled with ^{18}O atoms; therefore, upon fragmentation, y-fragment series (which retain the C-terminally labelled lysime or arginine residues) can easily be identified by their characteristic $^{16}O/^{18}O$ ratio (80). It has to be noted that in such experiments a 4-Da difference (corresponding to the incorporation of two ^{18}O) can be measured between labelled and unlabelled peptides (81). Since the de novo method does not rely on a match of the measured fragments to those generated in silico from a databank, the advantage of the de novo method is that it is particularly flexible and adaptable to the determination of posttranslationally modified peptides.

The new challenge in proteomics studies is to increase the biological interest of the collected data by addressing the crucial question of quantification, in order to compare two cellular populations in different physiological conditions. This is achieved by the ICAT and/or the DIGE methods, which are being actively developed in mass spectrometry laboratories.

Another challenging task is to increase the throughput of mass spectral analysis by developing non-gel-based techniques such as multidimensional chromatography analysis. This has to be done in relationship with an adequate data storage method, since proteome analysis is synonymous with a huge amount of collected data that must be archived and rationally handled (a single LC-MS/MS run can contain hundreds of fragmentation spectra and represent hundreds of megabytes).

Because proteomic studies by 2DE gel electrophoresis were characterized as spot collection a few years ago, the addition of protein names is generating "protein lists," and this is why proteomics is often not regarded as a biological science per se but rather as a tool to help the understanding of complex biological processes. This is becoming less and less true with the development of techniques such as SPR-MS that have produced more information on protein functions. These kinds of approaches may raise the interest of those who are not yet convinced by the use of systematic protein identification. In summary, the sensitivity, the accuracy, and the obvious potential of modern mass spectrometry to be interfaced with various sample preparations make it a unique technique for protein identification in the field of proteome research, and MS will surely seduce a growing number of biologists in the near future.

ABBREVIATIONS

2D-DIGE	2D difference gel electrophoresis
2DE	Two dimensional electrophoresis
ESI	Electrospray ionization
EDC	1-ethyl 3-(3 dimethylaminopropyl) carbodiimide
FTICR-MS	Fourier transform ion cyclotron resonance mass spectrometer
ICAT	Isotope coded affinity tag
IGOT	Isotope coded glycosylation-site-specific tagging
IMAC	Immobilized metal affinity chromatography
MALDI	Matrix-assisted laser desorption
TOF	Time of flight

NHS   N-hydroxysuccinimide
PTM   Posttranslational modification
QMS   Quantitative mass spectrometry
SELDI   Selective enhanced laser desorption ionization
TAP   Tandem affinity purification
TEA   Triethylamine

ACKNOWLEDGMENTS

The authors would like to thank J. L. Dacheux and A. P. Teixeira for their contributions to this chapter, and the French Institut National de la Recherche Agronomique (INRA) for financial support.

REFERENCES

1. Wilkins MR, Sanchez JC, Gooley AA, Appel RD, Humphery-Smith I, Hochstrasser DF, Williams KL. Progress with proteome projects: why all proteins expressed by a genome should be identified and how to do it.. Biotechnol Genet Eng Rev 1996; 13:19–50.
2. Biemann K. Sequencing of peptides by tandem mass spectrometry and high-energy collision-induced dissociation. Methods Enzymol 1990; 193:455–479.
3. Roepstorff P, Fohlman J. Proposal for a common nomenclature for sequence ions in mass spectra of peptides. Biomed Mass Spectrom 1984; 11:601.
4. Johnson RS, Martin SA, Biemann K, Stults JT, Watson JT. Novel fragmentation process of peptides by collision-induced decomposition in a tandem mass spectrometer: differentiation of leucine and isoleucine. Anal Chem 1987; 59:2621–2625.
5. Mann M, Wilm M. Error-tolerant identification of peptides in sequence databases by peptide sequence tags. Anal Chem 1994; 66:4390–4399.
6. Karas M, Hillenkamp F. Laser desorption ionization of proteins with molecular masses exceeding 10,000 daltons. Anal Chem 1988; 60:2299–2301.
7. Wattenberg A, Organ AJ, Schneider K, Tyldesley R, Bordoli R, Bateman RH. Sequence dependent fragmentation of peptides generated by MALDI quadrupole time-of-flight (MALDI Q-TOF) mass spectrometry and its implications for protein identification. J Am Soc Mass Spectrom 2002; 13:772–783.
8. Medzihradszky KF, Campbell JM, Baldwin MA, Falick AM, Juhasz P, Vestal ML, Burlingame AL. The characteristics of peptide collision-induced dissociation using a high-performance MALDI-TOF/TOF tandem mass spectrometer. Anal Chem 2000; 72:552–558.
9. Yergey AL, Coorssen JR, Backlund Jr PS, Blank PS, Humphrey GA, Zimmerberg J, Campbell JM, Vestal ML. De novo sequencing of peptides using MALDI/TOF-TOF. J Am Soc Mass Spectrom 2002; 13:784–791.
10. Fenn JB, Mann M, Meng CK, Wong SF, Whitehouse CM. Electrospray ionization for mass spectrometry of large biomolecules. Science 1989; 246:64–71.

11. Quemener B, Cabrera Pino JC, Ralet MC, Bonnin E, Thibault JF. Assignment of acetyl groups to O-2 and/or O-3 of pectic oligogalacturonides using negative electrospray ionization ion trap mass spectrometry. J Mass Spectrom 2003; 38:641–648.

12. Quemener B, Desire C, Debrauwer L, Rathahao E. Structural characterization by both positive and negative electrospray ion trap mass spectrometry of oligo-galacturonates purified by high-performance anion-exchange chromatography. J Chromatogr A 2003; 984:185–194.

13. Link AJ, Eng J, Schieltz DM, Carmack E, Mize GJ, Morris DR, Garvik BM, Yates JR 3rd. Direct analysis of protein complexes using mass spectrometry. Nat Biotechnol 1999; 17:676–682.

14. Le Blanc JC, Hager JW, Ilisiu AM, Hunter C, Zhong F, Chu I. Unique scanning capabilities of a new hybrid linear ion trap mass spectrometer (Q TRAP) used for high sensitivity proteomics applications. Proteomics 2003; 3:859–869.

15. Martin SE, Shabanowitz J, Hunt DF, Marto JA. Subfemtomole MS and MS/MS peptide sequence analysis using nano-HPLC micro-ESI fourier transform ion cyclotron resonance mass spectrometry. Anal Chem 2000; 72:4266–4274.

16. Lee SW, Berger SJ, Martinovic S, Pasa-Tolic L, Anderson GA, Shen Y, Zhao R, Smith RD. Direct mass spectrometric analysis of intact proteins of the yeast large ribosomal subunit using capillary LC/FTICR. Proc Natl Acad Sci U S A 2002; 99:5942–5947.

17. Smith RD, Anderson GA, Lipton MS, Pasa-Tolic L, Shen Y, Conrads TP, Veenstra TD, Udseth HR. An accurate mass tag strategy for quantitative and high-throughput proteome measurements. Proteomics 2002; 2:513–523.

18. Sutton CW, Wheeler CH, Sally U, Corbett JM, Cottrell JS, Dunn MJ. The analysis of myocardial proteins by infrared and ultraviolet laser desorption mass spectrometry. Electrophoresis 1997; 18:424–431.

19. Gerber SA, Rush J, Stemman O, Kirschner MW, Gygi SP. Absolute quantifica-tion of proteins and phosphoproteins from cell lysates by tandem MS. Proc Natl Acad Sci U S A 2003; 100:6940–6945.

20. Gygi SP, Rist B, Gerber SA, Turecek F, Gelb MH, Aebersold R. Quantitative analysis of complex protein mixtures using isotope-coded affinity tags. Nat Bio-technol 1999; 17:994–999.

21. Zhang R, Sioma CS, Wang S, Regnier FE. Fractionation of isotopically labeled peptides in quantitative proteomics. Anal Chem 2001; 73:5142–5149.

22. Zhang R, Regnier FE. Minimizing resolution of isotopically coded peptides in comparative proteomics. J Proteome Res 2002; 1:1391–47.

23. Tao WA, Aebersold R. Advances in quantitative proteomics via stable isotope tagging and mass spectrometry. Curr Opin Biotechnol 2003; 14:110–118.

24. Hansen KC, Schmitt-Ulms G, Chalkley RJ, Hirsch J, Baldwin MA, Burlingame AL. Mass spectrometric analysis of protein mixtures at low levels using cleava-ble 13C-ICAT and multi-dimensional chromatography. Mol Cell Proteomics May 23, 2003.

25. Griffin TJ, Han DK, Gygi SP, Rist B, Lee H, Aebersold R, Parker KC. Toward a high-throughput approach to quantitative proteomic analysis: expression-

dependent protein identification by mass spectrometry. J Am Soc Mass Spectrom 2001; 12:1238–1246.

26. Griffin TJ, Lock CM, Li XJ, Patel A, Chervetsova I, Lee H, Wright ME, Ranish JA, Chen SS, Aebersold R. Abundance ratio-dependent proteomic analysis by mass spectrometry. Anal Chem 2003; 75:867–874.

27. Gygi SP, Rist B, Griffin TJ, Eng J, Aebersold R. Proteome analysis of low-abundance proteins using multidimensional chromatography and isotope-coded affinity tags. J Proteome Res 2003; 1:47–54.

28. Goshe MB, Blonder J, Smith RD. Affinity labeling of highly hydrophobic integral membrane proteins for proteome-wide analysis. J Proteome Res 2003; 2:153–161.

29. Han DK, Eng J, Zhou H, Aebersold R. Quantitative profiling of differentiation-induced microsomal proteins using isotope-coded affinity tags and mass spectrometry. Nat Biotechnol 2001; 19:946–951.

30. Ranish JA, Yi EC, Leslie DM, Purvine SO, Goodlett DR, Eng J, Aebersold R. The study of macromolecular complexes by quantitative proteomics. Nat Genet 2003; 33:349–355.

31. Ficarro S, Chertihin O, Westbrook VA, White F, Jayes F, Kalab P, Marto JA, Shabanowitz J, Herr JC, Hunt DF, Visconti PE. Phosphoproteome analysis of capacitated human sperm. Evidence of tyrosine phosphorylation of a kinase-anchoring protein 3 and valosin-containing protein/p97 during capacitation. J Biol Chem 2003; 278:11579–11589.

32. Ficarro SB, McCleland ML, Stukenberg PT, Burke DJ, Ross MM, Shabanowitz J, Hunt DF, White FM. Phosphoproteome analysis by mass spectrometry and its application to Saccharomyces cerevisiae. Nat Biotechnol 2002; 20:301–305.

33. Oda Y, Nagasu T, Chait BT. Enrichment analysis of phosphorylated proteins as a tool for probing the phosphoproteome. Nat Biotechnol 2001; 19:379–382.

34. Zhou H, Watts JD, Aebersold R. A systematic approach to the analysis of protein phosphorylation. Nat Biotechnol 2001; 19:375–378.

35. Hayes BK, Greis KD, Hart GW. Specific isolation of O-linked N-acetylglucosamine glycopeptides from complex mixtures. Anal Biochem 1995; 228:115–122.

36. Greis KD, Hayes BK, Comer FI, Kirk M, Barnes S, Lowary TL, Hart GW. Selective detection and site-analysis of O-GlcNAc-modified glycopeptides by beta-elimination and tandem electrospray mass spectrometry. Anal Biochem 1996; 234:38–49.

37. Kaji H, Saito H, Yamauchi Y, Shinkawa T, Taoka M, Hirabayashi J, Kasai KI, Takahashi N, Isobe T. Lectin affinity capture, isotope-coded tagging and mass spectrometry to identify N-linked glycoproteins. Nat Biotechnol May 18, 2003.

38. Krishna RG, Wold F. Post-translational modification of proteins. Adv Enzymol Relat Areas Mol Biol 1993; 67:265–298.

39. MacCoss MJ, McDonald WH, Saraf A, Sadygov R, Clark JM, Tasto JJ, Gould KL, Wolters D, Washburn M, Weiss A, Clark JI, Yates JR 3rd. Shotgun identification of protein modifications from protein complexes and lens tissue. Proc Natl Acad Sci U S A 2002; 99:7900–7905.

40. Wu CC, MacCoss MJ, Howell KE, Yates JR. A method for the comprehensive proteomic analysis of membrane proteins. Nat Biotechnol 2003; 21:532–538.

41. Wilkins MR, Gasteiger E, Gooley AA, Herbert BR, Molloy MP, Binz PA, Ou K, Sanchez JC, Bairoch A, Williams KL, Hochstrasser DF. High-throughput mass spectrometric discovery of protein post-translational modifications. J Mol Biol 1999; 289:645–657.

42. Cooper CA, Gasteiger E, Packer N. GlycoMod - A software tool for determining glycosylation compositions from mass spectrometric Data. Proteomics 2001; 1:340–349.

43. Lopez F, Pichereaux C, Burlet-Schiltz O, Pradayrol L, Monsarrat B, Esteve JP. Improved sensitivity of biomolecular interaction analysis mass spectrometry for the identification of interacting molecules. Proteomics 2003; 3:402–412.

44. Rigaut G, Shevchenko A, Rutz B, Wilm M, Mann M, Seraphin BA. A generic protein purification method for protein complex characterization and proteome exploration. Nat Biotechnol 1999; 17:1030–1032.

45. Gavin AC, Bosche M, Krause R, Grandi P, Marzioch M, Bauer A, Schultz J, Rick JM, Michon AM, Cruciat CM, Remor M, Hofert C, Schelder M, Brajenovic M, Ruffner H, Merino A, Klein K, Hudak M, Dickson D, Rudi T, Gnau V, Bauch A, Bastuck S, Huhse B, Leutwein C, Heurtier MA, Copley RR, Edelmann A, Querfurth E, Rybin V, Drewes G, Raida M, Bouwmeester T, Bork P, Seraphin B, Kuster B, Neubauer G, Superti-Furga G. Functional organization of the yeast proteome by systematic analysis of protein complexes. Nature 2002; 415:141–147.

46. Taoka M, Ichimura T, Wakamiya-Tsuruta A, Kubota Y, Araki T, Obinata T, Isobe T. V-1, a protein expressed transiently during murine cerebellar development, regulates actin polymerization via interaction with capping protein. J Biol Chem 2003; 278:5864–5870.

47. Aphasizhev R, Aphasizheva I, Nelson RE, Gao G, Simpson AM, Kang X, Falick AM, Sbicego S, Simpson L. Isolation of a U-insertion/deletion editing complex from Leishmania tarentolae mitochondria. EMBO J 2003; 22:913–924.

48. Forler D, Kocher T, Rode M, Gentzel M, Izaurralde E, Wilm M. An efficient protein complex purification method for functional proteomics in higher eukaryotes. Nat Biotechnol 2003; 21:89–92.

49. Issaq HJ, Conrads TP, Prieto DA, Tirumalai R, Veenstra TD. SELDI-TOF MS for diagnostic proteomics. Anal Chem 2003; 75:148A–155A.

50. Zhukov TA, Johanson RA, Cantor AB, Clark RA, Tockman MS. Discovery of distinct protein profiles specific for lung tumors and pre-malignant lung lesions by SELDI mass spectrometry. Lung Cancer 2003; 40:267–279.

51. Cazares LH, Adam BL, Ward MD, Nasim S, Schellhammer PF, Semmes OJ, Wright Jr GL. Normal, benign, preneoplastic, and malignant prostate cells have distinct protein expression profiles resolved by surface enhanced laser desorption/ionization mass spectrometry. Clin Cancer Res 2002; 8:2541–2552.

52. Dare TO, Davies HA, Turton JA, Lomas L, Williams TC, York MJ. Application of surface-enhanced laser desorption/ionization technology to the detection and

identification of urinary parvalbumin-alpha: a biomarker of compound-induced skeletal muscle toxicity in the rat. Electrophoresis 2002; 23:3241–3251.

53. O'Farrell PH. High resolution two-dimensional electrophoresis of proteins. J Biol Chem 1975; 250:4007–4021.

54. Rabilloud T. Two-dimensional gel electrophoresis in proteomics: old, old fashioned, but it still climbs up the mountains. Proteomics 2002; 2:3–10.

55. Shevchenko A, Wilm M, Vorm O, Mann M. Mass spectrometric sequencing of proteins silver-stained polyacrylamide gels. Anal Chem 1996; 68:850–858.

56. Lauber WM, Carroll JA, Dufield DR, Kiesel JR, Radabaugh MR, Malone JP. Mass spectrometry compatibility of two-dimensional gel protein stains. Electrophoresis 2001; 22:906–18.

57. Sumner LW, Wolf-Sumner B, White SP, Asirvatham VS. Silver stain removal using H2O2 for enhanced peptide mass mapping by matrix-assisted laser desorption/ionization time-of-flight mass spectrometry. Rapid Commun Mass Spectrom 2002; 16:160–168.

58. Mortz E, Krogh TN, Vorum H, Gorg A. Improved silver staining protocols for high sensitivity protein identification using matrix-assisted laser desorption/ionization-time of flight analysis. Proteomics 2001; 1:1359–1363.

59. Patton WF. Detection technologies in proteome analysis. J Chromatogr B Analyt Technol Biomed Life Sci 2002; 771:3–31.

60. Matsui NM, Smith DM, Clauser KR, Fichmann J, Andrews LE, Sullivan CM, Burlingame AL, Epstein LB. Immobilized pH gradient two-dimensional gel electrophoresis and mass spectrometric identification of cytokine-regulated proteins in ME-180 cervical carcinoma cells. Electrophoresis 1997; 18:409–417.

61. Castellanos-Serra L, Proenza W, Huerta V, Moritz RL, Simpson RJ. Proteome analysis of polyacrylamide gel-separated proteins visualized by reversible negative staining using imidazole-zinc salts. Electrophoresis 1999; 20:732–737.

62. Katayama H, Satoh K, Takeuchi M, Deguchi-Tawarada M, Oda Y, Nagasu T. Optimization of in-gel protein digestion system in combination with thin-gel separation and negative staining in 96-well plate format. Rapid Commun Mass Spectrom 2003; 17:1071–1078.

63. Lopez MF, Berggren K, Chernokalskaya E, Lazarev A, Robinson M, Patton WF. A comparison of silver stain and SYPRO Ruby Protein Gel Stain with respect to protein detection in two-dimensional gels and identification by peptide mass profiling. Electrophoresis 2000; 21:3673–3683.

64. Tonge R, Shaw J, Middleton B, Rowlinson R, Rayner S, Young J, Pognan F, Hawkins E, Currie I, Davison M. Validation and development of fluorescence two-dimensional differential gel electrophoresis proteomics technology. Proteomics 2001; 1:3773–96.

65. Alban A, David SO, Bjorkesten L, Andersson C, Sloge E, Lewis S, Currie I. A novel experimental design for comparative two-dimensional gel analysis: Two-dimensional difference gel electrophoresis incorporating a pooled internal standard. Proteomics 2003; 3:36–44.

66. Bae SH, Harris AG, Hains PG, Chen H, Garfin DE, Hazell SL, Paik YK, Walsh BJ, Cordwell SJ. Strategies for the enrichment and identification of basic proteins in proteome projects. Proteomics 2003; 3:569–579.

67. Hunt DF, Henderson RA, Shabanowitz J, Sakaguchi K, Michel H, Sevilir N, Cox AL, Appella E, Engelhard VH. Characterization of peptides bound to the class I MHC molecule HLA-A2.1 by mass spectrometry. Science 1992; 255:1261–1263.

68. Pflieger D, Le Caer JP, Lemaire C, Bernard BA, Dujardin G, Rossier J. Systematic identification of mitochondrial proteins by LC-MS/MS. Anal Chem 2002; 74:2400–2406.

69. Gevaert K, Goethals M, Martens L, Van Damme J, Staes A, Thomas GR, Vandekerckhove J. Exploring proteomes and analyzing protein processing by mass spectrometric identification of sorted N-terminal peptides. Nat Biotechnol 2003; 21:566–569.

70. Larsen MR, Larsen PM, Fey SJ, Roepstorff P. Characterization of differently processed forms of enolase 2 from Saccharomyces cerevisiae by two-dimensional gel electrophoresis and mass spectrometry. Electrophoresis 2001; 22:566–575.

71. Washburn MP, Wolters D, Yates JR 3rd. Large-scale analysis of the yeast proteome by multidimensional protein identification technology. Nat Biotechnol 2001; 19:242–247.

72. Peng J, Elias JE, Thoreen CC, Licklider LJ, Gygi SP. Evaluation of multi dimensional chromatography coupled with tandem mass spectrometry (LC/LC-MS/MS) for large-scale protein analysis: the yeast proteome. J Proteome Res 2003; 2:43–50.

73. Santoni V, Molloy M, Rabilloud T. Membrane proteins and proteomics: un amour impossible? Electrophoresis 2000; 21:1054–1070.

74. Ferro M, Salvi D, Brugiere S, Miras S, Kowalski S, Louwagie M, Garin J, Joyard J, Rolland N. Proteomics of the chloroplast envelope membranes from Arabidopsis thaliana. Mol Cell Proteomics May 28, 2003.

75. Goshe MB, Blonder J, Smith RD. Affinity labeling of highly hydrophobic integral membrane proteins for proteome-wide analysis. J Proteome Res 2003; 2:153–161.

76. Luche S, Santoni V, Rabilloud T. Evaluation of nonionic and zwitterionic detergents as membrane protein solubilizers in two-dimensional electrophoresis. Proteomics 2003; 3:249–253.

77. van Montfort BA, Doeven MK, Canas B, Veenhoff LM, Poolman B, Robillard GT. Combined in-gel tryptic digestion and CNBr cleavage for the generation of peptide maps of an integral membrane protein with MALDI-TOF mass spectrometry. Biochim Biophys Acta 2002; 1555:111–115.

78. Wu CC, MacCoss MJ, Howell KE, Yates JR. A method for the comprehensive proteomic analysis of membrane proteins. Nat Biotechnol 2003; 21: 532–538.

79. Lu B, Chen T. A suboptimal algorithm for De Novo peptide sequencing via tandem mass spectrometry. J Comput Biol 2003; 10:1–12.

80. Shevchenko A, Chernushevich I, Ens W, Standing FG, Thomson B, Wilm M, Mann M. Rapid 'de novo' peptide sequencing by a combination of nanoelectrospray, isotopic labeling and a quadrupole/time-of-flight mass spectrometer. Rapid Commun Mass Spectrom 1997; 11:1015–1024.
81. Heller M, Mattou H, Menzel C, Yao X. Trypsin catalyzed (16)O-to-(18)O exchange for comparative proteomics: tandem mass spectrometry comparison using MALDI-TOF, ESI-QTOF, and ESI-ion trap mass spectrometers. J Am Soc Mass Spectrom 2003; 14(7):704–18.
82. Wilm M, Shevchenko A, Houthaeve T, Breit S, Schweigerer L, Fotsis T, Mann M. Femtomole sequencing of proteins from polyacrylamide gels by nanoelectrospray mass spectrometry. Nature 1996; 379:466–469.

19

Bioinformatics Tools

Simon M. Lin and Kimberly F. Johnson

Duke University Medical Center, Duke University, Durham,
North Carolina, U.S.A.

Seth W. Kullman

Nicholas School of the Environment, Duke University, Durham,
North Carolina, U.S.A.

1. BIOINFORMATICS AND NUTRITIONAL GENOMICS: AN INTRODUCTION

In this chapter we will discuss the application of genomics and bioinformatics as it applies to nutrition. Nutritional genomics ("Nutrigenomics") is the study of genome-wide influences of nutrition (1). From the molecular perspective, nutrients are dietary signals that are recognized by cellular sensors (often receptors), which influence gene expression, protein synthesis, and a resultant metabolome. Nutrigenomics seeks to examine the influence of dietary signals as it relates to cellular homeostasis and the effects of nutrition on health and disease. With the recent completion of several genome projects, we are now well poised to establish a linkage between gene expression and nutrition, the influence of micro and macronutrients on cellular homeostasis, and the relationship between diet, genetic predisposition, and human disease. The importance of diet to health is well established and many modern diseases are in part a result of chronic metabolic imbalances related to diet, including cancer, obesity, type 2 diabetes, hypertension atherosclerosis, osteoporosis, and inflammatory disease (2). Metabolic imbalance may be a result of a lack of essential nutrients and/or an abundance of nonessential components in the diet. Epidemiological and clinical studies have identified dietary determinants associated with these metabolic imbalances and have focused this knowledge toward disease cure and prevention.

The knowledge that diet influences gene expression was established long ago and, historically, this relationship was examined one gene at a time. This approach has led to numerous significant detailed discoveries in nutritional science and has been the basis for our initial understanding and insight into the mechanistic action of specific genes and the regulatory pathways through which diet influences cell homeostasis. A logical extension of "one gene at a time" has been the development of transgenic and knockout mice. The development of these mice has been instrumental in establishing the role of single genes in nutrition and disease and continues to be an important component relating gene function to animal health. With the development of new genomic tools and techniques including comparative genomics, expression profiling, genome association studies, and bioinformatics, we are now establishing a thorough knowledge of the complex interactions between genotype, phenotype, and environmental interaction. Equally important is the development of similar approaches for the analysis of nontranscriptional events occurring in the proteome (e.g., protein phosphorylation and glycosalation) and the analysis of resultant metabolomes. These new tools have facilitated a new paradigm in biological sciences based upon development of patterns in gene expression termed "systems biology." The use of such technology is rapidly changing the way in which nutrition science is conducted, emphasizing global expression differences at varying levels of biological organization.

Genomic sciences refer to mapping, sequencing, and functional analysis of individual genomes. As an emerging field, genomics incorporates aspects of structural, comparative, and functional biology. Structural genomics represents the construction of high-resolution genetic, physical, and transcript maps of a given genome including linkage analysis, physical mapping, genome sequencing, and genome organization. Comparative genomics allows the comparison of two or more genomes to identify the extent of similarity of specific features or a large-scale screening of a genome to identify the gene sequences present in another organism (3). Functional genomics is characterized by the development and application of genome-wide, high-throughput experimental approaches to assess gene function, combined with statistical and computational analysis. Functional genomics has been aided by the continual development of gene expression tools including microarray analysis, subtractive hybridization, Serial Analysis of Gene Expression (SAGE), and differential display.

The availability of draft sequences of human, mouse, rat, and other organisms' genomes extends the ability to study aspects of nutrition as it relates to the whole genome level. Integration of structural, comparative, and functional genomics through robust bioinformatics is facilitating a comprehensive view of the complex interactions between nutrition, diet, and genetics. The ultimate goal of this approach is to understand the relationship between nutrition and gene function as it relates to pathology and disease prevention. General questions to be addressed by nutrigenomics relate to how diet influences gene regulation, subsequent changes in metabolism along with the association to altered cellular homeostasis, and the influence of individual genetic variation on diet and nutrition.

In keeping with questions on the influence of specific dietary components on gene regulation, an emerging theme in nutrition science is the function of nuclear receptors as cellular sensors for nutrients and their metabolites (4). It is now understood that effects of many dietary compounds, including cholesterol, fatty acids, fat-soluble vitamins, and other lipids, are mediated by the action of nuclear receptors. It has been well established that nutritional lipid intake constitutes an important determinant of disease susceptibility and is often exacerbated by individuals with dislipidemia. Numerous studies in nutritional genomics are now addressing questions of the influence of dietary lipids on gene regulation and subsequent changes in metabolism with a goal of establishing how diets can be modified to improve health. Receptors including PPAR-fatty acids, LXR-oxysterols, FXR-bile acids, PXR/CAR-bile acids, and other hydrophobic dietary ingredients have been shown to coordinate metabolic gene networks associated with maintenance of cellular homeostasis by governing transcriptional regulation of genes involved with lipid metabolism, storage, transport, and elimination (4). The identification and activity of nuclear receptors as cellular sensors as they relate to nutrition is a subject of intensive study.

Genetic variation within populations and individuals is additionally a critical determinant of differences in nutritional requirements and disease susceptibilities. The most common form of genetic variation is the single nucleotide polymorphism (SNP), a single-base substitution within DNA. To date, over 2.8 million SNPs have been discovered and linked to discrete locations in the human genome. SNPs are highly stable and occur approximately once in every 1000–2000 nucleotides (5). With the recent development of genetic polymorphism databases, great strides have been made in understanding gene environmental interactions. Thus, it is now well substantiated that most major diseases including cardiovascular disease, diabetes, obesity, and cancers result from the interaction between genetic susceptibility and environmental factors, including diet. Several genetic polymorphisms of importance to nutrition have been identified including folate metabolism, iron homeostasis, lipid metabolism, and immune function (6). In the field of lipoprotein metabolism and cardiovascular disease, several gene polymorphisms for key proteins, such as apoproteins (apo) E, B, A-IV, and C-III, LDL receptor, microsomal transfer protein (MTP), fatty acid-binding protein (FABP), cholesteryl ester transfer protein (CETP), lipoprotein lipase, and hepatic lipase, have been identified and linked to variable responses to diets (7). Thus, the importance of genetic variation to nutrition continues to be established and highlights the relationships between genetic predisposition, physiological response, and disease susceptibility. Evidence now suggests that most common disorders are caused by the combined effects of multigenes and nongenetic environmental factors, i.e., they are multifactorial. SNP analysis provides an important component to understanding the relationship between nutrition, human health, and disease. Resolving the relative influence of gene environmental interactions will be a major challenge for nutritional and genomic scientists in the near future.

Integrative genomic approaches are now being conducted that coordinate dietary changes at gene (genomic), protein (proteomic), and metabolite (metabolomic) levels of organization. Combined systematic approaches are currently being employed to address the complexities of nutrient consumption, uptake, metabolism, and the resulting relationships to animal health and disease. Genomic technologies enable the nutritionist to simultaneously measure multiple biological events in molecular detail. The reliance of bioinformatics in these studies is paramount. Development in computerized data management has made much of genomics possible, and advancements in bioinformatics continue to broaden genomic approaches. In this chapter we will demonstrate the use of bioinformatics principles and application as applied to nutritional genomics. The computational strategy for gene sequence-related data is very different from that for numerical data. For example, we can easily define $20/10 = 2$, but sequence data are not so easily categorized. Analyzing sequence data involves many more complex mathematical operations. Thus, we need special computational models to solve sequence analysis problems. In the following sections, we will discuss computational tools, but also overview the models and assumptions used to develop these tools. In this way, readers can choose tools appropriately and interpret results correctly. Since it is impossible to list all bioinformatics tools in a chapter, we use examples from the nutritional research literature to illustrate key concepts. An overview of possible uses for bioinformatics tools is shown in Fig. 1. In addition, a web companion of this chapter is available at http://dbsr.duke.edu/pub/nutrition to provide updated web links to the tools and details of the worked-out examples.

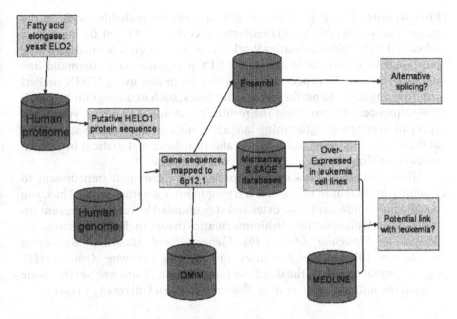

FIGURE 1 Bioinformatics tools used for the fictitious HELO1 cloning project.

Throughout this chapter, we use this example to illustrate the effective use of bioinformatics tools. We start with the yeast ELO2 protein, and search the human proteome database. A putative human homologue, HELO1, is identified. Then, we map this sequence to chromosome 6p12.1. The OMIM database is queried to determine if there are any human genetic diseases linked to this region. By using the Ensemble browser, we examine the genomic DNA structure of this gene, and identify a potential alternative splicing. A query of the gene expression databases indicates HELO1 is overexpressed in leukemia cell lines. We then search the MEDLINE literature database to see if this link is plausible. With these bioinformatics investigations prior to wet laboratory experiments, we can explore many possibilities before fine-tuning our hypothesis.

2. BIOINFORMATICS: A ROADMAP TO EDUCATE YOURSELF

Bioinformatics skills can be mastered at several levels. At an entry level, you should be familiar with the various bioinformatics database and tool websites, minimally the tool collections at the National Center for Biotechnology Information (NCBI) or European Bioinformatics Institute

(EBI) website. There are also integrated toolkits available, such as the commercially available GCG/SeqWeb (Accerlers, CA) and the academic tool called "The Bioinformatics Workbench" (8). At an intermediate level, you should be comfortable running UNIX programs via a command-line interface. You would expect to write some programs using UNIX or Perl scripting languages to perform repetitive tasks, such as a program to search 1000 sequences and parse out the results. At an advanced level, you would expect to master a programming language, such as C++ or Java; to know database constructions and queries; and to master a statistical or mathematical modeling language, such as S-plus or MatLab.

Bioinformatics tools evolve rapidly, and a common impediment to becoming proficient in this arena is trying to hit a moving target. Thus, you should study fundamental theories and stay updated by reviewing recent literature. Good introductions to bioinformatics theory include *Computational Methods in Molecular Biology* (9), *Computational Molecular Biology:an Introduction* (10), and *Bioinformatics: the Machine Learning Approach* (11). A good annual update is the database issue (January) and web service issue (July) of the journal, *Nucleic Acid Research* (Oxford University Press).

3. SCORING MATRIX AND SIMILARITY MATCH: BLAST THE DATABASE

Suppose you are interested in the elongation of saturated and monounsaturated fatty acids. As a first step, we want to find the human counterpart associated with this metabolic function. For purposes of this chapter, we have chosen to examine the ELO2p protein in *Saccharomyces cerevisiae* (12). To begin, we will screen the human genome database using the yeast sequence as a query. This database query should be based on protein sequence similarity. Such database queries are very different from most business applications, where the query is often looking at exact matches of certain criteria. How can we find and evaluate a similar match?

The Basic Local Alignment Search Tool (BLAST) is the primary utility to search similar sequences in a database. Based on the biological assumption that similar sequences usually have like functions, a BLAST search can quickly identify corresponding sequences and establish their putative function by analogy. Thus, BLAST is a fundamental bioinformatics tool that every experimental biologist should master. A simple BLAST search can either return an exciting array of possibilities to the bench work, or ruin your chance of making an important discovery (13). Readers are encouraged to read and learn more about BLAST. Good reviews and tutorials have been written by Altschul et al. (14) and Nicholas et al. (15).

To measure the degree of similarity, BLAST utilizes a scoring mechanism. The scoring schema is backed up by comparing physical and chemical properties of amino acids at each position (Fig. 2, groups I to VI). Some amino acids can be substituted with another one without affecting the structure and function of the protein. This is considered as a homologous substitution. Other amino acid substitutions can result in altered functionality. To quantify the compatibility of substitutions, Henikoff and Henikoff (16) compiled a distance matrix called BLOSUM62 (Fig. 2). In this matrix, a score larger than zero indicates that the two amino acids are essentially

	C	S	T	P	A	G	N	D	E	Q	H	R	K	M	I	L	V	F	Y	W
C	9																			
S	-1	4																		
T	-1	1	5																		
P	-3	-1	-1	7																
A	0	1	0	-1	4															
G	-3	0	-2	-2	0	6														
N	-3	1	0	-2	-2	0	6													
D	-3	0	-1	-1	-2	-1	1	6												
E	-4	0	-1	-1	-1	-2	0	2	5											
Q	-3	0	-1	-1	-1	-2	0	0	2	5										
H	-3	-1	-2	-2	-2	-2	1	-1	0	0	8									
R	-3	-1	-1	-2	-1	-2	0	-2	0	1	0	5								
K	-3	0	-1	-1	-1	-2	0	-1	1	1	-1	2	5							
M	-1	-1	-1	-2	-1	-3	-2	-3	-2	0	-2	-1	-1	5						
I	-1	-2	-1	-3	-1	-4	-3	-3	-3	-3	-3	-3	-3	1	4					
L	-1	-2	-1	-3	-1	-4	-3	-4	-3	-2	-3	-2	-2	2	2	4				
V	-1	-2	0	-2	0	-3	-3	-3	-2	-2	-3	-3	-2	1	3	1	4			
F	-2	-2	-2	-4	-2	-3	-3	-3	-3	-3	-1	-3	-3	0	0	0	-1	6		
Y	-2	-2	-2	-3	-2	-3	-2	-3	-2	-1	2	-2	-2	-1	-1	-1	-1	3	7	
W	-2	-3	-2	-4	3	-2	-4	-4	-3	-2	-2	-3	-3	-1	-3	-2	-3	1	2	11
	I		II				VIII			IV			V				VI			

Group I	C	sulfhydryl
Group II	STPAG	small hydrophilic
Group III	NDEQ	acid/acidamide and hydrophilic
Group IV	HRK	basic
Group V	MILV	small hydrophobic
Group VI	FYW	aromatic

FIGURE 2 The BLOSUM62 scoring matrix. Amino acids are grouped according to their chemical properties. Numbers in the matrix indicate the compatibility of the substitution in homologous proteins. A large number suggests the substitute is admissible.

```
Query sequence      . .  Q   E   C   K   E   C   D   Q   V   F  . .
Database match      . .  Y   R   C   E   D   C   K   Q   E   F  . .
                         ↑   ↑   ↑   ↑   ↑   ↑   ↑   ↑   ↑   ↑
Total score             (-1)+0 + 9 + 1 + 2 + 9 +(-1)+5+(-2)+ 6 = 28
```

FIGURE 3 Scoring a match of two similar protein sequences (a raw score).

inter-exchangeable in homologous proteins, whereas a score below zero indicates an unfavorable substitution. Note that the numbers on the diagonals of Fig. 2 are not the same. This reflects the idea that since residues such as cysteine (C) and tryptophan (W) are rare in sequences, a match should be given a higher significance.

By using BLOSUM62, we can score to what extent the two sequences match (Fig. 3). There are other scoring matrices, such as BLOSUM45 or PAM250 (17), which are designed for comparisons of distantly related proteins. To make the scores using different scoring matrices comparable, the raw scores are usually adjusted into a bit score that is reported in BLAST searches (Fig. 4). We use the bit score in conjunction with the E-value (discussed later) to evaluate sequence similarity.

Using our ELO2 protein as an example, we run a BLAST search of computationally identified proteins in the human genome and find a homologous

```
Score = 80.9 bits (198), Expect = 2e-15
 Identities = 61/216 (28%), Positives = 101/216 (46%), Gaps = 37/216 (17%)

Query: 128 VQHGLYFAICN----IGAWTQPLVTLYYMNYIVKFIEFIDTFFLVLK--HKKLTFLHTYH 181
            V GY C        G    ++ + +  Y  K IEF+DTFF +L+  + ++T LH YH
Sbjct: 134 VWEGKYNFFCQGTRTAGESDMKIIRVLWWYYFSKLIEFMDTFFFILRKNNHQITVLHVYH 193

Query: 182 HGATALLCYTQLMGTTSISWVPI-------SLNLGVHVVMYWYYFLAA-RGIR--VWWKE 231
            H +  + + +    +WVP        +LN +HV+MY YY L++   +R  +WWK+
Sbjct: 194 HASMLNIWWFVM------NWVPCGHSYFGATLNSFIHVLMYSYYGLSSVPSMRPYLWWKK 247

Query: 232 WVTRFQIIQFVLDIGFIYFAVYQKAVHLYFPILPHCGDCVGSTTATFAGCAIISSYLVLF 291
            ++T+ Q++QFVL     + Q + +P      C        +  + S + LF
Sbjct: 248 YITQGQLLQFVL-------TIIQTSCGVIWP-------CTFPLGWLYFQIGYMISLIALF 293

Query: 292 ISFYINVYKRKGTKTSRVVKRAH-GGVAAKVNEYVN 326
            +FYI  Y +KG   +  + H G  A VN + N
Sbjct: 294 TNFYIQTYNKKGASRRKDHLKDHQNGSMAAVNGHTN 329
```

FIGURE 4 BLAST search result. The yeast ELO2 protein was used to search the human proteome database. In this example, the yeast ELO2 protein is called the Query sequence; the matching sequence in the human database is called the subject (Sbjct) sequence. The alignment of these two sequences is shown. "-" indicates a gap. On the central line of the alignment, identical amino acids are indicated by characters; similar ones that receive a positive score are identified as "+" signs. The number of matches with a score equivalent to or better than this one is expected to be 2e-15 (2×10^{-15}) by chance.

protein as shown in Fig. 4. This match received a bit score of 80.9 from a raw score of 198. The bit score is a numerical measurement of similarity, but it does not take the size of the database into consideration. Further evaluation of the result is indicated by the expected number of matches which is given an expect value, or E-value. This value indicates how often there is a similar match to proteins in the database with no homology to the query sequence. This is why a smaller E-value generally indicates a more significant sequence match. For the BLAST search of ELO2, the combined positive and identical matches is 74%; together with the E-value of 2E-15, we can conclude that this search is a significant one.

The BLAST search is essentially a template-matching model that uses a single input sequence to find similar ones in the database. This searching strategy is more sensitive if we can specify a better template from several related sequences in the protein family as an input. Thus, iterative BLAST tools were developed such as PSI-BLAST (18). The PSI-BLAST tool utilizes one input sequence as a starting query to search the databases. Resulting sequences from the first round of the search are used to construct the possibilities of different amino acids at each position, which is represented by a position-specific scoring matrix (PSSM). The PSI-BLAST tool is used most appropriately when searching for distantly related proteins. Otherwise, a BLAST search is adequate. Often BLAST and PSI-BLAST searches provide many sequences as an output. In the following section, we will discuss tools to deal with multiple sequences, and then introduce a more mathematically powerful tool to model sequences in general.

4. MULTIPLE SEQUENCE ALIGNMENTS: PILE THEM UP

Besides a BLAST search, sequence alignment is one of the most commonly used bioinformatics tools. The conserved sequences from multiple species can help us design PCR primers to amplify a gene from the species of interest. Furthermore, the conserved domains suggest a primary sequence basis for certain biological functions and from which mutations can be made to confirm their importance. In addition, a multiple sequence alignment is the starting point for phylogenetic analysis of the evolutionary relationships.

To cover a broader range of examples in nutritional research, here we temporarily depart from the ELO1 example and introduce another study of C_2H_2 zinc finger-containing proteins (Fig. 5) in this and the following section. Zinc finger domains are characterized by incorporation of a single zinc ion bound by two pairs of cysteine (Cys) residues, or two cysteine and two histidine (His) residues in a tetrahedral arrangement. This unique structure forms a finger-like motif that is highly conserved among DNA-binding proteins. In the multiple sequence alignment of Fig. 5, we can clearly see the

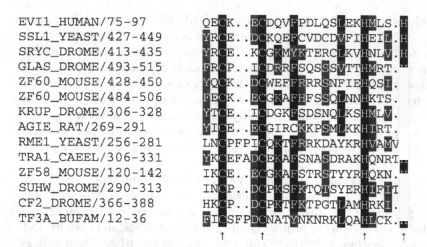

EVI1_HUMAN/75-97   QECK..ECDQVFPDLQSLEKHMLS.H
SSL1_YEAST/427-449   YRCE..DCKQEFCVDCDVFIHEIL.H
SRYC_DROME/413-435   YRCE..KCGKMYKTERCLKVHNLV.H
GLAS_DROME/493-515   FRCP..ICDRRFSQSSSVTTHMRT.
ZF60_MOUSE/428-450   YQCK..DCWEFFRRRSNFIEHQSI.
ZF60_MOUSE/484-506   FECK..ECGKAFHFSSQLNNHKTS.
KRUP_DROME/306-328   YTCE..ICDGKFSDSNQLKSHMLV.
AGIE_RAT/269-291   YICE..ECGIRCKKPSMLKKHIRT.
RME1_YEAST/256-281   LNCPFPICQKTFRRKDAYKRHVAMV
TRA1_CAEEL/306-331   YKCEFADCEKAFSNASDRAKHQNRT.
ZF58_MOUSE/120-142   IKCE..ECGKAFSTRSTYYRHQKN.
SUHW_DROME/290-313   INCP..DCPKSFKTQTSYERHIFIT
CF2_DROME/366-388   HKCP..DCPKTFKTPGTLAMHRKI.
TF3A_BUFAM/12-36   FICSFPDCNATYNKNRKLQAHLCK.

                          ↑   ↑   ↑   ↑

FIGURE 5   Alignment of zinc finger proteins. Conserved residues are highlighted using BOXSHADE (http://www.ch.embnet.org/software/BOX_form.html). "." indicates a gap. Due to the highly conserved Cys and His residues (indicated by arrows), the domain is called C_2H_2.

conserved Cys and His residues. The biological significance of zinc fingers is that they usually mediate the protein/DNA interactions of transcription factors. Zinc finger proteins thus regulate a variety of cellular activities, such as development, differentiation, and tumor suppression (19). It has been shown that dietary zinc deficiency can implicate zinc finger signal transduction proteins (20).

The best alignment of two sequences can be solved by dynamic programming algorithms (21,22). However, finding the optimal alignment of multiple sequences is computationally intractable in theory (10). Thus, a heuristic algorithm is utilized to speed it up. A progressive pairwise method (23) transforms the difficult multiple alignment problem into a series of pairwise alignment problems. Briefly, the most similar pairs of sequences are aligned and then merged into a consensus. This consensus sequence is aligned again with the rest of other sequences, and the procedure continues until done. In such an iterative manner, a multiple sequence alignment is achieved.

Two frequently used programs align sequences in this manner: PILEUP (in GCG package) and CLUSTAL (24). The alignment result depends on several factors: first, the scoring matrix used for gauging similarity among substitutes; second, the gap penalties for nonaligning stretches; and third, the alignment method being either global (PILEUP program) or local (CLUSTAL program). A global alignment optimizes the overlap from the beginning to the end of the sequences, whereas a local alignment searches for the best similar segments. Thus, a local alignment is more appropriate for assessing domain structures.

Due to the computational complexity and the occasional misalignment between the mathematically optimal alignment vs. the biologically significant alignment, sequence alignment results are usually subject to manual adjustments. A human expert can force the alignment of critical residues that are identified by experimental mutations. To this end, a multiple alignment editor program, such as SeqLab in GCG or CINEMA (25), is handy. Multiple sequence alignments can also be optimized by the hidden markov model (HMM) discussed in the next section (26).

5. HIDDEN MARKOV MODELS AND THEIR APPLICATIONS: A BEAUTIFUL MIND

Once multiple alignments of a protein family are produced, we can create a fingerprint that represents the multiple alignment and its variations. This fingerprint leverages the observed variations from the alignment, and can be used itself as a query to identify protein with a similar pattern when constructing a database search. This is a powerful method to identify unknown proteins with similar structural/functional motifs that can be missed by using a BLAST search. To implement this idea, here we will discuss two formal methodologies from mathematical linguistics: regular expressions and HMMs. A readable introduction of linguistic modeling of sequences has been written by Searls (27).

The following is a regular expression describing the matching portions of strings by using a generic template to describe the likely amino acids and their prescribed order. For example, the zinc finger domain can be expressed as

```
C- x(2,4)- C- x(3)-[LIVMFYWC] - x(8)-H-x(3,5) - H
↑                  ↑              ↑            ↑
Cysteine           Choice from    Eight amino  Histine
at this position;  any of         acids of any at this position;
                   LIVMFYWC;      kind;
```

which indicates that there is a gap of two to four amino acids of any kind between the first and the second cysteine; a gap of three to five amino acids between the two histidines; and a middle position that always contains a small hydrophobic or aromatic amino acid (LIVMFYWC). A compilation of common protein motifs such as the one just shown are available in the PROSITE database (28). The MOTIFS program in the GCG package can search databases using PROSITE motifs to identify potential members in the protein family. In addition, MOTIFS search can be used to see if there are any functional domains in a newly identified protein and thus infer the protein's function.

Manually writing a regular expression as illustrated in the preceding paragraph from a multiple sequence alignment not only is time-consuming but also does not lend itself to quantifying the likelihood of each amino acid at each position. Thus, we can use HMMs in sequence analysis (29–31). An HMM elegantly models substitutions, insertions, and deletions of sequences by "states" and "state transitions." It assumes there is a hidden process (unobservable) that generates a sequence of amino acid residues (observed). Each observed protein sequence comes from this hidden process with some realization of the probability distribution of the HMM. Fig. 6 shows part of an HMM model of the zinc finger motif. Sequences such as "CE. DC" and "CKFPDC" can all be derived from this model by traveling through the nodes (states) with different paths (state transitions) from left to right.

Similar to regular expressions, HMMs are used to identify new family members from database searches and to characterize whether a new protein contains a certain functional domain. Such models of proteins can be constructed by the HMMER program (33). The database Pfam contains precompiled HMM models. Using an HMM model of zinc finger proteins,

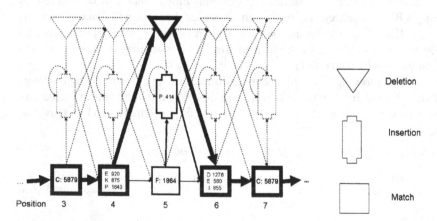

FIGURE 6 HMM representation of the zinc finger motif. Boxes in different shapes represent different states: match, insertion, and deletion. Arrows indicate state transitions. For clarity, we do not show the full probability distribution of each residue at each state, but indicate only the dominant ones. The number is proportional to the likelihood of each residue at a given state. For the transitions, a thicker line indicates a higher probability. This figure is constructed according to the HMM model (# PF00096) in the pfam database (From Ref. 32.).

Clarke and Berg (34) tried to find all zinc finger proteins in *Caenorhabditis elegans*, and Manning et al.(35) identified a new cyclin-dependent kinase in the human genome. The HMM is also a powerful tool to model sequences in general. It can also be used to find genes (36), CpG islands (37), RNA secondary structures, and to align sequences (31).

6. THE HUMAN GENOME: KNOWING OURSELVES

The promise of the genomic era is evidenced by the complete sequencing of several model organisms, including yeast, worm, zebrafish, fly, mouse, rat, and human. Utilizing the wealth of these data in everyday research is a necessity for molecular biologists. To provide integrated access to these data, several browsers were designed at NCBI, UCSC, and the Sanger Center.

Returning to our ELO2 example, we demonstrate how a comparative genomic approach can help identify human homologues. Having run the BLAST search against the proteins identified in the human genome, we were able to locate HELO1, which is the human homologue of yeast ELO2. Based on the direct mapping of computationally translated proteins to genomic DNA sequences, we are able to map the exact location of HELO1 on chromosome 6p12.1 (Fig. 7). We are also able to determine the intron-exon structure, and inspect the potential alternative splicing variants. With the experimental indication of alternative splicing of a closely related enzyme, ELOVL6 (38), we are interested in the splicing of HELO1. Based on ab initio (meaning from the beginning) gene boundary predictions and expressed sequence tags (ESTs), some possible gene exon-intron structures of HELO1 are suggested (Fig. 7C). The true gene transcription structure is subject to experimental biology verification.

To identify the relevance of HELO1 to human disease, we first query the Online Mendelian Inheritance in Man (OMIM) database (39). We find chromosomal 6p12 is implicated in diabetes (OMIM # 125853), epilepsy (OMIM # 606904), and Char syndrome (OMIM # 169100).

Other databases of interest include transcription factor databases (40); pathway databases, such as the Kyoto Encyclopedia of Genes and Genomes (KEGG) (41) and the Encyclopedia of *Escherichia coli* (EcoCyc) (42); and microarray databases. In the following section, we further illustrate how to generate additional biological hypotheses through data mining of the gene expression databases.

7. DATABASE OF CHIPS: THE POWER OF EXPRESSION

With the rapid accumulation of microarrays and Serial Analysis of Gene Expression (SAGE) data from profiling the transcriptomes, it is now possible

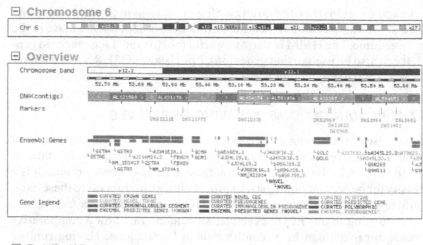

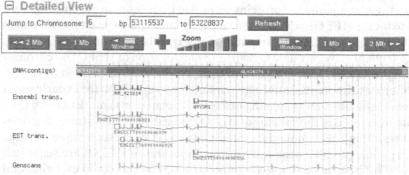

FIGURE 7 A view of the genomic region containing HELO1 (NM_021814) using the Ensembl browser. (Top) Cytochromal bands on chromosome 6. The 6p12.1 band where HELO1 resides is highlighted in the red box. (Middle) An overview of 6p12.1. The genes surrounding HELO1 are FBOX9, GCM1, GCLC, and other computationally found genes. (Bottom) A detailed view showing the ensemble gene model (Ensembl trans), an ab initio computational prediction (Genscans), and alignments with ESTs (EST trans). Vertical bars indicates exon-inton boundaries. This computational evidence suggests a potential alternative splicing of HELO1.

to mine these databases and generate new hypotheses without doing preliminary experiments. Such databases include the Gene Expression Omnibus (GEO) (43) at NCBI and ArrayExpress (44) at EBI. There are also commercial expression database providers such as GeneLogic (Gaithersburg, MD).

In this example, we have used the gene expression database at Novartis GNF (45) and the NCBI SAGE database (46) to continue exploring our fictitious HELO1 cloning project. First, we conducted a BLAST search to see if HELO1 is represented in the Novartis GNF database. Our results suggest that HELO is represented as an EST ('33821_at') on the Affymetrix U74A chip. This enables us to follow expression information of HELO1 in various tissues. In Fig. 8, the GNF database query of HELO1 is in agreement with the experimental data of Leonard et al. (12) where a significant amount of expression is found in spinal cord, testis, prostate, and the adrenal gland. In addition, we identify this gene as highly expressed in various types of blood

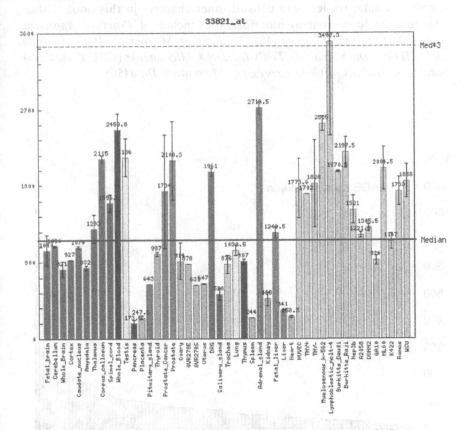

FIGURE 8 Tissue distribution of HELO1 expression from the Novartis GNF microarray database. The abundance of HELO1 expression in various tissues is plotted. Data are from the GNF survey of 47 human tissues using the Affymetrix U95A chip. The EST 33821_at is the identifier of HELO1 on the Affymetrix chip. The original chip design of 33821_at is based on a genomic DNA contig sequence of 6p12.1–21.1 (Genbank AL034374) that contains computationally identified unknown genes. This unknown gene matches HELO1. (From Ref. (45).)

cells, especially in the lymphoblastic MOLT-4 cell line. This observation is further supported by another microarray study of acute lymphoblastic leukemia at St. Jude Children's Research Hospital (47) in which HELO1 was identified as a potential marker protein.

The SAGE data (Fig. 9) from the NCBI SAGE database suggests HELO1 is significantly overexpressed in ovarian and breast cancer samples. All of these bioinformatics studies point to a new direction of investigation on the relevance of HELO1 in various types of cancers.

Due to space limitations, we will not discuss the statistical analysis of microarray data; readers can consult other chapters in this book. Other introductions to microarray bioinformatics include *A Practical Approach to Microarray Data Analysis* (48), *Methods of Microarray Data Analysis vol. I-III* (49), *Data Analysis Tools for DNA Microarrays* (CRC Press), and *Statistical Analysis of Gene Expression Microarray Data* (50).

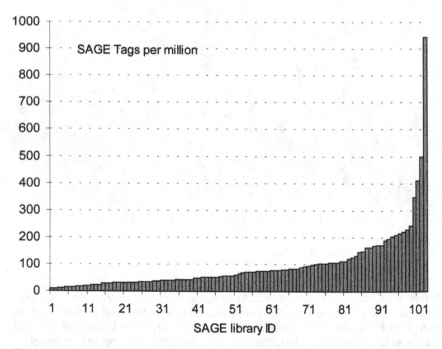

FIGURE 9 Expression of HELO1 in 103 SAGE libraries including kidney, prostate, mammary gland, ovary, brain, and other tissues. The right-most four libraries are ovary carcinoma and three mammary gland carcinomas.

8. ONTOLOGIES: THE LINGO OF SMARTER REASONING

With microarrays, we dramatically extend our investigation spectrum to all genes in the genome. Such experimental results require the development of a corresponding computational model to handle the relationship among the genes. An ontology is a formal specification of concepts and relationships in the domain of interest (51). This powerful tool comes from decades of artificial intelligence research. Usually, it is presented as a vocabulary of terms. For example, phosphorylation is a kind of interaction (Fig. 10) and fatty acid is a kind of lipid (Fig. 11A). Gene ontology is an emerging conceptual model to define gene functions (52). This instrumental development helps to interpret gene lists in terms of biochemical function, biological process, and cellular compartments. Practical tools using gene ontology include Onto-Express (53), MAPPFinder (54), FunSpec (55), TreeMaps (56), and FatiGO (57). Herrero et al.(57) demonstrated the use of gene ontology to interpret both microarray and proteomic data sets using a publicly available web server.

Because an ontology captures the knowledge structure of a particular scientific domain, it becomes an important tool for intelligent database searches. In particular, the machine can answer queries beyond the primary facts deposited into the database by applying known rules to the facts. For example, if we tell the computer that "GABA-R is phosphorylated by PKA" and "the components participating in phosphorylation must be in the same

Fact:
- "Phosphorylation of GABA-R (Q) by PKA (P) at Ser892"

Ontology and Rules:
- Phosphorylation is a kind of interaction.

- PKA is a kinase; A kinase is a protein.

- If there is an interaction between P and Q, then P and Q are in the same cellular compartment.

Machine-deduced knowledge:
- GABA-R interacts with PKA.

- GABA-R and PKA can be in the same cellular compartment.

- PKA is a protein.

FIGURE 10   Ontology and logic extend the database query capability. Based on ontology and rules, the computer can answer intelligent questions beyond the literal facts entered as parameters.

(A)

Select	Subject Heading	Hits :
− Chemicals and Drugs (Non MeSH)		
+ ⌐ Inorganic Chemicals		264
+ ⌐ Organic Chemicals		1897
+ ⌐ Heterocyclic Compounds		3060
+ ⌐ Polycyclic Hydrocarbons		3449
+ ⌐ Environmental Pollutants, Noxae, and Pesticides		0
+ ⌐ Hormones, Hormone Substitutes, and Hormone Antagonists		0
+ ⌐ Reproductive Control Agents		226
+ ⌐ Enzymes, Coenzymes, and Enzyme Inhibitors		0
+ ⌐ Carbohydrates and Hypoglycemic Agents		0
− ⌐ Lipids and Antilipemic Agents		0
+ ⌐ Antilipemic Agents		5922
− ⌐ Lipids		78112
⌐ Ceroid		249
+ ⌐ Fats		3198
− ⌐ Fatty Acids		39863
+ ⌐ Caprylates		766
+ ⌐ Decanoic Acids		701
+ ⌐ Eicosanoic Acids		575
⌐ Fatty Acids, Nonesterified		19965
− ☑ Fatty Acids, Unsaturated		10922
+ ⌐ Arachidonic Acids		13070
+ ⌐ Eicosanoids		2289
+ ⌐ Fatty Acids, Essential		3340
+ ⌐ Fatty Acids, Monounsaturated		1888
− ⌐ Fatty Acids, Omega-3		2608
⌐ alpha-Linolenic Acid		984
⌐ Docosahexaenoic Acids		1969
⌐ 5,8,11,14,17-Eicosapentaenoic Acid		1884

(B) PubMed ID: 11597788

Agatha G., Hafer R., and Zintl F. (2001). Fatty acid composition of lymphocyte membrane phospholipids in children with acute leukemia. Cancer Lett 173: 139–44.

The composition of phospholipid fatty acids (PLFA) of separated mononuclear blood cells (MNC) from patients with leukemia was established by high-resolution gas chromatography. Abnormal fatty acid concentrations are detected in the MNC membrane phospholipids in patients with acute lymphoblastic leukemia (ALL) without a deficiency of essential fatty acids (EFA). Significantly reduced relative levels of linoleic acid (4.35 vs. 7.82%; P < 0.001) are found in the MNC-PL in patients with ALL as compared to a healthy control group. Moreover, the Delta6-desaturated fatty acids are increased: gamma-linoleic acid (3.56 vs. 0.17%; P < 0.001), arachidonic acid (21.82 vs. 16.27%; P < 0.05), docosatetraenoic acid (3.52 vs. 1.56%; P < 0.001), docosapentaenoic acid (0.34 vs. 0.04%; P < 0.001), octadecatetraenoic acid (0.53 vs. 0.23%; P < 0.05), eicosatetraenoic acid (1.83 vs. 0.08%; P < 0.001) and docosahexaenoic acid (2.77 vs. 1.54%; P < 0.001). A increased Delta(6)-desaturase activity is postulated as the cause for the increased level of desaturate products or the increased Delta6-activity index (Ratio of gamma-linoleic acid+dihomogamma-linolenic acid to linoleic acid) (1.21 vs. 0.27; P < 0.001). The Delta6-enzyme activities measured using linoleic acid and alpha-linoleic acid as substrate underscore these findings (Delta6(n-6); 2.49 vs. 0.65 and Delta6(n-3); 2.75 vs. 1.12 nmol x h(−1)/10(8) MNC). In contrast, patients with acute myeloid leukemia (AML) do not show any significant differences in the lymphocyte membrane PLFA and no Delta6-desaturase abnormalities.

(C) PubMed ID: 12423658

Lima T. M., Kanunfre C. C., Pompeia C., Verlengia R., and Curi R. (2002). Ranking the toxicity of fatty acids on Jurkat and Raji cells by flow cytometric analysis. Toxicol In Vitro 16: 741–7.

The fatty acids have an important role in the control of leukocyte metabolism and function. Higher concentrations of certain fatty acids, particularly polyunsaturated fatty acids (PUFAs) . . .

FIGURE 11   MeSH ontology for literature search. (A) The hierarchical structure of chemical compounds, indicating the parenting concepts of unsaturated fatty acids. (B) and (C) are resultant abstracts by searching "fatty acids, unsaturated" and "leukemia." Note that although "unsaturated fatty acids" never appears as a keyword in (B), we still get this abstract due to semantic mapping of ontology terms.

physical location," then the computer automatically knows GABA-R should be in the same cellular compartment with PKA (Fig. 10) when the interaction happens. This example might sound trivial, but in the genomic exploration of interconnected gene products and small molecules, this artificial intelligence capability can help answer questions involving thousands of facts and rules.

A mature application of ontologies is MeSH, a medical ontology (58). It has been extensively used in MEDLINE to search the medical literature. For example, after we get some hint of HELO1 being associated with leukemia from microarrays, we want to search the literature and see if there are any links between unsaturated fatty acid and leukemia. The two abstracts obtained by a conceptual search in Fig. 11 further shape our hypothesis for future investigation.

9.   Medline: Knowledge Is Power

Genomic and proteomic screening usually results in a long list of genes. Harnessing the power of existing biological knowledge to interpret these findings has been an active research topic in bioinformatics. A frequently encountered problem is: Given such a list of statistically significantly changed genes, how do we find the pathways involved? How do we discover the potential interconnection between the genes?

To answer these questions, we first need a vast repository of all current knowledge with regard to the gene products. MEDLINE can be such a knowledge base. A rudimentary approach could be simply looking at the co-occurrence of a pair of genes in a MEDLINE abstract, and thus construct a gene-gene interaction network. Inpharmix (Greenwood, IN) and Pubgene (59) implemented this idea. Figure 12 shows a resultant biological network generated by using the gene list identified in Fig. 2 of the fibroblast micro-

Gene	Name	Ratio (1h/ Ref)
IL6	interleukin-6	5.08
IL8	interleukin-8	3.75
ICAM1	intercellular adhesion molecule 1	1.41
PBEF	pre-B-cell colony-enhancing factor	1.92
PTGS2	prostaglandin-endoperoxide synthase 2	11.9
EDN1	endothelin 1	3.88

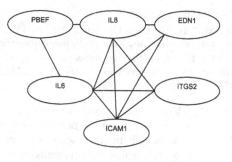

FIGURE 12 Using Pubgene to infer a potential genetic network from a list of genes. MEDLINE abstracts were searched for co-occurrence from a list of 517 genes. Shown is one of the resultant sub-networks and the corresponding gene expression ratio (1 h/common reference). Each node indicates a gene; each line indicates a connection in the literature.

array study (60). This subnet includes genes for immunological responses of activating B-cells and T-cells (IL6, IL8, ICAM1, and PBEF), inflammation and mitogenesis (PTGS2), and fibroblast proliferation (EDN1). The microarray data at 1 hour after serum stimulation are all upregulated, which suggests this network of genes is acting in concordance. This kind of analysis can greatly help experimental biologists digest the microarray observations in terms of biological knowledge, and then formulate a new hypothesis to test.

10. The Fun Has Just Started: A Conclusion

With the yeast ELO2/human HELO1 protein cloning example in this chapter, we have shown how bioinformatics and functional genomic databases can transform molecular biology research. With the existing blueprint of the human genome and the transcriptome databases, we first located the potential human homologue on chromosome 6p12.1, and then surveyed its expression distribution in human tissues. In addition, we explored potential alternative splicing, and its link to human diseases. This rich information will help to guide the wet laboratory experimental design and hypothesis testing.

With the maturation of genomic technologies, experimental biologists are more capable of acquiring large data sets. "Data mining" is a commercial description of the process of extracting actionable knowledge from these data. Bioinformatics tools will continue to be indispensable to mining the realm of genomic and proteomic data to come.

REFERENCES

1. Muller M, Kersten S. Nutrigenomics: goals and strategies. Nature Reviews Genetics 2003; 4:315–322.
2. German JB, Roberts MA, Watkins SM. Genomics and metabolomics as markers for the interaction of diet and health: Lessons from lipids. J Nutr 2003; 133:2078S–2083S.
3. Seda O, Sedova L. New apolipoprotein A-V: Comparative genomics meets metabolism. Physiological Research 2003; 52:141–146.
4. Chawla A, Repa JJ, Evans RM, Mangelsdorf DJ. Nuclear receptors and lipid physiology: Opening the X-files. Science 2001; 294:1866–1870.
5. Sachidanandam R, Weissman D, Schmidt SC, Kakol JM, Stein LD, Marth G, Sherry S, Mullikin JC, Mortimore BJ, Willey DL, Hunt SE, Cole CG, Coggill PC, Rice CM, Ning ZM, Rogers J, Bentley DR, Kwok PY, Mardis ER, Yeh RT, Schultz B, Cook L, Davenport R, Dante M, Fulton L, Hillier L, Waterston RH, McPherson JD, Gilman B, Schaffner S, Van Etten WJ, Reich D, Higgins J, Daly MJ, Blumenstiel B, Baldwin J, Stange-Thomann NS, Zody MC, Linton L, Lander ES, Altshuler D. A map of human genome sequence variation containing 1.42 million single nucleotide polymorphisms. Nature 2001; 409:928–933.
6. Elliott R, Ong TJ. Science, medicine, and the future - Nutritional genomics. British Medical Journal 2002; 324:1438–1442.
7. Vincent S, Planells R, Defoort C, Bernard MC, Gerber M, Prudhomme J, Vague P, Lairon D. Genetic polymorphisms and lipoprotein responses to diets. Proc Nutr Soc 2002; 61(4):427–434.
8. Subramaniam S. The Biology Workbench–A seamless database and analysis environment for the biologist. Proteins 1998; 32:1–2.
9. Salzberg SL, Searls DB, Kasif S. Computational methods in molecular biology. Amsterdam New York: Elsevier, 1998.
10. Clote P, Backofen R. Computational molecular biology an introduction. Chichester, New York: John Wiley, 2000.
11. Baldi P, Brunak S. Bioinformatics: the machine learning approach. 2nd Cambridge, Mass: MIT Press, 2001.
12. Leonard AE, Bobik EG, Dorado J, Kroeger PE, Chuang LT, Thurmond JM, Parker-Barnes JM, Das T, Huang YS, Mukerji P. Cloning of a human cDNA encoding a novel enzyme involved in the elongation of long-chain polyunsaturated fatty acids. Biochem J 350 Pt 2000; 3:765–770.
13. Pertsemlidis A, Fondon JW, III Having a BLAST with bioinformatics (and avoiding BLASTphemy). Genome Biol 2, 2001.
14. Altschul SF, Boguski MS, Gish W, Wootton JC. Issues in searching molecular sequence databases. Nat Genet 1994; 6:119–129.
15. Nicholas, HB, Jr. Deerfield DW, II, Ropelewski AJ. Strategies for searching sequence databases. Biotechniques 2000; 28:1174–1178.
16. Henikoff S, Henikoff JG. Amino acid substitution matrices from protein blocks. Proc Natl Acad Sci U S A 1992; 89:10915–10919.
17. Pearson WR. Comparison of Methods for Searching Protein-Sequence Databases. Protein Science 1995; 4:1145–1160.

18. Jones DT, Swindells MB. Getting the most from PSI-BLAST. Trends Biochem Sci 2002; 27:161–164.
19. Evans RM, Hollenberg SM. Zinc fingers: gilt by association. Cell 1988; 52:1–3.
20. Lepage LM, Giesbrecht JA, Taylor CG. Expression of T lymphocyte p56(lck), a zinc-finger signal transduction protein, is elevated by dietary zinc deficiency and diet restriction in mice. J Nutr 1999; 129:620–627.
21. Needleman SB, Wunsch CD. A general method applicable to the search for similarities in the amino acid sequence of two proteins. J Mol Biol 1970; 48:443–453.
22. Smith TF, Waterman MS. Identification of common molecular subsequences. J Mol Biol 1981; 147:195–197.
23. Feng DF, Doolittle RF. Progressive sequence alignment as a prerequisite to correct phylogenetic trees. J Mol Evol 1987; 25:351–360.
24. Higgins DG, Thompson JD, Gibson TJ. Using CLUSTAL for multiple sequence alignments. Methods Enzymol 1996; 266:383–402.
25. Parry-Smith DJ, Payne AW, Michie AD, Attwood TK. CINEMA–a novel colour INteractive editor for multiple alignments. Gene 1998; 221:GC57–63.
26. Tanaka H, Ishikawa M, Asai K, Konagaya A. Hidden Markov models and iterative aligners: study of their equivalence and possibilities. Proc Int Conf Intell Syst Mol Biol 1993; 1:395–401.
27. Searls DB. The Linguistics of DNA. American Scientist 1992; 80:579–591.
28. Sigrist CJ, Cerutti L, Hulo N, Gattiker A, Falquet L, Pagni M, Bairoch A, Bucher P. PROSITE: a documented database using patterns and profiles as motif descriptors. Brief Bioinform 2002; 3:265–274.
29. Baldi P, Chauvin Y, Hunkapiller T, McClure MA. Hidden Markov models of biological primary sequence information. Proc Natl Acad Sci USA 1994; 91:1059–1063.
30. Krogh A, Brown M, Mian IS, Sjolander K, Haussler D. Hidden Markov models in computational biology. Applications to protein modeling. J Mol Biol 1994; 235:1501–1531.
31. Durbin R. Biological sequence analysis : probabilistic models of proteins and nucleic acids. Cambridge, New York, U.K: Cambridge University Press, 1998.
32. Bateman A, Birney E, Cerruti L, Durbin R, Etwiller L, Eddy SR, Griffiths-Jones S, Howe KL, Marshall M, Sonnhammer EL. The Pfam protein families database. Nucleic Acids Res 2002; 30:276–280.
33. Eddy SR. Profile hidden Markov models. Bioinformatics 1998; 14:755–763.
34. Clarke ND, Berg JM. Zinc fingers in Caenorhabditis elegans: finding families and probing pathways. Science 1998; 282:2018–2022.
35. Manning G, Whyte DB, Martinez R, Hunter T, Sudarsanam S. The protein kinase complement of the human genome. Science 2002; 298:1912–1934.
36. Dong S, Searls DB. Gene structure prediction by linguistic methods. Genomics 1994; 23:540–551.
37. Dasgupta N, Lin S, Carin L. Sequential modeling for identifying CpG island locations in human genome. Ieee Signal Processing Letters 2002; 9:407–409.
38. Matsuzaka T, Shimano H, Yahagi N, Yoshikawa T, Amemiya-Kudo M, Hasty AH, Okazaki H, Tamura Y, Iizuka Y, Ohashi K, Osuga J, Takahashi A, Yato S,

Sone H, Ishibashi S, Yamada N. Cloning and characterization of a mammalian fatty acyl-CoA elongase as a lipogenic enzyme regulated by SREBPs. J Lipid Res 2002; 43:911–920.

39. Hamosh A, Scott AF, Amberger J, Bocchini C, Valle D, McKusick VA. Online Mendelian Inheritance in Man (OMIM), a knowledgebase of human genes and genetic disorders. Nucleic Acids Res 2002; 30:52–55.

40. Krull M, Voss N, Choi C, Pistor S, Potapov A, Wingender E. TRANSPATH: an integrated database on signal transduction and a tool for array analysis. Nucleic Acids Res 2003; 31:97–100.

41. Kanehisa M, Goto S, Kawashima S, Nakaya A. The KEGG databases at GenomeNet. Nucleic Acids Res 2002; 30:42–46.

42. Karp PD, Riley M, Saier M, Paulsen IT, Collado-Vides J, Paley SM, Pellegrini-Toole A, Bonavides C, Gama-Castro S. The EcoCyc Database. Nucleic Acids Res 2002; 30:56–58.

43. Edgar R, Domrachev M, Lash AE. Gene Expression Omnibus: NCBI gene expression and hybridization array data repository. Nucleic Acids Res 2002; 30:207–210.

44. Brazma A, Parkinson H, Sarkans U, Shojatalab M, Vilo J, Abeygunawardena N, Holloway E, Kapushesky M, Kemmeren P, Lara GG, Oezcimen A, Rocca-Serra P, Sansone SA. ArrayExpress–a public repository for microarray gene expression data at the EBI. Nucleic Acids Res 2003; 31:68–71.

45. Su AI, Cooke MP, Ching KA, Hakak Y, Walker JR, Wiltshire T, Orth AP, Vega RG, Sapinoso LM, Moqrich A, Patapoutian A, Hampton GM, Schultz PG, Hogenesch JB. Large-scale analysis of the human and mouse transcriptomes. Proc Natl Acad Sci U S A 2002; 99:4465–4470.

46. Lash AE, Tolstoshev CM, Wagner L, Schuler GD, Strausberg RL, Riggins GJ, Altschul SF. SAGEmap: a public gene expression resource. Genome Res 2000; 10:1051–1060.

47. Yeoh EJ, Ross ME, Shurtleff SA, Williams WK, Patel D, Mahfouz R, Behm FG, Raimondi SC, Relling MV, Patel A, Cheng C, Campana D, Wilkins D, Zhou X, Li J, Liu H, Pui CH, Evans WE, Naeve C, Wong L, Downing JR. Classification, subtype discovery, and prediction of outcome in pediatric acute lymphoblastic leukemia by gene expression profiling. Cancer Cell 2002; 1:133–143.

48. Berrar DP, Dubitzky W, Granzow M. A practical approach to microarray data analysis. Boston: Kluwer Academic Publishers, 2003.

49. Lin SM, Johnson KF. Methods of microarray data analysis II : papers from CAMDA'01. Boston: Kluwer Academic Publishers, 2002.

50. Speed TP. Statistical analysis of gene expression microarray data. Boca Raton, Fla: Chapman & Hall/CRC, 2003.

51. Uschold M, King M, Moralee S, Zorgios Y. The Enterprise Ontology. Knowledge Engineering Review 1998; 13:31–89.

52. Ashburner M, Ball CA, Blake JA, Botstein D, Butler H, Cherry JM, Davis AP, Dolinski K, Dwight SS, Eppig JT, Harris MA, Hill DP, Issel-Tarver L, Kasarskis A, Lewis S, Matese JC, Richardson JE, Ringwald M, Rubin GM,

Sherlock G. Gene ontology: tool for the unification of biology. The Gene Ontology consortium. Nat Genet 2000; 25:25–29.

53. Khatri P, Draghici S, Ostermeier GC, Krawetz SA. Profiling gene expression using onto-express. Genomics 2002; 79:266–270.

54. Doniger SW, Salomonis N, Dahlquist KD, Vranizan K, Lawlor SC, Conklin BR. MAPPFinder: using gene ontology and GenMAPP to create a global gene-expression profile from microarray data. Genome Biol 4, 2003.

55. Robinson MD, Grigull J, Mohammad N, Hughes TR. FunSpec: a web-based cluster interpreter for yeast. BMC Bioinformatics 2002; 3:35.

56. McConnell P, Johnson K, Lin S. Applications of Tree-Maps to hierarchical biological data. Bioinformatics 2002; 18:1278–1279.

57. Herrero J, Al-Shahrour F, Diaz-Uriarte R, Mateos A, Vaquerizas JM, Santoyo J, Dopazo J. GEPAS: a web-based resource for microarray gene expression data analysis. Nucleic Acids Research 2003; 31:3461–3467.

58. Lowe HJ, Barnett GO. Understanding and using the medical subject headings (MeSH) vocabulary to perform literature searches. Jama 1994; 271:1103–1108.

59. Jenssen TK, Laegreid A, Komorowski J, Hovig E. A literature network of human genes for high-throughput analysis of gene expression. Nat Genet 2001; 28:21–28.

60. Iyer VR, Eisen MB, Ross DT, Schuler G, Moore T, Lee JC, Trent JM, Staudt LM, Hudson J, Jr. Boguski MS, Lashkari D, Shalon D, Botstein D, Brown PO. The transcriptional program in the response of human fibroblasts to serum. Science 1999; 283:83–87.

20

Statistical Principles for Analysis of Array Experiments

Arnold M. Saxton
Department of Animal Science, University of Tennessee,
Knoxville, Tennessee, U.S.A.

E. Barry Moser
Department of Experimental Statistics, Louisiana State University,
Baton Rouge, Louisiana, U.S.A.

1. INTRODUCTION

Statistical analysis is generally defined as the process of making scientific inferences from data containing variability. Microarray experiments certainly illustrate this definition, with scientists trying to decide which genes show differential expression in the face of substantial biological and technical variation (1). Methods for statistical analysis of array data have been widely presented, as in Chapter 7 (St Onge et al.) and elsewhere (2,3). These presentations generally assume a working knowledge of statistics and focus on methodology rather than on underlying statistical concepts. Even if scientists collecting array data will not themselves statistically analyze the data, some knowledge of how statistical principles apply to these experiments would enhance collaboration with statisticians.

This chapter will consider only the question: Does average gene expression across biological groups differ? This question is answered by hypothesis testing. Other scientific questions are addressed by different statistical methods, such as clustering. These methods are beyond the scope of this chapter. Hypothesis testing for comparing means is commonly conducted by t-tests or analysis of variance, appropriate when the measured response variable is normally distributed.

In this chapter we present basic statistical principles needed during analysis of array data, in the hopes that scientists will be able to make better decisions with their data. Example data are presented, so researchers can use them to practice their statistical analysis skills using their choice of software. We have used SAS software to produce the results presented here.

2. ANALYSIS OF VARIANCE CONCEPTS

An example is taken from Chapter 4 (Urs et al.), where adipose tissue from six patients was sampled and processed to yield two cell types. Each cell type was assigned a dye color (red and green) for this cDNA microarray experiment. Experimental data consist of six patients, one array per patient, two cell types (or dyes) per array, and two spots per cell type. The arrays actually measured over 10,000 genes, but just Perilipin results are initially considered. Raw data collected by measuring fluorescent intensity are given in Table 1.

To statistically process these data, and answer whether red intensity differs from green, the most comprehensive and flexible approach is to use a linear model. For example, a t-test to compare two means can be done with the linear model

$$y_{ij} = \mu + d_i + e_{ij} \tag{1}$$

Model (1) explains y, the intensity measures, with an overall mean (μ), an effect of the ith dye (i = red or green), and residual unexplained error (e). If the dye effects differ greatly, then we conclude there is differential expression of the Perilipin gene. An objective decision on what is "greatly" is obtained from the significance probability, or P-value. For this model, applied to Table 1 data, the F statistic is 2.26, indicating that dye differences are over twice as large as error differences. Statistical tests reflect the signal-to-noise concept, testing whether the dye signal is greater than the error noise. This F ratio has a P-value of 0.1469, which tells us that if red and green intensities are truly equal, we would have a 14.7% chance of observing data at least as different as Table 1. This chance is large, and only if the P-value goes below 5% do scientists generally consider that there is sufficient evidence that red and green intensities differ. If the 0.05 significance level cutoff is used, then

TABLE 1   Raw Data for the Perilipin Gene, Comparing Two Cell Types (Dyes) Using
Six Microarrays with Duplicate Spots

Array	Signal	Background	Dye
40	277	257	Green
40	248	235	Green
43	192	189	Green
43	198	168	Green
55	2,292	4,010	Green
55	2,130	4,268	Green
69	1,007	350	Green
69	978	404	Green
70	663	585	Green
70	792	536	Green
72	1,161	820	Green
72	1,287	816	Green
40	1,090	984	Red
40	690	732	Red
43	113	60	Red
43	101	59	Red
55	467	297	Red
55	478	639	Red
69	7,901	240	Red
69	7,220	330	Red
70	632	169	Red
70	735	143	Red
72	3,615	933	Red
72	2,972	890	Red

it should be kept in mind that there will be a 5% chance of declaring a difference when in truth there is none, the false positive result of a Type I error.

Model (1) does not address any sources of variation except for the dye effect. We know that all arrays are not equal. In Model (1), any array differences will be left in the residual error term, making it too large, in turn making the F ratio too small and leading to an inability to detect true differences. This gives a false negative, a consequence of low statistical power. One way to increase power is to correctly address known sources of variation in the linear model, thus making the error variation smaller. This model,

$$y_{ij} = \mu + a_j + d_i + e_{ij} \tag{2}$$

uses both array and dye effects to explain observed intensities. Model (2) is a "mixed model" because it includes fixed and random effects. Dye effects are "fixed effects" because they are modeled as additive constants. Suppose the red dye effect is 100, and the green dye effect is −100. Defining dye as fixed means that whenever red is observed, we expect a 100-unit increase in intensity, whereas green produces a constant 100-unit decrease. In contrast, arrays are generally modeled as random effects. They produce variation in the observations, but on average do not change the mean. In short, arrays are modeled as having a zero mean. Statistically, array in this model is an example of a "block" effect, with a block being a group of observations that are similar and have several treatments (usually all) applied within the block. For experiments with two treatments, this model is equivalent to the possibly more familiar paired t-test.

Correct statistical analysis of a mixed model requires software that addresses the random effects differently from the fixed effects, something that a standard two-way analysis of variance will not do. Be sure to understand the capabilities of the statistical software you are using. Applying Model (2) to Table 1 data gives an F of 3.66 and a P-value of 0.0726. Thus, addressing array variation did decrease error, making the F ratio larger. Some would interpret this as being suggestive evidence for differential expression, since the P-value is close to the 0.05 cutoff generally used. But strict interpretation would conclude that the gene is not differentially expressed between cell types ($P > 0.05$).

The statistical analysis still is not correct. Statistical tests must use error terms that reflect the variation among replicates. This is why replication is so important in experiments. For microarray experiments, replicates are arrays, as these are the units to which treatments can be independently applied. As a counter example, duplicate spots within an array are not replicates, as they will measure the same experimental application of a treatment. Statistically spots are called samples, multiple observations on the replicate. In Model (2), the error term has both replicate and sample variation. To separate out sampling variation, the correct error term must be added, producing the model

$$y_{ijk} = \mu + a_j + d_i + a \times d_{ij} + e_{ijk} \tag{3}$$

The complexity of subscripts may give scientists headaches, but they are needed to identify all components in the model, here an observation on the jth array for the ith dye and kth spot. The new term, array × dye, is a random term that captures replicate variation. Running a mixed model analysis of variance with this model produces an F of 1.09 ($P = 0.3439$). Correctly using replicate variation among arrays shows that evidence for differential gene expression is much weaker than previous models suggested.

Most common microarray designs use the blocking concept. For example, reference designs have arrays with a common control and one other treatment. By including block (array) in the model, data are adjusted so that effectively all observations are equalized by the control value. Then correct comparisons among the other treatments can be made. If there are at least three treatments including the control, then the statistical design is called an incomplete block design. Incomplete refers to the fact that all treatment conditions are not included or could not fit within the same block. Loop designs with more than two treatments also produce incomplete block designs. It is critical that incomplete block designs be analyzed with statistical software designed for mixed models. Only mixed model analysis can extract all information about treatment differences from such designs, producing the most precise results.

The example in Table 1 does not include dye reversals, where replicate arrays are run but with red and green dyes assigned to opposite treatments. Designs with reversals allow dye effects to be estimated separately from treatment effects, providing more information (4). Interestingly, arguments against extensive use of dye swapping have been made (5). The smallest possible design for dye reversals has no replicate arrays, i.e., multiple arrays with the same dye-treatment assignments. For example, with two treatments the array-dye-treatment combinations would be 1-R-1, 1-G-2, 2-R-2, and 2-G-1. Such a small experiment permits only estimation of the linear model

$$y_{ijk} = \mu + a_j + d_i + t_k + e_{ijk} \tag{4}$$

Dye and treatment effects can be addressed separately, but no other information is provided. It is better to include replicate arrays, allowing dye by treatment interactions to be estimated. These interactions, symbolized as $D \times T$, measure how much treatment differences change for the red dye vs. green. If an interaction does occur, it means measuring differential gene expression is affected by dye, and interpretation must be done more carefully. A typical linear model, including duplicate spots on each array, would look like

$$y_{ijkl} = \mu + a_j + d_i + t_k + d \times t_{ik} + a \times d \times t_{ijk} + e_{ijkl} \tag{5}$$

This model can be viewed as another incomplete block design, with the dye-treatment combinations incompletely represented in each array block. An important concept researchers should see from this section is how the appropriate linear model changes depending on how the experiment was conducted. Of primary importance is to choose an experimental design that permits all effects of interest, such as treatments, and of concern, such as arrays, to be considered in the model. The design should assign treatments to arrays in such a way that the questions of interest can be answered with the highest precision for a given experiment size (5).

3. DATA QUALITY ISSUES

Model (3) is the correct linear model for analysis of Table 1 data, but there are several other concepts that may impact statistical analysis results. The first issue is to consider background noise on the microarray slide. This intensity, not associated with real gene expression, could be added to the linear model as a "covariate," producing the model

$$y_{ijk} = \mu + a_j + d_i + a \times d_{ij} + \beta N_{ijk} + e_{ijk} \tag{6}$$

Covariates are regression variables that remove variation from the error term, thus increasing the ability to detect dye differences. This approach opens up even more complex possibilities, such as modeling the spatial fluctuations in background across the microarray slide. Such approaches could benefit from using all the background data, instead of correcting each spot separately for its local background. However, it is much more common to simply subtract the local background from each spot's signal, and analyze the resulting corrected intensities with Model (3). This can produce negative intensities, which are best set to a small positive number (2). Note that setting $\beta = 1$ in Model (6) produces Model (3) for the corrected data. After background correction of the example data, an F ratio of 2.02 ($P = 0.2140$) is produced, suggesting that background correction has removed noise from the data, allowing dye differences to be more clearly seen. However, differences are not large from the statistical significance viewpoint.

The P values for testing mean differences are based on a normal distribution. Specifically the residual errors (e) in linear models must have the symmetric, single-peak, bell-shaped curve characteristic of the normal distribution. This requirement should be checked during every statistical analysis. Figure 1A shows the residuals for Model (3). They have a distribution with a strong central peak and long tails, giving a Shapiro-Wilk W statistic of 0.87 (W = 1 indicates perfect normality). This information is not strongly indicative of nonnormality (due to the small number of observations), but general experience with microarray intensity data suggests a log transformation provides better normality characteristics. Instead of analyzing signal intensities, log base 2 values are used. Choice of base 2 gives a convenient scale for expression ratios, as a two-fold increase will have a value of 1, and so forth. For achieving statistical normality, any base will perform similarly. Figure 1B shows background corrected signal residuals after log transformation. This distribution appears more bell-shaped, but still has a W of only 0.89, not a large improvement. The statistical test for dye differences now has an F of 3.43 ($P = 0.1233$), an example of the influence that deviations from normality can have. Depending on the observed distribution, it is possible to make either false positive or false negative conclusions when the data are

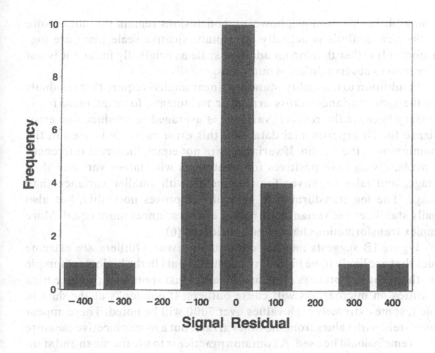

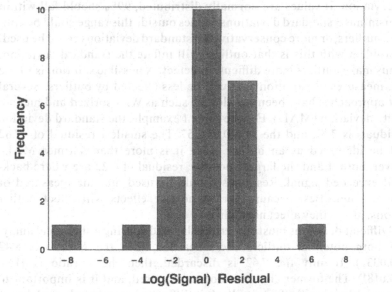

Figure 1 Frequency distribution of residuals from Model (3) for background corrected (A) signal intensity and (B) log base 2 signal intensity.

not normal. In this example, however, conclusions remain the same. Note that the new analysis is actually on a multiplicative scale (data are log-transformed) rather than on an additive scale as originally fit, and it is not uncommon to observe different outcomes.

In addition to normality, standard linear models require that residuals have the same variance across arrays or treatments, for example. This is necessary because the residual variance is averaged to produce one error variance for all experimental data, and this error variance is used in the denominator of the F ratio. If variances are not equal, incorrect inferences are made, giving false positives for treatments with larger variance than average, and false negatives for treatments with smaller variance than average. The log transformation generally improves normality, but also usually stabilizes the variance, or makes error variances more equal. More complex transformations have been considered (6).

Figure 1B suggests another data quality issue. Outliers are extreme values that are likely to be biased by unusual events in the biological sample or in the technical process of measurement. Dust spots that produce high intensities on microarrays will cause outliers. If one looks at the data in Table 1, some extremely high values over 7000 will be noted. These appear inconsistent with values from the other arrays, but a more objective measure of "extreme" should be used. A common practice is to use the mean and standard deviation. If values are normally distributed, 99% should fall within plus-or-minus 3 standard deviations. Values outside this range could be considered outliers, or more conservatively 4 standard deviations could be used. One problem with this is that outliers will inflate the standard deviation, and thus make outliers more difficult to detect. A possible solution is to use robust measures of variation that will be less affected by outliers. Several robust approaches have been developed, such as Windsorized and median absolute deviation (MAD). For the Table 1 example, the standard deviation of residuals is 2.57, and the MAD is 1.57. The smallest residual of -7.62 would be identified as an outlier since it is more than 4 times MAD. However, both it and the largest positive residual of 6.22 are where background exceeded signal. Residuals should be used, not the measured or corrected intensities, because array and dye effects will affect outlier decisions, just as they affect normality.

Difficult decisions must sometimes be made during a statistical analysis. If both potential outliers are discarded, then the F ratio is 8.42 ($P = 0.0337$). If only the 7.62 is discarded, then the F ratio is 11.94 ($P = 0.0181$). The former choice is commonly used, and it is important to realize this choice will affect results. It is difficult to remain unbiased, but in our judgment, none of the data should be discarded in this example. Although the 7.62 meets a common criterion for identifying outliers, the small number of observations and the shape of Figure 1B do not strongly support this observation being "extremely" unusual.

Adding to the difficulty is the realization that this is just one gene of over 10,000, and it is not realistic to make careful judgments such as those just presented for each individual gene. An automated process must be programmed, and the results interpreted in light of the possibility that data quality issues affected the analysis.

We can now complete the analysis of differential expression for this gene. Using all background-corrected data and Model (3), least squares mean log expressions are 4.83 and 7.06 for green and red dyes. The difference is 2.2 ($P = 0.12$) with a standard error of 1.2. Mathematically, $\log(R)-\log(G) = \log(R/G)$, so the difference can be back-transformed by raising 2 (the log base) to the difference power, giving 4.59. Interpretation is thus that the red dye shows over a four-fold increase in gene expression compared to green. That this large observed difference is not statistically different suggests that a false negative has occurred. There was too much variability among arrays (or among individual subjects since the biological sample for each array is one person), causing low statistical power or low ability to detect true differences. If this difference is scientifically important, then future experiments should attempt to reduce variation, or the number of observations must be increased. These are the two variables that researchers can control to increase statistical power. Estimation of required sample size is a fairly complex process, with more details given elsewhere (1,7). But a reasonable approximation can be made with the formula

Number per treatment = 25 × Variance(difference × difference).

Variance is the error variance, which can be obtained from the standard error of the difference by

$$\text{Variance} = N \times SE \times SE/2 \tag{7}$$

where N is the number of observations for the mean. For the example data, this gives $12 \times 1.2 \times 1.2/2 = 8.64$. Difference in Eq. (7) is the treatment difference of interest, and the observed difference can be used. Doing so here gives 45 observations, or 23 arrays considering duplicate spots per array. If variation can not be reduced, then a substantially larger experiment will be needed.

4. CONCEPTS FOR MULTIPLE GENE ANALYSIS

The value of microarrays is the ability to measure thousands of genes, but analysis of such data requires extension of the above concepts. For linear models, gene becomes a factor that needs to be included in the analysis.

However, this produces a complex model with many interaction terms involving gene, very difficult to interpret scientifically. Also computationally, the statistical analysis is difficult because of the thousands of genes being considered. Wolfinger et al. (3) suggested a two-step modeling process that avoids these difficulties. First, a model such as

$$y_{ijk} = \mu + a_j + e_{ijk} \tag{8}$$

is fit to all gene data, so the k subscript now represents genes and duplicate spots. The purpose of this analysis is to remove average array effects, averaged over all genes. Dye is not included because for the Table 1 example, dye is equivalent to treatment. If a dye swap experiment were used, then dye effects would be included in this first step. Then for the second step, residuals from the first analysis are used as the y-variable in a second model, run on each gene separately.

$$e_{ijk} = \mu + a_j + d_i + a \times d_{ij} + e'_{ijk} \tag{9}$$

This model must have correct error terms and treatments of interest, as discussed earlier, since it will produce the scientific results to be interpreted.

For data quality issues, analysis of all genes allows an additional correction to be made. Scientifically, it is assumed that most genes will not be over- or underexpressed. Thus, if expression ratios are plotted against intensity, a flat line should be observed (8). In reality, each array may have a different pattern, and corrections such as the loess smoothing can be used to remove these array differences.

Finally, there is the concept of multiple testing. Analysis of each gene produces a P-value test for differential expression. Each statistical test has a chance of false positives. For example, if $P < 0.05$ is used, then each test has a 5% chance of a Type I error, or false positive. If all tests are independent, then these 5% chances accumulate. After 200 tests, the chance of at least one false positive is essentially 100%. Microarray analysis generally involves over 10,000 tests, so there is real concern that test results will include too many false positives. The Bonferroni correction is mathematically proven to prevent increases in false positives for independent tests, simply by dividing the critical P-value by the number of tests (7). Thus we would use a significance level of $0.05/10000$, for example. However, almost certainly tests for microarrays are not independent, since genes operating in the same pathway are likely to be over- or underexpressed to a somewhat similar degree. The Bonferroni correction then becomes too conservative, or the significant P value is too small and we make false negative errors too often. There is growing acceptance for an alternative method, called the False Discovery Rate. Details can be found in (1,9), but basically an attempt is made to balance false positives and false negatives.

5. ANALYSIS FOR OLIGONUCLEOTIDE ARRAYS

Oligonucleotide arrays provided commercially as GeneChips[TM] from Affymetrix (Santa Clara, CA) add their own set of challenges for the analysis of microarrays. In these experiments, each gene is represented by a set of (usually) 20 oligonucleotide sequences called probes. In addition, each probe has a control probe that is identical except for a single nucleotide mismatch at position 13. These pairs of probes are referred to as perfect match (PM) and mismatch (MM) probes, respectively.

The design of experiments using oligonucleotide arrays differs from cDNA arrays in that only a single set of treatment conditions is applied to an array, rather than having two treatment conditions applied as with the cDNA array. Thus, the array no longer serves as a block (or incomplete block when more than two treatments are considered) in the design, but as an experimental unit. Traditional completely randomized designs with factorial treatment arrangements apply to these experiments. This distinction is very important because it has implications on the models appropriate for analysis of these experiments.

Since each array is calibrated and read independently of the others, and since the arrays no longer serve as blocks in the experiment, normalization may be required to put the arrays on a comparable basis. Chu et al. (10) recommend a log base 2 transformation of the intensities, then centering these values at zero separately for each array. In this way the geometric means of the arrays are equilibrated. The assumption is that centering of the data removes calibration noise, but does not remove treatment effects. One also has to make decisions regarding whether or not to use the mismatch data in the analyses. The differences PM-MM are supposed to indicate the degree to which the PM probes are expressed by the treatments above the noise generated by nonspecific binding as measured with the MM probes. Unfortunately, these differences can be negative. If a log base 2 transformation of the differences is desired, then a constant could be added to the differences to make them all positive, or the negative values may be truncated to 1 or some other small positive number prior to transformation. Alternatives include using the MM intensity as a covariate in the model or using the log base 2-transformed MM intensities as a covariate when the PM data are log base 2-transformed.

Chu et al. (10) recommend a mixed-models approach to the analysis of oligonucleotide arrays. Their approach fits traditional split-plot models to the data where the array is the whole-plot experimental unit, and probe locations on an array are the subplot experimental unit. This design assumes that arrays on which treatment factors are applied, such as cell lines or other treatment conditions, are independent, whereas the probe measurements made on a given array may be correlated. This is the approach that we recommend. However, we treat the model in a repeated measures context, so that if use of the spatial information about the probe locations in the analysis is desired, then it becomes a straightforward extension of our approach.

We will illustrate the analysis of an oligonucleotide experiment using data taken from an experiment by Tusher et al. (11) to examine the ionizing radiation response on two cell lines. The experiment was a completely randomized design with a two-factor factorial arrangement of treatment (irradiated or unirradiated) and cell line (1 or 2) with each combination replicated on two separate arrays making a total of 8 arrays in the experiment. These combinations are referred to as I1A, I1B, I2A, I2B, U1A, U1B, U2A, and U2B, with the three-letter codes indicating treatment, cell line, and replicate, respectively. Again there are 20 PM probes and 20 MM probes for each gene on each array. A model corresponding to this experiment should consider treatment, cell line, probe, and array effects. A gene-specific model for the PM data is

$$y_{ijkl} = \mu + \tau_i + \varphi_j + \tau\varphi_{ij} + a_{ijk} + \rho_l + \tau\rho_{il} + \varphi\rho_{jl} + e_{ijkl} \tag{10}$$

where τ_i is the effect due to ionizing radiation, φ_j is the effect due to the jth cell line, ρ_l is the probe effect, and a_{ijk} and e_{ijkl} are random effects due to variation among arrays and probe locations within arrays, respectively. This is the split-plot model proposed for these data by Chu et al. (10). Note that they have not allowed for a three-factor interaction of treatment × cell line × probe in this model. The repeated measures (multivariate) model differs from model 10 by having separate equations for each probe, but then linking those equations by specifying a correlation structure (12). The coding for the mixed-model analysis in SAS using the repeated measures model approach matches that for the split-plot model, so we will use the split-plot model specification. In addition, we have selected a compound symmetry covariance structure that will give results identical to the traditional split-plot model (12). Given the spatial coordinates for the probes on the arrays, more explicit spatial correlation structures can be fit. Results are reported for the gene numbered 170 in the data set with PM and MM data shown in Tables 2 and 3, respectively. Prior to analysis, all of the PM data are log base 2-transformed and centered at zero for each array. The log base 2 geometric means for each array are included in Tables 2 and 3. To center the data, compute the log base 2 transformation for each value including the means, and then subtract the transformed array mean from each of the transformed probe values.

A large treatment effect ($F = 125.7$; df $= 1,4$; $P > F = 0.0004$) is observed on the log base 2-transformed and centered PM data for gene 170 using the repeated measures Model (10). In addition, significant effects are

TABLE 2 Perfect Match (PM) Data for Gene 170 from an Oligonucleotide Experiment

Probe	I1A	I1B	I2A	I2B	U1A	U1B	U2A	U2B
1	1,652.0	1,805.0	1,875.5	1,970.3	1,442.5	1,636.5	2,628.0	2,152.0
2	2,330.0	2,450.0	2,351.3	2,478.5	1,939.3	2,001.5	2,867.3	2,525.5
3	2,761.5	2,732.3	2,554.8	2,687.0	2,058.3	2,259.3	3,333.5	2,730.3
4	1,505.0	1,573.8	1,432.8	1,587.0	1,111.0	1,304.0	1,962.3	1,657.8
5	529.5	469.5	440.3	498.5	398.3	437.0	554.0	521.8
6	972.0	958.8	976.0	1,044.5	899.5	918.3	1,243.8	1,086.0
7	799.3	747.0	835.5	729.5	652.3	699.3	1,006.3	812.8
8	1,308.5	1,420.5	1,441.8	1,481.5	1,097.0	1,200.5	1,612.5	1,578.0
9	931.8	938.3	850.0	896.3	734.5	841.8	1,088.8	971.0
10	1,151.3	1,201.5	1,294.5	1,279.8	1,049.3	1,113.3	1,623.5	1,355.3
11	1,206.3	1,175.8	1,204.3	1,198.8	1,083.8	1,013.8	1,438.3	1,234.5
12	1,995.3	2,018.8	2,142.5	2,164.8	1,538.5	1,632.3	2,253.5	2,152.8
13	1,048.3	948.5	1,093.8	1,059.0	884.8	959.3	1,327.0	1,071.8
14	2,374.0	2,453.3	2,399.8	2,494.0	2,074.8	2,208.5	2,631.8	2,521.3
15	2,635.8	2,614.5	2,804.5	2,944.5	2,249.8	2,334.0	3,391.3	2,956.0
16	1,415.5	1,267.0	1,470.5	1,406.8	1,116.8	1,308.0	1,850.8	1,542.3
17	1,720.0	1,656.8	1,728.3	1,842.0	1,339.8	1,474.0	2,214.0	1,822.0
18	2,026.8	2,045.0	2,259.8	2,067.5	1,729.0	1,790.3	2,651.5	2,233.5
19	589.8	570.0	588.0	592.0	458.3	547.5	733.8	636.3
20	1,833.5	1,669.0	1,929.5	1,813.0	1,377.3	1,547.5	2,109.0	2,009.0
Array geometric mean	980.6	951.7	1,014.2	1,009.7	886.1	950.9	1,301.9	1,136.2

(Data from Ref. 11.)

observed for treatment × cell line ($F = 8.4$; df $= 1,4$; $P > F = 0.0441$), probe ($F = 997.8$; df $= 19,95$; $P > F = 0.0001$), and cell line × probe ($F = 2.1$; df $= 19,95$; $P > F = 0.0093$). Thus, there may be some evidence that the probes do not respond the same way for each cell line, and that the treatment may affect the cell lines differently for this gene. The estimated (least-squares) means from the model for treatment, cell line, and treatment × cell line are given in Table 4. One can examine the treatment × cell line means and observe that the difference between cell lines means appears greater for the unirradiated than the irradiated treatment. The simple effects or interaction slices provide a more convenient way to investigate this inter-action. Simple effects compare the means of one factor at fixed levels of another factor. These simple effect tests are reported in Table 5. These indi-cate that the cell lines do not differ for the irradiated treatment, but do differ in the unirradiated treatment, which is consistent with an interaction of treatment with cell line. The probe effect is very large and indicates that the intensities for a given experimental condition vary considerably among

TABLE 3 Mismatch (MM) Data for Gene 170 from an Oligonucleotide Experiment

Probe	I1A	I1B	I2A	I2B	U1A	U1B	U2A	U2B
1	692.5	679.0	734.3	819.0	607.0	618.0	876.5	826.8
2	933.8	927.8	997.5	1,102.8	824.5	869.8	1,157.5	1,013.0
3	666.8	670.8	658.0	694.0	603.5	629.0	835.3	694.3
4	597.5	611.5	620.0	620.5	599.0	576.3	809.5	674.8
5	453.8	447.5	447.0	429.0	381.3	378.8	534.8	492.0
6	596.3	558.3	582.5	580.3	484.5	522.8	728.0	597.8
7	471.8	413.8	480.5	456.0	401.5	411.0	527.8	525.0
8	515.8	543.8	578.0	575.3	532.3	548.3	767.0	605.8
9	475.8	514.3	503.0	517.8	415.3	507.3	665.5	594.5
10	643.5	767.5	759.8	725.3	636.3	635.0	944.0	754.5
11	544.3	539.3	575.8	619.8	506.0	485.0	662.8	621.3
12	590.5	646.5	696.8	698.0	539.5	628.8	736.0	703.0
13	544.8	569.3	589.8	569.0	566.3	526.0	702.3	655.8
14	842.0	962.8	955.3	1,104.0	935.3	848.0	1,092.8	929.3
15	954.3	1,101.5	1,128.8	1,212.3	908.8	880.0	1,381.5	1,022.8
16	526.3	502.3	480.3	513.0	406.0	463.3	643.0	524.5
17	954.3	1,025.0	1,026.3	1,145.5	977.0	984.8	1,332.0	1,199.3
18	819.8	821.8	871.8	917.3	737.3	754.5	1,120.0	863.8
19	525.3	446.0	463.3	498.0	429.5	447.5	649.5	548.8
20	496.0	479.0	514.5	545.8	489.8	538.0	621.3	565.3
Array geometric mean	829.6	831.2	874.4	877.9	763.6	808.6	1,097.7	970.0

(Data from Ref. 11.)

TABLE 4 Least-Squares Means[a](LSMean) of Treatment, Cell Line, and Treatment × Cell Line Effects for Gene 170

Effect	Treatment	Cell line	LSMean	Standard error
Treatment	I		0.5114	0.007362
Treatment	U		0.3946	0.007362
Cell Line		1	0.4474	0.007362
Cell Line		2	0.4586	0.007362
Treatment × Cell Line	I	1	0.5209	0.01041
Treatment × Cell Line	I	2	0.5018	0.01041
Treatment × Cell Line	U	1	0.3740	0.01041
Treatment × Cell Line	U	2	0.4153	0.01041

[a]Means are of centered log base 2 PM data from an oligonucleotide experiment
(Data from Ref. 11.)

TABLE 5 Treatment × Cell Line Interaction Simple Effect Slices for the PM Data of Gene 170 from an Oligonucleotide Experiment

Effect	Treatment	Cell line	Numerator DF	Denominator DF	F Value	Prob > F
Treatment × Cell Line	I		1	4	1.67	0.2655
Treatment × Cell Line	U		1	4	7.88	0.0485
Treatment × Cell Line		1	1	4	99.56	0.0006
Treatment × Cell Line		2	1	4	34.54	0.0042

(Data from Ref. 11)

the probes for this gene. These centered log base 2 mean intensities are graphed as a function of probe number in Fig. 2 and suggest that some probe intensities are 4 times larger than others (mean difference of 2 on a log base 2 scale).

Again, the usual goal is to identify the more important, distinguishing genes, so all genes are analyzed using the selected model, and then

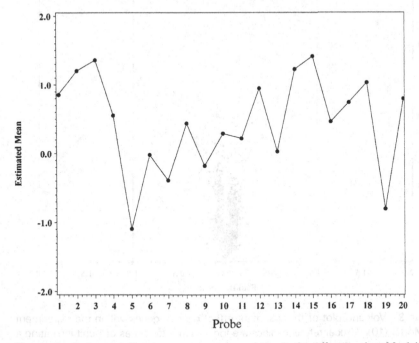

FIGURE 2 Estimated (least-squares) means of the probe effects using Model (10) plotted as a function of probe number.

techniques to identify the important genes are applied as discussed for the cDNA array experiments. One graphical aid to help with this decision is the volcano plot (10), in which the effect differences and associated hypothesis test P-values are presented together. Rather than being plotted directly, the P-values are rescaled so that small P-values appear near the top of the plot, and so that judgments relative to their importance can be made using the P-values. We have transformed the P-values using $-2 \log P$-value, resulting in an approximate chi-squared-distributed random variable. The fold-differences are plotted on the x-axis and the chi-square values are plotted on the y-axis. The resulting volcano plot for the treatment effect for the Tusher experiment is given in Fig. 3. Notice that just because the P-value is very small, it does not necessarily imply that the fold-difference is large. The P-value not only is a function of the fold-difference, but also depends upon the variability in the data. In this experiment, many small fold-differences have been assigned small P-values. Thus, using a cut-off value of

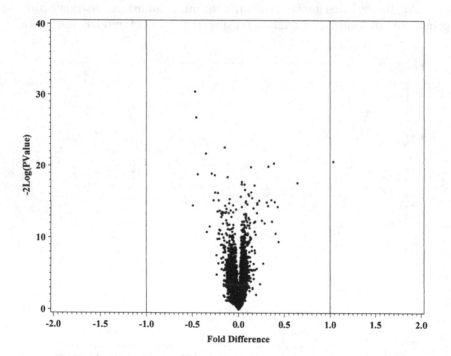

FIGURE 3 Volcano plot of the treatment effect over all genes within the experiment for Model (10). Vertical reference lines are located at differences of 1 unit indicating a two-fold change. The $-2 \log(\text{PValue})$ are natural logarithms of the treatment effect P-values and are approximately chi-squared distributed.

$P < 0.05$ is not very informative, as many of the genes would be selected as informative. If the volcano plot is used, genes with points near the top and to the left or right extremes would be more closely examined.

The residuals from the above mentioned analysis are examined for normality and consistency with the model using a quantile-quantile plot. The observed residual quantiles appear consistent with those that would be expected if the residuals are normally distributed (Fig. 4), following closely along the straight reference line. Unfortunately, examining such a plot for each gene in the experiment is overwhelming (thousands of such plots), so techniques of combining residuals over all genes have been proposed. Chu et al. (10) suggest standardizing the residuals from each gene's analysis by the estimated standard deviation of the residuals. For the split-plot or repeated measures Model (10), the standardization is to compute the estimate of $\sigma_a^2 + \sigma^2$ for each gene, where σ_a^2 and σ^2 are the variance components for the whole-plot and subplot units, respectively, using the estimates from each gene's analysis. For gene 170, the estimated variance components are $\hat{\sigma}_a^2 = 0.000018$ and $\hat{\sigma}^2 = 0.003979$, yielding an estimated standard deviation

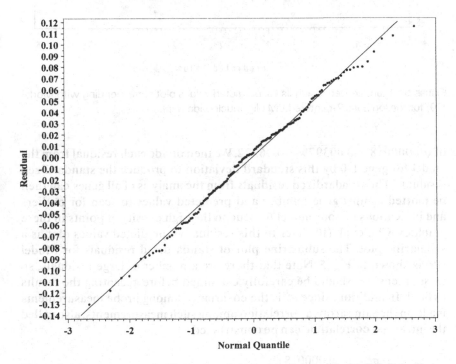

FIGURE 4 Quantile-quantile (QQ) plot of the residuals from Model (10) fit to the log base 2-centered PM oligonucleotide data.

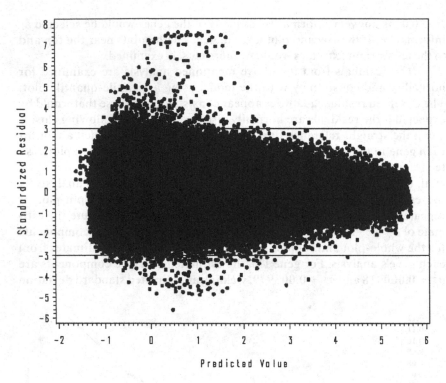

FIGURE 5 Standardized residuals vs. predicted values plot corresponding with Model (10) for the log base 2-centered PM oligonucleotide data.

of $\sqrt{0.000018 + 0.003979} = 0.06322$. We then divide each residual from the model for gene 170 by this standard deviation to produce the standardized residuals. The standardized residuals from the analysis of all genes can then be plotted against gene number and predicted values to scan for outliers and indications of poor model fit. Due to the high density of points in these graphics, Chu et al. (10) refer to this residual vs. predicted values plot as a submarine plot. The submarine plot of standardized residuals for Model (10) is shown in Fig. 5. Note that there are a number of large residuals, so these intensities should be carefully examined before accepting the results as final. In addition, since σ_a^2 is the covariance among probe measurements made on the same array, a correlation among such measurements, also called the intraclass correlation, can be constructed as

$$r = \frac{\sigma_a^2}{\sigma_a^2 + \sigma^2} = \frac{0.000018}{0.003997} = 0.0045 \tag{11}$$

indicating that there is very little spatial association among the probe measurements from the same array. One could also use the actual spatial coordinates of the probe locations on the arrays to fit more explicit spatial correlation models.

6. CONCLUSIONS

The analysis of both cDNA and oligonucleotide microarrays requires an understanding of the experimental designs that are used for each, along with the accompanying statistical models that account for or explain the various sources of experimental variation from those designs. For the oligonucleotide experiments, probe-level information must also be dealt with. Normalization of the data may be required to adjust for overall differences in intensity readings from array to array. Various diagnostics are required to examine overall model fit, and to identify outliers. Due to the large number of genes examined, new techniques and graphics are required to accommodate the multiplicity in testing and to aid in the identification of important, discriminating genes. The mixed-models approach that we advocate here is quite flexible and accommodates a wide number of experimental designs permitting a consistent approach to the analysis of such data.

REFERENCES

1. Simon R, Radmacher MD, Dobbin K. Design of studies using DNA microarrays. Genetic Epidemiology 2002; 23:21–36.
2. Craig BA, Black MA, Doerge RW. Gene expression data: the technology and statistical analysis. J Agricultural, Biological and Environmental Statistics 2003; 8:1–28.
3. Wolfinger RD, Gibson G, Wolfinger E, Bennett L, Hamadeh H, Bushel P, Afshari C, Paules RS. Assessing gene significance from cDNA microarray expression data via mixed models. J Computational Biology 2001; 8:625–637.
4. He YD, Dai H, Schadt EE, Cavet G, Edwards SW, Stepaniants SB, Duenwald S, Kleinhanz R, Jones AR, Shoemaker DD, Stoughton RB. Microarray standard data set and figures of merit for comparing data processing methods and experiment designs. Bioinformatics 2003; 19:956–965.
5. Dobbin K, Shih JH, Simon R. Questions and Answers on Design of Dual-Label Microarrays for Identifying Differentially Expressed Genes. J National Cancer Institute 2003; 95:1362–1365.
6. Rocke DM, Durbin B. Approximate variance-stabilizing transformations for gene-expression microarray data. Bioinformatics 2003; 19:966–972.
7. Zar JH. Biostatistical Analysis. 2nd ed. Englewood Cliffs, New Jersey: Prentice-Hall, 1984.

8. Yang YH, Dudoit S, Luu P, Lin DM, Peng V, Ngai J, Speed TP. Normalization for cDNA microarray data: a robust composite method addressing single and multiple slide systematic variation. Nucleic Acids Research 2002; 30:e15.
9. Efron B, Tibshirani R. Empirical Bayes methods and false discovery rates for microarrays. Genetic Epidemiology 2002; 23:70–86.
10. Chu T, Weir B, Wolfinger R. A systematic statistical linear modeling approach to oligonucleotide array experiments. Math Biosci 2002; 176:35–51.
11. Tusher VG, Tibshirani R, Chu G. Significance analysis of microarrays applied to the ionizing radiation response. Proc Nat Acad Sci 2001; 98:5116–5121.
12. Moser EB, Saxton AM, Pezeshki SR. Repeated measures analysis of variance: application to tree research. Can J For Res 1990; 20:524–535.

Index

Printed in the United States
by Baker & Taylor Publisher Services